Nanomaterials for Tumor Targeting Theranostics

A Proactive Clinical Perspective

Nanomaterials for Tumor Targeting Theranostics

A Proactive Clinical Perspective

Editors

Mingqian Tan
Aiguo Wu

Chinese Academy of Sciences, China

高等教育出版社
HIGHER EDUCATION PRESS

World Scientific

Published by

Higher Education Press Limited Company
4 Dewai Dajie, Beijing 100120, P. R. China

and

World Scientific Publishing Co. Pte. Ltd.
5 Toh Tuck Link, Singapore 596224

Library of Congress Cataloging-in-Publication Data
Nanomaterials for tumor targeting theranostics : a proactive clinical perspective / editors Mingqian Tan & Aiguo Wu.
p. ; cm.
Includes bibliographical references and index.
ISBN 978-9814635417 (hardcover : alk. paper)
I. Tan, Mingqian, editor. II. Wu, Aiguo, editor.
[DNLM: 1. Nanostructures. 2. Neoplasms--diagnosis. 3. Neoplasms--therapy. QZ 241]
RC261
616.99'4--dc23

2015031988

British Library Cataloguing-in-Publication Data
A catalogue record for this book is available from the British Library.

Printed in P. R. China

Contents

Foreword 1

This excellent reference text comes at a time when the research on theranostics is reaching a larger and larger audience. Theranostics is a contraction of therapy and diagnostic, meaning that the action of a drug on a patient can be followed in real time. Theranostics is presently a cutting-edge topic in biomedicine and requires both extended knowledge in genomics, proteomics, metabolomics, and suitable advanced techniques in bioanalysis and bioimaging. It is expected that its developments will lead to personalized medicine within the next decade, especially in the field of cancer therapy and early detection. There are several ways of taking advantage of this new array of medical technologies. One is to identify the patients for whom a given drug is not working; another one is to apply diagnosis and therapy in sequence for distinguishing variants of diseases in order to apply the best suited therapy to a given patient; the later one is the so-called co-development in which two drugs are applied leading to a synergistic effect.

It is now well recognized that nanotechnology permits a combination of functionalities within a single particle and therefore is ideal for taking diagnosis from the medical laboratories to the point-of-care by making use of lab-on-a-chip devices. In this respect, a whole range of nanomaterials are at hand, facilitating the work of biochemists and medical doctors.

This book precisely informs chemists, biochemists, biologists, and medical doctors of the strengths and applications of nanomaterials in this fast developing field. The various chapters perfectly illustrate that scientists involved in the development of nanomaterials are finding better, faster and more cost-effective technologies. All aspects are covered, from magnetic nanomaterials, to quantum dots, mesoporous silica nanoparticles, surface enhanced Raman scattering nanomaterials, polymeric nanostructures, carbon nanomaterials, lipopolyplex nanoparticles and microbubbles. Upconversion nanomaterials, functional dendrimers, and protein-based nanoparticles are also included.

The editors are leading figures in the field and their competences have led to a well-balanced text with two introductory chapters, 11 topical reviews, and a perspective chapter to conclude the opus. I certainly recommend this book for teaching at the master and doctorate levels and for any scientists involved in either nanoscience or theranostics and, of course in both.

Jean-Claude G. Bünzli FRSC

Honorary Professor, Swiss Federal
Institute of Technology (EPFL) and
University of Lausanne (Switzerland)
WCU Professor, Korea University (ROK)

Lausanne, September 24, 2013

Foreword 2

The integration of nanotechnology with medicine has led to a significant advancement in the medical sciences. Nanomaterials can be used for both diagnostic and therapeutic purposes at the same time, which can't be achieved by small molecules or other microscale tools. Nanotechnology has given rise to a new theranostic tool. It provides the promising agents that propel the biomedical practice towards personalized medicine, and paves a way towards the development of innovative strategies in the cancer treatment.

This book is an excellent introduction of the nanomaterials based platforms for tumors theranostics. It will explicate the mechanism and application of this platform in targeting diagnostic imaging, such as magnetic resonance, optical, positron emission tomography and surface enhanced Raman scattering; and in therapeutics including hyperthermia, photodynamics and photothermal therapy, as well as gene and drug delivery. It covers all the major materials from magnetic nanomaterials, quantum dots, noble metals and carbon nanomaterials to polymeric nanoarchitectures, proteins, *etc.*

The editors of this book are young elites in the field, who have carried out extensive research projects and accomplished many remarkable achievements in the nanomaterials based tumor targeting theranostics. They devoted a great deal of precious time and effort to present useful information in this book. I am delighted to see that besides them, there are lots of young scholars in China and around the world who dedicate themselves to the research of targeting, imaging, and therapeutics at the cellular and molecular level. They are becoming the drive force of biomedical science development in recent years.

This is a timely book for introduction of nano-based theranostics. I, therefore, strongly recommend this book to all of the students and scholars in academe or industry who are interested in this rapidly expanding field. It is also helpful for the clinicians to acquire cutting-edge knowledge and to get a better understanding of nanomaterials and tumor targeting theranostics.

Shen, Baozhong M.D., Ph.D.

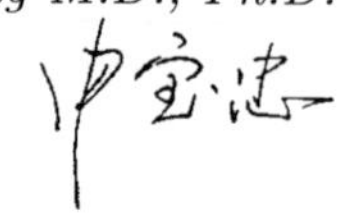

Professor of Medical Imaging and Nuclear Medicine
Chair, Department of Radiology
Director, Molecular Imaging Center of Harbin Medical University
President, Fourth Hospital of Harbin Medical University
Vice Chancellor, Heilongjiang Academy of Medical Sciences

Harbin, December 12, 2013

Preface

At the beginning of the twenty-first century, cancer is still a major leading cause of death worldwide because of the unpredictable and numerous challenges for early diagnosis and effective treatment. The remarkable advance about "nanoworld" during the past two decades has encouraged hopes of nanotechnology to be a game changer for the diagnosis and treatment of cancer. However, most efforts to improve the cancer treatment through nanotechnology are still in its infancy and there has been no routine nanomedicine for tumor theranostics in clinical practice. Indeed, the use of nanomaterials in cancer diagnosis and treatment offers great potential for effective detection and inhibition of tumor growth at the early stage, as well as destroying cancer tumors with minimal damage to healthy tissue and organs. On the basis of this, the idea for writing a book to summarize the nanomaterials for tumor targeting theranostics came to us. We hope that the book may be helpful for the readers to get a more realistic understanding of both the advantages and limitations of nanomaterials for potential tumor targeting theranostics.

Theranostics is a term coined to describe a strategy that combines diagnostic and therapeutic modalities into one platform to develop all-in-one therapeutic agents. For the cancer treatment, theranostics agents may offer the possibility to monitor and treat the tumor according to the disease progresses in one dosage. Multiple diagnostic and therapeutic techniques are integrated into a functional platform to provide tumor patients with selectable treatments for improved prognoses. Examples of designing and evaluating various types of nanomaterials for tumor targeting theranostics are found throughout the book, and we hope thereby to give a profile to the entire development. The book is organized as follows: Chapter 1 introduces the concept of theranostics for tumor therapy based on the major nanomaterials discussed throughout the rest of the text. Chapter 2 discusses the basic ideas of tumor theranostics. Chapter 3 covers the advances of magnetic nanomaterials in cancer theranostics with an emphasis on magnetic resonance imaging, hyperthermia, photodynamic therapy, photothermal therapy and drug/gene delivery. Chapter 4 surveys the synthesis, surface functionalization, properties, and biomedical applications of different types of semiconductor quantum dots (QDs) for tumor imaging, and targeted drug delivery and cancer therapy, both *in vitro* and *in vivo*. Chapter 5 highlights the progression of mesoporous silica nanoparticles with diverse functional structure and control morphology for tumor theranostics. Chapter 6 moves up to surface enhanced Raman scattering technique based on silver/gold nanoparticles for tumor theranostics. Chapter 7 describes the critical issues confronted in the design, synthesis, and characterization of polymeric nanosystem and their applications in theranostics. Chapter 8 introduces carbon nanomaterials including carbon nanotubes, graphene, carbon dots, fullerene and nanodiamonds, used for tumor targeting theranostics for improving tumor imaging, earlier detection and specific drug delivery. Chapter 9 discusses the development of cationic nonviral lipopolyplex for targeted gene delivery combined with diagnosis and therapy. Chapter 10 considers the possibility of nano/microbubbles for imaging and therapeutic purpose in tumor theranostics. Chapter 11 discusses the mechanism of upconversion fluorescence and bio-applications of upconversion nanomaterials in bioimaging, tumor photodynamic therapy and drug/gene delivery. Chapter 12 introduces the functional dendrimers as theranostic vehicles for the cancer treatment. Chapter 13 describes the nontoxic, non-immunogenic, biocompatible and biodegradable protein-based nanoparticles for tumor targeting theranostics. Chapter 14 makes a summary of this book and gives a perspective of nanomaterials for targeting theranostics.

We hope that the publication of this book will accelerate the spread of ideas that are currently trickling through the scientific literature. Also a greater understanding of the potential and challenge of nanomaterials for tumor targeting theranostics is highly expected for the practical clinical use. We thank all the authors and reviewers who made valuable contributions to this book for helpful discussions, comments, and corrections during the writing and revision process.

Dr. Mingqian Tan
Professor of Division of Biotechnology,
Dalian Institute of Chemical Physics,
Chinese Academy of Sciences,
Dalian 116023, China
Dr. Aiguo Wu
Professor of Ningbo Institute of Materials Technology and
Engineering, Chinese Academy of Sciences,
Ningbo 315201, China

Chapter 1

Nanomaterials as Therapeutic/Imaging Agent Delivery Vehicles for Tumor Targeting Theranostics

Mingqian Tan[1,2,3*], *Yanfang Wang*[1], *Xiaojie Song*[1], *and Yaqi Wu*[1]

1 Division of Biotechnology, Dalian Institute of Chemical Physics, Chinese Academy of Sciences, Dalian 116023, China

2 Liaoning Key Laboratory of Food Biological Technology, School of Food Science and Technology, Dalian Polytechnic University, Dalian 116034, China

3 SENESCYT/Proyecto Prometeo, Instituto de Investigación Científica, Universidad Técnica de Manabí, Portoviejo, Ecuador

Malignant tumor is a significant cause of death worldwide and the effective tumor treatment remains a challenging task. The successful cancer treatment is highly dependent on accurate diagnosis and specific therapy approaches. Nanomaterials are of great scientific interest as effective carriers for antitumor drug or imaging agents due to their large surface area to volume ratio or large nanocavity. Over recent decades, a variety of nanomaterials including magnetic nanomaterials, quantum dots, mesoporous silica nanomaterials, surface enhanced Raman scattering (SERS) gold/silver nanomaterials, carbon nanomaterials, lipopolyplex nanoparticles, nano/microbubbles, upconversion nanomaterials, dendrimers and protein-based nanomaterials have been developed as nanocarriers for tumor imaging and therapy. Unlike the traditional low molecule weight contrast media or therapeutic drugs, these new nanomaterial-based medicines enable the incorporation of tumor targeting moiety, imaging agents and anticancer drug in one system. The tumor theranostics combining diagnostic and therapeutic agents in one nanocarrier system has the benefit of early detection of tumors, simultaneous monitoring the disease progression and targeting tumor therapy with minimum side effect. This chapter highlights the advantages of nanomaterials and the emerging concept of theranostics for the effective tumor treatment in the next generation nanomedicine.

1.1 Introduction

Malignant tumors are cancerous masses of tissue created by uncontrolled cells that can spread throughout the body and often cause death unless receiving adequate treatment. They have become one of the leading causes of death worldwide.[1] The National Cancer Institute (NCI) estimated that, in 2010, there

* Corresponding author: mqtan@dicp.ac.cn

were more than 1.5 million new cases of cancer in the United States and more than half a million will die of it.[2] The early diagnosis and effective treatment of malignant tumors pose a special challenge due to cancer metastasis to other organs or tissue.[3] Surgery, radiotherapy, chemotherapy, immunotherapy and combination therapy are all well accepted treatment techniques of malignant tumors. However, all these methods are currently carried out for palliative purposes or to induce remission and the survival rate is significantly lower when the malignant tumor is detected in its late stage, *i.e.* after it metastasizes to another non-adjacent organ or part.[4] Although remarkable advances have been made towards accurate tumor diagnosis and therapy, treatment options are still limited and the prognosis is poor for the metastatic cancer patients in many cases due to the lack of specificity and sensitivity, along with the serious off-target side effect.[5] Therefore, much effort in malignant tumor research has been focused on improving the sensitivity and accuracy of the diagnosis for the purpose of early diagnosis and better treatment.

Over the last few decades, considerable efforts have been put forth to the development of injectable drug delivery platforms as effective carriers for the anticancer drug or imaging media.[6–8] Nanomaterials are particularly advantageous for this purpose and have served a new role in malignant tumor research by providing a variety of nanotechnology platforms.[9] Nanomaterials are chemical substances or materials that are with at least one external dimension in the size ranging from approximately 1–100 nanometers,[10] which hold great potential in tumor diagnostics and therapeutics. Despite the size restriction, nanomaterials commonly refer to the structures that are up to several hundred nanometers in size and that are prepared by top-down or bottom-up methods with starting reactants.[11] In view of their huge surface to volume ratios and unique nanostructures, nanomaterials can carry a variety of diagnostic or therapeutic agents *via* chemical conjugation or physical encapsulation. When intravenously injected, nanomaterial platforms are known to have the ability of targeting tumors *via* passive accumulation and/or active-targeting approaches.[4] Applications of the nanomaterials for tumor treatment include early diagnosis of cancer, molecular imaging and targeted therapy.[12] The nanomaterial-based diagnosis (imaging) and therapy have been studied separately before and the combined delivery of the therapy drug and imaging media is able to achieve better therapeutic efficacy and high-quality images without repeat severe toxicity to normal tissue. This can be defined that the nanomaterials can co-deliver therapeutic and imaging agents for simultaneous tumor diagnosis and therapy. To this end, theranostic nanomaterials for both therapeutic and diagnostic purposes have emerged as new-generation delivery systems for tumor treatment (Fig.1.1).

Theranostics is a term originally coined in 2002 by Picard to define a material that combines the dual functions of therapy and diagnostic imaging.[13]

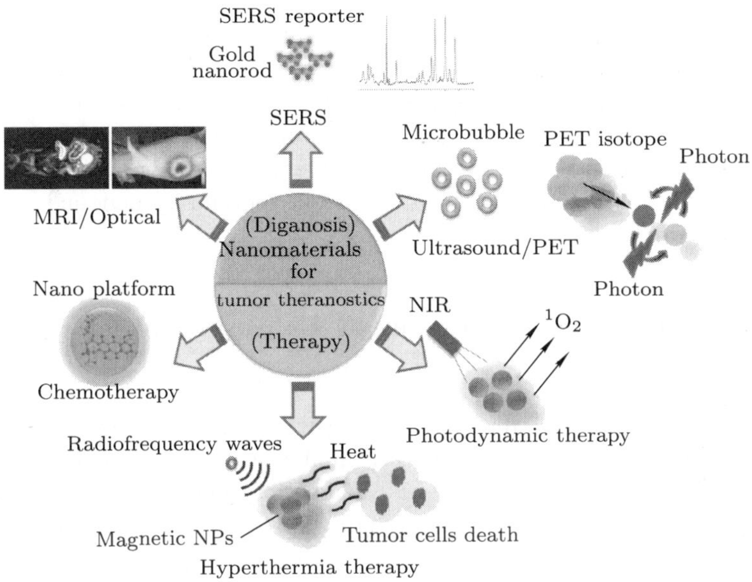

Fig. 1.1 Schematic illustration of nanomaterials for tumor theranostics.

A more simple definition of theranostics made by Warner is diagnostics plus therapy.[14] The concept describes the simultaneous delivery of therapeutic drugs and diagnostic imaging media using the same delivery vehicle. Compared with the traditional separate materials for these dual-purpose deliveries, theranostics combines these features into one dosage and has the possibility to realize same biodistribution and selectivity for both imaging and therapeutic agents. The advantage of theranostic technology includes gaining the ability to monitor and tune the therapy according to the disease progresses, providing seamless integration of diagnosis, therapy, follow-up and different therapeutic modalities. It is anticipated with a great hope that theranostic agents will be able to provide malignant tumor patients with various diagnostic and treatment options that are suitable for individuals. Improved prognoses are highly expected as the theranostic agents that can monitor real-time therapeutic efficacy after tumor treatment and help the clinician to make the next step therapeutic decisions.

In this chapter, we will summarize various types of nanomaterials including magnetic nanomaterials, quantum dots, mesoporous silica nanoparticles, surface enhanced Raman scattering (SERS) gold/silver nanomaterials, polymeric nanosystems (dendrimers), carbon nanomaterials, lipopolyplex nanoparticles, nano/microbubbles, upconversion nanomaterials, and protein-based nanomaterials, and discuss their applications for tumor targeted theranostics. The chapter will begin with strategies for nanomaterials as delivery vehicles for tumor theranostics, followed by a short introduction of targeting effect of nanomaterials including passive targeting and active targeting effects. We will focus on the discussion of theranostic applications of above-mentioned nanomaterials delivery systems with the emphasis of ① strategies for nanomaterials as delivery vehicles for theranostics, ② targeting effects of nanomaterials, and ③ multi-functional theranostic systems, including the molecular imaging modalities as well as various nanomaterials current available for tumor theranostics. Finally, the limitations and future challenges of current theranostic nanomaterials systems will be addressed.

1.2 Strategies for Nanomaterials as Delivery Vehicles for Tumor Theranostics

The unique properties of nanomaterials, such as the size-dependent effect, good physical stability, ultra-small sizes, tunable surface chemistry, and special photo, electronic, heat, ultrasound and magnetic properties, have made them good candidates as therapeutic and imaging agent delivery vehicles for tumor targeting theranostics. The most interesting characteristic of nanomaterials for tumor targeting theranostics is their tunable sizes that are small enough to avoid opsonization but big enough for carrying imaging and therapeutic agents to reach the tumor sites.[15] Their nanosize has a particular advantage of getting injected intravenously during the process of tumor theranostics. Depending on their different preparation methods (self-assembling or not), the nanomaterials can be categorized by liposomal delivery systems including liposomes, polymeric micelles, polysaccharide nanomaterials, and lipopolyplex nanoparticles, dendrimers, as well as nanoparticles delivery systems, such as magnetic nanomaterials, quantum dots, mesoporous silica nanoparticles, gold nanomaterials, carbon nanomaterials and protein-based nanomaterials. We summarized the representative nanomaterials for tumor theranostics in Table 1.1.

Nanomaterials used for tumor theranostics can simultaneously deliver diagnostic and therapeutic agents in a single procedure. They must be designed to interact with the biological components, such as proteins, genes and cells without interfering their functions. After surface modification, they must maintain their physical properties during the period of staying in the body. The nanomaterials must be nontoxic for the applications in tumor theranostics. The size, surface characteristics, stability and shape of nanomaterials play a key role in its pharmacokinetic properties *in vivo.* Their size in nanometer dimension is advantageous for the purpose of intravenous injection. Nanomaterials with size less than 5 nm are cleared from the circulation by rapid renal clearance through kidney, and large microsized particles are likely removed by the reticuloendothelial system (RES) of liver and spleen.[40] To this regard, large particles can accumulate in the circulation for an extended period of time when injected intravenously. In addition, the particle shape is also known to influence the biodistribution of nanomaterials. For example, filomicelles with flexible filaments shape were found persisting in the circulation up to one week after the intravenous injection into rodents. The residence time in the body is about ten times longer than their spherical counterparts and is more persistent than any known synthetic nanoparticle.[41] The prolonged residence of filomicelles can help the effective delivery of the anticancer drug paclitaxel and shrinkage human-derived tumors in mice.

Table 1.1 Representative nanomaterials for tumor theranostics.

Nanomaterials	Imaging modalities	Therapeutic drug	Targeting	Ref.
Liposomes-QDs	OI (QDs)	DOX	N	16
Ferri-liposomes	MRI	Biotherapeutics	N	17
Liposome-siRNA	MRI, OI	siRNA	N	18
PEG-PLGA-NPs	N	Paclitaxel	N	19
Chitosan NPs	N	DOX, DNA	RGD	20
Lipoplexes NPs	BL	siRNA	Folate	21
Phospholipid micelle	OI (QDs)	Doxorubicin	N	22
Polymeric micelle	MRI	Doxorubicin	N	23,24
Magnetic NPs	MRI/OI	siRNA/HT	RGD	25–28
QDs	OI	DOX/PDT	Folate	29–31
Mesoporous silica	OI	siRNA	Aptamer	32
SERS gold NPs	OI	DOX	Antibody	33
Dendrimers	MRI	N	Peptide	34
Carbon NPs	OI	PDT	N	35
Lipopolyplex NPs	OI	DNA	Peptide	36
Nano/Microbubbles	US	DOX	N	37
Upconversion NPs	OI	PDT	Folate	38
Protein NPs	OI	DOX	Peptide	39

Abbrev.: MRI – magnetic resonance imaging; OI – optical imaging; PDT – photodynamic therapy; US – ultrasound; DOX – doxorubicin; HT – hyperthermia treatment; BL – bioluminescence.

Bottom-up, top-down and/or combinational strategies of design and preparation for nanomaterials have been employed to improve the physicochemical properties including solubility, permeability, shape, morphology, and biopharmaceutical properties, such as enhanced selectivity, stimuli-sensitivity, bioavailability and multifunctionality.[42] The bottom-up strategies require the ability to predict the shape of nanostructures and to synthesize nanomaterials in a self-assembling or highly controlled manner. The resulting platforms mimic biological systems and have the potential for encapsulating and delivering diagnostic and therapeutic agents.[43] The top-down approaches enable the production of monodispersed nanoparticles with simultaneous and independent control over particle size, shape, composition and surface functionality.[43] Using nanomaterials as the carriers of diagnostic and therapeutic agents for tumor targeting theranostics, it may be possible to achieve ① simultaneous delivery of diagnostic and therapeutic agents for combination therapy, ② improved delivery efficacy of hydrophobic drugs, ③ targeted delivery of drugs or diagnostic agents in a cell- or tissue-specific manner, ④ transcytosis of drugs across tight epithelial and endothelial barriers, ⑤ visualization of drug delivery process and delivery sites by combining therapeutic agents with imaging modalities, and ⑥ real-time read on the *in vivo* efficacy of a therapeutic agent.[11] Various nanomaterials may be able to provide a more rational and fundamental basis for the effective and safe therapeutic nanomedicine development.

1.3 Targeting Effects of Nanomaterials

Tumor targeting functionality is a vital segment of nanomaterials for tumor theranostics. The most of current clinically approved low molecular weight diagnostic and therapeutic agents are relatively simple and generally lack targeting moieties. Since the first targeted liposome drug was reported in the literature, much effort has been devoted towards improving therapeutic effects and reducing side effects for normal tissues.

1.3.1 *Passive Targeting*

Tumors consist of some normal tissue and cancer cells and their blood vessels are fenestrated because of the rapid vascularization. These tumor vessels are more permeable or leakier than normal blood vessels in other organs or tissues. This phenomenon has been identified as enhanced permeability and retention (EPR) effect or passive targeting (Fig.1.2).[44,45] Passive targeting of nanomaterial delivery vehicles takes advantage of the permeability of tumor tissues and allows the nanomaterials staying with the tumor interstitium. As reported by Maeda *et al.*[46], the EPR phenomenon is based on two factors: ① the more disorder of the capillary endothelium in a tumor tissue, the more permeable of nanomaterial vehicles for tumor theranostics; ② the resulted nanomaterial vehicles accumulation is due to the lack of tumor lymphatic drainage in the tumor tissue.[47]

1.3.2 *Active Targeting*

Although many studies on nanomaterials have shown the passive targeting effects on tumor imaging or therapeutics, the *in vivo* applications have revealed that the tumor specificity was only slightly improved compared with the controls in many cases.[40,48] In order to overcome such issues, a ligand-based active targeting has been developed for nanomaterials in tumor theranostics (Fig.1.2). Active targeting is usually achieved by conjugating nanomaterials to a targeting moiety, such as folic acid (FA) molecules, peptide, aptamer or antibody, thereby allowing preferential accumulation of the delivery vehicles in the tumor tissue. This approach is used to direct nanomaterial delivery vehicles to cell surface carbohydrates, receptors, and antigens as well as the extra cellular matrix.[47] Almost any nanomaterial delivery vehicle can be targeted to a tumor tissue by conjugating them to a targeting ligand with specificity for a

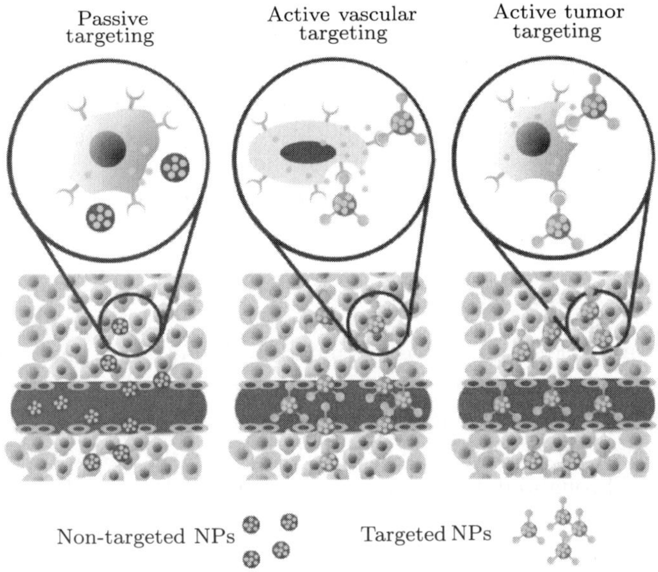

Fig. 1.2 Schematic illustration of passive and active targeting. Image is adapted with permission from Ref. 11. Copyright (2009) American Chemical Society.

pathologic cell type.

The targeting moieties play an important role in adhering nanomaterial delivery vehicles to the target site. In immunohistochemistry, antigens in tissue sections are localized by the use of labeled antibodies as specific targeting reagents through antigen-antibody interactions. The bonds formed between a targeted nanomaterial and the tumor receptor may not be strong enough to stop the motion of delivery vehicles, especially for the bigger size nanoparticles. As a result, the targeted nanomaterial delivery vehicles may not be anchored to the specific site. Therefore, the number of targeting groups should be well designed and investigated for the bigger size nanomaterial delivery vehicles. To guarantee the adhesive power of the nanomaterial delivery vehicles, it is necessary to attach enough targeting moieties to the nanomaterials.[49]

1.3.2.1 *Folic Acid (FA)*

FA (MW = 441 Da①) is a vitamin that is essential for the proliferation and maintenance of all cells.[50] Over-expressed receptor of FA is often observed on some cancer cells[51–53], activated macrophages[54,55], and the proximal tubule cell of the kidney[56]. Ovary, lung, kidney, endometrium, breast, brain, and colon cancers commonly up-regulate the folate receptors.[50] The FA binds to

① 1 Da = 1 u = $1.660\ 54 \times 10^{-27}$ kg.

folate receptors on these cells with high affinity ($K_d \approx 10^{-9}$ M②) and enters FA-expressing cells by receptor-mediated endocytosis.[57] When FA is used as a targeting moiety, it is stable, inexpensive, and non-immunogenic compared with the monoclonal antibodies. Many applications of FA for targeting a variety of therapeutic agents have been reported for the treatment of malignant diseases. There have been about seven folate-based targeting therapeutic agents tested in clinic for human cancers.[58–64]

1.3.2.2 *Monoclonal Antibodies*

Monoclonal antibodies are a class of monospecific proteins that are produced by one type of immune cell derived from a single parent cell and can recognize a specific target. Monoclonal antibodies were first reported by Milstein in 1975,[65] followed by the development in 1986 of the murine anti-CD3 mAb for therapies.[66] Throughout the progression regarding the development of monoclonal antibodies, there have been four major antibody types that were developed: murine, chimeric, humanized and human. Murine monoclonal antibodies were obtained by hybridoma technology, which were easy to cause allergic reactions due to the dissimilarity between murine and human immune systems.[49] Advances in biotechnology reduced this problem with the production of high-quality chimeric, humanized, or complete human monoclonal antibodies. Examples include natalizumab, a humanized monoclonal antibody against the cellular adhesion molecule α_4-integrin that is used in monoclonal antibody therapy for multiple sclerosis and Crohn's disease.[67,68] Although monoclonal antibodies possess a high affinity to the target, their relatively large size (~25 nm) may hinder diffusion into the interstitium of blood vessel.[69]

1.3.2.3 *Antibody Fragments*

Generally, the molecular weight of antibodies is approximately 150 kDa and they have a slow blood clearance. This increases the opportunities of accumulation in normal tissues and limits the delivery efficiency to the tumor sites.[70] Therefore, use of small size antibody fragments, such as Fabs (molecular weight ~50 kDa) and Fvs (molecular weight ~50 kDa) can facilitate the penetration into target tissues and are cleared much faster from the blood. This can significantly improve the targeting efficiency.

1.3.2.4 *Peptides*

Peptides are short polymerized α-amino acids in a defined order. Unlike antibodies and antibody fragments, they have relatively small sizes, they are cleared rapidly from the blood and they are less likely to be immunogenic.

② 1 M = 1 mol/dm^3.

Small peptides are easily synthesized from the constituent amino acids. Now, peptides with hundreds of amino acids in length can be made, including full proteins like ubiquitin (8.5 kDa).[71] The synthetic small peptides can be incorporated into nanomaterials as targeting groups. For example, the use of RGD peptide facilitates specific binding to $\alpha_v\beta_3$-integrin,[72] while CLT1 peptide can promote binding to fibrin-fibronectin complexes.[73] There are a bulk of literatures that describes RGD peptide as targeting groups for tumor imaging and therapy.[74–76]

1.3.2.5 *Aptamer*

Aptamers are factitious oligonucleotides selected by the process of cell-based systematic evolution of ligands through exponential enrichment (SELEX).[77] The obtained aptamers have the ability to bind tightly and specifically to the tumor cell surface protein or biomarkers produced by tumor endothelial cells.[78,79] In the early 1990s, the SELEX method to select aptamer by an iterative *in vitro* selection procedure was first reported independently by two research groups, Ellington's and Tuerk's.[80,81] Aptamers are generally nontoxic, and have low molecular weights (8–15 kDa), which can rapidly diffuse into tissues and organs, thus permitting faster targeting in drug delivery.[78] Other advantage of aptamers includes their high stability against heat sterilization. They can be easily modified onto the nanomaterials surface and aptamer-conjugated nanomaterials may thus be expected to allow loading diagnostic and therapeutic agents to generate targeted drugs for tumor theranostics.

1.4 Multifunctional Theranostic Systems

1.4.1 *Medical Imaging Modalities*

Different medical imaging modalities, including X-ray computerized tomography (X-CT), magnetic resonance imaging (MRI), ultrasound imaging (UI), optical imaging (OI) and nuclear imaging (NI), have been developed to view the human body in order to diagnose, monitor or treat the tumors.[82] Each type of modality has different strengths and weaknesses and gives different information about the area of interest being diagnosed or treated.

The selection of an appropriate imaging modality is highly dependent on the biochemical process one wishes to visualize and the type of imaging data that is needed to be obtained.[83]

X-CT scanner was introduced in 1972 by G. N. Hounsfield at the Annual Congress of the British Institute of Radiology.[84] It is the greatest advancement in radiology since the clinical applications of X-ray. X-CT can produce thin two-dimensional (2D) sections of the body, approximately 1 mm in thickness and result in a three-dimensional (3D) anatomic image of the subject being scanned, achieving sub-millimetre spatial resolution.

MRI is a powerful diagnostic tool that uses magnetic waves and the spin of protons by a computer to display tissues as thin topographic slices with high spatial and temporal resolution.[85] Compared with X-CT, patients imaged by MRI are complete non-invasiveness and have very low risk. The MRI scans can provide high spatial resolution (sub-mm) and are very adept at morphological imaging and functional imaging. However, MRI is not safe for patients with metallic devices or implanted electrical devices.

UI exploits the properties and behavior of the high-frequency sound waves as they travel through the body.[86] The transducer of ultrasound scanner can emit ultrasound waves that enter the body, and some sound waves are bounced back to the transducer once striking an object and converted into electrical signals. The computer then processes these signals and displays them as images on the screen. This is useful for noninvasively real-time detecting differences between solid and liquid materials in the body.[87,88] Ultrasound imaging is a unique clinical tool that can be used both for diagnostic imaging and as a therapeutic tool.

OI is a sensitive imaging modality without strong safety concerns like X-CT or nuclear imaging.[89] Its penetration depth highly depends on the fluorescent probes and the absorption/reflectance of the body. For example, the absorption of endogenous chromophores in living tissue results in penetration depths of only several millimeters when the excitation source is below 700 nm. In the near infrared (NIR) region of 700–900 nm, light can penetrate more deeply, to depths of several centimeters.[90]

NI includes single photon emission computed tomography (SPECT) and positron emission tomography (PET). SPECT uses a radioactive material as a tracer to produce gamma rays then converts the information into a 3D image to show how blood flows to tissues and organs.[91,92] **PET-CT** is also an NI technology that can produce a 3D image or picture of functional processes in the body.[93,94] A molecule tagged with a positron emitting isotope is used in a PET scan. When the positrons annihilate with nearby electrons, two photons are emitted and directed 180 degrees apart in opposite directions. The scanner can detect these two photons and convert the signals into 2D or 3D images.[95] The foremost advantage of PET is its high sensitivity, which can reach as low as 10^{-12} mol/L of radio labeled substrate.[96,97] However, the major disadvantage of PET is that most of the short-lived (in hours) probes must be made with a cyclotron on site. This factor can make PET prohibitively expensive.

1.4.2 *Magnetic Nanomaterials*

Magnetic nanomaterials like iron oxide nanoparticles are an important class of biomaterials for various biomedical applications, such as imaging, magnetic separation, hyperthermia, drug/gene delivery and cell control.[98] In principle, magnetic nanomaterials contain one or more than one type of magnetic elements, such as iron, cobalt, nickel, manganese, gadolinium, and their alloys, possessing substantial saturation magnetization values at room temperature.[99] The interest in utilizing magnetic nanoparticles for tumor theranostics has increased very recently.[25,27,100,101] Figure 1.3 shows the schematic diagram of the magnetic nanoparticles as versatile platform materials for tumor theranostics. The reasons are mainly because ① they are the most popular agents that have the ability to shorten T_2 relaxation time resulting in negative contrast for MRI;[101] ② functional surface is suitable for carrying the diagnostic and therapeutic agents; ③ easy physiochemical property controlled by the well-developed surface modification methods; ④ they are attractive platforms that show the potential for both MRI diagnosis and hyperthermia therapy; ⑤ the major element of iron oxide nanoparticles, ferrum, is hardly considered as toxic substance for *in vivo* applications because they can be degraded and metabolized into the serum Fe pool to form hemoglobin or to enter other metabolic processes.[4]

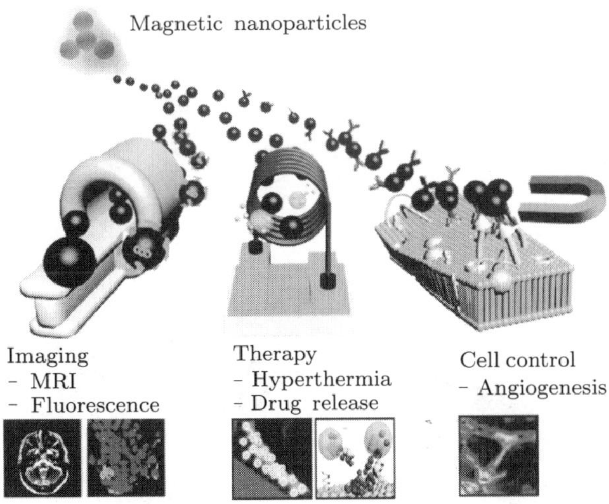

Fig. 1.3 Magnetic nanoparticles with the functions of tumor targeting, MRI imaging, thermal therapy, precise drug release, and cell signaling controls can be used as versatile platform materials for tumor theranostics. Image is adapted with permission from Ref. 102. Copyright (2011) American Chemical Society.

As an efficient T_2 contrast agent, magnetic nanoparticles with higher relaxivity (r_2) coefficient are very helpful for detecting small sized pathogenic targets precisely at an early stage. Jiang *et al.*[103] synthesized $(Zn_{0.4}Mn_{0.6})Fe_2O_4$ nanoparticles by substituting Fe^{2+} with Zn^{2+} and Mn^{2+} as the contrast agents for MRI. The r_2 reaches up to 676 $mM^{-1}{\cdot}s^{-1}$ which is 6 times higher than that of Feridex, one of the most representative iron-oxide based contrast agents.[98] Transmission electron microscopy (TEM) image of $(Zn_{0.4}Mn_{0.6})Fe_2O_4$ nanoparticles shows that the nanoparticles are monodispersed with a particle size about 15 nm in diameter. In addition, these $(Zn_{0.4}Mn_{0.6})Fe_2O_4$ nanoparticles can continuously emit heat for hyperthermia therapy *via* Néel and Brownian pathways [Fig.1.4(a)] upon exposure to the alternating external magnetic field. The hyperthermia device is shown in Fig.1.4(b), which consists of resistor, inductor, and capacitor circuit causing the induction of an AC field with a water-jacketed coil. The measured specific loss power (SLP) of $(Zn_{0.4}Mn_{0.6})Fe_2O_4$ is about 4 times larger than that of Feridex [Fig.1.4(c)]. Consequently, the application of such magnetic nanoparticles for hyperthermia in cancer cells resulting in more death of HeLa cancer cells which of those were treated with $(Zn_{0.4}Mn_{0.6})Fe_2O_4$ [Fig.1.4(d)]. The fluorescence microscope images of the viability of HeLa cells upon hyperthermia treatment with $(Zn_{0.4}Mn_{0.6})Fe_2O_4$ further confirm the result [Fig.1.4(e,f)]. These results suggest that magnetic nanoparticles can be

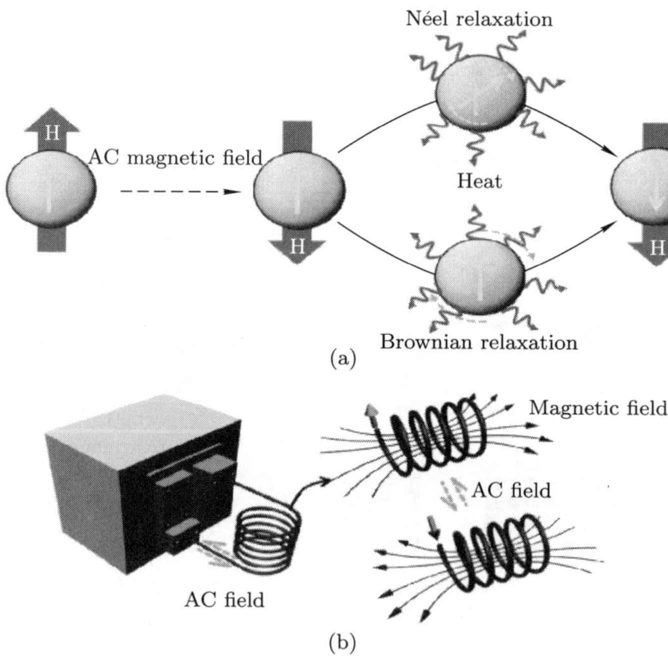

Fig. 1.4 $(Zn_{0.4}Mn_{0.6})Fe_2O_4$ magnetic nanoparticles for HeLa cell hyperthermia. (a) These nanoparticles can emit heat for hyperthermia therapy *via* Néel and Brownian pathways. (b) Experimental apparatus. (c) Specific loss power (SLP) values for $(Zn_{0.4}Mn_{0.6})Fe_2O_4$ and Feridex in AC magnetic field. (d) Percentage of HeLa cells killed after hyperthermic treatment with $(Zn_{0.4}Mn_{0.6})Fe_2O_4$ and Feridex. Fluorescence microscopy images of HeLa cells treated with (e) $(Zn_{0.4}Mn_{0.6})Fe_2O_4$ and (f) Feridex. Image is adapted with permission from Ref. 98. Copyright (2011) American Chemical Society.

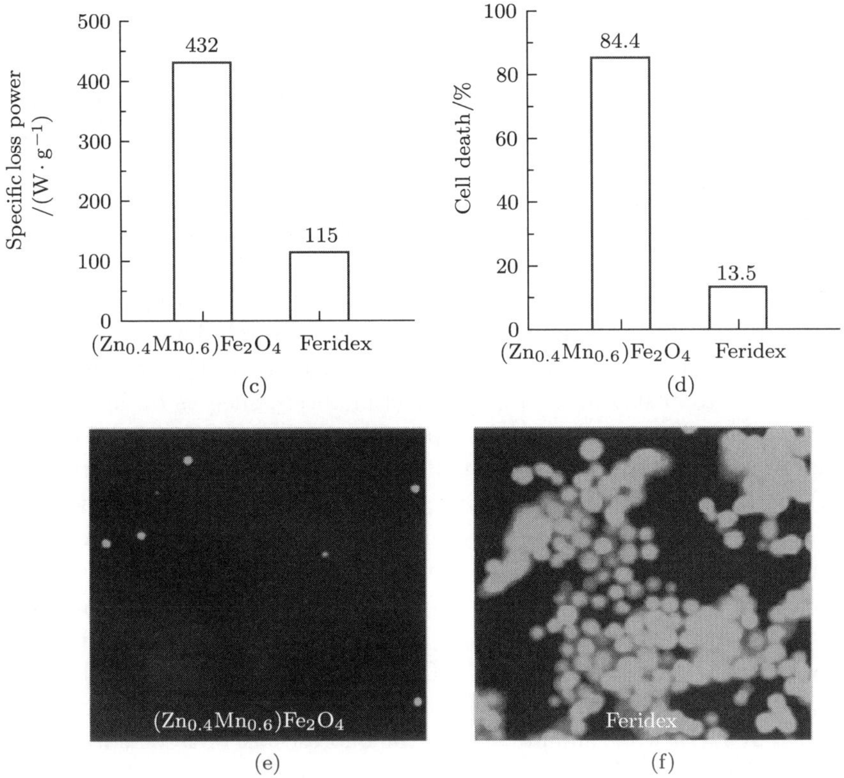

Fig. 1.4 (Continued).

used as an efficient platform for noninvasive tumor MR imaging and hyperthermia therapy. Other applications of these nanoparticles including targeted imaging of specific biological objects, remote actuation for controlled drug release, cell signaling for therapeutics and monitoring of the efficacy of the treatments, have been well described in the literature.[100,102]

1.4.3 *Quantum Dots*

Quantum dots (QDs) are a class of semiconductor nanocrystals that confine the motion of conduction band electrons, valence band holes, or excitons (bound pairs of conduction band electrons and valence band holes) in three dimensions.[104] They were first discovered in 1981 by Alexei Ekimov in transparent dielectric matrix[105] and the term was coined by Mark Reed[106]. There are several characteristics that distinguish QDs from traditional organic dyes: ① QDs are nanocrystals with the size of a few nanometers and have composition- and size-dependent absorption and light emission;[107] ② size of QDs are smaller than the so-called Bohr exciton radius (several nanometers) and the energy levels are quantized with values directly related to the QDs' size (quantum confinement effect); ③ lifetime (>10 ns) of QDs is longer than most of the organic dyes due to the radiative recombination of an exciton. This makes it possible for their applications in time-gated fluorescence detection by separating the signal from that of the short-lived background autofluorescence or other species.[108]

A variety of quantum-sized nanomaterials, including group II-VI QDs (CdS, CdSe, and CdTe)[109,110], group III-V QDs (InP, InAs, GaAs, GaN)[111], and recently developed group IV QDs (C, Si, and Ge)[112–114], have been developed and extensively explored for both biomedical imaging and drug delivery, especially for tumor diagnosis and therapy studies. The well-developed chemistry enables the conjugation of drugs or targeting ligands on QDs resulting in an "all-in-one" nanoplatform with multi-functionalities, such as targeting, therapeutic and imaging modalities.[115] A representative example of QD-based tumor theranostics is the QD-aptamer(Apt)-doxorubicin (DOX) conjugate (QD-Apt-DOX) for the synchronous cancer imaging and the traceable drug delivery.[29] The CdSe/ZnS core-shell QDs were surface modified with A10 RNA aptamer which recognized the extracellular domain of the prostate specific membrane antigen (PSMA) and served as drug vehicles. After carrying the anticancer drug, DOX, the fluorescence of both QDs and DOX were quenched due to the interaction of DOX within aptamer-QDs conjugated through a Bi-FRET (fluorescence resonance energy transfer) mechanism: the fluorescence of QD was quenched by DOX while simultaneously the fluorescence of DOX was quenched by intercalation within the RNA aptamer resulting in the "OFF" state. After the uptake of QD-Apt-DOX by the targeted tumor cells, DOX gradually released from the conjugate, turning both QD and DOX fluorescence back "ON". Thus, this smart system can be used for delivery DOX to the targeted tumor cells and monitor the release of DOX by switching on the fluorescence, thereby enabling the synchronous fluorescent localization and killing of cancer cells.

1.4.4 *Mesoporous Silica Nanoparticles*

Mesoporous silica nanoparticles (MSNs) have been successfully synthesized by the groups of Cai[116] and Mann[117]. The inherent properties of small size effect, high surface areas, large pore sizes and volumes have led to the rapid development of MSNs for various applications.[118–121] Given that MSNs are bio-inert, biocompatible and easy to be modified with controllable morphology, they have been used as delivery vehicles for tumor theranostic purpose.[122,123]

Liong *et al.*[124] demonstrated the synthesis of multifunctional MSNs for cancer cell-specific delivery of hydrophobic anticancer drugs with dual-imaging capability of optical and MR imaging (Fig.1.5A). The size of MSNs loaded superparamagnetic iron oxide nanocrystals in the center was about 100–200 nm as displayed in Fig.1.5B and Fig.1.5C. After surface attachment with hydrophilic groups the stability of the nanoparticle dispersion was increased in aqueous solution. MSNs were further surface modified with fluorescent molecule WGA-Alexa Fluor 594 and folate as targeting ligand, and the pores were filled with drugs. Fluorescence microscopy images of cancer cells PANC-1 and human foreskin fibroblasts (HFF) treated with camptothecin (CPT)-loaded folate-NP (FA-NP) revealed that the FA-NP can selectively increase the delivery of drugs to the PANC-1 cells that overexpress α-FA receptor (Fig.1.5D). The FA-modified CPT-loaded NPs showed a considerable increase in the cytotoxicity to PANC-1 cells, but not to HFF cells. Therefore, MSNs have shown great potential as an attractive nanoplatform for tumor targeting drug delivery and bio-imaging applications.

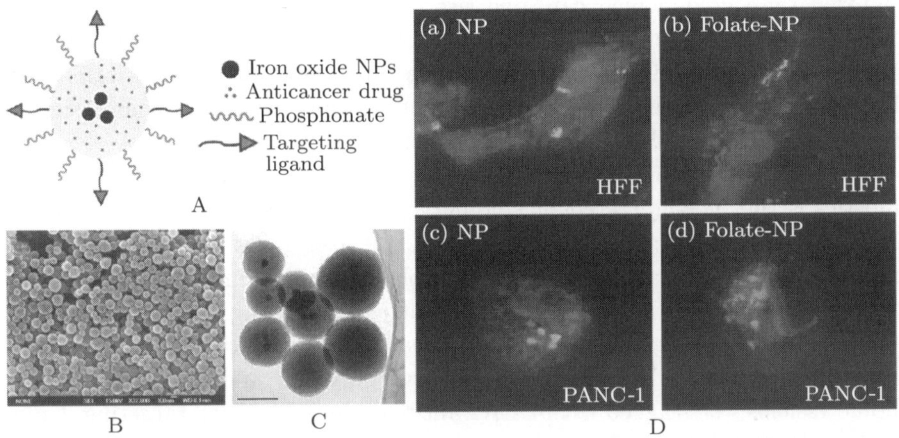

Fig. 1.5 A. Schematic illustration of multifunctional MSNs doped with iron oxide NPs, hydrophobic anticancer drugs, and surface modifications with phosphonate and FA. SEM (B) and TEM (C) images of the iron oxide NPs incorporated MSNs. D. Fluorescence microscopy images of the cancer cells PANC-1 and human foreskin fibroblasts (HFF) treated with camptothecin (CPT)-loaded FA-NP and NPs (green fluorescence). HFF cells treated with (a) NPs and (b) folate-NP, PANC-1 cells treated with (c) NPs and (d) folate-NP. The blue fluorescence shows the cell nuclei stained with DAPI, and red fluorescence indicates the membranes stained with WGA. Image is adapted with permission from Ref. 124. Copyright (2008) American Chemical Society. (color figure at the end of the book)

1.4.5 *SERS Gold/Silver Nanomaterials*

Gold/silver nanomaterials, such as nanospheres, nanorods, nanoshells, nanostars, and nanocages, have become attractive diagnostic and therapeutic agents for tumor research owing to their unique physicochemical properties, versatility of their morphologies, flexibility in functionalization, efficiency in drug delivery and their remarkable capacity to absorb and scatter light *i.e.* due to the localized surface plasmon resonance effects.[125,126] Recent advances in plasmonics have stimulated much research efforts in surface enhanced Raman scattering (SERS), which is emerging as a powerful tool for cancer cells theranostics.[33,127,128] SERS is an ultrasensitive vibrational spectroscopic technique that enhances Raman scattering by molecules adsorbed on rough metal surface.[129] It is one of the most powerful analytical techniques with the potential for identification of single molecule under ambient conditions.

SERS intensity is highly dependent on the resonance frequency of the noble metal substrate. One can obtain the greatest field enhancement when the plasmon frequency of metal substrate is in resonance with the laser radiation. Gold and silver nanospheres are the most widely used Raman enhancing substrates. By comparison, silver is a much more efficient Raman signal-enhancing material than gold, which can give rise to SERS signals 10-fold to 100-fold higher than that of gold.[130] However, silver nanomaterials have poor biocompatibility, uncontrollable size distributions, and only short-term stability.[128] Therefore, gold nanomaterials are still most commonly used in the design of SERS probes, especially those for *in vivo* tumor theranostics studies.

A good example of SERS-based gold nanoparticles for cancer-targeted theranostics is the bioconjugated plasmonic vesicles, reported by Song *et al.*[33] in 2012 (Fig.1.6). This new type of SERS-active assemblies as anticancer drug carriers can be simultaneously tracked by plasmonic imaging and Raman spectroscopy to monitor the targeted cancer intracellular drug delivery. As shown in Fig.1.6A, the gold nanoparticles were coated with the Raman reporter, amphiphilic polymer of PEG and pH-sensitive polymer, and drug-loaded plasmonic vesicles were tagged with HER2 antibody for cancer cell targeting. Upon tuning pH from 7.4 to 5.0, the dissociation of plasmonic vesicles was triggered due to the dissolution of the hydrophobic polymer brushes. Strong fingerprint SERS signal of the vesicle dropped dramatically and became very weak upon the pH change. The plasmonic vesicles incubated with HER2-positive SKBR-3 breast cancer cells experiment demonstrated that the plasmonic vesicles can quickly bind to the SKBR-3 cancer cells. These results were confirmed both by the dark-field plasmonic imaging and Raman spectroscopy. While HER2-targeted vesicles labeled SKBR-3 cells displayed the strong fingerprint signal of the Raman reporter, significantly weaker Raman signals were detected from SKBR-3 cells exposed to nontargeted vesicles and HER2-nagetive MCF-7 cells incubated with targeted vesicles [Fig.1.6 B(a)]. SKBR-3 cells labeled with HER2-targeted vesicles also showed gradually reduced SERS intensity in the same timeline [Fig.1.6 B(b)]. The dual-modality imaging of DOX-loaded plasmonic vesicles in live SKBR-3 cells is shown in Fig.1.6C. The strong fluorescence of DOX at 590 nm and the light scattering of plasmonic vesicles allow independently tracking the cargo and carriers using fluorescence and dark-field imaging, respectively. The images illustrated the cellular binding and intracellular distribution of plasmonic vesicles and DOX molecules during the uptake process of cancer cells. After incubating with the cells for 30 min, the drug-loaded vesicles firstly recognized the cells through the targeting effect of the conjugated HER2 antibodies and primarily accumulated at the cell membrane region. Highly colocalized signals from DOX and plasmonic vesicles were observed at this stage, suggesting that DOX was still retained in the vesicles. After 60 min post-incubation, the scattering color of the plasmonic vesicles changed from orange-red to orange-yellow, indicating an increase in interparticle spacing. At the same time, the fluorescence of DOX was not colocalized with the vesicles and strong signal appeared in the cytosol, suggesting that DOX was released from the vesicles into the cytoplasm. After 90 min post-incubation, overlapped fluorescence of DOX and nuclei was found, indicating that the DOX molecules penetrated into the nuclei. The scattering color of the vesicle further changed from orange-yellow to green-yellow and the signal became very weak because of the disruption of plasmonic vesicles upon the pH change.

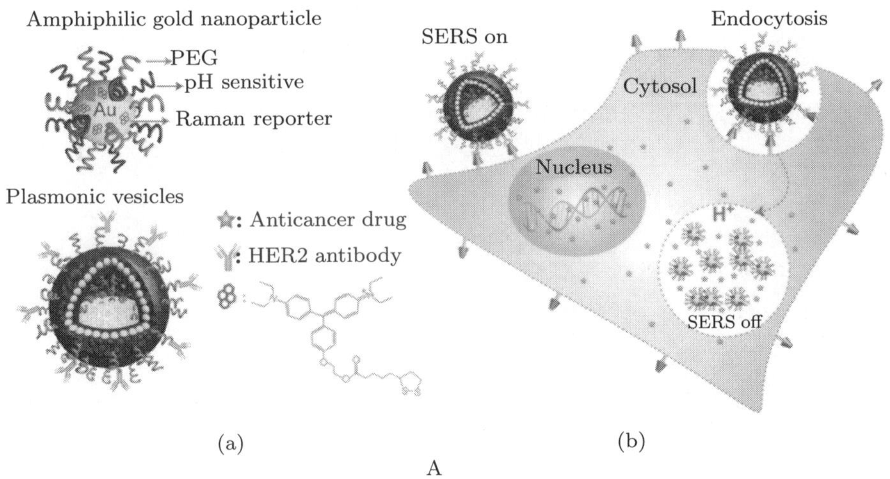

Fig. 1.6 A. (a) Schematic illustration of gold NPs modified with Raman reporter, PEG and pH-sensitive polymer grafts and the drug-loaded plasmonic vesicle conjugated with HER2 antibody for cancer cell targeting. (b) Schematic illustration of cellular binding, uptake, and intra-organelle disruption of SERS-encoded pH-sensitive plasmonic vesicles. B. (a) SERS spectra of SKBR-3 cells treated with HER2-targeted vesicles (1) and untargeted vesicle (2), MCF-7 cells treated with targeted vesicles (3), and SKBR-3 control cells (4). (b) SERS spectra of SKBR-3 cells treated incubated with targeted vesicles at 30 min, 60 min and 90 min, respectively. C. Dark-field (a, e, i), fluorescence (c, g, k), and the overlaid images (b, d, f, h, j, l) of SKBR-3 cells treated with DOX-loaded pH-sensitive plasmonic vesicles after 30 min incubation (a–d) and the post incubation images of the cells at 60 min (e–h) and 90 min (i–l). The red fluorescence shows DOX, blue fluorescence indicates cell nuclei. Image is adapted with permission from Ref. 33. Copyright (2012) American Chemical Society. (color figure at the end of the book)

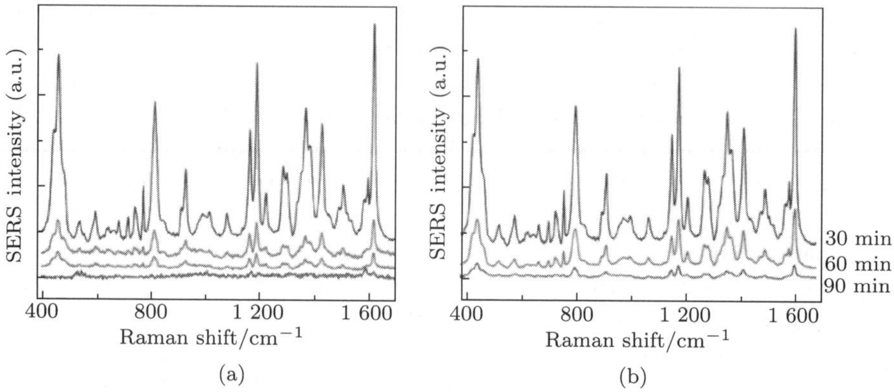

B

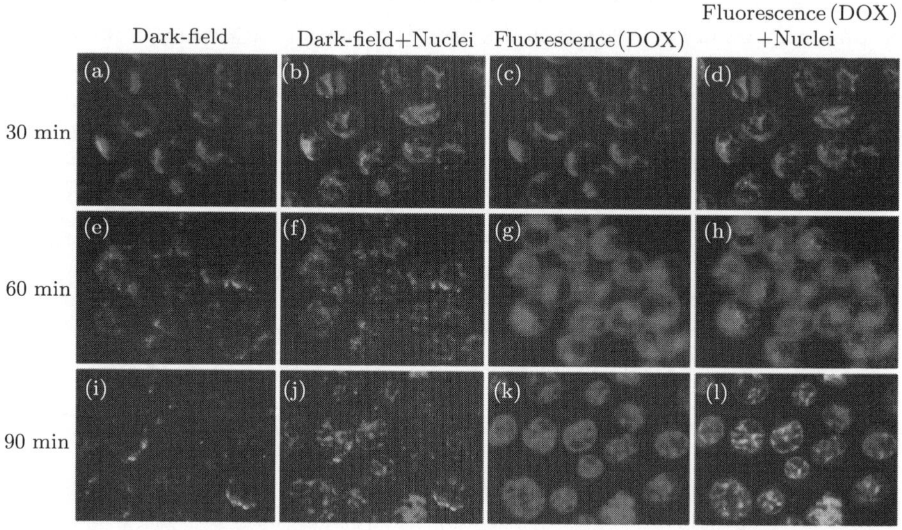

C

Fig. 1.6 (Continued).

SERS gold/silver nanomaterials can also be used for *in vivo* Raman imaging with pronounced advantages. For example, the SERS imaging technology is able to detect multiple targets simultaneously and use the NIR laser excitation for fluorescence *in vivo* imaging in deep tissues. Moreover, SERS gold/silver nanomaterials are superiorly photostable, making them suitable for the long-term imaging. The successful use of gold nanoparticles as SERS probes for *in vivo* tumor targeting was reported by Nie and co-workers in 2008.[131] A highly resolved SERS signal was obtained from subcutaneous as well as deep muscular injections of pegylated gold nanoparticles in live animals. The *in vivo* SERS spectra measured by using a 785 nm laser beam on the tumor site showed significant differences between the targeted and non-targeted nanoparticles in tumor signal intensities, while SERS signals from non-specific liver uptake were similar. Multiplexed *in vivo* SERS imaging in mice using four types of SERS gold nanoparticles was described by Gambhir and co-workers in 2008.[132] The concentration of each gold nanoparticle could be determined by using the component analysis of different Raman spectra. The encouraging result of this study was that up to ten different subcutaneously injected and five intravenously injected gold probes tags were identified from the Raman spectral fingerprints. Potentially, the *in vivo* multiplex-imaging ability may be used for high-throughput drug screening in living subjects. Therefore, the development of gold/silver nanomaterials for the use of SERS-active vesicles for tumor targeted drug delivery, *in vivo* Raman imaging and multiplex-imaging strongly imply that these gold/silver nanomaterials have a potential as a versatile theranostic platforms for tumor targeted theranostics.

1.4.6 *Polymeric Nanosystems*

Polymeric nanosystems, including micelle, liposome, polymer-drug conjugate, dendrimers, polymer-protein conjugate and polymer-DNA conjugate, are a class of nanomaterials which have been extensively used as therapeutic carriers and imaging medium.[133,134] Among them, dendrimers are highly branched polymers which have well-defined compact morphology with tunable and precisely controllable structures and sizes.[135] The hydrodynamic volume of a dendrimer is much smaller than that of a linear polymer with similar molecular weight. The common dendrimers include polyamidoamine (PAMAM) and poly (propyleneimine) (PPI) dendrimers.[136] As compared with the traditional anticancer drug, the polymeric nanosystems offer many advantages such as easy functionality, high drug solubility, good biocompatibility and stability, and improved delivery efficiency upon administration into the body. They can be functionalized *via* the terminal, main chain/side groups through either covalent chemical reaction like amide reaction, ester formation, click chemistry, or physical interaction including hydrophobic interaction, coordination, stereocomplex, supramolecular chemistry. National Cancer Institute (NCI, the U.S.) has performed extensive research on the translational development of polymeric cancer drugs to promote their advancement into clinical trials, such as demonstrating efficacy in a biological system, anticipated benefits of strategy on clinical cancer therapeutics and/or diagnosis, physicochemical characterization, compatibility with scale-up, as well as inherent toxicity testing of polymeric nanosystems. For example, albumin-bound paclitaxel nanoparticles, called Abraxane® with significantly lower toxic side effects in comparison to free paclitaxel have been approved by FDA (Food and Drug Administration) for clinical use in breast cancer therapy.[137]

With the emergency and development of the concept of theranostics, polymeric nanosystems have been considered as important carriers for the design of theranostic medicine because of their good biodegradability, biocompatibility, structural flexibility and functional versatility. The incorporation of therapeutic agents (drug) and diagnosis medium (imaging) into a polymeric system allows the real-time monitoring of drug distribution and assessment of therapeutic efficacy, which may gain much better treatment effects. A variety of theranostic polymeric nanosystems have been developed for an image-guided drug or gene delivery based on the different imaging modalities, such as MRI, ultrasound imaging and fluorescent imaging.[4,5,8,9,40,133] For example, Li *et al.*[138] have developed a multifunctional hybrid micelles based on amphiphilic block copolymers Pluronic F127 and peptide-amphiphile with shell-embedded magnetic nanoparticles for theranostic applications. The MR imaging agents, iron oxide nanoparticles, were *in situ* synthesized in the shell of the complex micelles under an ambient condition without using any organic solvents, followed by efficient loading the anticancer drug DOX into the hybrid micelles. Interestingly, the DOX release rate can be regulated by the magnetic nanoparticles in the shell of micelles. MR imaging based on these hybrid micelles showed clear contrast enhancement. Other examples include the optical imaging-based drug or gene delivery, the ultrasound-based drug or gene delivery, and the image-guided phototherapy and hyperthermia.

Recently, a new class of poly-L-lysine nanoglobules derived from a symmetric octa (3-aminopropyl) silsesquioxane (OAS) cubic core have been developed and assessed as nanocarriers for MR vascular imaging and molecular cancer imaging [Fig.1.7(a)].[139–143] The nanoglobular MRI agents have a relatively high payload of a stable Gd(Ⅲ) chelate and their effectiveness for molecular MRI of liver fibrosis was investigated in an experimental mouse model of liver fibrosis at 7 T.[34] Importantly, the stroma (fibrin-fibronectin complexes) specific peptides conjugated on the surface of the agents *via* click chemistry have showed an enhanced contrast for molecular MRI. As shown in Fig.1.7(b), Gd-P [peptide-targeted Gd(Ⅲ) nanoglobular MRI agent] produced higher contrast enhancement than Gd-C [non-targeted Gd(Ⅲ) nanoglobular MRI agent control] in fibrotic liver up to 54 min postinjection.[34] This study demonstrates that the peptide-targeted nanoglobular contrast agent may be valuable in detecting and characterizing liver fibrosis by probing the accumulation of fibronectin in fibrotic livers with molecular MRI. The effective contrast enhancement of the agents at reduced doses may reduce the dose-related toxic side effects. The MRI guided drug delivery based on the nanoglobule is now in progress and will further expand the nanoglobule applications for tumor theranostics.

Polymeric nanosystems provide a reliable and inexpensive substrate where both the therapeutic and imaging components can be attached, and thus more functionalities including tumor targeting moieties can also be conjugated to the tumor targeted theranostics. Moreover, they can be easily tailored to improve the biocompatibility and water solubility of the theranostic conjugate. There-

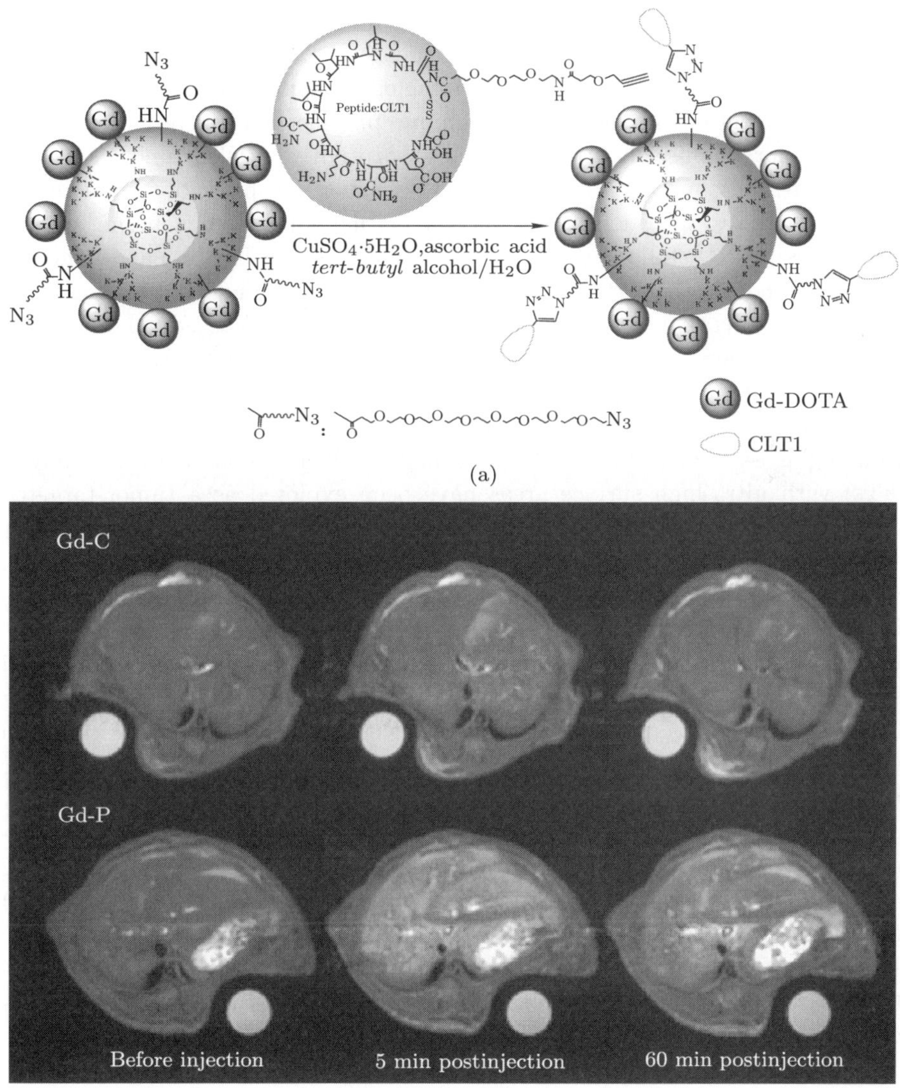

(b)

Fig. 1.7 (a) Schematic illustration of nanoglobular MRI contrast agents for tumor targeted molecular imaging. (b) T_1-weighted spin-echo images of fibrotic liver before, 5 min and 60 min after injection of peptide-targeted nanoglobular contrast agent (Gd-P) and its control (Gd-C) at 0.03 mmol (Gd)/kg.

fore, the polymeric nanosystems may offer great potential for tumor targeted theranostics.

1.4.7 *Carbon Nanomaterials*

The research of carbon nanomaterials in biological applications has received the extensive attention because of their unique physicochemical properties, such as multicolored emission, strong optical absorptivity, excellent photo-stability, biocompatibility and low toxicity.[112,144,145] Carbon is capable of forming many allotropes due to its valency. Various carbon nanomaterials including carbon dots (CDs)[145–147], fullerenes[148], carbon nanotubes (CNTs)[149], nanodiamonds (NDs)[150] and graphene[151] have been discovered for biomedical applications. The unique electrical or fluorescence properties of carbon-based nanomaterials are advantageous for use as diagnostic and therapeutic agent carriers in tumor targeting theranostics. For instance, CNTs and graphene have excellent electrical properties, which have been used as biosensing platforms for the detection of glucose[152] or lactate oxidase[153]. Single-walled carbon nanotubes (SWCNTs) with ultra-high surface areas have been explored as a tumor-targeted drug delivery system (DDS) for the efficient drug delivery in the treatment of cancer cells.[154,155] The hollow carbon nanomaterials, fullerene and CNTs, with inside cavity may be used as diagnostic and therapeutic agent carriers for tumor targeting theranostics.[156–158] Nanosized active carbon[159] and carbon nanotubes[160], with strong optical absorptivity in the NIR region have been studied for the photo-thermal cancer therapy. Moreover, unlike QDs containing heavy metals, carbon nanomaterials composed of intrinsically nontoxic carbon element are relatively safe and more environmentally friendly for the tumor targeting theranostics.[161]

Huang *et al.*[35] demonstrated a novel theranostics based on photosensitizer-conjugated carbon dots (CD-Ce6) with excellent imaging and tumor-homing ability for NIR fluorescence imaging monitored photodynamic therapy (PDT) treatment. The synthetic scheme for preparation CD-Ce6 is shown in Fig.1.8. PEG-coated CDs were covalently bonded with the photosensitizer Ce6 for photosensitizer fluorescence detection and PDT. The fluorescence emission of CDs overlaps with the absorption of Ce6 in the range of 450–650 nm. Therefore, the fluorescence emission of CDs can be used to excite Ce6 indirectly through the FRET mechanism. Enhanced fluorescence of CD-Ce6 ($\sim$35-fold) in PBS was observed than that of free Ce6 due to indirect excitation through FRET from CDs to Ce6 leading to improved solubility of Ce6 in aqueous solution. The CDs can enhance the efficiency of singlet oxygen production of Ce6 on CD-Ce6. The PDT effect of CD-Ce6 evaluated on MGC803 cells by CCK-8 assay is stronger than that of Ce6 alone due to the enhanced efficiency of singlet oxygen production of Ce6 by CDs. The *in vivo* fluorescence imaging-guided PDT on nude mice bearing subcutaneous MGC803 gastric cancer xenograft demonstrated that CD-Ce6 had excellent imaging and tumor-homing ability and the CD-Ce6 administration/irradiation group showed partial tumor regression. These results suggest that CDs are good candidates as carriers for the effective NIR fluorescence imaging guided PDT treatment, which may have a potential for applications in the clinic to treat patients with gastric cancer or other tumors in the future.[35]

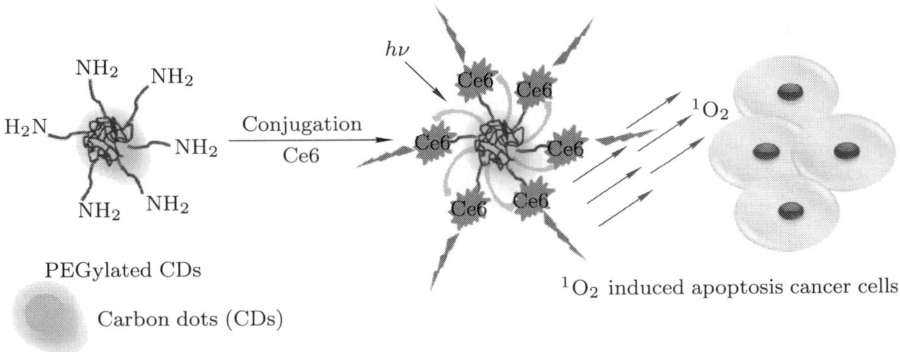

Fig. 1.8 Schematic illustration of photosensitizer-conjugated carbon dots for PDT treatment of cancer cells through the fluorescence resonance energy transfer (FRET) process between CDs and Ce6.

1.4.8 *Lipopolyplex Nanoparticles*

Nonviral vectors are recently regarded as an important approach for the therapeutic delivery of DNA into cells because the delivery of DNA using nonviral vectors may not cause undesirable immune responses.[162,163] The first generation of nonviral vectors, including lipoplex[164] based on plasmid DNA complexes and polyplexes[165] constituted with polycationic polymers such as polyethylenimine (PEI), peptides, and dendrimers, initially lacked any mechanisms to overcome the poor *in vivo* stabilities and frequently exhibited low transfection efficiencies compared with viral vectors.[166] A new generation of lipopolyplex (LPD) vectors has been developed for both *in vitro* and *in vivo* delivery applications.[166] LPD nanoparticles are a class of nanosized nonviral vectors consisting of one or more amphipathic molecules and one or more ionic polymers or proteins, used to deliver vaccines, drugs, enzymes, nucleic acid, or other substances to target cells or organs.[162] There are several advantages of using LPD for DNA delivery: ① LPD nanoparticles can provide improved gene transfer activity; ② they have unlimited packaging capacity; ③ their structures can be easily functionalized; ④ they can be produced in large-scale; and ⑤ they have few of the toxicity and safety issues associated with viral gene delivery vectors.[162,167,168]

Kudsiova *et al.*[36] demonstrated the design of novel LPD vectors prepared using a series of structurally related C-14 cationic lipids together with a bifunctional peptide for *in vitro* gene transfection studies in breast cancer cells (Fig.1.9). Targeting peptides were introduced in this study to improve biological stability, targeting efficiency and cellular internalization. The results of using LPD vectors in DNA transfection revealed that ternary formulations of lipid-peptide-DNA vectors possessed highly synergistic transfection activity up to 2 500-fold higher than their respective lipid-DNA or peptide-DNA counterparts. The significant increase in transfection was attributable to a synergistic activity between the lipids and the peptide. The localization and intracellular trafficking of the targeting peptide in relation to pDNA was determined by the combination of rhodamine green-labeled peptide (green) and rhodamine-labeled pDNA (red) in LPD delivery vectors. The results indicated that pDNA was incorporated into the nucleus. The authors thought that the targeting peptide possibly facilitated pDNA entry into the cell nucleus by acting as a nuclear localization agent. Therefore, the targeting lipid-peptide-DNA vectors show a great potential for the effective gene transfection and treatment of cancer.

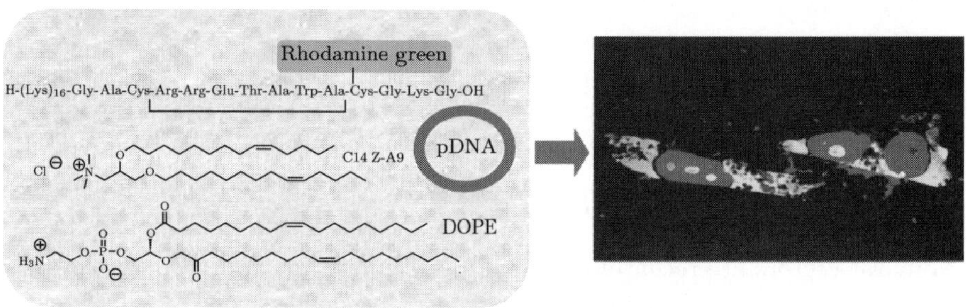

Fig. 1.9 Lipopolyplex vectors consisting of lipid-peptide-pDNA for *in vitro* DNA transfection in breast MDA-MB-231cancer cells. Image is adapted with permission from Ref. 36. Copyright (2010) American Chemical Society. (color figure at the end of the book)

1.4.9 *Nano/Microbubbles*

The ultrasound imaging has become one of the most widespread diagnostic modalities used in clinics. Although most of the ultrasound examinations are performed without contrast agents, gas-filled microbubbles (or nanobubbles) have been proven to be highly beneficial as contrast agents in imaging the liver lesions, tumors, intravascular diseases in soft tissues and the assessment of therapy response.[169] Nano/microbubble structure is comprised of an encapsulating shell and an inner cavity core, where gas or drug can be loaded for diagnosis and therapeutic effects. Generally, microbubbles for the ultrasound contrast enhancement usually have a diameter in the range of 1 to 4 μm, which are much larger than MRI or CT contrast agents.[169] However, these larger ultrasound imaging contrast agents are easily trapped in the blood pool after intravenous injection when they are used in diagnostic ultrasound. Consequently, fabrication of nanobubbles with various shells such as block polymers, phospholipids is promising for the development of tumor targeting contrast agents.

By oscillating in an acoustic field, these nano/microbubbles can provide increased sensitivity to produce nonlinear responses which can be distinguished from the surrounding tissues. Moreover, the functionalized shell surface enables nano/microbubbles to be conjugated with the variety of targeting modalities for tumor targeting theranostics. Thus, in this sense, nano/microbubbles combined with acoustics may be one of the potential modality for tumor targeting theranostics.

The possibility of using microbubbles for tumor theranostics has been evaluated by Rapoport *et al.* in 2007.[37] They prepared multifunctional nanoparticles composed of DOX-loaded polymeric micelles and perfluoropentane (PFP) nano/microbubbles stabilized by the same biodegradable block copolymer. The nanosized polymeric micelles function as DOX carriers and nano/microbubbles serve as drug carriers and ultrasound contrast agents. The *in vitro* experiment of incubation of DOX-loaded PFP-based nanobubbles with breast cancer MDA MB231 cells demonstrated that DOX was strongly retained by the nanobubbles in the absence of ultrasound and DOX was released and internalized by the cells under the action of ultrasound. The combination chemotherapy by DOX-loaded nanobubbles and ultrasound induced drug release in mice bearing MDA MB231 breast tumors revealed that the sonicated tumors didn't grow due to the DOX release from the nanobubbles, which were destroyed by the ultrasound, eventually resulting highly effective tumor growth inhibition.

Very recently, several different drug-loaded nano/microbubbles were reported for *in vivo* visualizing tumor and ultrasound triggered drug release tumor therapy.[170–172] Bin *et al.*[170] described the application of ultrasound-targeted microbubbles destruction to increase the extravasation of Evens blue in spinotrapezius muscles of rats without transient nanopores. Li *et al.*[171] demonstrated a new strategy to overcome the low drug payload of microbubbles for the ultrasound triggered drug release from 10-hydroxycamptothecin-loaded phospholipid microbubbles for targeted tumor therapy in mice. Tang *et al.*[172] reported the use of ultrasound microbubbles destruction to augment site-targeted engraftment of bone marrow mesenchymal stem cells to kidney tissue and promote kidney recovery in acute kidney injury. Due to the unique advantages of the nano/microbubbles in their structure, it can be concluded that the ultrasound image-guided therapy in combination with nano/microbubbles will become a versatile tool for tumor molecular imaging, therapy, and theranostics.

1.4.10 *Upconversion Nanomaterials*

Rare-earth ions-doped upconversion nanoparticles (UCNPs) have attracted extensive attention in recent years because of their unique photo-physical properties, such as the ability of transfering near-infrared (NIR) light into visible light, long luminescence lifetime, and narrow emission band widths, which allows their potential applications as imaging agents and drug delivery carriers.[173–176] Significantly, UCNPs can emit visible light under near-infrared irradiation, leading to low autofluorescence, less scattering and absorption, and deep penetration in biological tissues. Upconversion is a phenomenon related to nonlinear optical process in which the sequential absorption of two or more photons leads to the emission of light at shorter wavelength than the excitation wavelength (anti-Stokes type emission) first observed by F. Auzel in 1966.[177,178] Detailed mechanisms will be discussed in Chapter 11. During the past ten years, high-quality UCNPs have been successfully synthesized and are becoming more prominent in the biological field for bioimaging.[179–181] Up to now, $NaYF_4$: Yb, Er(Tm) UCNP is one of the most efficient upconversion nanomaterials explored as imaging agent for tumor imaging and therapy.

Zhang and coworkers[38] demonstrated the application of mesoporous silica-coated UCNPs for *in vivo* tumor-targeted PDT. $NaYF_4$(Yb:Er) UCNPs were coated with mesoporous silica and co-loaded with photo-sensitizers, ZnPc and merocyanine 540 (MC540). The absorption bands of ~540 nm (ZnPc) and ~660 nm (MC540) match well with the emission peaks of UCNPs excitated by 980 nm laser, resulting in efficient $^1O_2$ generation in the live cells under the illustration of NIR light. This is valuable for fully utilization of upconverted energy to maximize the PDT efficiency. After further modification with PEG and FA on the surface of UCNPs, they have carefully assessed the therapeutic efficacy of PEG-FA-ZnPc/MC540 co-loaded UCNPs in a melanoma-bearing mouse model *via* intravenous tail vein injection. A great tumor growth inhibition was observed as compared with the control mice injected with PBS buffer. This is the first report of using UCNPs for *in vivo* tumor-targeted PDT, which demonstrates that UCNPs may be useful for tumor targeted theranostics.

1.4.11 *Protein Nanomaterials*

Proteins are important biological components consisting of one or more chains of amino acids. Protein-based nanoparticles are particularly interesting because they have certain advantages such as good stability, being non-toxic and non-antigenic, greater biocompatibility and biodegradability to either enhance pharmacodynamic efficiency in systemic circulation or reduce side effects.[182–185] To this regard, protein-based nanoparticles are promising for the development of imaging and therapeutic agents for tumor theranostics. They hold the possibility of loading multiple functionalities including targeting moieties, imaging or therapeutic agents onto a single platform, which are useful for tumor-targeted therapeutic and/or diagnostic purposes. Some animal proteins including albumin, gelatin, collagen, elastin, casein, white protein and plant proteins such as gliadin and zein, *etc.*, have been explored for the preparation of protein-based nanoparticles.[186,187] Among different protein-based nanoparticle delivery systems, albumin nanoparticles have attracted considerable attention owing to their low immunogenicity, good biocompatibility, great drug loading capacity and minimized side-effect.[188,189]

Shen *et al.*[190] demonstrated the design and preparation of DOX-loaded FA-conjugated albumin nanospheres specific for HeLa cells, which showed greater drug delivery efficiency over the DOX-loaded albumin nanospheres. Dubey *et al.*[191] described RGD peptide-anchored sterically stabilized bovine serum albumin (BSA) nanospheres bearing 5-fluorouracil for tumor vasculature targeting. Zensi *et al.*[192] reported the synthesis of human serum albumin (HSA) nanoparticles with covalently bound apolipoprotein E as target for the delivery of drugs across the blood-brain barrier (BBB) after intravenously injected in SV 129 mice. Particularly, Xu *et al.*[39] constructed innovative protein and genes carriers, RGD-PEG-albumin-DOX conjugates (RPA-DOX), for the delivery of amphipathic drugs. The synthetic scheme is shown in Fig.1.10(a). Alternatively, PEGylated albumin-DOX conjugate attached 2-mercaptoethanol (MPA-DOX)

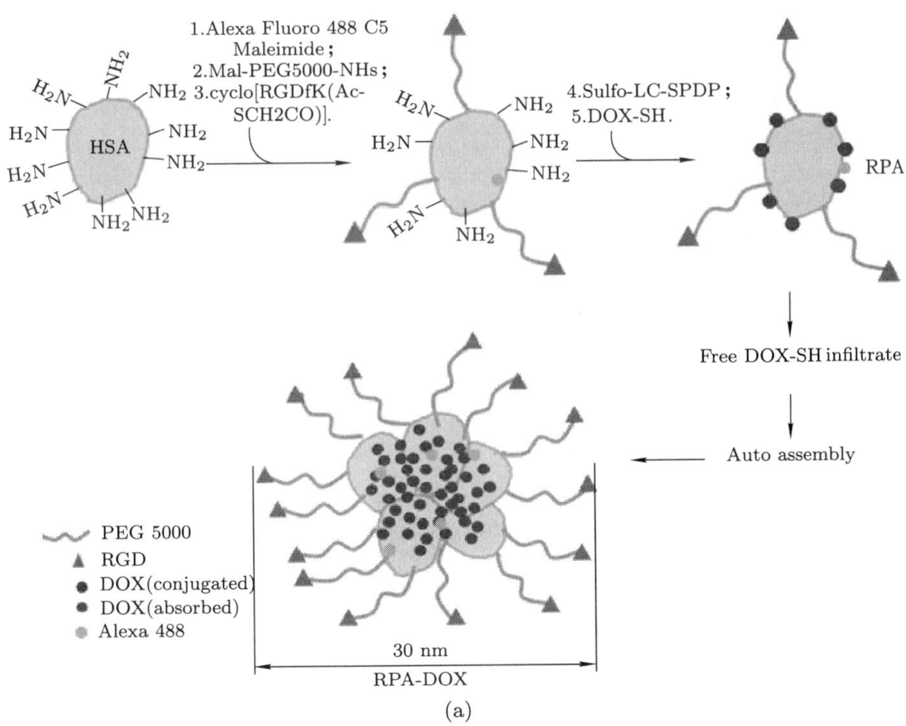

Fig. 1.10 Synthetic scheme of RPA-DOX preparation (a), cytotoxicity of RPA-DOX, MPA-DOX and free DOX on M21+ human melanoma cells (b), and confocal microscopy image of DOX uptake in M21+ cells after 12 h incubation (c). Arrow points to the DOX subcellular distributions. Image is adapted with permission from Ref. 39. Copyright (2011) American Chemical Society.

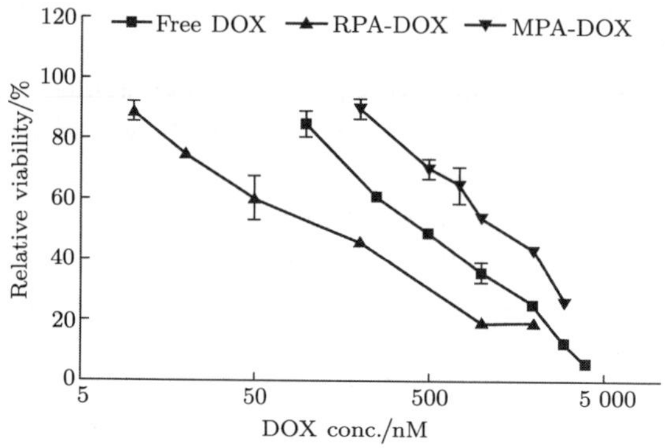

(b)

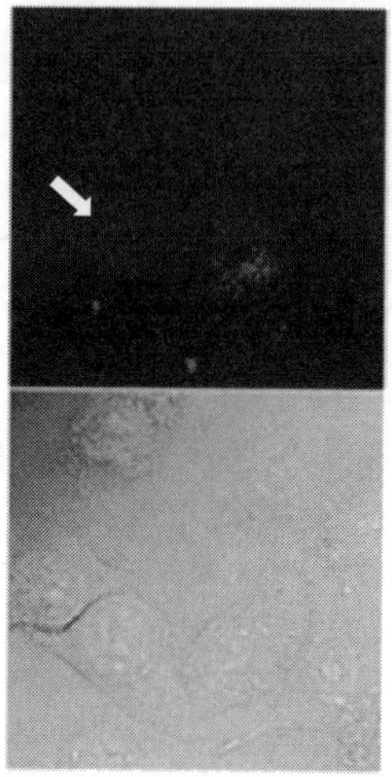

(c)

Fig. 1.10 (Continued).

was used as a control. The obtained RPA-DOX had approximately 24.6 chemically conjugated DOX per albumin as determined by a ninhydrin test. The size of RPA-DOX is about 30 nm in diameter. As shown in Fig.1.10(b), RPA-DOX exhibited the highest potency to M21+ human melanoma cells, followed by free DOX and then MPA-DOX. It is difficult to attribute the cytotoxic effect to the improved delivery efficacy of the albumin carrier because RPA-DOX contains both bound DOX and conjugated DOX. However, the actual drug release may be quicker due to contributions of endosomal proteases. Figure 1.10(c) shows the subcellular distribution of RPA-DOX determined by confocal microscopy, indicating that a greater amount of DOX was delivered to the M21+ cells by using the protein-based nanoparticles. All the *in vitro* and *in vivo* studies have demonstrated that the protein-based nanoparticles have a bright future for tumor targeting theranostics.

1.5 Conclusions and Future Perspective

In this chapter, we have highlighted the typical nanomaterials with multifunctional structure and components for tumor targeting theranostics. The advances in this field have shown that the theranostic vehicles based on nanomaterials hold great benefits for image-guided therapy, ultrasound triggered drug release, imaging monitored photodynamic therapy and assessment of therapy response, *etc.* Nanoplatforms for tumor targeting theranostics provide multifunctions by combination of the diagnosis and therapy into one dosage and have the ability to monitor and adjust the therapy during the disease progression. It is highly desired to realize seamless integration of tumor diagnosis, therapy and follow-up by using a suitable theranostic nanoplatform with targeting, diagnosis and therapy functions. A wide variety of theranostic vehicles based on diverse nanomaterials including magnetic nanomaterials, quantum dots, mesoporous silica, silver/gold, polymeric (dendrimers), carbon, lipopolyplex, upconversion, protein-based nanomaterials and nano/microbubbles, have showed great potential for tumor targeting theranostics. However, the studies of the theranostic nanomaterials for tumor targeting theranostics are still in its infancy and theranostic nanomaterials have yet to be clinically used in the world. Much efforts are required for systematic research of the safety issue, side-effect, lower efficiency of drug delivery and diagnosis before these theranostic nanoplatforms reach the clinical trials.

Acknowledgments

This work was supported by the National Nature Science Foundation of China (91227126, 81441052), National Special Fund for Key Scientific Instrument and Equipment Development (2013YQ17046307), Nature Science Foundation of Liaoning Province, China (2013020177), and Proyecto Prometeo and Secretaría de Educación Superior, Ciencia, Tecnología e Innovación, Ecuador.

References

1. Pisani, P.; Bray, F.; Parkin, D. M. Estimates of the world-wide prevalence of cancer for 25 sites in the adult population. *Int J Cancer* **2002**, *97* (1), 72-81.
2. http://www.cancer.gov/statistics.
3. Swapp, R. E.; Shon, W.; Peethambaram, P. P.; Moran, S. L.; Reynolds, C. Cutaneous presentation of a distant metastasis of malignant phyllodes tumor. *Int J Dermatol* **2012**, *51* (1), 72-74.
4. Choi, K. Y.; Liu, G.; Lee, S.; Chen, X. Y. Theranostic nanoplatforms for simultaneous cancer imaging and therapy: Current approaches and future perspectives. *Nanoscale* **2012**, *4* (2), 330-342.
5. Fernandez-Fernandez, A.; Manchanda, R.; McGoron, A. J. Theranostic applications of nanomaterials in cancer: Drug delivery, image-guided therapy, and multifunctional platforms. *Appl Biochem Biotech* **2011**, *165* (7-8), 1628-1651.
6. Prabhakar, U.; Maeda, H.; Jain, R. K.; Sevick-Muraca, E. M.; Zamboni, W.; Farokhzad, O. C.; Barry, S. T.; Gabizon, A.; Grodzinski, P.; Blakey, D. C. Challenges and key considerations of the enhanced permeability and retention effect for nanomedicine drug delivery in oncology. *Cancer Res* **2013**, *73* (8), 2412-2417.
7. Stegh, A. H. Toward personalized cancer nanomedicine: Past, present, and future. *Integr Biol-UK* **2013**, *5* (1), 48-65.
8. Goncalves, H. M.; Duarte, A. J.; Davis, F.; Higson, S. P.; Esteves da Silva, J. C. Layer-by-layer immobilization of carbon dots fluorescent nanomaterials on single optical fiber. *Anal Chim Acta* **2012**, *735*, 90-95.

9. Kateb, B.; Chiu, K.; Black, K. L.; Yamamoto, V.; Khalsa, B.; Ljubimova, J. Y.; Ding, H.; Patil, R.; Portilla-Arias, J. A.; Modo, M.; Moore, D. F.; Farahani, K.; Okun, M. S.; Prakash, N.; Neman, J.; Ahdoot, D.; Grundfest, W.; Nikzad, S.; Heiss, J. D. Nanoplatforms for constructing new approaches to cancer treatment, imaging, and drug delivery: What should be the policy? *Neuroimage* **2011**, *54*, S106-S124.
10. Borm, P.; Klaessig, F. C.; Landry, T. D.; Moudgil, B.; Pauluhn, J.; Thomas, K.; Trottier, R.; Wood, S. Research strategies for safety evaluation of nanomaterials, part V: Role of dissolution in biological fate and effects of nanoscale particles. *Toxicol Sci* **2006**, *90* (1), 23-32.
11. Farokhzad, O. C.; Langer, R. Impact of nanotechnology on drug delivery. *ACS Nano* **2009**, *3* (1), 16-20.
12. Ahmed, N.; Fessi, H.; Elaissari, A. Theranostic applications of nanoparticles in cancer. *Drug Discov Today* **2012**, *17* (17-18), 928-934.
13. Picard, F. J.; Bergeron, M. G. Rapid molecular theranostics in infectious diseases. *Drug Discov Today* **2002**, *7* (21), 1092-1101.
14. Warner, S. Diagnostics plus therapy = theranostics. *Scientist* **2004**, *18* (16), 38-39.
15. Rivera Gil, P.; Parak, W. J. Composite nanoparticles take aim at cancer. *ACS Nano* **2008**, *2* (11), 2200-2205.
16. Al-Jamal, W. T.; Al-Jamal, K. T.; Tian, B.; Lacerda, L.; Bornans, P. H.; Frederik, P. M.; Kostarelos, K. Lipid-quanturn dot bilayer vesicles enhance tumor cell uptake and retention *in vitro* and *in vivo*. *ACS Nano* **2008**, *2* (3), 408-418.
17. Mikhaylov, G.; Mikac, U.; Magaeva, A. A.; Itin, V. I.; Naiden, E. P.; Psakhye, I.; Babes, L.; Reinheckel, T.; Peters, C.; Zeiser, R.; Bogyo, M.; Turk, V.; Psakhye, S. G.; Turk, B.; Vasiljeva, O. Ferri-liposomes as an MRI-visible drug-delivery system for targeting tumours and their microenvironment. *Nat Nanotechnol* **2011**, *6* (9), 594-602.
18. Kenny, G. D.; Kamaly, N.; Kalber, T. L.; Brody, L. P.; Sahuri, M.; Shamsaei, E.; Miller, A. D.; Bell, J. D. Novel multifunctional nanoparticle mediates siRNA tumour delivery, visualisation and therapeutic tumour reduction *in vivo*. *J Control Release* **2011**, *149* (2), 111-116.
19. Danhier, F.; Vroman, B.; Lecouturier, N.; Crokart, N.; Pourcelle, V.; Freichels, H.; Jerome, C.; Marchand-Brynaert, J.; Feron, O.; Preat, V. Targeting of tumor endothelium by RGD-grafted PLGA-nanoparticles loaded with paclitaxel. *J Control Release* **2009**, *140* (2), 166-173.
20. Wang, J. Q.; Tao, X. Y.; Zhang, Y. F.; Wei, D. Z.; Ren, Y. H. Reversion of multidrug resistance by tumor targeted delivery of antisense oligodeoxynucleotides in hydroxypropyl-chitosan nanoparticles. *Biomaterials* **2010**, *31* (15), 4426-4433.
21. Kolli, S.; Wong, S. P.; Harbottle, R.; Johnston, B.; Thanou, M.; Miller, A. D. pH-triggered nanoparticle mediated delivery of siRNA to liver cells *in vitro* and *in vivo*. *Bioconjugate Chem* **2013**, *24* (3), 314-332.
22. Kumar, R.; Kulkarni, A.; Nagesha, D. K.; Sridhar, S. *In vitro* evaluation of theranostic polymeric micelles for imaging and drug delivery in cancer. *Theranostics* **2012**, *2* (7), 714-722.

23. Nasongkla, N.; Bey, E.; Ren, J. M.; Ai, H.; Khemtong, C.; Guthi, J. S.; Chin, S. F.; Sherry, A. D.; Boothman, D. A.; Gao, J. M. Multifunctional polymeric micelles as cancer-targeted, MRI-ultrasensitive drug delivery systems. *Nano Lett* **2006**, *6* (11), 2427-2430.
24. Kaida, S.; Cabral, H.; Kumagai, M.; Kishimura, A.; Terada, Y.; Sekino, M.; Aoki, I.; Nishiyama, N.; Tani, T.; Kataoka, K. Visible drug delivery by supramolecular nanocarriers directing to single-platformed diagnosis and therapy of pancreatic tumor model. *Cancer Res* **2010**, *70* (18), 7031-7041.
25. Yoo, H.; Moon, S. K.; Hwang, T.; Kim, Y. S.; Kim, J. H.; Choi, S. W.; Kim, J. H. Multifunctional magnetic nanoparticles modified with polyethylenimine and folic acid for biomedical theranostics. *Langmuir* **2013**, *29* (20), 5962-5967.
26. Singh, S. P. Multifunctional magnetic quantum dots for cancer theranostics. *J Biomed Nanotechnol* **2011**, *7* (1), 95-97.
27. Hayashi, K.; Nakamura, M.; Sakamoto, W.; Yogo, T.; Miki, H.; Ozaki, S.; Abe, M.; Matsumoto, T.; Ishimura, K. Superparamagnetic nanoparticle clusters for cancer theranostics combining magnetic resonance imaging and hyperthermia treatment. *Theranostics* **2013**, *3* (6), 366-376.
28. Medarova, Z.; Pham, W.; Farrar, C.; Petkova, V.; Moore, A. *In vivo* imaging of siRNA delivery and silencing in tumors. *Nat Med* **2007**, *13* (3), 372-377.
29. Bagalkot, V.; Zhang, L.; Levy-Nissenbaum, E.; Jon, S.; Kantoff, P. W.; Langer, R.; Farokhzad, O. C. Quantum dot - aptamer conjugates for synchronous cancer imaging, therapy, and sensing of drug delivery based on bi-fluorescence resonance energy transfer. *Nano Lett* **2007**, *7* (10), 3065-3070.
30. Hoepfner, M.; Kuzyniak, W.; Sekhosana, E.; Hoffmann, B.; Litwinski, C.; Pries, A.; Ermilov, E. Semiconductor quantum dots for innovative approaches in photodynamic diagnosis and therapy. *Cardiovasc Res* **2012**, *93*, S123.
31. Samia, A. C. S.; Chen, X. B.; Burda, C. Semiconductor quantum dots for photodynamic therapy. *J Am Chem Soc* **2003**, *125* (51), 15736-15737.
32. Li, X.; Chen, Y. J.; Wang, M. Q.; Ma, Y. J.; Xia, W. L.; Gu, H. C. A mesoporous silica nanoparticle - PEI - fusogenic peptide system for siRNA delivery in cancer therapy. *Biomaterials* **2013**, *34* (4), 1391-1401.
33. Song, J. B.; Zhou, J. J.; Duan, H. W. Self-assembled plasmonic vesicles of SERS-encoded amphiphilic gold nanoparticles for cancer cell targeting and traceable intracellular drug delivery. *J Am Chem Soc* **2012**, *134* (32), 13458-13469.
34. Chow, A. M.; Tan, M. Q.; Gao, D. S.; Fan, S. J.; Cheung, J. S.; Qiao, Z. W.; Man, K.; Lu, Z. R.; Wu, E. X. Molecular MRI of liver fibrosis by a peptide-targeted contrast agent in an experimental mouse model. *Invest Radiol* **2013**, *48* (1), 46-54.
35. Huang, P.; Lin, J.; Wang, X. S.; Wang, Z.; Zhang, C. L.; He, M.; Wang, K.; Chen, F.; Li, Z. M.; Shen, G. X.; Cui, D. X.; Chen, X. Y. Light-triggered theranostics based on photosensitizer-conjugated carbon dots for simultaneous enhanced-fluorescence imaging and photodynamic therapy. *Adv Mater* **2012**, *24* (37), 5104-5110.
36. Kudsiova, L.; Fridrich, B.; Ho, J.; Mustapa, M. F. M.; Campbell, F.; Welser, K.; Keppler, M.; Ng, T.; Barlow, D. J.; Tabor, A. B.; Hailes, H. C.; Lawrence, M. J. Lipopolyplex ternary delivery systems incorporating c14 glycerol-based lipids. *Mol Pharmaceut* **2011**, *8* (5), 1831-1847.

37. Rapoport, N.; Gao, Z. G.; Kennedy, A. Multifunctional nanoparticles for combining ultrasonic tumor imaging and targeted chemotherapy. *J Natl Cancer I* **2007**, *99* (14), 1095-1106.
38. Idris, N. M.; Gnanasammandhan, M. K.; Zhang, J.; Ho, P. C.; Mahendran, R.; Zhang, Y. *In vivo* photodynamic therapy using upconversion nanoparticles as remote-controlled nanotransducers. *Nat Med* **2012**, *18* (10), 1580-1585.
39. Xu, R. Z.; Fisher, M.; Juliano, R. L. Targeted albumin-based nanoparticles for delivery of amphipathic drugs. *Bioconjugate Chem* **2011**, *22* (5), 870-878.
40. Petros, R. A.; DeSimone, J. M. Strategies in the design of nanoparticles for therapeutic applications. *Nat Rev Drug Discov* **2010**, *9* (8), 615-627.
41. Geng, Y.; Dalhaimer, P.; Cai, S. S.; Tsai, R.; Tewari, M.; Minko, T.; Discher, D. E. Shape effects of filaments versus spherical particles in flow and drug delivery. *Nat Nanotechnol* **2007**, *2* (4), 249-255.
42. Youan, B. B. C. Impact of nanoscience and nanotechnology on controlled drug delivery. *Nanomedicine* **2008**, *3* (4), 401-406.
43. Tu, R. S.; Tirrell, M. Bottom-up design of biomimetic assemblies. *Adv Drug Deliver Rev* **2004**, *56* (11), 1537-1563.
44. Torchilin, V. Tumor delivery of macromolecular drugs based on the EPR effect. *Adv Drug Deliver Rev* **2011**, *63* (3), 131-135.
45. Maeda, H. The enhanced permeability and retention (EPR) effect in tumor vasculature: The key role of tumor-selective macromolecular drug targeting. *Adv Enzyme Regul* **2001**, *41*, 189-207.
46. Maeda, H.; Matsumura, Y. Tumoritropic and lymphotropic principles of macromolecular drugs. *Criti Rev Ther Drug* **1989**, *6* (3), 193-210.
47. Sinha, R.; Kim, G. J.; Nie, S. M.; Shin, D. M. Nanotechnology in cancer therapeutics: Bioconjugated nanoparticles for drug delivery. *Mol Cancer Ther* **2006**, *5* (8), 1909-1917.
48. Koo, H.; Huh, M. S.; Sun, I. C.; Yuk, S. H.; Choi, K.; Kim, K.; Kwon, I. C. *In vivo* targeted delivery of nanoparticles for theranostics. *Accounts Chem Res* **2011**, *44* (10), 1018-1028.
49. Debbage, P.; Jaschke, W. Molecular imaging with nanoparticles: Giant roles for dwarf actors. *Histochem Cell Biol* **2008**, *130* (5), 845-875.
50. Xia, W.; Low, P. S. Folate-targeted therapies for cancer. *J Med Chem* **2010**, *53* (19), 6811-6824.
51. Parker, N.; Turk, M. J.; Westrick, E.; Lewis, J. D.; Low, P. S.; Leamon, C. P. Folate receptor expression in carcinomas and normal tissues determined by a quantitative radioligand binding assay. *Anal Biochem* **2005**, *338* (2), 284-293.
52. Maltzman, J.; O'Shannessy, D.; Somers, E.; Kamen, B. Folate receptor alpha (FRα) expression in triple negative breast cancer (TNBC). *Tumor Biol* **2012**, *33*, 30-30.
53. Nunez, M. I.; Behrens, C.; Woods, D. M.; Lin, H.; Suraokar, M.; Kadara, H.; Minna, J. D.; Hofstetter, W.; Kalhor, N.; Hong, W. K.; Lee, J. J.; Franklin, W.; Stewart, D.; Wistuba, I. I. High expression of folate receptor alpha in lung cancer correlates with adenocarcinoma histology and EGFR mutation. *Modern Pathol* **2011**, *24*, 420a-421a.

54. Shen, J. Y.; Chelvam, V.; Cresswell, G.; Low, P. S. Use of folate-conjugated imaging agents to target alternatively activated macrophages in a murine model of asthma. *Mol Pharmaceut* **2013**, *10* (5), 1918-1927.
55. Piscaer, T. M.; Muller, C.; Mindt, T. L.; Lubberts, E.; Verhaar, J. A. N.; Krenning, E. P.; Schibli, R.; De Jong, M.; Weinans, H. Imaging of activated macrophages in experimental osteoarthritis using folate-targeted animal single-photon-emission computed tomography/computed tomography. *Arthritis Rheum-US* **2011**, *63* (7), 1898-1907.
56. Morshed, K. M.; Ross, D. M.; McMartin, K. E. Characteristics of folate receptor (FR) and reduced-folate carrier (RFC)-mediated bidirectional 5-methyltetrahydrofolate transport in cultured human proximal tubule (HPT) cells. *Faseb J* **1996**, *10* (3), 2680.
57. Antony, A. C. The biological chemistry of folate receptors. *Blood* **1992**, *79* (11), 2807-2820.
58. Leamon, C. P.; Reddy, J. A.; Vlahov, I. R.; Westrick, E.; Dawson, A.; Dorton, R.; Vetzel, M.; Santhapuram, H. K.; Wang, V. Preclinical antitumor activity of a novel folate-targeted dual drug conjugate. *Mol Pharmaceut* **2007**, *4* (5), 659-667.
59. Lu, Y. J.; Xu, L. C.; Parker, N.; Westrick, E.; Reddy, J. A.; Vetzel, M.; Low, P. S.; Leamon, C. P. Preclinical pharmacokinetics, tissue distribution, and antitumor activity of a folate-hapten conjugate-targeted immunotherapy in hapten-immunized mice. *Mol Cancer Ther* **2006**, *5* (12), 3258-3267.
60. Betzel, T.; Muller, C.; Groehn, V.; Muller, A.; Reber, J.; Fischer, C. R.; Kramer, S. D.; Schibli, R.; Ametamey, S. M. Radiosynthesis and preclinical evaluation of 3′-Aza-2′-[(18)F]fluorofolic acid: A novel PET radiotracer for folate receptor targeting. *Bioconjugate Chem* **2013**, *24* (2), 205-214.
61. Al Jammaz, I.; Al-Otaibi, B.; Amer, S.; Al-Hokbany, N.; Okarvi, S. Novel synthesis and preclinical evaluation of folic acid derivatives labeled with (18)F-[FDG] for PET imaging of folate receptor-positive tumors. *Nucl Med Biol* **2012**, *39* (6), 864-870.
62. Fani, M.; Wang, X. J.; Nicolas, G.; Medina, C.; Raynal, I.; Port, M.; Maecke, H. R. Development of new folate-based PET radiotracers: Preclinical evaluation of Ga-68-DOTA-folate conjugates. *Eur J Nucl Med Mol I* **2011**, *38* (1), 108-119.
63. Reddy, J. A.; Dorton, R.; Westrick, E.; Dawson, A.; Smith, T.; Xu, L. C.; Vetzel, M.; Kleindl, P.; Vlahov, I. R.; Leamon, C. P. Preclinical evaluation of EC145, a folate-vinca alkaloid conjugate. *Cancer Res* **2007**, *67* (9), 4434-4442.
64. Muller, C.; Hohn, A.; Schubiger, P. A.; Schibli, R. Preclinical evaluation of novel organometallic 99mTc-folate and 99mTc-pteroate radiotracers for folate receptor-positive tumour targeting. *Eur J Nucl Med Mol I* **2006**, *33* (9), 1007-1016.
65. Milstein, G. K. C. Continuous cultures of fused cells secreting antibody of predefined specificity. *Nature* **1975**, *256*, 495-497.
66. Goldstein, G.; Fuccello, A.; Norman, D.; Shield, C.; Colvin, R.; Cosimi, A. OKT3 monoclonal-antibody plasma-levels during therapy and the subsequent development of host antibodies to OKT3. *Transplantation* **1986**, *42* (5), 507-511.
67. Stuve, O.; Gold, R.; Chan, A.; Mix, E.; Zettl, U.; Kieseier, B. Alpha 4-integrin antagonism with natalizumab effects and adverse effects. *J Neurol* **2008**, *255*,

58-65.

68. Engelhardt, B.; Kappos, L. Natalizumab: Targeting alpha(4)-integrins in multiple sclerosis. *Neurodegener Dis* **2008**, *5* (1), 16-22.
69. Hilger, I.; Kaiser, W. A. Iron oxide-based nanostructures for MRI and magnetic hyperthermia. *Nanomedicine* **2012**, *7* (9), 1443-1459.
70. Colcher, D.; Pavlinkova, G.; Beresford, G.; Booth, B.; Choudhury, A.; Batra, S. Pharmacokinetics and biodistribution of genetically-engineered antibodies. *Q J Nucl Med* **1998**, *42* (4), 225-241.
71. Ramage, R.; Green, J.; Muir, T.; Ogunjobi, O.; Love, S.; Shaw, K. Synthetic, structural and biological studies of the ubiquitin system: The total chemical synthesis of ubiquitin. *Biochem J* **1994**, *299*, 151-158.
72. Ruoslahti, E. RGD and other recognition sequences for integrins. *Annu Rev Cell Dev Bi* **1996**, *12*, 697-715.
73. Pilch, J.; Brown, D.; Komatsu, M.; Jarvinen, T.; Yang, M.; Peters, D.; Hoffman, R.; Ruoslahti, E. Peptides selected for binding to clotted plasma accumulate in tumor stroma and wounds. *P Natl Acad Sci* **2006**, *103* (8), 2800-2804.
74. Liu, Z.; Niu, G.; Wang, F.; Chen, X. Ga-68-labeled NOTA-RGD-BBN peptide for dual integrin and GRPR-targeted tumor imaging. *Eur J Nucl Med Mol I* **2009**, *36* (9), 1483-1494.
75. Dubey, P.; Mishra, V.; Jain, S.; Mahor, S.; Vyas, S. Liposomes modified with cyclic RGD peptide for tumor targeting. *J Drug Target* **2004**, *12* (5), 257-264.
76. Burtea, C.; Laurent, S.; Murariu, O.; Rattat, D.; Toubeau, G.; Verbruggen, A.; Vansthertem, D.; Elst, L.; Muller, R. Molecular imaging of alpha(v)beta(3) integrin expression in atherosclerotic plaques with a mimetic of RGD peptide grafted to Gd-DTPA. *Cardiovasc Res* **2008**, *78* (1), 148-157.
77. Wang, C. H.; Kang, S. T.; Lee, Y. H.; Luo, Y. L.; Huang, Y. F.; Yeh, C. K. Aptamer-conjugated and drug-loaded acoustic droplets for ultrasound theranostics. *Biomaterials* **2012**, *33* (6), 1939-1947.
78. Hicke, B. J.; Stephens, A. W.; Gould, T.; Chang, Y. F.; Lynott, C. K.; Heil, J.; Borkowski, S.; Hilger, C. S.; Cook, G.; Warren, S.; Schmidt, P. G. Tumor targeting by an aptamer. *J Nucl Med* **2006**, *47* (4), 668-678.
79. Ara, M. N.; Hyodo, M.; Ohga, N.; Hida, K.; Harashima, H. Development of a novel DNA aptamer ligand targeting to primary cultured tumor endothelial cells by a cell-based selex method. *Plos One* **2012**, *7* (12).
80. Ellington, A. D.; Szostak, J. W. *In vitro* selection of RNA molecules that bind specific ligands. *Nature* **1990**, *346* (6287), 818-822.
81. Tuerk, C.; Gold, L. Systematic evolution of ligands by exponential enrichment - RNA ligands to bacteriophage-T4 DNA-polymerase. *Science* **1990**, *249* (4968), 505-510.
82. Cherry, S. R. Multimodality imaging: Beyond PET/CT and SPECT/CT. *Semin Nucl Med* **2009**, *39* (5), 348-353.
83. Cogbill, T. H.; Ziegelbein, K. J. Computed tomography, magnetic resonance, and ultrasound imaging: Basic principles, glossary of terms, and patient safety. *Surg Clin N Am* **2011**, *91* (1), 1-14.
84. Raju, T. N. K. The Nobel chronicles - 1979: Allan MacLeod Cormack (b 1924) and Sir Godfrey Newbold Hounsfield (b 1919). *Lancet* **1999**, *354* (9190), 1653.

85. Vlaardingerbroek MT, d. B. J. *Magnetic Resonance Imaging : Theory and Practice.* 3rd ed. Physics and astronomy online library. Springer: New York, **2002**.
86. Ophir, J.; Maklad, N. F. Digital scan converters in diagnostic ultrasound imaging. *P IEEE* **1979**, *67* (4), 654-664.
87. Coatney, R. W.; Zhao, S. F.; Woods, T.; Legos, J. Non-invasive imaging in animal research: Principles and applications of ultrasound imaging in rodents. *J Pharmacol Sci* **2003**, *91*, 18.
88. Athanasiou, A.; Tardivon, A. Ultrasound elastography in cancer imaging: Principles and current applications. *Oncologie* **2010**, *12* (3), 208-212.
89. Solomon, M.; Liu, Y.; Berezin, M. Y.; Achilefu, S. Optical imaging in cancer research: Basic principles, tumor detection, and therapeutic monitoring. *Med Prin Pract* **2011**, *20* (5), 397-415.
90. Ntziachristos, V. Going deeper than microscopy: The optical imaging frontier in biology. *Nat Methods* **2010**, *7* (8), 603-614.
91. Gullberg, G. T.; Reutter, B. W.; Sitek, A.; Maltz, J. S.; Budinger, T. F. Dynamic single photon emission computed tomography-basic principles and cardiac applications. *Phys Med Biol* **2010**, *55* (20), R111-R191.
92. Grunder, G.; Vernaleken, I.; Bartenstein, P. Use of PET and SPECT in psychiatry. *Nervenarzt* **2010**, *81* (1), 97-107.
93. Townsend, D. W. Physical principles and technology of clinical PET imaging. *Ann Acad Med Singap* **2004**, *33* (2), 133-145.
94. Wada, Y. PET measurement: Principles, theories, and practice. *J Physiol Sci* **2009**, *59*, 118.
95. Seemann, M. D.; Nekolla, S.; Ziegler, S.; Bengel, F.; Schwaiger, M. PET/CT: Fundamental principles. *Eur J Med Res* **2004**, *9* (5), 241-246.
96. Richard, J. C.; Zhou, Z. H.; Ponde, D. E.; Dence, C. S.; Factor, P.; Reynolds, P. N.; Luker, G. D.; Sharma, V.; Ferkol, T.; Piwnica-Worms, D.; Schuster, D. P. Imaging pulmonary gene expression with positron emission tomography. *Am J Resp Crit Care* **2003**, *167* (9), 1257-1263.
97. Phelps, M. E. PET: A biological imaging technique. *Neurochemical Res* **1991**, *16* (9), 929-940.
98. Yoo, D.; Lee, J. H.; Shin, T. H.; Cheon, J. Theranostic magnetic nanoparticles. *Accounts Chem Res* **2011**, *44* (10), 863-874.
99. Sun, S. H.; Zeng, H.; Robinson, D. B.; Raoux, S.; Rice, P. M.; Wang, S. X.; Li, G. X. Monodisperse MFe_2O_4 (M = Fe, Co, Mn) nanoparticles. *J Am Chem Soc* **2004**, *126* (1), 273-279.
100. Cole, A. J.; Yang, V. C.; David, A. E. Cancer theranostics: The rise of targeted magnetic nanoparticles. *Trends Biotechnol* **2011**, *29* (7), 323-332.
101. Xie, J.; Jon, S. Magnetic nanoparticle-based theranostics. *Theranostics* **2012**, *2* (1), 122-124.
102. Yoo, D.; Lee, J. H.; Shin, T. H.; Cheon, J. Theranostic magnetic nanoparticles. *Account Chem Res* **2012**, *45* (9), 1622.
103. Jiang, J. T.; Nah, H.; Lee, J. H.; Moon, S. H.; Kim, M. G.; Cheon, J. Critical enhancements of MRI contrast and hyperthermic effects by dopant-controlled magnetic nanoparticles. *Angew Chem Int Edit* **2009**, *48* (7), 1234-1238.

104. Alivisatos, A. P. Semiconductor clusters, nanocrystals, and quantum dots. *Science* **1996**, *271* (5251), 933-937.
105. Ekimov, A. I. Onushchenko, A. A. Quantum size effect in three-dimensional microscopic semiconductor crystals. *JETP Lett* **1981**, *34*, 345-349.
106. Reed, M. A.; Randall, J. N.; Aggarwal, R. J.; Matyi, R. J.; Moore, T. M.; Wetsel, A. E. Observation of discrete electronic states in a zero-dimensional semiconductor nanostructure. *Phys Rev Lett* **1988**, *60* (6), 535-537.
107. Michalet, X.; Pinaud, F. F.; Bentolila, L. A.; Tsay, J. M.; Doose, S.; Li, J. J.; Sundaresan, G.; Wu, A. M.; Gambhir, S. S.; Weiss, S. Quantum dots for live cells, *in vivo* imaging, and diagnostics. *Science* **2005**, *307* (5709), 538-544.
108. Mitchell, A. C.; Dad, S.; Morgan, C. G. Selective detection of luminescence from semiconductor quantum dots by nanosecond time-gated imaging with a colour-masked CCD detector. *J Microsc-Oxford* **2008**, *230* (2), 172-176.
109. Lesoine, M. D.; Bhattacharjee, U.; Guo, Y. J.; Vela, J.; Petrich, J. W.; Smith, E. A. Subdiffraction, iuminescence-depletion imaging of isolated, giant, CdSe/CdS nanocrystal quantum dots. *J Phys Chem C* **2013**, *117*, 3662-3667.
110. Smith, A. M.; Duan, H.; Mohs, A. M.; Nie, S. Bioconjugated quantum dots for *in vivo* molecular and cellular imaging. *Adv Drug Deliv Rev* **2008**, *60* (11), 1226-1240.
111. Zrazhevskiy, P.; Sena, M.; Gao, X. H. Designing multifunctional quantum dots for bioimaging, detection, and drug delivery. *Chem Soc Rev* **2010**, *39* (11), 4326-4354.
112. Cao, L.; Meziani, M. J.; Sahu, S.; Sun, Y. P. Photoluminescence properties of graphene versus other carbon nanomaterials. *Accounts Chem Res* **2013**, *46* (1), 171-180.
113. Erogbogbo, F.; Yong, K. T.; Roy, I.; Xu, G. X.; Prasad, P. N.; Swihart, M. T. Biocompatible luminescent silicon quantum dots for imaging of cancer cells. *ACS Nano* **2008**, *2* (5), 873-878.
114. Dashiell, M. W.; Denker, U.; Muller, C.; Costantini, G.; Manzano, C.; Kern, K.; Schmidt, O. G. Photoluminescence of ultrasmall Ge quantum dots grown by molecular-beam epitaxy at low temperatures. *Appl Phys Lett* **2002**, *80* (7), 1279-1281.
115. Porat, G.; Suchowski, H.; Silberberg, Y.; Arie, A. Tunable upconverted optical parametric oscillator with intracavity adiabatic sum-frequency generation. *Opt Lett* **2010**, *35* (10), 1590-1592.
116. Cai, Q.; Luo, Z. S.; Pang, W. Q.; Fan, Y. W.; Chen, X. H.; Cui, F. Z. Dilute solution routes to various controllable morphologies of MCM-41 silica with a basic medium. *Chem Mater* **2001**, *13* (2), 258-263.
117. Fowler, C. E.; Khushalani, D.; Lebeau, B.; Mann, S. Nanoscale materials with mesostructured interiors. *Adv Mater* **2001**, *13* (9), 649-652.
118. Bore, M. T.; Pham, H. N.; Switzer, E. E.; Ward, T. L.; Fukuoka, A.; Datye, A. K. The role of pore size and structure on the thermal stability of gold nanoparticles within mesoporous silica. *J Phys Chem B* **2005**, *109* (7), 2873-2880.
119. Das, D. D.; Sayari, A. Applications of pore-expanded mesoporous silica: Novel synthesis of monodispersed supported palladium nanoparticles and their catalytic activity for suzuki reaction. *J Catal* **2007**, *246* (1), 60-65.

120. Gu, J. L.; Shi, J. L.; You, G. J.; Xiong, L. M.; Qian, S. X.; Hua, Z. L.; Chen, H. R. Incorporation of highly dispersed gold nanoparticles into the pore channels of mesoporous silica thin films and their ultrafast nonlinear optical response. *Adv Mater* **2005**, *17* (5), 557-560.

121. Zhao, W. R.; Zhang, H. T.; Chang, S.; Gu, J. L.; Li, Y. S.; Li, L.; Shi, J. L. An organosilane route to mesoporous silica nanoparticles with tunable particle and pore sizes and their anticancer drug delivery behavior. *Rsc Adv* **2012**, *2* (12), 5105-5107.

122. Lee, J. E.; Lee, N.; Kim, T.; Kim, J.; Hyeon, T. Multifunctional mesoporous silica nanocomposite nanoparticles for theranostic applications. *Accounts Chem Res* **2011**, *44* (10), 893-902.

123. Chen, N. T.; Cheng, S. H.; Souris, J. S.; Chen, C. T.; Mou, C. Y.; Lo, L. W. Theranostic applications of mesoporous silica nanoparticles and their organic/inorganic hybrids. *J Mater Chem B* **2013**, *1* (25), 3128-3135.

124. Liong, M.; Lu, J.; Kovochich, M.; Xia, T.; Ruehm, S. G.; Nel, A. E.; Tamanoi, F.; Zink, J. I. Multifunctional inorganic nanoparticles for imaging, targeting, and drug delivery. *ACS Nano* **2008**, *2* (5), 889-896.

125. Panchapakesan, B.; Book-Newell, B.; Sethu, P.; Rao, M.; Irudayaraj, J. Gold nanoprobes for theranostics. *Nanomedicine-UK* **2011**, *6* (10), 1787-1811.

126. Zhang, Z. J.; Wang, J.; Chen, C. Y. Gold nanorods based platforms for light-mediated theranostics. *Theranostics* **2013**, *3* (3), 223-238.

127. Lin, J.; Wang, S. J.; Huang, P.; Wang, Z.; Chen, S. H.; Niu, G.; Li, W. W.; He, J.; Cui, D. X.; Lu, G. M.; Chen, X. Y.; Nie, Z. H. Photosensitizer-loaded gold vesicles with strong plasmonic coupling effect for imaging-guided photothermal/photodynamic therapy. *ACS Nano* **2013**, *7* (6), 5320-5329.

128. Wang, Y. Q.; Yan, B.; Chen, L. X. SERS tags: Novel optical nanoprobes for bioanalysis. *Chem Rev* **2013**, *113* (3), 1391-1428.

129. Alvarez-Puebla, R. A.; Liz-Marzan, L. M. SERS-based diagnosis and biodetection. *Small* **2010**, *6* (5), 604-610.

130. Abalde-Cela, S.; Aldeanueva-Potel, P.; Mateo-Mateo, C.; Rodriguez-Lorenzo, L.; Alvarez-Puebla, R. A.; Liz-Marzan, L. M. Surface-enhanced Raman scattering biomedical applications of plasmonic colloidal particles. *J R Soc Interface* **2010**, *7*, S435-S450.

131. Qian, X. M.; Peng, X. H.; Ansari, D. O.; Yin-Goen, Q.; Chen, G. Z.; Shin, D. M.; Yang, L.; Young, A. N.; Wang, M. D.; Nie, S. M. *In vivo* tumor targeting and spectroscopic detection with surface-enhanced Raman nanoparticle tags. *Nat Biotechnol* **2008**, *26* (1), 83-90.

132. Keren, S.; Zavaleta, C.; Cheng, Z.; de la Zerda, A.; Gheysens, O.; Gambhir, S. S. Noninvasive molecular imaging of small living subjects using Raman spectroscopy. *P Natl Acad Sci USA* **2008**, *105* (15), 5844-5849.

133. Abeylath, S. C.; Ganta, S.; Iyer, A. K.; Amiji, M. Combinatorial-designed multifunctional polymeric nanosystems for tumor-targeted therapeutic delivery. *Accounts Chem Res* **2011**, *44* (10), 1009-1017.

134. Wang, Z.; Niu, G.; Chen, X. Polymeric materials for theranostic applications. *Pharm Res* **2013**, *31*, 1358-1376.

135. Villaraza, A. J. L.; Bumb, A.; Brechbiel, M. W. Macromolecules, dendrimers, and nanomaterials in magnetic resonance imaging: The interplay between size, function, and pharmacokinetics. *Chem Rev* **2010**, *110* (5), 2921-2959.

136. Svenson, S. Dendrimers as versatile platform in drug delivery applications. *Eur J Pharm Biopharm* **2009**, *71* (3), 445-462.

137. Miele, E.; Spinelli, G. P.; Miele, E.; Tomao, F.; Tomao, S. Albumin-bound formulation of paclitaxel (Abraxane ABI-007) in the treatment of breast cancer. *Int J Nanomed* **2009**, *4* (1), 99-105.

138. Li, Y. Y.; Ma, J. P.; Zhu, H. Y.; Gao, X. L.; Dong, H. Q.; Shi, D. L. Green synthetic, multifunctional hybrid micelles with shell embedded magnetic nanoparticles for theranostic applications. *ACS Appl Mater Inter* **2013**, *5* (15), 7227-7235.

139. Kaneshiro, T. L.; Wang, X.; Lu, Z. R. Synthesis, characterization, and gene delivery of poly-l-lysine octa(3-aminopropyl)silsesquioxane dendrimers: Nanoglobular drug carriers with precisely defined molecular architectures. *Mol Pharm* **2007**, *4* (5), 759-768.

140. Tan, M. Q.; Wu, X. M.; Jeong, E. K.; Chen, Q. J.; Lu, Z. R. Peptide-targeted nanoglobular Gd-DOTA monoamide conjugates for magnetic resonance cancer molecular imaging. *Biomacromolecules* **2010**, *11* (3), 754-761.

141. Tan, M. Q.; Wu, X. M.; Jeong, E. K.; Chen, Q. J.; Parker, D. L.; Lu, Z. R. An effective targeted nanoglobular manganese(II) chelate conjugate for magnetic resonance molecular imaging of tumor extracellular matrix. *Mol Pharmaceut* **2010**, *7* (4), 936-943.

142. Tan, M. Q.; Ye, Z.; Jeong, E. K.; Wu, X. M.; Parker, D. L.; Lu, Z. R. Synthesis and evaluation of nanoglobular macrocyclic Mn(II) chelate conjugates as non-gadolinium(III) MRI contrast agents. *Bioconjugate Chem* **2011**, *22* (5), 931-937.

143. Tan, M.; Ye, Z.; Lindner, D.; Lu, Z-R. Synthesis and evaluation of a targeted nanoglobular dual-modal imaging agent for MR imaging and image-guided surgery of prostate cancer. *Pharm Res* **2014**, *31* (6), 1469-1476.

144. Zhu, S. J.; Meng, Q. N.; Wang, L.; Zhang, J. H.; Song, Y. B.; Jin, H.; Zhang, K.; Sun, H. C.; Wang, H. Y.; Yang, B. Highly photoluminescent carbon dots for multicolor patterning, sensors, and bioimaging. *Angew Chem Int Edit* **2013**, *52* (14), 3953-3957.

145. Baker, S. N.; Baker, G. A. Luminescent carbon nanodots: Emergent nanolights. *Angew Chem Int Edit* **2010**, *49* (38), 6726-6744.

146. Tan, M. Q.; Zhang, L. X.; Tang, R.; Song, X.; Li, Y.; Wu, H.; Wang, Y. F.; Lv, G.; Liu, W. F.; Ma, X. J. Enhanced photoluminescence and characterization of multicolor carbon dots using plant soot as a carbon source. *Talanta* **2013**, *115*, 950-956.

147. Yan, H. P.; Tan, M. Q.; Zhang, D. M.; Cheng, F. S.; Wu, H.; Fan, M. K.; Ma, X. J.; Wang, J. H. Development of multicolor carbon nanoparticles for cell imaging. *Talanta* **2013**, *108*, 59-65.

148. Krishna, V.; Singh, A.; Sharma, P.; Iwakuma, N.; Wang, Q. A.; Zhang, Q. Z.; Knapik, J.; Jiang, H. B.; Grobmyer, S. R.; Koopman, B.; Moudgil, B. Polyhydroxy fullerenes for non-invasive cancer imaging and therapy. *Small* **2010**, *6* (20), 2236-2241.

149. Antaris, A. L.; Robinson, J. T.; Yaghi, O. K.; Hong, G. S.; Diao, S.; Luong, R.; Dai, H. J. Ultra-low doses of chirality sorted (6,5) carbon nanotubes for

simultaneous tumor imaging and photothermal therapy. *ACS Nano* **2013**, *7* (4), 3644-3652.

150. Kaur, R.; Badea, I. Nanodiamonds as novel nanomaterials for biomedical applications: Drug delivery and imaging systems. *Int J Nanomed* **2013**, *8*, 203-220.
151. Nurunnabi, M.; Khatun, Z.; Reeck, G. R.; Lee, D. Y.; Lee, Y. K. Near infra-red photoluminescent graphene nanoparticles greatly expand their use in noninvasive biomedical imaging. *Chem Commun* **2013**, *49* (44), 5079-5081.
152. Parra-Alfambra, A. M.; Casero, E.; Ruiz, M. A.; Vazquez, L.; Pariente, F.; Lorenzo, E. Carbon nanotubes/pentacyaneferrate-modified chitosan nanocomposites platforms for reagentless glucose biosensing. *Anal Bioanal Chem* **2011**, *401* (3), 883-889.
153. Casero, E.; Alonso, C.; Vazquez, L.; Petit-Dominguez, M. D.; Parra-Alfambra, A. M.; de la Fuente, M.; Merino, P.; Alvarez-Garcia, S.; de Andres, A.; Pariente, F.; Lorenzo, E. Comparative response of biosensing platforms based on synthesized graphene oxide and electrochemically reduced graphene. *Electroanal* **2013**, *25* (1), 154-165.
154. Chen, S. Y.; Chen, J. Y.; Zhao, X. R.; Wong, S. S.; Ojima, I. Medi 328-synthesis and evaluation of functionalized SWNT as transporter for tumor-targeting drug delivery. *Abstr Pap Am Chem S* **2007**, *234*.
155. Chen, J. Y.; Chen, S. Y.; Zhao, X. R.; Kuznetsova, L. V.; Wong, S. S.; Ojima, I. Functionalized single-walled carbon nanotubes as rationally designed vehicles for tumor-targeted drug delivery. *J Am Chem Soc* **2008**, *130* (49), 16778-16785.
156. Liang, X. J.; Meng, H.; Wang, Y. Z.; He, H. Y.; Meng, J.; Lu, J.; Wang, P. C.; Zhao, Y. L.; Gao, X. Y.; Sun, B. Y.; Chen, C. Y.; Xing, G. M.; Shen, D. W.; Gottesman, M. M.; Wu, Y.; Yin, J. J.; Jia, L. Metallofullerene nanoparticles circumvent tumor resistance to cisplatin by reactivating endocytosis. *P Natl Acad Sci USA* **2010**, *107* (16), 7449-7454.
157. de la Zerda, A.; Kim, J. W.; Galanzha, E. I.; Gambhir, S. S.; Zharov, V. P. Advanced contrast nanoagents for photoacoustic molecular imaging, cytometry, blood test and photothermal theranostics. *Contrast Media Mol I* **2011**, *6* (5), 346-369.
158. Judkins, J.; Lee, H. H.; Tung, S.; Kim, J. W. Diffusion of single-walled carbon nanotube under physiological conditions. *J Biomed Nanotechnol* **2013**, *9* (6), 1065-1070.
159. Chu, M. Q.; Peng, J. L.; Zhao, J. J.; Liang, S. L.; Shao, Y. X.; Wu, Q. Laser light triggered-activated carbon nanosystem for cancer therapy. *Biomaterials* **2013**, *34* (7), 1820-1832.
160. Yu, J. G.; Jiao, F. P.; Chen, X. Q.; Jiang, X. Y.; Peng, Z. G.; Zeng, D. M.; Huang, D. S. Irradiation-mediated carbon nanotubes use in cancer therapy. *J Cancer Res Ther* **2012**, *8* (3), 348-354.
161. Liu, Z.; Liang, X. J. Nano-carbons as theranostics. *Theranostics* **2012**, *2* (3), 235-237.
162. Mustapa, M. F. M.; Grosse, S. M.; Kudsiova, L.; Elbs, M.; Raiber, E. A.; Wong, J. B.; Brain, A. P. R.; Armer, H. E. J.; Warley, A.; Keppler, M.; Ng, T.; Lawrence, M. J.; Hart, S. L.; Hailes, H. C.; Tabor, A. B. Stabilized integrin-targeting ternary LPD (lipopolyplex) vectors for gene delivery designed to disassemble within the target cell. *Bioconjugate Chem* **2009**, *20* (3), 518-532.

163. Munoz-Ubeda, M.; Misra, S. K.; Barran-Berdon, A. L.; Aicart-Ramos, C.; Sierra, M. B.; Biswas, J.; Kondaiah, P.; Junquera, E.; Bhattacharya, S.; Aicart, E. Why is less cationic lipid required to prepare lipoplexes from plasmid DNA than linear DNA in gene therapy? *J Am Chem Soc* **2011**, *133* (45), 18014-18017.

164. Uyechi, L. S.; Gagne, L.; Thurston, G.; Szoka, F. C. Mechanism of lipoplex gene delivery in mouse lung: Binding and internalization of fluorescent lipid and DNA components. *Gene Ther* **2001**, *8* (11), 828-836.

165. Gebhart, C. L.; Kabanov, A. V. Evaluation of polyplexes as gene transfer agents. *J Control Release* **2001**, *73* (2-3), 401-416.

166. Whitmore, M.; Li, S.; Huang, L. LPD lipopolyplex initiates a potent cytokine response and inhibits tumor growth. *Gene Ther* **1999**, *6* (11), 1867-1875.

167. Somia, N.; Verma, I. M. Gene therapy: Trials and tribulations. *Nat Rev Genet* **2000**, *1* (2), 91-99.

168. Hacein-Bey-Abina, S.; Von Kalle, C.; Schmidt, M.; McCcormack, M. P.; Wulffraat, N.; Leboulch, P.; Lim, A.; Osborne, C. S.; Pawliuk, R.; Morillon, E.; Sorensen, R.; Forster, A.; Fraser, P.; Cohen, J. I.; de Saint Basile, G.; Alexander, I.; Wintergerst, U.; Frebourg, T.; Aurias, A.; Stoppa-Lyonnet, D.; Romana, S.; Radford-Weiss, I.; Gross, F.; Valensi, F.; Delabesse, E.; Macintyre, E.; Sigaux, F.; Soulier, J.; Leiva, L. E.; Wissler, M.; Prinz, C.; Rabbitts, T. H.; Le Deist, F.; Fischer, A.; Cavazzana-Calvo, M. LMO2-associated clonal T cell proliferation in two patients after gene therapy for SCID-X1. *Science* **2003**, *302* (5644), 415-419.

169. Kiessling, F.; Fokong, S.; Koczera, P.; Lederle, W.; Lammers, T. Ultrasound microbubbles for molecular diagnosis, therapy, and theranostics. *J Nucl Med* **2012**, *53* (3), 345-348.

170. Bin, J. P.; Xiu, J. C.; Lao, Y.; Huang, W. F.; Zha, D. G.; Liu, Y. L. Ultrasound-targeted microbubbles destruction increases the extravasation of evens blue in spinotrapezius muscle of rats without transient nanopores. *Eur Heart J* **2008**, *29*, 377.

171. Li, P.; Zheng, Y. Y.; Ran, H. T.; Tan, J. X.; Lin, Y. J.; Zhang, Q. X.; Ren, J. L.; Wang, Z. G. Ultrasound triggered drug release from 10-hydroxycamptothecin-loaded phospholipid microbubbles for targeted tumor therapy in mice. *J Control Release* **2012**, *162* (2), 349-354.

172. Tang, H. L.; Wang, Z. G.; Li, Q.; Ran, H. T.; Zheng, Y. Y.; Ren, J. L.; Ling, Z. Y.; Li, A.; Zhao, B. W. Targeted delivery of bone mesenchymal stem cells by ultrasound destruction of microbubbles promotes kidney recovery in acute kidney injury. *Ultrasound Med Biol* **2012**, *38* (4), 661-669.

173. Jia, X. K.; Yin, J. J.; He, D. G.; He, X. X.; Wang, K. M.; Chen, M.; Li, Y. H. Polyacrylic acid modified upconversion nanoparticles for simultaneous ph-triggered drug delivery and release imaging. *J Biomed Nanotechnol* **2013**, *9* (12), 2063-2072.

174. Tian, G.; Gu, Z. J.; Zhou, L. J.; Yin, W. Y.; Liu, X. X.; Yan, L.; Jin, S.; Ren, W. L.; Xing, G. M.; Li, S. J.; Zhao, Y. L. Mn^{2+} dopant-controlled synthesis of NaYF4: Yb/Er upconversion nanoparticles for *in vivo* imaging and drug delivery. *Adv Mater* **2012**, *24* (9), 1226-1231.

175. Wang, C.; Cheng, L. A.; Liu, Z. A. Drug delivery with upconversion nanoparticles for multi-functional targeted cancer cell imaging and therapy. *Biomaterials*

2011, *32* (4), 1110-1120.

176. Yang, D. M.; Dai, Y. L.; Ma, P. A.; Kang, X. J.; Cheng, Z. Y.; Li, C. X.; Lin, J. One-step synthesis of small-sized and water-soluble $NaReF_4$ upconversion nanoparticles for *in vitro* cell imaging and drug delivery. *Chem-Eur J* **2013**, *19* (8), 2685-2694.

177. Auzel, F. Compteur quantique par transfert denergie DeYb^{3+} a Tm^{3+} Dans un tungstate mixte et dans un verre germanate. *Cr Acad Sci B Phys* **1966**, *263* (14), 819-821.

178. Auzel, F. Compteur quantique par transfert denergie entre deux ions de terres rares dans un tungstate mixte et dans un verre. *Comptes Rendus Hebdomadaires Des Seances De L Academie Des Sciences Serie B* **1966**, *262* (15), 1016-1019.

179. Chien, Y. H.; Chou, Y. L.; Wang, S. W.; Hung, S. T.; Liau, M. C.; Chao, Y. J.; Su, C. H.; Yeh, C. S. Near-infrared light photocontrolled targeting, bioimaging, and chemotherapy with caged upconversion nanoparticles *in vitro* and *in vivo*. *ACS Nano* **2013**, *7* (10), 8516-8528.

180. Cao, T. Y.; Yang, Y.; Gao, Y. A.; Zhou, J.; Li, Z. Q.; Li, F. Y. High-quality water-soluble and surface-functionalized upconversion nanocrystals as luminescent probes for bioimaging. *Biomaterials* **2011**, *32* (11), 2959-2968.

181. Liu, Y.; Chen, M.; Cao, T. Y.; Sun, Y.; Li, C. Y.; Liu, Q.; Yang, T. S.; Yao, L. M.; Feng, W.; Li, F. Y. A cyanine-modified nanosystem for *in vivo* upconversion luminescence bioimaging of methylmercury. *J Am Chem Soc* **2013**, *135* (26), 9869-9876.

182. Langer, K.; Balthasar, S.; Vogel, V.; Dinauer, N.; von Briesen, H.; Schubert, D. Optimization of the preparation process for human serum albumin (HSA) nanoparticles. *Int J Pharm* **2003**, *257* (1-2), 169-180.

183. Rubino, O. P.; Kowalsky, R.; Swarbrick, J. Albumin microspheres as a drug-delivery system: Relation among turbidity ratio, degree of cross-linking, and drug-release. *Pharmaceut Res* **1993**, *10* (7), 1059-1065.

184. Kommareddy, S.; Amiji, M. Preparation and evaluation of thiol-modified gelatin nanoparticles for intracellular DNA delivery in response to glutathione. *Bioconjugate Chem* **2005**, *16* (6), 1423-1432.

185. Fuchs, S.; Coester, C. Protein-based nanoparticles as a drug delivery system: Chances, risks, perspectives. *J Drug Deliv Sci Tec* **2010**, *20* (5), 331-342.

186. Elzoghby, A. O.; Samy, W. M.; Elgindy, N. A. Protein-based nanocarriers as promising drug and gene delivery systems. *J Control Release* **2012**, *161* (1), 38-49.

187. MaHam, A.; Tang, Z. W.; Wu, H.; Wang, J.; Lin, Y. H. Protein-based nanomedicine platforms for drug delivery. *Small* **2009**, *5* (15), 1706-1721.

188. Vandelli, M. A.; Rivasi, F.; Guerra, P.; Forni, F.; Arletti, R. Gelatin microspheres crosslinked with d,l-glyceraldehyde as a potential drug delivery system: Preparation, characterisation, *in vitro* and *in vivo* studies. *Inter J Pharmac* **2001**, *215* (1-2), 175-184.

189. Wilson, B.; Ambika, T. V.; Patel, R. D. K.; Jenita, J. L.; Priyadarshini, S. R. B. Nanoparticles based on albumin: Preparation, characterization and the use for 5-flurouracil delivery. *Int J Biol Macromol* **2012**, *51* (5), 874-878.

190. Shen, Z. Y.; Li, Y.; Kohama, K.; Oneill, B.; Bi, J. X. Improved drug targeting of cancer cells by utilizing actively targetable folic acid-conjugated albumin nanospheres. *Pharmacol Res* **2011**, *63* (1), 51-58.

191. Dubey, P. K.; Singodia, D.; Verma, R. K.; Vyas, S. P. RGD modified albumin nanospheres for tumour vasculature targeting. *J Pharm Pharmacol* **2011**, *63* (1), 33-40.

192. Zensi, A.; Begley, D.; Pontikis, C.; Legros, C.; Mihoreanu, L.; Wagner, S.; Buchel, C.; von Briesen, H.; Kreuter, J. Albumin nanoparticles targeted with Apo E enter the CNS by transcytosis and are delivered to neurones. *J Control Release* **2009**, *137* (1), 78-86.

Chapter 2

Basics of Theranostics in Tumor

Huichao Zou[1], *Yaohua Liu*[2], *and Shiguang Zhao*[2*]
1 Department of Pain Medicine, the Third Affiliated Hospital, Harbin Medical University, Harbin 150081, China
2 Department of Neurosurgery, the First Affiliated Hospital, Harbin Medical University, Harbin 150081, China

Theranostic nanomedicine is a promising treatment strategy for cancer. In this chapter, we will discuss the developments in theranostic research and potential application in clinical practice. Some of the currently used nanoplatforms, nanoprobes and nanodiagnostic tools will be listed. At the end of this chapter, the safety issues and future research direction of theranostic nanomedicine will be discussed.

2.1 Introduction

Cancer is a serious incurable global health threat. It is one of the leading causes of mortality worldwide.[1] There are over 10 million new patients diagnosed with cancers every year.[2] Surgical resection, radiation and chemotherapy are commonly used treatment options for cancers. Despite of extensive research, the heterogeneity and resistance of tumor cells and poor specificity of current therapies limit the efficacy of current cancer treatments.[3] Most of the chemotherapeutic agents are given systemically. Therapeutic agents for chemotherapy are toxic to both cancer cells and normal cells which limit the allowed dosage and leads to severe systemic side-effects and treatment failure. Therefore, novel therapies which can improve the therapeutic index and decrease toxicity to normal tissues are urgently needed.

Theranostics (diagnostics + therapeutics) is a process of diagnostic therapy which can tailor a treatment for individual patients based on the test results (Fig.2.1).[4,5] Theranostic nanomedicine is a nanotherapeutic system which combines the diagnostic tests with targeted therapies and real-time monitoring to the therapies. Theranostic nanomedicine is an effective approach to recognize the molecular heterogeneity of the cancer cells.[6] Theranostic nanomedicine integrates cancer biology, diagnostic imaging and materials science. It applies nanobiotechnology at the nanoscale for diagnosing, curing or preventing diseases and involves the creation and application of nano-biomaterials and devices

* Corresponding author: guangsz@hotmail.com

at the molecular level for personalized therapy. It has several potential advantages over conventional therapies for cancer treatment, including improvement of the sensitivity of cancer diagnosis, and reducing both toxicity and the duration of the treatment.[7–9]

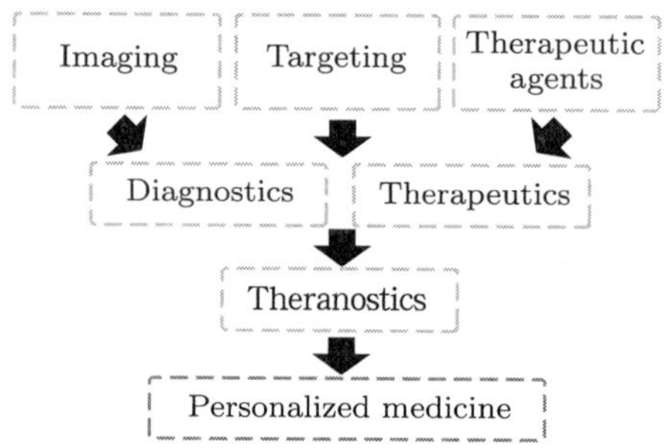

Fig. 2.1 Schematic of the basics of theranostics.

Theranostic nanomedicine utilizes molecular imaging to characterize the cellular phenotypes of the tumor, thus providing specific information of tumor markers and guidance to the following therapies.[10] The tumor can be viewed in the molecular level by giving different cancer target contrasts in one nanoplatform. The following chemotherapy after molecular imaging will eradicate the tumor cells by targeting the identified molecular markers on the cellular surface.

One advantage of theranostic nanomedicine over the conventional small molecular therapies is that nanomedicine can target multiple tumor markers and deliver multiple chemotherapeutic agents simultaneously.[11] The main mechanism of current chemotherapeutic agents is either to kill cancer cell or inhibit cancer cell proliferation. However, because of the unclear distinction between cancer cells and normal cells, the efficacy of current chemotherapy is limited by its toxicity. Novel targeted agents have been used in cancer treatment to achieve higher binding affinity, higher specificity and lower systemic toxicity. For example, Herceptin has been routinely used in breast cancer treatment. It is a monoclonal antibody to Her2/Neu receptor in breast cancer.[12,13] However, breast cancer cells can develop resistance to this single targeted agent by proliferating through other pathways instead of HER2/neu pathway,[14,15] which leads to even stronger resistance. The shortcoming of single target agents in chemotherapy can be overcome by administrating multiple monoclonal antibodies against different molecular markers. This synergistic effect can be achieved by giving a single nanomedicine formulation which will provide a platform to combine multiple antibodies and small peptides together.[16–18]

Another specific advantage of theranostic nanomedicine is that by co-delivering of imaging contrast agents and chemotherapeutic drugs, a physician can get real-time information about the efficacy of the treatment strategy by monitoring the change of expression level of the biomarkers.[19,20] These markers can either be the downstream effector molecules of the drugs or key proteins in cell death or proliferation. This approach will provide early feedback of therapeutic efficacy before detection of the traditional end points, such as tumor shrinkage (Fig.2.2).

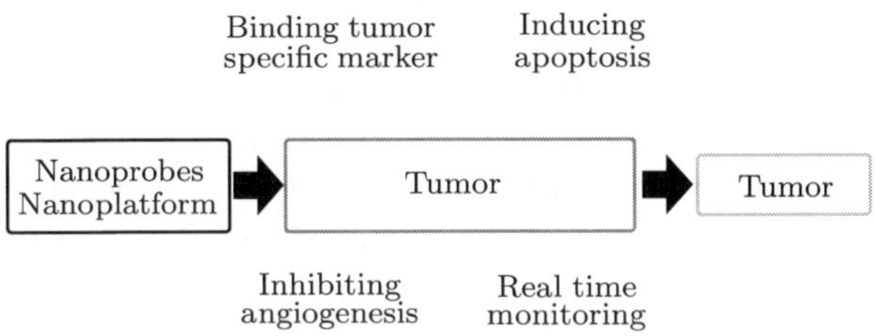

Fig. 2.2 Schematic of the mechanisms of theranostics.

Current research on theranostic nanomedicine is being conducted on developing a system which has higher sensitivity and specificity that can detect cancer at early stage, improve the efficiency of therapy with less toxicity and monitor the response of tumor to various therapies, thus the overall prognosis of the cancer patients will be improved.

2.2 Nanoplatforms

The basis of successful development for theranostic nanomedicine is the development of nanomaterials. An ideal nanomedicine platform should be small in size, efficient in providing high drug-loading densities and specific in targeting the tumor tissues with minimal nonspecific uptake.[16]

Several nanomedicine platforms have been widely developed which include polymer-based nanomedicine templates, liposome-based nanomedicine platform, ferritin/apoferritin-based nanomedicine platform, and viral capsids-based nanomedicine.[21–24] Nanomedicine platforms have promising potential for drug delivery. For example, small interfering RNA (siRNA) has been widely used in several diseases such as cancer, infectious diseases, and metabolic disturbances by silencing specific genetic expression. However, the low penetration ability of naked siRNA limits the efficacy of siRNA therapy[25–27] and the nanomedicine platforms can overcome this obstacle effectively. One recent study has established a perfluorocarbon-based nanoparticle system as a tumor vascular targeting system that can simultaneously deliver drug molecules and can act as ultrasound and MRI contrast agents.[28,29] Fluorescent probes will detect rapid visualization of the tumor and vascular-constrained nanoparticles will more selectively detect cancer-induced angiogenesis. The combination of fluorescent probes and vascular-constrained nanoparticles provides a better strategy to monitor the angiogenesis of tumor.

2.3 Nanoprobes

Nanoprobes are nanomaterial-based molecular probes which can target tumors by binding with specific tumor markers or be utilized to target imaging biological processes such as apoptosis, autophagy, angiogenesis *etc.*[30,31] It can also be applied to monitor the response of tumors to the therapies. Nanoprobes can detect tumors at early stage, guide receptor-targeted sentinel node biopsies and customize individualized treatment options. Nanoprobes can also improve the sensitivity of detection and accuracy of biomarker analysis. Examples of probes include gold nanoparticles, single wall carbon nanotubes CNTs, magnetic iron oxide nanoparticles for MRI, and fluorescent nanoparticles for *in vivo* fluorescence imaging.[32,33]

Glioma is the most common primary malignant tumor in central nervous system. The existence of blood-brain barrier limits the available chemotherapeutic agents. In the studies done by Mani and Koo, a nanoprobe has been developed that can cross the blood-brain barrier to specifically target glioma and remain in tumors sustainably.[33,34]

Another recent study done by Veiseh *et al.* developed a multifunctional nanoprobe which binds specifically with chlorotoxin, a glioma specific targeting molecule.[35] The nanoprobe can be detected by both magnetic resonance imaging and fluorescence microscope and can be used in preoperative and postoperative diagnostic imaging with MRI and real-time intraoperative visualization of tumor margins with optical devices.

In the work of Cui *et al.*[31] a fluorescent superparamagnetic nanoprobes conjugated with an antibody against gamma-seminoprotein showed highly selective targeting, fluorescent imaging, and magnetic resonance imaging. The most interesting fact in that study is that this nanoprobe has cytotoxic effects on the prostate cancer cells under *in vitro* alternating magnetic field irradiation. It was also observed that the size of tumors implanted in nude mice was significantly reduced with thermal treatment.

Immunoassay involves another application of nanoprobes. The traditional immunoassay used for detection of tumor markers is usually time-consuming. Nanoprobes-based immunoassay can detect biomarker proteins at very low concentration. For example, prostate specific antigen (PSA) is an established biomarker for the prostate cancer diagnosis. By using gold nanoparticle-based immunoassay, the detection limit is improved to 0.02 pM for gold nanoparticles, which is nine orders of magnitudes lower than the assays using proteins or DNA molecules.[36]

Gold nanoprobe coupling with dynamic light scattering (DLS) measurements was developed to detect free-PSA.[36] Compared to the traditional plate-based immunoassay, this nanoprobe assay has the advantage of both the large scattering cross section of nanoparticles and high sensitivity of DLS measurement. These advantages make the biomarker detection faster and more sensitive.

Molecular nanodiagnostics applied to cancer treatment may provide rapid and sensitive detection of cancer related molecular alterations, which would enable early detection even when those alterations occur only in a small percentage of cells. For example, an Au-nanoparticles based nanoprobe was developed and used to recognize and quantify the expression level of the Bcr-abl mRNA. The sensitivity of the Au-nanoprobes allows differential gene expression at 10 ng/μL of total RNA.[37]

DNAzyme-functionalized gold nanoparticles are used as catalytic labels in new immunoassay strategy for the amplified detection of α-fetoprotein.[38] As gold nanoparticles can carry a large number of DNAzyme units per protein binding event, they lead to substantial amplification of the detection efficiency.

Another antibody-conjugated hydrophilic magnetic nanoprobe conjugated with monoclonal Trastuzumab antibody was developed to detect human epidermal growth factor receptor (HER2/neu) in breast cancer *via* magnetic resonance imaging.[39]

2.4 Nanocarriers

Conventional chemotherapeutic agents affect both cancer cells and normal cells simultaneously. Thus, the systemic toxicities limit the dose of chemotherapeutic drugs which resulted in suboptimal or failed treatment. Nanocarrier based therapy overcomes this shortcoming of traditional chemotherapeutic agents by specifically attacking the target cancer cells.

Nanocarriers have the capability to deliver drugs to the tumor site specifically (Fig.2.3). This specificity significantly reduces the toxicity of the drugs to normal cells. Ideally, such carriers should be able to decrease drug concentration in normal tissue, minimize drug degradation, prevent undesirable side effects, increase the solubility of drug, increase drug bioavailability and release of the drug at the targeted site.[40–42] Several nanocarriers (nanoparticles, liposomes, water-soluble polymers, micelles and dendrimers) have been used for targeted delivery of cancer diagnostic and therapeutic agents.[43]

Several promising formulations of nanocarriers have been used in clinical trials. Doxil is a liposomal adriamycin which has therapeutic effects on Kaposi sarcoma and ovarian cancers.[44] Genexol®-PM is paclitaxel using polymeric micelles as carrier for breast, lung, pancreatic cancer and recurrent breast cancer.[45] Abraxane® is albumin-paclitaxel using nanoparticles as carrier for metastatic breast cancer.[46] Polymer-drug conjugated Paclitaxel Xyotax® is paclitaxel using polymer-drug conjugates for breast and ovarian cancers.[47] Combining several therapeutic agents into one delivery vehicle may enhance the effectiveness of cancer interventions. This may include delivery of multiple

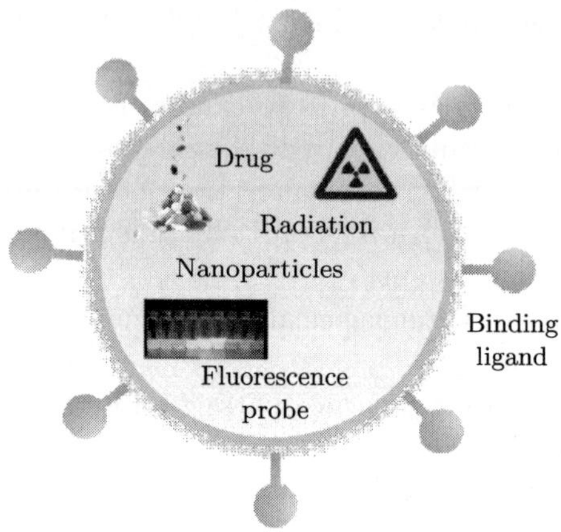

Fig. 2.3 Illustration of nanocarriers.

chemotherapeutic drugs, co-delivery of chemotherapeutic and antiangiogenic agents, co-delivery of drugs and genes, and co-delivery of drugs and siRNA, among others.

Nanocarriers are able to carry large amount of therapeutic agents, increase aqueous solubility and elevate the bioavailability. In addition, nanocarriers have the capability to overcome multidrug resistance, which is a frequently encountered problem in cancer treatment.[48,49]

A new strategy is the development of activated nanocarriers.[50] When the nanocarriers arrive at the target site, the unactivated nanocarriers will be transformed to activated nanocarriers by internal or external stimuli. While the internal stimuli consists of the pH of the tumor microenvironment and tumor specific enzymes, the external stimuli includes temperature, irradiation, ultrasound, *etc.*[51]

Enormous research efforts have been invested to develop innovative nanocarriers. Multi-functional targeted nanocarriers are one of the novel strategies to extend the potential application of nanocarriers.[52] Multi-functional nanocarriers combine different therapeutic approaches to achieve additional or synergistic efficacy. One example is the conjugation of magnetic nanoparticles with anti-VEGF monoclonal antibody for the radioimmunotherapy of liver cancer with an external magnetic field.[52] Another advantage of multi-functional nanocarriers is the capacity to overcome the resistance of cancer cells to chemotherapeutic agents. Multi-functional nanocarriers are designed to deliver resistance modulators, agents that lower the cell death threshold and agents that regulate the pH in the tumor microenvironment. All these properties contribute to the enhanced effectiveness of anticancer agents in refractory cancers.[49]

2.5 Nanodiagnostics

A simple, sensitive and high-throughput diagnostic tool having higher sensitivity and precise recognition capacity is ideal for cancer diagnostics. Nanodiagnostics can be defined as the usage of nanobiotechnology in molecular diagnostics. The sizes of tumor markers are mostly within the nanometer scale. The nanoparticles used for nanodiagnostics are within the size domain as most of the molecules like protein *etc.* Nanodiagnostic tools can collect the individual information in a heterogeneous population of the tumor. Thus, nanotechnology has clear advantage compared to the conventional diagnostic tools in the aspect of increase of sensitivity and specificity.[53]

Nanodiagnostics can detect cancers at the single-cell and molecular levels. The main categories of the nanodiagnostics are: nanoscale visualization, nanoparticle biolabels, nanotechnology-based biochips/microarrays, nanoparticle-based nucleic acid diagnostics, nanoproteomic-based diagnostics, biobarcode assays, nanopore technology, DNA nano-machines for molecular diagnostics, nanoparticle-based immunoassays, nanobiosensors and combinations of multiple diagnostic technologies.[54] Nanoparticles, such as gold magnetic nanoparticles and quantum dots, are all widely used in nanodiagnostics.[55]

Magnetic nanoparticles binding with specific antibody are used to label molecules for bioscreening and cell-tracking.[56–58] Magnetic nanoparticles can be detected by a magnetometer or MRI to reveal small lymph-node metastases which can't be detected by traditional imaging tools. A gold nanoparticle can be tagged with a short segment of DNA to detect millions of different DNA sequences of the tumor sample simultaneously. QDs and luminescent can be used to visualize cancer cells under fluorescent microscope at high resolution in living animals.[53]

In the study of Sokolov *et al.*, the affinity of gold nanoparticles conjugated to anti-epidermal growth factor receptor monoclonal antibodies to cancer cells is much higher compared to the affinity to normal cells.[59] One advantage of gold nanoparticles is that it can also convert the absorbed light to heat which can be used to kill cancer cells. This strategy can link diagnosis to therapeutics by detecting tumor cells and kill them.

2.6 Safety Issues and Future Direction

There are some potential toxic effects with the use of nanoparticles for *in vivo* diagnostics. Nanoparticles with particular sizes may have toxic effects, although no conclusions can be made currently. There have been a large volume of works about the potential application of theranostic nanomedicine in cancer therapy. However, only a few clinical trials of viable theranostic nanomedicine products are reported. It is a clear fact that further preclinical, clinical and long term toxicity studies need to be done to understand the limitation and side effects of the theranostic nanomedicine. Future efforts should be focused on translating the *in vitro* study to *in vivo* clinical application.

In the near future, the development of theranostic nanomedicine can make the detection and treatment of cancers faster, more effective and less expensive. Future trends in theranostic nanomedicine will be linking diagnosis with treatment and development "whole in one approach" for individualized treatment. The theranostic tools can detect cancers at an early stage, release anticancer agents in the targeted location and monitor the response of cancers to the therapies. The ideal theranostic agents should be able to deliver therapeutics (genetic and/or chemotherapeutic) and provide real-time feedback of therapy response by molecular imaging.

References

1. Pisani, P.; Bray, F.; Parkin, D. M. Estimates of the world-wide prevalence of cancer for 25 sites in the adult population. *Int J Cancer* **2002**, *97* (1), 72-81.
2. Jemal, A.; Siegel, R.; Xu, J.; Ward, E. Cancer statistics, 2010. *CA Cancer J Clin* **2010**, *60* (5), 277-300.
3. Somasundaram, R.; Villanueva, J.; Herlyn, M. Intratumoral heterogeneity as a therapy resistance mechanism: Role of melanoma subpopulations. *Adv Pharmacol* **2012**, *65*, 335-359.
4. Bardhan, R.; Lal, S.; Joshi, A.; Halas, N. J. Theranostic nanoshells: From probe design to imaging and treatment of cancer. *Acc Chem Res* **2011**, *44* (10), 936-946.
5. Kumar, R.; Kulkarni, A.; Nagesha, D. K.; Sridhar, S. *In vitro* evaluation of theranostic polymeric micelles for imaging and drug delivery in cancer. *Theranostics* **2012**, *2* (7), 714-722.
6. Blanco, E.; Kessinger, C. W.; Sumer, B. D.; Gao, J. Multifunctional micellar nanomedicine for cancer therapy. *Exp Biol Med* **2009**, *234* (2), 123-131.
7. Allegra, A.; Penna, G.; Alonci, A.; Rizzo, V.; Russo, S.; Musolino, C. Nanoparticles in oncology: The new theranostic molecules. *Anticancer Agents Med Chem* **2011**, *11* (7), 669-686.
8. Kryza, D.; Taleb, J.; Janier, M.; Marmuse, L.; Miladi, I.; Bonazza, P.; Louis, C.; Perriat, P.; Roux, S.; Tillement, O.; Billotey, C. Biodistribution study of nanometric hybrid gadolinium oxide particles as a multimodal SPECT/MR/optical imaging and theragnostic agent. *Bioconjugate Chem* **2011**, *22* (6), 1145-1152.
9. Shapira, A.; Livney, Y. D.; Broxterman, H. J.; Assaraf, Y. G. Nanomedicine for targeted cancer therapy: Towards the overcoming of drug resistance. *Drug Resist Updat* **2011**, *14* (3), 150-163.
10. Abeylath, S. C.; Ganta, S.; Iyer, A. K.; Amiji, M. Combinatorial-designed multifunctional polymeric nanosystems for tumor-targeted therapeutic delivery. *Acc Chem Res* **2011**, *44* (10), 1009-1017.
11. Zhang, X. Q.; Xu, X.; Bertrand, N.; Pridgen, E.; Swami, A.; Farokhzad, O. C. Interactions of nanomaterials and biological systems: Implications to personalized nanomedicine. *Adv Drug Deliv Rev* **2012**, *64* (13), 1363-1384.

12. Izumi, Y.; Xu, L.; Di Tomaso, E.; Fukumura, D.; Jain, R. K. Tumour biology: Herceptin acts as an anti-angiogenic cocktail. *Nature* **2002**, *416* (6878), 279-280.
13. Baselga, J.; Norton, L.; Albanell, J.; Kim, Y.-M.; Mendelsohn, J. Recombinant humanized anti-HER2 antibody (Herceptin^{TM}) enhances the antitumor activity of paclitaxel and doxorubicin against HER2/neu overexpressing human breast cancer xenografts. *Cancer Res* **1998**, *58* (13), 2825-2831.
14. Kute, T.; Lack, C. M.; Willingham, M.; Bishwokama, B.; Williams, H.; Barrett, K.; Mitchell, T.; Vaughn, J. P. Development of Herceptin resistance in breast cancer cells. *Cytometry Part A* **2004**, *57* (2), 86-93.
15. Nahta, R.; Esteva, F. J. Herceptin: Mechanisms of action and resistance. *Cancer Lett* **2006**, *232* (2), 123-138.
16. Sumer, B.; Gao, J. Theranostic nanomedicine for cancer. *Nanomedicine* **2008**, *3* (2), 137-140.
17. Ferrari, M. Cancer nanotechnology: Opportunities and challenges. *Nat Rev Cancer* **2005**, *5* (3), 161-171.
18. Lammers, T.; Kiessling, F.; Hennink, W. E.; Storm, G. Nanotheranostics and image-guided drug delivery: Current concepts and future directions. *Mol Pharmceut* **2010**, *7*, 1899-1912.
19. Zhang, L.; Gu, F.; Chan, J.; Wang, A.; Langer, R.; Farokhzad, O. Nanoparticles in medicine: Therapeutic applications and developments. *Clin Pharmacol Therapeutics* **2007**, *83* (5), 761-769.
20. Pridgen, E. M.; Langer, R.; Farokhzad, O. C. Biodegradable, polymeric nanoparticle delivery systems for cancer therapy. *Nanomedicine* **2007**, *2* (5), 669-680.
21. MaHam, A.; Tang, Z.; Wu, H.; Wang, J.; Lin, Y. Protein - based nanomedicine platforms for drug delivery. *Small* **2009**, *5* (15), 1706-1721.
22. Al-Jamal, W. T.; Kostarelos, K. Liposomes: From a clinically established drug delivery system to a nanoparticle platform for theranostic nanomedicine. *Acc Chem Res* **2011**, *44* (10), 1094-1104.
23. Huynh, N.; Passirani, C.; Saulnier, P.; Benoit, J. Lipid nanocapsules: A new platform for nanomedicine. *Int J Pharmaceut* **2009**, *379* (2), 201-209.
24. Khemtong, C.; Kessinger, C. W.; Gao, J. Polymeric nanomedicine for cancer MR imaging and drug delivery. *Chem Commun* **2009**, (24), 3497-3510.
25. Minko, T. Multifunctional nanomedicine platform for cancer specific delivery of siRNA by superparamagnetic iron oxide nanoparticles-dendrimer complexes. *Cur Drug Delivery* **2011**, *8*, 59-69.
26. Ozpolat, B.; Sood, A.; Lopez-Berestein, G. Nanomedicine based approaches for the delivery of siRNA in cancer. *J Int Medicine* **2009**, *267* (1), 44-53.
27. Xiong, X.-B.; Lavasanifar, A. Traceable multifunctional micellar nanocarriers for cancer-targeted co-delivery of MDR-1 siRNA and doxorubicin. *ACS Nano* **2011**, *5* (6), 5202-5213.
28. Lanza, G. M.; Winter, P. M.; Caruthers, S. D.; Hughes, M. S.; Cyrus, T.; Marsh, J. N.; Neubauer, A. M.; Partlow, K. C.; Wickline, S. A. Nanomedicine opportunities for cardiovascular disease with perfluorocarbon nanoparticles. *Nanomedicine* **2006**, *1* (3), 321-329.
29. Winter, P. M.; Cai, K.; Caruthers, S. D.; Wickline, S. A.; Lanza, G. M. Emerging nanomedicine opportunities with perfluorocarbon nanoparticles. *Expert Rev Med Devic* **2007**, *4* (2), 137-145.

30. Wang, H.-N.; Vo-Dinh, T. Multiplex detection of breast cancer biomarkers using plasmonic molecular sentinel nanoprobes. *Nanotechnol* **2009**, *20* (6), 065101.
31. Cui, D.; Han, Y.; Li, Z.; Song, H.; Wang, K.; He, R.; Liu, B.; Liu, H.; Bao, C.; Huang, P. Fluorescent magnetic nanoprobes for *in vivo* targeted imaging and hyperthermia therapy of prostate cancer. *Nano Biomed Eng* **2009**, *1* (1), 61-74.
32. Zhang, J.-J.; Gu, M.-M.; Zheng, T.-T.; Zhu, J.-J. Synthesis of gelatin-stabilized gold nanoparticles and assembly of carboxylic single-walled carbon nanotubes/Au composites for cytosensing and drug uptake. *Anal Chem* **2009**, *81* (16), 6641-6648.
33. Mani, V.; Chikkaveeraiah, B. V.; Patel, V.; Gutkind, J. S.; Rusling, J. F. Ultrasensitive immunosensor for cancer biomarker proteins using gold nanoparticle film electrodes and multienzyme-particle amplification. *ACS Nano* **2009**, *3* (3), 585-594.
34. Koo, Y.-E. L.; Reddy, G. R.; Bhojani, M.; Schneider, R.; Philbert, M. A.; Rehemtulla, A.; Ross, B. D.; Kopelman, R. Brain cancer diagnosis and therapy with nanoplatforms. *Adv Drug Deliv Rev* **2006**, *58* (14), 1556-1577.
35. Veiseh, O.; Sun, C.; Gunn, J.; Kohler, N.; Gabikian, P.; Lee, D.; Bhattarai, N.; Ellenbogen, R.; Sze, R.; Hallahan, A. Optical and MRI multifunctional nanoprobe for targeting gliomas. *Nano Lett* **2005**, *5* (6), 1003-1008.
36. Liu, X.; Dai, Q.; Austin, L.; Coutts, J.; Knowles, G.; Zou, J.; Chen, H.; Huo, Q. A one-step homogeneous immunoassay for cancer biomarker detection using gold nanoparticle probes coupled with dynamic light scattering. *J Am Chem Soc* **2008**, *130* (9), 2780-2782.
37. Diagaradjane, P.; Orenstein-Cardona, J. M.; Colón-Casasnovas, N. E.; Deorukhkar, A.; Shentu, S.; Kuno, N.; Schwartz, D. L.; Gelovani, J. G.; Krishnan, S. Imaging epidermal growth factor receptor expression *in vivo*: Pharmacokinetic and biodistribution characterization of a bioconjugated quantum dot nanoprobe. *Clin Cancer Res* **2008**, *14* (3), 731-741.
38. Zhou, W.-H.; Zhu, C.-L.; Lu, C.-H.; Guo, X.; Chen, F.; Yang, H.-H.; Wang, X. Amplified detection of protein cancer biomarkers using dnazyme functionalized nanoprobes. *Chem Commun* **2009**, (44), 6845-6847.
39. Lee, J.; Yang, J.; Seo, S.-B.; Ko, H.-J.; Suh, J.-S.; Huh, Y.-M.; Haam, S. Smart nanoprobes for ultrasensitive detection of breast cancer *via* magnetic resonance imaging. *Nanotechnol* **2008**, *19* (48), 485101.
40. Torchilin, V. P. Targeted pharmaceutical nanocarriers for cancer therapy and imaging. *Aaps Journal* **2007**, *9* (2), 128-147.
41. Emerich, D. F.; Thanos, C. G. Targeted nanoparticle-based drug delivery and diagnosis. *J Drug Targeting* **2007**, *15* (3), 163-183.
42. Emerich, D. F.; Thanos, C. G. The pinpoint promise of nanoparticle-based drug delivery and molecular diagnosis. *Biomol Eng* **2006**, *23* (4), 171-184.
43. Cho, K.; Wang, X.; Nie, S.; Shin, D. M. Therapeutic nanoparticles for drug delivery in cancer. *Clin Cancer Res* **2008**, *14* (5), 1310-1316.
44. Peer, D.; Margalit, R. Tumor-targeted hyaluronan nanoliposomes increase the antitumor activity of liposomal doxorubicin in syngeneic and human xenograft mouse tumor models. *Neoplasia* **2004**, *6* (4), 343.

45. Kim, S. C.; Kim, D. W.; Shim, Y. H.; Bang, J. S.; Oh, H. S.; Wan, K. S.; Seo, M. H. *In vivo* evaluation of polymeric micellar paclitaxel formulation: Toxicity and efficacy. *J Control Release* **2001**, *72* (1-3), 191.
46. Moreno-Aspitia, A.; Perez, E. A. North central cancer treatment group N0531: Phase Ⅱ trial of weekly albumin-bound paclitaxel (ABI-007; Abraxane®) in combination with gemcitabine in patients with metastatic breast cancer. *Clin Breast Cancer* **2005**, *6* (4), 361-364.
47. Vicent, M. J.; Duncan, R. Polymer conjugates: Nanosized medicines for treating cancer. *Trends Biotechnol* **2006**, *24* (1), 39-47.
48. Peer, D.; Karp, J. M.; Hong, S.; Farokhzad, O. C.; Margalit, R.; Langer, R. Nanocarriers as an emerging platform for cancer therapy. *Nat Nanotechnol* **2007**, *2* (12), 751-760.
49. Jabr-Milane, L. S.; van Vlerken, L. E.; Yadav, S.; Amiji, M. M. Multi-functional nanocarriers to overcome tumor drug resistance. *Cancer Treatment Rev* **2008**, *34* (7), 592-602.
50. Gullotti, E.; Yeo, Y. Extracellularly activated nanocarriers: A new paradigm of tumor targeted drug delivery. *Mol pharmaceut* **2009**, *6* (4), 1041-1051.
51. Danhier, F.; Feron, O.; Préat, V. To exploit the tumor microenvironment: Passive and active tumor targeting of nanocarriers for anti-cancer drug delivery. *J control release* **2010**, *148* (2), 135-146.
52. Torchilin, V. P. Multifunctional nanocarriers. *Adv Drug Deliv Rev* **2012**, *64*, 302-315.
53. Jain, K. K. Applications of nanobiotechnology in clinical diagnostics. *Clin Chem* **2007**, *53* (11), 2002-2009.
54. He, J.; VanBrocklin, H. F.; Franc, B. L.; Seo, Y.; Jones, E. F. Nanoprobes for medical diagnosis: Current status of nanotechnology in molecular imaging. *Curr Nanosci* **2008**, *4* (1), 17-29.
55. Jain, K. K. Nanodiagnostics: Application of nanotechnology in molecular diagnostics. *Expert Rev Mol Diagn* **2003**, *3* (2), 153-161.
56. Pankhurst, Q. A.; Connolly, J.; Jones, S. K.; Dobson, J. Applications of magnetic nanoparticles in biomedicine. *J Phys D Appl Phys* **2003**, *36* (13), R167-R181.
57. Tiefenauer, L. X.; Kuehne, G.; Andres, R. Y. Antibody-magnetite nanoparticles: *In vitro* characterization of a potential tumor-specific contrast agent for magnetic resonance imaging. *Bioconjugate Chem* **1993**, *4* (5), 347-352.
58. Gruttner, C.; Muller, K.; Teller, J.; Westphal, F.; Foreman, A.; Ivkov, R. Synthesis and antibody conjugation of magnetic nanoparticles with improved specific power absorption rates for alternating magnetic field cancer therapy. *J Magn Magn Mater* **2007**, *311* (1), 181-186.
59. Sokolov, K.; Follen, M.; Aaron, J.; Pavlova, I.; Malpica, A.; Lotan, R.; Richards-Kortum, R. Real-time vital optical imaging of precancer using anti-epidermal growth factor receptor antibodies conjugated to gold nanoparticles. *Cancer Res* **2003**, *63* (9), 1999-2004.

Chapter 3

Magnetic Nanomaterials for Tumor Targeting Theranostics

*Leyong Zeng, Zheyu Shen, and Aiguo Wu**

Division of Functional Materials and Nano Devices, Key Laboratory of Magnetic Materials and Devices, Ningbo Institute of Materials Technology and Engineering (NIMTE), Chinese Academy of Sciences, Ningbo 315201, China

Magnetic nanomaterials are the most promising biomaterials as theranostic nanoprobes for the diagnosis and therapy of cancers. Commercial magnetic resonance imaging contrast agents, based on magnetic nanomaterials, have been widely used in clinical medicine. In this chapter, we will focus on the applications of magnetic nanomaterials including paramagnetic and superparamagnetic nanomaterials as diagnostic and/or therapeutic agents in biomedicine, commonly known as theranostics. Therefore, the applications of the magnetic nanomaterials in cancer research are mainly targeted for MRI, hyperthermia, photodynamic and photothermal therapy, and drug/gene delivery. In addition, the chapter concludes with the brief discussion on future perspective of this field.

3.1 Introduction

Magnetic nanomaterials are the materials composed of magnetic elements, such as iron, gadolinium (Gd^{3+}), cobalt, manganese, nickel, their corresponding oxides or alloys, and their polymeric cationic polyplexes among others. Magnetic nanomaterials show ferromagnetism, paramagnetism or even superparamagnetism depending on their element types, crystal structures, sizes, and chemical composition of the nanomaterials, synthetic methodologies, *etc.* In general, besides the four basic effects of nanomaterials, such as the quantum size effect, the surface effect, the small size effect and the macroscopic quantum tunneling effect, superparamagnetism, high coercivity, low Curie temperature and high magnetic susceptibility are the four special properties of the magnetic nanomaterials.[1]

The physical and chemical properties of the magnetic nanomaterials are attractive and, therefore, these nanomaterials are utilized extensively in biomedical fields, environmental sciences, catalysis, mineralogy, data storage in informatics, and nanofluids, *etc.* In biomedicine, the applications of the magnetic nanomaterials include magnetic separation, magnetic transfection, magnetic

∗ Corresponding author: aiguo@nimte.ac.cn

hyperthermia, magnetic resonance imaging (MRI), and nanodrug delivery under the influence of external magnetic fields.[1] In this chapter, we will focus on the applications of magnetic nanomaterials and their complexes/composites as diagnostic and/or therapeutic agents in biomedicine, that is, magnetic nanomaterials-based theranostics. While the paramagnetic nanomaterials include those of Gd^{3+} and manganese, the superparamagnetic nanomaterials are based on iron oxide.

3.2 Paramagnetic Nanomaterials for Theranostics

3.2.1 *Introduction to Magnetic Resonance Imaging and Paramagnetic Nanomaterials*

In the last 30 years, MRI has been probably the most used tool for the diagnosis of various diseases. MRI was explored by Paul Lauterbur in 1973 and approved by Food and Drug Administration (FDA) for clinical application in 1985.[2] Soon after its discovery, MRI has been one of the most common diagnostic tools in hospitals worldwide. For clinical MRI detection, the body needs to be placed in a magnetic field. In this protocol, the hydrogen (proton) in the body could emit nuclear magnetic resonance signals for MRI detection. Because MRI is very useful for monitoring and diagnosis of many human diseases including cancers, the Nobel Prize in physiology and medicine was awarded in 2003 to two scientists, Paul Lauterbur and Sir Peter Mansfield, for their contribution to MRI.[3,4]

The magnetic resonance images are originated from the nuclear magnetic resonance signals of H_2O protons. Therefore, the resolution of the images basically lies on three aspects: ① the density of H_2O protons, ② the longitudinal relaxation time (T_1), and ③ the transverse relaxation time (T_2) of the H_2O protons.[5] In order to increase the resolution of the image, a contrast agent, with a high longitudinal relaxivity (r_1) or transverse relaxivity (r_2), is used.[6] The definition of relaxivity (r) is the ratio of the inverse relaxation time to the concentration of contrast agent. It is usually reported as the parameter r_2/r_1, that indicates whether contrast agents (CAs) are positive or negative CAs. For paramagnetic nanomaterials-based CAs, the r_2/r_1 values are relatively low. When the r_2/r_1 value is below 2, the observed T_1-weighted images are brightened. Such CAs are called positive CAs or T_1-weighted MRI CAs.[7,8]

Paramagnetic nanomaterials, preferably rare-earth-containing particles (chelates, oxides, hydroxides, fluorides, *etc*), are widely used as MRI CAs. At room temperature, the paramagnetic nanoparticles are purely paramagnetic in nature.[9] Their magnetization follows a classical Langevin function, but the magnetic saturation is not expected at the magnetic fields used for MRI. Their magnetization simply increases linearly with the field, as predicted by the Langevin function at lower fields. When approaching magnetic saturation, all of the paramagnetic ions constituting the paramagnetic nanoparticles will finally align in the field, resulting in a very large magnetic moment for the particle.[10] Their paramagnetic behavior opens many perspectives in MRI technology.[11,12]

3.2.2 *Synthesis of Paramagnetic Nanomaterials*

Naturally, metal ions having unpaired electrons are essential paramagnetic components in MRI CAs. These metal ions involve transition and lanthanide metals including manganese (Mn^{2+}, Mn^{3+}), Gd^{3+}, lanthanide (La^{3+}), iron (Fe^{2+}, Fe^{3+}), dysprosium (Dy^{3+}) and other metals in the *d*- and *f*-blocks.[13] Because of seven unpaired electrons, symmetric electronic states and high relaxivity, Gd^{3+} is an excellent paramagnetic metal ion. Nonetheless, the paramagnetic metal ions can rarely be applied as CAs in the ionic state because of their unacceptable biodistribution and high cytotoxicities. Consequently, chelates of paramagnetic metal ions and ligands have been synthesized as the chelates may effectively stabilize the ions until they are completely excreted from the body, thus reducing their cytotoxicities effectively. Polyaminocarboxylates, including diethylene triamine pentaacetic acid (DTPA), 1,4,7,10-tetraazacyclododecane-1,4,7,10-tetraacetic acid (DOTA), 1,4,7,10-tetraazacyclododecane-1-(2-hydroxypropyl)-4,7,10-triacetic acid (DO3A) and their derivatives, have been mainly used as the stable chelating ligands. The current clinically used Gd^{3+} chelates are Gd-DTPA (Magnevist, Schering AG, Germany), Gd-DTPA-bismethylamide (Gd-DTPA-BMA) (Omniscan, Amersham-Nycomed, Norway), Gd-DO3A-HP (ProHance, Bracco Imaging, Italy), and Gd-DOTA (Dotarem, Guerbet SA, France).[14]

In addition, nanoparticles of paramagnetic metal oxides, *e.g.*, Gd_2O_3, have also been synthesized to overcome the low biodistribution and high cytotoxicities. Small Gd_2O_3 nanoparticles with 20–40 nm of diameters were investigated by McDonald and Watkin who have shown that the relaxivity of the Gd_2O_3 nanoparticles is comparable to that of Magnevist (Schering AG, Germany).[15] It was also reported that the relaxivity of smaller Gd_2O_3 nanoparticles with the diameters of 3–10 nm is twice higher than that of Magnevist.[16] Such Gd_2O_3 nanoparticles can be synthesized by a polyol method in which $GdCl_3$ (1.582 g) and NaOH (300 mg) were respectively added into two DEG solvents (30 mL), followed by magnetic stirring of the resulting mixture and/or solution and heating to 140 °C using an oil bath. The reaction was kept at this temperature for 1.0 h and subsequently at 180 °C for another 4.0 h. Finally, the synthesized Gd_2O_3 nanoparticles were purified by dialysis.[17]

In addition, multifunctional hybrid nanoparticles of paramagnetic metal oxides have also been synthesized by Petoral *et al.* who have further confirmed the magnetic characteristics of Gd_2O_3 nanoparticles and fluorescent properties of Tb^{3+} ion. The core of the Gd_2O_3 nanoparticles were doped with Tb^{3+} ion, stabilized with some organic acids, and then functionalized with polyethylene glycol.[18]

3.2.3 *Paramagnetic Nanomaterials for Tumor Targeting Diagnosis*

The MRI technique has been widely used for clinical diagnosis during the last two decades. The paramagnetic nanomaterials as T_1-weighted CAs are used to improve the contrast MRI images and play an important role for increasing the MRI utility. The current paramagnetic MRI CAs in clinical use mainly include Magnevist (Gd-DTPA), Omniscan (Gd-DTPA-BMA), ProHance (Gd-DO3A-HP), Dotarem (Gd-DOTA), OptiMARK (Gd-DTPA-BMEA), Gadovist (Gd-DO3A-Butriol), Multihance (Gd-BOPTA, Gadobenate), and so on.[14] Currently, the Gd^{3+}-chelates and Gd-based inorganic nanoparticles are the paramagnetic CAs used in MRI.[19,20]

3.2.3.1 *Chelates*

Gd-chelates are the most commonly used MRI CAs because of their high magnetic moment and asymmetric electronic ground state.[4,21] The relaxivity of CAs is the efficiency of changing the relaxation time, that mainly depends on three factors: ① the content of H_2O in the coordination shell, ② the exchange rate of the coordinated H_2O with the bulk H_2O and, ③ the rotational correlation time of the molecule (τ_R).[22–24] The Gd-chelates mainly include small-molecular weight Gd-complexes and their macromolecular analogues as discussed below.

To date, the most frequently used Gd-chelates in clinical diagnosis imaging is Gd-DTPA, Gd-ethoxybenzyl-DTPA (Gd-EOB-DTPA) and Gd-DTPA-bismethylamide (Gd-DTPA-BMA). In 1995, Timo Kurki *et al.* reported the effect of Gd-DTPA *in vivo* by measuring the spin-lattice relaxation time and relaxation parameters in MR imaging of intracranial tumors before and after injection of Gd-DTPA.[25] Papadopoulos *et al.* studied the labyrinthine enhancement on MRI as a special signal of labyrinthine disease. The MRI with enhanced contrast by Gd appeared to be not sensitive in demonstrating labyrinthine disease, and the normal examination findings in a patient with sudden sensorineural hearing loss cannot exclude damage at the cochlear level.[26] In addition, for MRI of brain and central nerves system, the paramagnetic CAs including Gd-DTPA have been successfully utilized to increase the contrast.[4,5] But when the T_R (intrinsic rotational correlation time) is 100 ps, the r_1 and r_2 of these small-molecular Gd-chelates are lower than 10 $s^{-1}\cdot mM^{-1}$. However, the maximal theoretical value is higher than 100 $s^{-1}\cdot mM^{-1}$.[27,28] Furthermore, the blood circulation and tissue retention time of these small-molecular CAs are relatively short (<30 min), which limits some applications of MRI for those who needs longer time for data collection.[29,30] Therefore, to obtain the best diagnostic effect of the CAs, it is essential to enable them to target specific tissues and cells. Some small molecular CAs with liver targetability have been explored and marketed, such as Gd-EOB-DTPA[31], Gd-benzyloxypropionictetraacetate (Gd-BOPTA)[32]. Additionally, functional Gd-chelates have also been developed as new tissue-specific CAs by complexing the Gd-DTPA or Gd-DOTA with biomacromolecules or by synthesizing the corresponding polymers with albumin[33], polysaccharide[34–37], polylysine[38] and polyamidoamine dendrimer[39]. Incorporation of the biomacromolecules or polymers may decrease the T_1 relaxation of the surrounding H_2O protons, but can increase the tissue targetability while enhancing the blood circulation time.[40]

Recently, Chandrasekharan *et al.* explored a new Gd-chelate, *i.e.*, Gd (DO3A-xy-TPEP), as a promising targetable paramagnetic MRI contrast agent for diagnostic imaging. The results of MTT assay show that the cytotoxicity of this Gd-chelate is low at low Gd concentrations. The experimental results of cellular uptake indicate that this Gd-chelate is highly targetable to cancer cells. The r_1 of the Gd (DO3A-xy-TPEP) is around 50 % larger than that of the clinical CAs including Gd-DTPA (Magnevist) and Gd-DOTA (Dotarem). MRI results studied on tumor-bearing mice illustrate that the signal at solid tumors is significantly enhanced.[41] Ye *et al.* designed and explored a tumor targetable dendritic contrast agent (DCA) [(FA-PEG)-G2-(DTPA-Gd)], which is prepared by conjugating Gd-chelates and PEG [with folic acid (FA) at the end of chains] onto the surface of a biodegradable polyester dendrimer. The longitudinal relaxivity of DCA is very large up to 17.1 $mM^{-1}\cdot s^{-1}$, which is 4 times larger than that of Magnevist (the most commonly used contrast agent in clinic). The MRI spotted the tumor with (FA-PEG)-G2-(DTPA-Gd) contrast agent. Moreover, the contrast enhancement and the retention time in solid tumor are respectively much higher and longer than those of Magnevist. When compared with non-degradable DCAs, the Gd retention in biodegradable (FA-PEG)-G2-(DTPA-Gd) is very low in all of the normal organs or tissues. Therefore, the proposed biodegradable (FA-PEG)-G2-(DTPA-Gd) should be a promising contrast agent for tumor diagnosis.[42]

In order to specifically deliver MRI CAs to tumors, Han *et al.*[43] explored a tumor targetable dendrimer, *i.e.*, peptide HAIYPRH (T7)-conjugated PEG-modified polyamidoamine dendrimer (PAMAM-PEG-T7). The MRI efficiency of the PAMAM-PEG-T7 was evaluated in liver cancer and early brain glioma model (subcutaneous Bel-7402 xenograft). Gd-DTPA-PAMAM-PEG-T7 was synthesized by conjugating PAMAM-PEG-T7 with DTPA and then chelating with Gd. The ICP-AES results shows that around 92 Gd^{3+} can be loaded by one molecular dendrimer. The longitudinal relaxivity r_1 of the Gd-DTPA-

PAMAM-PEG-T7 is calculated to be 10.7 $mM^{-1}\cdot s^{-1}$, which is much larger than that of Gd-DTPA (4.8 $mM^{-1}\cdot s^{-1}$). The targeting capacity of PAMAM-PEG-T7 towards malignant hepatoma is better than that of PAMAM-PEG. However, PAMAM-PEG-T7 is not specific for glioma cells. Furthermore, Gd-DTPA-PAMAM-PEG-T7 can increase the MRI contrast of malignant hepatoma with the enhanced signal (187 %), which is 130 % for PAMAM-PEG and 121 % for Gd-DTPA. The results show that Gd-DTPA-PAMAM-PEG-T7 is not selective for brain glioma, but can target malignant hepatoma with high specificity. Therefore, the paramagnetic Gd-DTPA-PAMAM-PEG-T7 could be used to diagnose tumors that have no natural barriers.[43]

Shiraishi *et al.*[44] applied a micelle as carrier of MRI CAs to obtain the targetability. The polymer PEG-b-poly (L-lysine) was synthesized and conjugated to DOTA using the $—NH_2$ of L-lysine. DOTA was then used for chelating the gadolinium ions. The synthesized polymer PEG-b-poly(L-lysine-DOTA) can be used to prepare micelle. The polymeric micelle could be stable when the loading content of Gd^{3+} (mass ratio of Gd^{3+} to DOTA) was lower than 40%. The experimental results in a mouse model show that the Gd-loaded polymeric micelle exhibits stable blood circulation, and a considerable amount of Gd-loaded micelle accumulates at tumor site *via* enhanced permeability and retention (EPR) effect after 24 h of intravenous injection. After using the above-indicated Gd-loaded micelle, the intensity of MRI signal at the tumor site is enhanced 200 %.[44]

In conclusion, various Gd-chelates have been widely researched and commercialized as paramagnetic CAs. But Gd-chelates are cytotoxic to some extent and their relaxivity r_1 is very low. In addition, they are not very specific to tumors. Consequently, it is very necessary to explore new and more efficient (more sensitive and specific to tumors) CAs with lower cytotoxicity.

3.2.3.2 *Inorganic Nanoparticles*

It is now clear that an advanced paramagnetic MRI contrast agent, which is more sensitive (*i.e.*, has larger r_1) and more specific to tumors than Gd-chelates, needs to be explored. Gd-based inorganic nanoparticles, such as Gd oxide nanoparticles[45] and Gd fluorides nanoparticles,[46] could be one of such candidates because they can be passively or actively targeted to tumor sites and have larger r_1 than Gd-chelates, but depending on the particle sizes.

Synthetic methods of the Gd oxide (Gd_2O_3) nanoparticles are given below. The Gd_2O_3 nanoparticles with 2–15 nm size can be synthesized from $GdCl_3$ and NaOH at high temperatures.[18,47,48] The Gd_2O_3 nanoparticles with particle size of about 10 nm can also be synthesized from the combustion of $Gd(NO_3)_3$ and amino acid glycin.[49] Other Gd inorganic nanoparticles with different compositions can also be synthesized. For example, reaction between $Gd(NO_3)_3$ and $(NH_4)_2HPO_4$ could be utilized to prepare $GdPO_4$ nanorods with diameter of 20–30 nm (major axis) and 6–15 nm (minor axis).[8] Evanics *et al.*[46] synthesized GdF_3 and GdF_3/LaF_3 nanoparticles with diameter of 10–110 nm. Although the r_1 values of these Gd-based inorganic nanoparticles depend on the sizes, they are, however, much larger than that of the Gd-chelates.

Paramagnetic ultrasmall Gd oxide nanoparticles with sizes of about 1.0 nm can be prepared from the reaction of Gd (Ⅲ) ion precursors and O_2 flow in tripropylene glycol. The r_1 was estimated to be 9.9 $mM^{-1}\cdot s^{-1}$, which is much higher than that of the Gd (Ⅲ)-chelates (Fig.3.1). Moreover, the Gd oxide nanoparticles were applied to diagnose the brain tumor of a rat and high-contrast T_1 MRI's were observed. The higher r_1 can be ascribed to the tremendous specific surface area of the ultrasmall Gd_2O_3 nanoparticles associated with the cooperative induction of surface Gd^{3+} for the longitudinal relaxation of a water proton. Because the r_1 is dependent on the particle size, the diameter with a highest r_1 can be optimized to be 1–2.5 nm. The Gd_2O_3 nanoparticles around 1–2.5 nm is a kind of promising T_1 MRI CAs.[45]

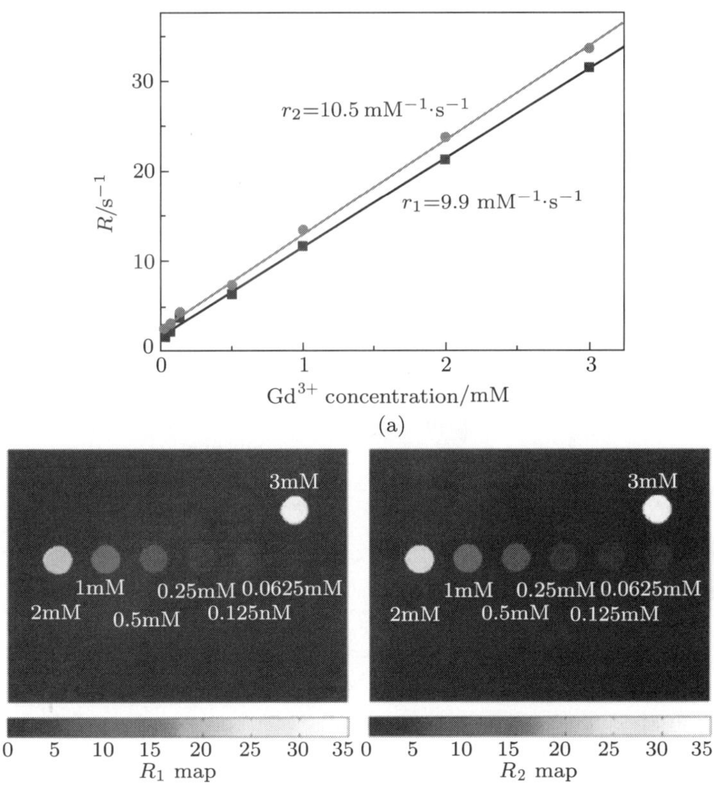

Fig. 3.1 (a) Plots of the R_1 and R_2 versus Gd^{3+} concentration. r_1 and r_2 are calculated from the slopes. (b) R_1 and (c) R_2 map images as a function of Gd^{3+} concentration. Image is adapted with permission from Ref. 45. Copyright (2009) American Chemical Society. (color figure at the end of the book)

A new design of PEGylated Gd oxide nanoparticles as a positive MRI contrast agent was also reported by Ahren *et al.* (Fig.3.2).[17] The results for functionalized nanoparticles (after dialyzed sufficiently) show a considerable relaxivity increase. The r_1 and r_2 are around 4 times those of Magnevist. The microscopy results show that the PEGylated Gd oxide nanoparticles do not activate neutrophils compared with normal Gd oxide nanoparticles without capping. Moreover, Rhodamine is also used to modify the Gd oxide nanoparticles and the results indicate that the PEGylated Gd oxide nanoparticles are applicable for further coupling chemistry and for association with targeted molecules. Therefore, it is promising to explore a potential MRI contrast agent with specific targetability for early diagnosis of the disease.[17]

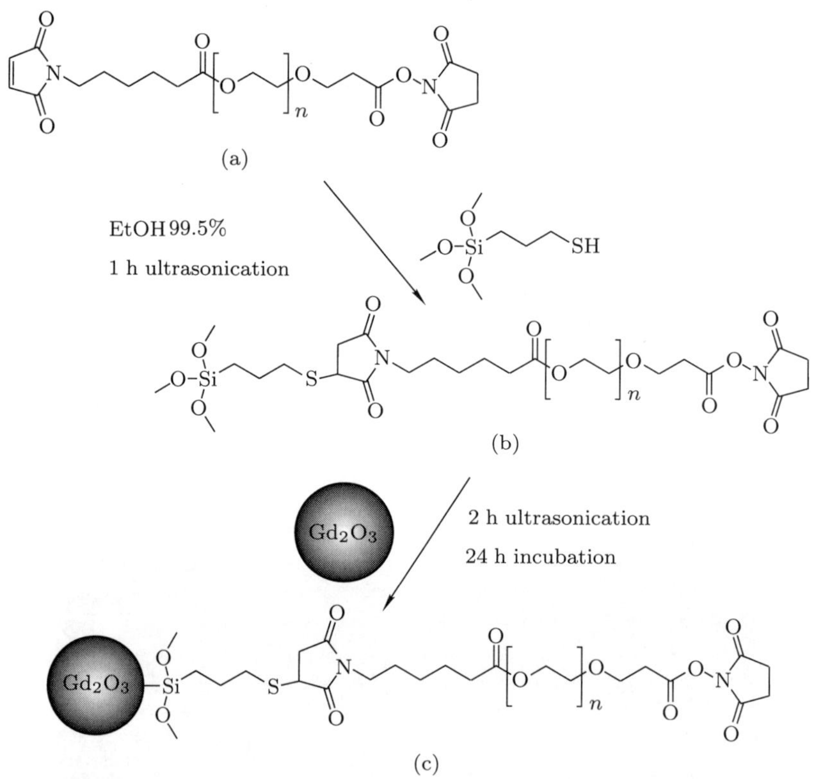

Fig. 3.2 Schematic illustration for the synthesis of PEGylated Gd_2O_3 nanoparticles. (a) Synthesis of α-Maleinimido-ω-carboxysuccinimidyl ester poly (ethylene glycol) (Mal-PEG-NHS). (b) Association of 3-mercaptopropyl trimethoxysilane (MPTS) with Mal-PEG-NHS. (c) PEGylation of the Gd_2O_3 nanoparticles through the reaction between —OH on the surface of Gd_2O_3 nanoparticles and methoxysilane groups in MPTS-Mal-PEG-NHS. Image is adapted with permission from Ref. 17. Copyright (2010) American Chemical Society.

Very recently, Wu and his research group have designed a new type of MRI

CAs based on gadolinium oxide nanocrystals (GON) for targeted imaging and early diagnosis of cancer.[50] The GON were prepared using a polyol method and then encapsulated into albumin nanoparticles (AN), which are cross-linked with glutaraldehyde and they were found to exhibit bright and stable auto- fluorescence without conjugation to any fluorescent agent. Subsequently, the target molecule, FA, was conjugated onto the surface of the GON-loaded AN (GON-AN) to fabricate a GON-AN-FA composite. The results of MRI relaxation studies show that the longitudinal relaxivities (r_1) of GON-AN (11.6 $mM^{-1}\cdot s^{-1}$) and GON-AN-FA (10.8 $mM^{-1}\cdot s^{-1}$) are much larger than those of traditional positive MRI CAs, such as Magnevist (4.7 $mM^{-1}\cdot s^{-1}$). The results of cell viability assays indicate that GON-AN and GON-AN-FA are almost noncytotoxic. Furthermore, the specificities of GON-AN and GON-AN-FA were evaluated with two kinds of cancer cells that overexpress folate receptor alpha (FRα). The results reinforce that the autofluorescent GON-AN-FA is able to target cancer cells *via* recognition of the ligand FA and the receptor FRα. Therefore, this type of autofluorescent GON-AN-FA possessing a large longitudinal relaxivity and good biocompatibility represents a significant advance for the targeted imaging and early diagnosis of cancer.[50]

3.2.4 *Paramagnetic Nanomaterials for Tumor Targeting Theranostics*

In 2002, Funkhouser created the term "theranostic" whose definition is combination of the therapeutic function and diagnostic function in one kind of material. Therefore, theranostics simultaneously deliver therapeutic drugs and diagnostic CAs using one bioformulation.[51] There are numerous clinical benefits for the cancerous early diagnosis and therapy. If the tumor is found at an early stage, therapies could be highly localized, the side effects could be reduced and the damage to normal tissues could be minimized compared to conventional chemotherapy and invasive surgery.

The example of the localized therapy combined with paramagnetic MRI diagnosis may be PDT, chemotherapy, gene therapy and hyperthermia.[51]

PDT uses photosensitizing agents as drugs that can create singlet oxygen to kill cancer cells when exposing to a laser light with an appropriate wavelength.[52] Spernyak *et al.*[52] explored a kind of conjugate, which combined 3-(1′-hexyloxyethyl)-3-devinyl pyropheophorbide (HPPH) with multiple Gd (Ⅲ) aminobenzyl-diethylene triamine pentacetic acid (ADTPA) for cancer diagnositic imaging and PDT. This conjugate of HPPH-3Gd-ADTPA is soluble in pH 7.4 of water. The experimental results on mice and rat tumors show that MRI signal of tumors is significantly enhanced at 24 h post-injection compared with surrounding tissues. Moreover, the HPPH-3Gd-ADTPA conjugate can be significantly uptaken by tumor at 24 h post-injection and induce good therapeutic efficiency through PDT. Toxicity of HPPH-3Gd (Ⅲ)-ADTPA administered at or above diagnostic/therapeutic doses was also evaluated and, evidently, no toxicity was found for organs. Therefore, this particular conjugate is a novel theranostic agent which is efficient for tumor diagnosis and PDT.[52]

Another example of the localized therapy combined with paramagnetic MRI diagnosis is the targeted delivery with precise release of chemotherapeutants. Howell *et al.* reported a novel theranostic strategy, that is loading manganese oxide (MnO, MRI contrast agent) and DNA and/or drug (for disease treatment) using a kind of lipid-micellar nanoparticles (LMNPs). The oleic acid-capped MnO nanoparticles were loaded in LMNPs prepared from polyethylene glycol (PEG-2000), phosphatidyl ethanolamine (PE), DC-cholesterol, and dioleoyl phosphatidyl ethanolamine (DOPE).[53] The nanoparticles are almost non-cytotoxic and well-taken up by human embryonic kidney (HEK293), Lewis lung carcinoma (LLC1), and human lung adenocarcinoma epithelial cells (A549). The nanoparticles were also intra-nasally administered to mice. The results show that most of them accumulated in the lung sites. *In vitro* phantom and *ex vivo* lung MRI results confirm that MnO-LMNPs can increase T_1 MRI contrast. In addition, MnO-LMNPs encapsulated with pDNA and/or doxorubicin are also significantly uptaken by HEK293 cells *in vitro* and by target cells *in vivo*. Consequently, the above results reinforce that MnO-LMNPs could deliver MRI CAs and DNA and/or drugs at the same time to the lung tissues and cells and thus it should be a promising lung tumor theranostic agent.[53]

The localized therapy combined with paramagnetic MRI diagnosis can also be gene therapy. Bae *et al.* explored hollow manganese oxide nanoparticles (MnO-NPs) capped with polyethylenimine (PEI), which could be used to load siRNA drugs and MRI CAs.[54] It was found that the nanoparticles loaded with therapeutic siRNA and MRI CAs were efficiently delivered into human breast cancer cells, and enhanced the T_1 signal intensity of MRI. Therefore, the nanoparticles may be potential and promising multifunctional theranostic agents.[54]

Finally, hyperthermia can also be associated with paramagnetic MRI diagnosis. Hossann *et al.* proposed thermo-sensitive liposomes (TSL) loaded with paramagnetic MRI CAs for MRI assisted interventional hyperthermia in tumor sites. They tried to encapsulate 6 CAs on the market (Gd-DTPA, Gd-DTPA-BMA, Gd-DOTA, Gd-BOPTA, Gd-HP-DO3A and Gd-BT-DO3A) into the TSL. CAs were passively encapsulated with 323 mOs/kg into 1,2-dipalmitoyl-sn-glycero-3-phosphocholine /1,2-distearoyl-sn-glycero-3-phosphocholine / 1,2-dipalmitoyl-sn-glycero-3-phosphodiglycerol 50/20/30 (mol/mol) TSL (DPPG2-TSL) to get stable bioformulations. All preparations are mono-dispersed with particle sizes of around 135 nm. The encapsulation of CAs has no influence on phase transition temperature (T_m) and zeta potential. The r_1 of the bioformulations increases in the temperature ranging from 38 °C to 44 °C, which could be ascribed to the phase transition. The r_1 at a temperature lower than T_m

and the r_1 change [$\Delta r_1 = r_1$ (45.3 °C) − r_1 (37.6 °C)] depend on the concentration of CAs loaded in the TSL. All CAs except Gd-DTPA-BMA can cause hydrolysis of the phospholipid and induce undesired leakage of the CAs. The results indicate that the concentration of CAs loaded in TSL should be high to obtain an enough high Δr_1.[55]

3.2.5 *Overview*

MRI technology is much more sensitive than X-ray technologies for soft tissues in human body. Moreover, some new MRI technologies requiring paramagnetic MRI nanomaterials have been frequently used in clinical protocols including MR angiography (MRA), perfusion MRI, diffusion-weighted MRI, functional MRI (fMRI), BOLD (blood oxygenation level-dependent) imaging, dynamic contrast enhanced (DCE) MRI and molecular imaging. Therefore, future exploration of paramagnetic MRI nanomaterials might meet some challenges. The perfect paramagnetic MRI nanomaterials should meet the following requirements: ① their relaxivity r_1 should be high; ② they should specifically target tumors; ③ their cytotoxicity should be low; ④ the intravascular duration should be long enough and the excretion time should be suitable; ⑤ they should highly enhance the contrast of MRI at low doses; and ⑥ should be inexpensive.

3.3 Superparamagnetic Nanomaterials for Theranostics

3.3.1 *Introduction to Superparamagnetic Nanomaterials*

Superparamagnetic nanomaterials are a kind of important magnetic materials with zero coercive force and residual magnetism, and high saturation magnetization under room temperature. While saturation magnetization is similar to that of ferromagnetic species, superparamagnetic materials have smaller sizes. However, when compared to paramagnetic materials, superparamagnetic materials had much higher magnetism. Typical superparamagnetic nanomaterials included metal-iron oxide nanoparticles, which had important applications in biomedical imaging and therapy due to their high biocompatibility and low cytotoxicity. In addition, iron oxide nanoparticles were the safest candidate as nanoprobes and carriers for MRI, multimodal imaging, magnetic hyperthermia, PDT and PTT, drug delivery, gene delivery, *etc.*

3.3.2 *Synthesis of Superparamagnetic Nanomaterials*

Generally, three types of methods are used for synthesizing superparamagnetic nanomaterials and they are high-temperature thermal decomposition, solvothermal and co-precipitation.

Using high-temperature thermal decomposition method, monodisperse metal-iron oxide nanoparticles could be synthesized in which the size distribution of nanoparticles was narrow. As shown in Fig.3.3, Sun *et al.*[56] synthesized monodisperse Fe_3O_4 nanoparticles that had good dispersibility in organic solvents. Similarly, other monodipsperse MFe_2O_4 (M= Co, Mn) nanoparticles were also synthesized.[56] In 2004, Hyeon and coworkers reported the large-scale synthesis of monodipsperse magnetic nanoparticles by thermal decomposition of metal-oleate complex in different solvents with high boiling points in order to control the size of nanoparticles precisely from 5 nm to 22 nm.[57] Nanoparticles of 1 nm-scale size could also be synthesized by controlling the growth conditions from 6 nm to 13 nm.[58] However, the high-temperature thermal decomposition method was expensive and the prepared nanoparticles had poor water-dispersibility and high cytotoxicity. Therefore, the phase transfer and surface modification of nanoparticles are necessary before their application in biomedical fields can be explored.

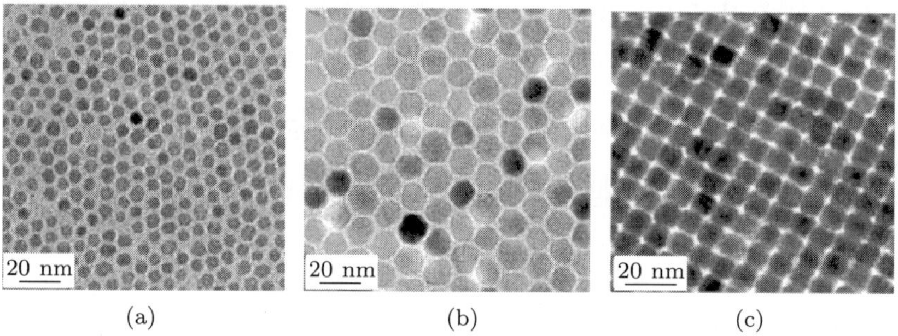

Fig. 3.3 TEM and 3D superlattice images of Fe_3O_4 nanoparticles with different sizes. (a) TEM image of 6 nm Fe_3O_4 nanoparticles dispersed in hexane. (b) TEM image of 12 nm Fe_3O_4 nanoparticles dispersed in hexane. (c) 3D superlattice image of 10 nm Fe_3O_4 nanoparticles dispersed in octane. Image is adapted with permission from Ref. 56. Copyright (2004) American Chemical Society.

Solvothermal method is also important for the synthesis of monodisperse supermagnetic nanoparticles in which water or other organic solvents are used.[59] Similar to the high-temperature thermal decomposition method, the prepared magnetic nanoparticles also have regular morphology and size uniformity. Using solvothermal method, the iron oxide nanoparticles of 1 nm-scale could be synthesized in precise sizes of 4 nm, 5 nm and 6 nm with excellent monodispersibility in organic solvents.[60] However, similar to high-temperature thermal decomposition, the water-dispersibility of nanoparticles is also poor. Generally, using water as solvent in the solvothermal method, the isolated superparamagnetic nanoparticles were with the size appropriate for better water-dispersibility, but in the organic solvent, much bigger sized nanoparticles were synthesized.

In the biomedical application of superparamagnetic nanomaterials, the nanoparticles are needed to be well dispersed in water, PBS and FBS. Co-precipitation is the low-cost method that is convenient for the synthesis of water-soluble superparamagnetic nanomaterials. Many commercial MRI CAs were prepared by co-precipitation method, such as Feridex, Resovist, *etc.* However, using co-precipitation method, the isolated nanoparticles were easily aggregated, and did not possess regular morphology and uniform size in order to qualify for monodisperse superparamagnetic nanoparticles. Nonetheless, the superparamagnetic nanoparticles, based on nanocomplexes, were difficult to be synthesized by co-precipitation method. Therefore, the high-temperature thermal decomposition and solvothermal methods were usually selected to synthesize superparamagnetic nanomaterials-based nanocomposites.

3.3.3 *Superparamagnetic Nanomaterials for Bioimaging*

Superparamagnetic nanoparticles are important nanomaterials for MRI CAs due to their high biocompatibility, magnetism and low cytotoxicity. Many metal-iron oxides, such as $ZnFe_2O_4$, $NiFe_2O_4$, $MnFe_2O_4$, $CoFe_2O_4$ and Fe_3O_4, as nanoparticles, are all typical superparamagnetic nanomaterials. Commercial MRI CAs based on iron oxides have been used for early diagnosis of cancers in clinical medicine. Superparamagnetic nanoparticles could be commonly used as T_2-weighted MRI CAs, but also sometimes as T_1-weighted MRI CAs. In addition, superparamagnetic nanomaterials-based nanocomposites could be multifunctional agents of dual-modal or multimodal imaging.

3.3.3.1 T_2-*weighted MRI*

Generally, superparamagnetic nanomaterials could be used as T_2-weighted MRI CAs due to high superparamagnetism. Typical superparamagnetic nanomaterials, such as metal-iron oxide nanoparticles, have been investigated greatly for the early diagnosis of cancers *in vitro* and *in vivo*.[61–64] Cheon and coworkers developed innovative magnetic nanoprobes to evaluate the magnetic spin, size and type of spinel metal ferrites, that showed the $MnFe_2O_4$ nanoparticles to posses stronger r_2 relaxivity than that of Fe_3O_4, $CoFe_2O_4$ and $NiFe_2O_4$ nanoparticles.[65] Moreover, the size of iron oxide nanoparticles also had important effect on the MRI performance.[66] In the past decades, orbicular metal-iron oxides were commonly investigated as MRI CAs. Recently, Hyeon and coworkers reported the synthesis of iron oxide nanocubes for sensitve T_2-weighted MRI, in which iron oxide nanocubes had excellent biocompatibility, water-dispersibility and low cytotoxicity. In addition, the r_2 relaxivity was extremely high with 761 $mM^{-1}\cdot s^{-1}$.[67] Iron oxide nanoparticles were further investigated for the MRI of stem cells to track their biodistribution and migration *in vivo*, which may also offer a new promising approach in stem cell-based cancer therapy.[68–70]

3.3.3.2 T_1-weighted MRI

Although superparamagnetic nanoparticles have been investigated widely as T_2-weighted MRI CAs, the T_1-weighted MRI signal could be triggered when the size of nanoparticles is sufficiently small due to change of magnetic spin and magnetic moment of ultrasmall iron oxide nanoparticles.[56,71–73] Hyeon and coworkers[74] reported the controllable synthesis of ultrasmall-sized iron oxide nanoparticles with 1.5 nm, 2.2 nm, 3 nm and 3.7 nm for T_1-weighted MRI CAs. As shown in Fig.3.4, the r_1 relaxivity and r_2/r_1 ratio were 4.78 $mM^{-1} \cdot s^{-1}$ and 6.12, respectively, in which the T_1-weighted MRI signal was attributed to large number of Fe^{3+} ions with 5 unpaired electrons on the surface of the nanoparticles. Compared to conventional T_1-weighted Gd-based MRI CAs, superparamagnetic metal-iron oxide-based T_1-weighted MRI CAs exhibit good biocompatibility and low cytotoxicity. These results are exciting as new MRI CAs, other than the current Gd-based ones, are introduced for early diagnosis of cancers.

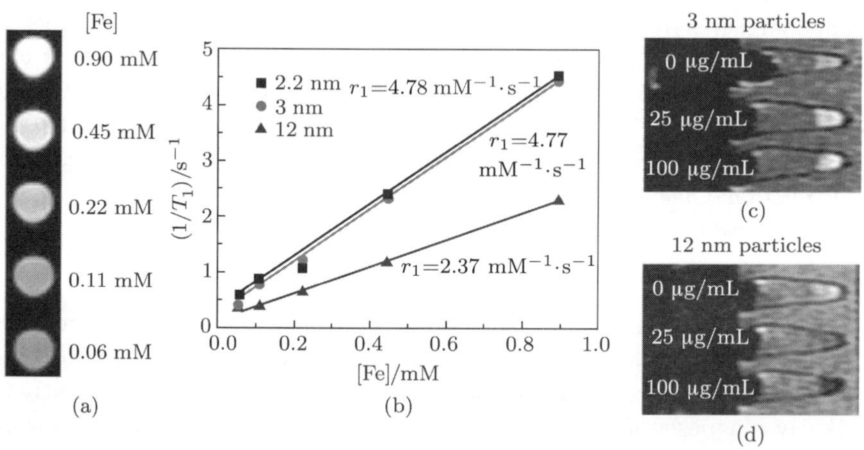

Fig. 3.4 (a) T_1-weighted MRI of Fe_3O_4 nanoparticles with the size of 3 nm. (b) Longitudinal MR relaxivity of Fe_3O_4 nanoparticles with the sizes of 2.2 nm, 3 nm, and 12 nm, respectively. (c, d) T_1-weighted MRI's of MCF-7 cells incubated with Fe_3O_4 nanoparticles with diameters for 24 h, in which the size of Fe_3O_4 nanoparticles was (c) 3 nm and (d) 12 nm, respectively. Image is adapted with permission from Ref. 74. Copyright (2011) American Chemical Society.

3.3.3.3 T_1- and T_2-weighted MRI

Multimodal imaging technique is important for improving the diagnosis efficiency of cancers. Especially, dual-modal T_1- and T_2-weighted MRI's could enhance the resolution of cancers with positive and negtive MRI signals. When nanocompexes of superparamagnetic and paramagnetic materials were synthesized, dual-modal T_1- and T_2-weighted MRI could be achieved. Generally, iron oxide and Gd-based complex/oxide nanocomposites were synthesized as the dual-modal T_1- and T_2-weighted MRI CAs.[75–77] Cheon and coworkers[78] proposed a "magnetically decoupled" core-shell design concept to develop a dual mode nanoparticle contrast agent, as shown in Fig.3.5. The nanocomposites showed strong $T_1$and T_2 signal, and led to greater diagnostic accuracy.[78] In addition, iron/manganese oxide nanocomposites could also be used as T_1- and T_2-weighted MRI CAs due to high r_1 relaxivity of manganese oxides. Lee and coworkers reported Fe_3O_4/MnO hybrid nanocrystals as a dual contrast agent of T_1- and T_2-weighted MRI for liver.[79]

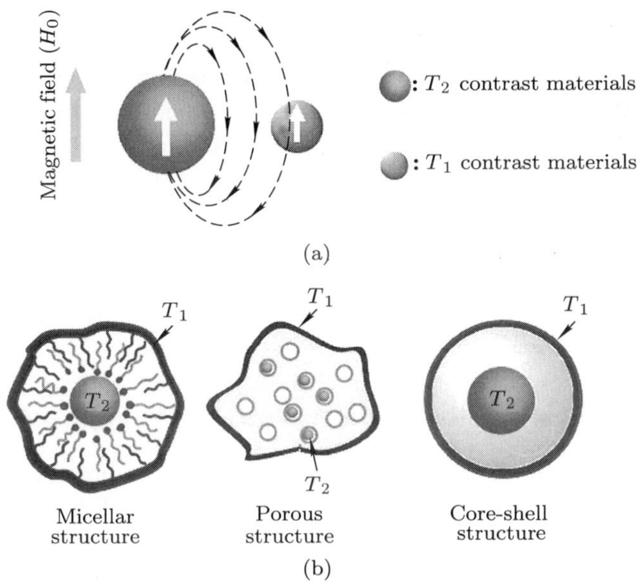

Fig. 3.5 (a) Magnetic coupling interaction between T_1 and T_2 contrast materials, in which electronic spins of paramagnetic T_1 contrast materials are affected by a magnetic field from superparamagnetic T_2 contrast materials when they are in close proximity. (b) Three possible structures of a dual mode CAs preventing magnetic coupling between different layers. Image is adapted with permission from Ref. 78. Copyright (2010) American Chemical Society.

3.3.3.4 *Multimodal Imaging*

All kinds of imaging techniques, such as computerized tomography (CT), ultrasound (US), positron emission tomography (PET) and MRI, are all important diagnostic tools in clinical medicine. Other imaging techniques, such as fluorescence and optical imaging, have also been investigated thoroughly. However, they are not complimenting with each other and thus the diagnostic accuracy of cancers is minimum, while nanoprobes-based multimodal imaging is promising only in the early diagnosis of cancers. It has also been demonstrated that the metal-iron oxide nanoparticles are excellent T_1- and T_2-weighted MRI CAs for early diagnosis due to their high superparamagnetism and non-invasive of MRI. Therefore, dual-modal or multimodal imaging, based on superparamagnetic nanomaterials, is a promising protocol for clinical diagnosis. Magnetic nano-in-micro materials were commonly designed and used as dual-modal CAs of MR/US. For example, both the superparamagnetic PLA-iron oxide nanoparticles and PLGA-iron oxide microcapsules are good candidates as dual-modal CAs of MR/US.[80,81] The poly (butyl cyanoacrylate) microbubbles containing iron oxide were also used as MR/US dual-modality imaging due to their biodegradability and biocompatibility.[82] Rare-earth doped upconversion luminescence (UCL) nanoparticles are also popular in bioimaging, while iron oxide-UCL nanocomposites could be applied as dual-modal CAs of MRI/UCL and even for multimodal imaging of MRI/UCL/CT due to high X-ray absorption of some rare-earth elements.[83,84] Other multimodal imaging, such as MRI/PET[85], MRI/CT/fluorescence[86], MRI/PET/fluorescence[87], and MRI/PET/optical[88], could also be achieved when superparamagnetic nanomaterials-based multifunctional nanocomposites are synthesized. The noble metal nanoparticles, such as gold or silver, are important substrate materials for surface enhanced Raman scattering (SERS) that could be used for SERS diagnosis of cancers. Medarova and coworkers reported dual-modal imaging agents of MRI/SERS, based on gold-iron oxide nanostructures, for "Live Imaging" of mice.[89]

3.3.4 *Superparamagnetic Nanomaterials for Theranostics*

Multifunctional agents with imaging and therapy are becoming interesting and have attracted great attention worldwide. Other than MRI diagnosis and magnetic hyperthermia, superparamagnetic-based nanocomposites could also be applied in therapy of cancers as carriers due to their biocompatibility, such as drug delivery, gene delivery, PDT and PTT. Therefore, superparamagnetic nanomaterials could be good candidates as theranostic agents for diagnosis and therapy of cancers.

3.3.4.1 *MRI and Magnetic Hyperthermia*

Superparamagnetic nanoparticles are conventional magnetic hyperthermia agents that have been investigated widely in the past decades as MRI CAs.[90–93] However, the dual-function of MRI and hyperthermia were not thoroughly investigated.[94,95] Using superparamagnetic nanoparticles, MRI and magnetically-guided therapy of cancers could be achieved simultaneously. In 2009, Cheon and coworkers[95] reported the Zn^{2+}-modulated metal-iron oxide nanoparticles for MRI and hyperthermia in which the performance were improved for eight to fourteen times and four times, respectively, than the conventional nanoparticles.

3.3.4.2 *MRI and Photodynamic Therapy (PDT)*

Multifunctional nanoprobes of MRI and PDT have potential applications in early diagnosis and therapy of cancers due to their non-invasive and high selectivity. Generally, photosensitizers are coupled with superparamagnetic nanoparticles and special light was used as excitation source. Some researchers have reported the MRI/PDT, based on photosensitizers-conjugated iron oxide nanocomposites, under the excitation of visible laser.[96–98] Due to deep penetration and non-invasive of near-infrared lights, UCL nanoparticles could be good carriers for PDT. Therefore, photosensitizers-conjugated magnetic-UCL nanocomposites were designed and synthesized as multifunctional agents of MRI and PDT, and even for UCL imaging. Multifunctional agents of T_1-weighted MR/UCL imaging and PDT, such as photosensitizers-conjugated Gd-based UCL nanoparticles, were reported extensively.[99,100] However, photosensitizers-conjugated iron oxide-UCL nanocomposites for T_2-weighted MR/UCL imaging and PDT were rarely investigated.[101] Inorganic photosensitizers, such as TiO_2 nanoparticles, are also important in PDT of cancers due to their stability and long-circulating time in body.[102–104] Recently, Wu and coworkers[105] reported Fe_3O_4-TiO_2 nanocomposites for MRI and inorganic PDT of cancer cells in which the viability of MCF-7 cells could be decreased about 60 % with the irradiation of UV lights for 15 min.

3.3.4.3 *MRI and Photothermal Therapy (PTT)*

PTT is another technique of light therapy, in which the local heating would kill cancer cells. In general, gold nanoparticles are excellent PTT carrier due to high absorption of special lights. Therefore, iron oxides-gold nanocomposites have been investigated as multifunctional agents of MRI and PTT.[106–108] Liu and coworkers[109] reported UCL-iron oxides-Au nanocomplexes for dual-modal imaging of UCL/MR and imaging-guided magnetically targeted PTT of cancers under the irradiation of 808 nm laser.

3.3.4.4 *MRI and Drug Delivery*

Anticancer drugs, such as doxorubicin (DOX), were used widely in cancer therapy, but had severe negative effect, and drug efficiency was poor. Drug delivery is an important technique in cancer therapy, and many nanomaterials could be used as carrier of drugs. Due to their biocompatibility and multifunction of superparamagnetic nanomaterials, the efficiency of iron oxides for the delivery of the therapeutics to tumor was investigated and further explored.[110–115] Gong and coworkers[116] reported multifunctional iron oxide/DOX-loaded worm-like polymer vesicles for MRI and DOX delivery. In this work, iron oxide nanoclusters were encapsulated into aqueous cores to improve MR relaxivity, and hydrophobic DOX was stably loaded into the hydrophobic membrane of vesicles.[116] Hyeon and coworkers[117] reported biodegradable PLGA polymer encapsulated iron oxide/DOX for MRI and magnetically-targeted drug delivery, and the synergetic targeting efficiency was increased due to iron oxide facilitated the magnetic guiding of the polymer particles. In addition to DOX, other drugs could also be loaded into iron oxide nanoparticles, suggesting that the superparamagnetic nanomaterials are good drug carriers for cancer therapy.[118,119] Recent reports described the multimodal imaging and drug delivery, based on superparamagnetic nanoparticles.[120–122]

3.3.4.5 *MRI and Gene Delivery*

Small interfering RNA (siRNA) has received significant attention as potential therapeutic agents for treating therapy of a kind of malignant, infectious, and autoimmune diseases, because they could make silence target genes in a specific manner. Many efforts and significant advances have been made to develop gene delivery vehicles. However, the lack of non-invasive delivery monitoring remained major hurdles towards clinical application. Superparamagnetic nanomaterials are good carriers for non-viral DNA delivery.[123] Moreover, using the MRI performance of superparamagnetic nanoparticles, the transfection of siRNA could be monitored. Zhang and coworkers[124] developed a new multifunctional magnetic nanovector, which is comprised of Fe_3O_4 nanoparticle core, PEG-coupled chitosan, and PEI shell. The results showed that the synthesized multifunctional magnetic nanovectors could effectively deliver sufficient dose of siRNA to target cells to induce gene silencing. Moreover, the multifunctional magnetic nanovectors could also monitor the carrier through MRI.[124] Recently, Cheon and coworkers[125] reported multifunctional "all in one" magnetic nanoparticles, in which cell-specific targeting moiety, fluorescent dye, and therapeutic siRNA were included in one system. In addition, the intracellular transfection of siRNA was monitored by MRI fluorescent dye, which provided sensitive subcellular visualization.[125]

3.3.5 *Overview*

Superparamagnetic nanomaterials are the most promising biomaterials as theranostic nanoprobes for the diagnosis and therapy of cancers. Commercial MRI CAs based on iron oxide nanoparticles have been used in clinical medicine. Moreover, multimodal imaging and multifunctional imaging and therapy based on superparamagnetic nanomaterials are being investigated at great extent. However, the surface modification of nanoparticles is important for the biocompatibility and the stability of nanoprobes, while targeting coupling is also important for the selectivity and sensitivity of the nanoprobes. Therefore, many efforts must be made for the improvement of theranostic performance, also including safety evaluation in cells and animals. In future, with the development of medical equipments and techniques, the multimodal and multifunctional agents, based on superparamagnetic nanomaterials, would play an important role in early diagnosis and therapy of cancers.

3.4 Outlook

Although some progress of magnetic nanomaterials for diagnosis and therapy in cancers is shown in this chapter, there are still many big challenges to be resolved before these magnetic nanomaterials is implemented for cancer theranostics in the near future. These challenges are as following:

(1) How to control the sizes, components and shapes of magnetic nanomaterials precisely to adjust the magnetic properties of them?

(2) How to combine the dual or multimodalities of bioimaging within one entity of magnetic nanomaterials with good enhancement for all modalities during the diagnosis *in vitro* or *in vivo*?

(3) How to make the magnetic nanomaterials to possess the diagnostic capability, such as MRI enhancement along with good therapeutic effects, for use in hyperthermia and PDT *in vitro*, *in vivo* or even in clinical trials?

(4) Could we manipulate the cell life, particularly for the cancer cells using the magnetic nanomaterials, based on excitation or trigger of their paramagnetic or superparamagnetic properties?

If so, the cancer could be imaged and killed simultaneously during MRI scanning or the cancer will be eliminated by the magnetic nanomaterials swallowed, injected, or administrated in the body while the patients stay in a suitable magnetic field area.

These ideas are sometimes naive, but they are the immediate expectations from the current clinic point of view for the cancer diagnosis and therapy based on magnetic nanomaterials.

Acknowledgments

This work was supported by the National Nature Science Foundation of China (51102251, 51203175, U1332117, and U1432114), Hundred Talents Program of the Chinese Academy of Sciences (2010-735), and Zhejiang Provincial Natural Science Foundation of China (Grants No. LQ13E030004).

References

1. Wu, A.; Ou, P.; Zeng, L. Biomedical applications of magnetic nanoparticles. *Nano* **2010**, *5* (5), 245-270.
2. Lauterbur, P. C. Image formation by induced local interactions: Examples employing nuclear magnetic resonance. *Clin Orthop Relat R* **1989**, *244*, 3-6.

3. Gallez, B.; Swartz, H. M. *In vivo* EPR: When, how and why? *NMR Biomed* **2004**, *17* (5), 223-225.
4. Lauffer, R. B. Paramagnetic metal complexes as water proton relaxation agents for NMR imaging: Theory and design. *Chem Rev* **1987**, *87* (5), 901-927.
5. Caravan, P.; Ellison, J. J.; McMurry, T. J.; Lauffer, R. B. Gadolinium (Ⅲ) chelates as MRI contrast agents: Structure, dynamics, and applications. *Chem Rev* **1999**, *99* (9), 2293-2352.
6. Tiefenauer, L. X.; Tschirky, A.; Kuhne, G.; Andres, R. Y. *In vivo* evaluation of magnetite nanoparticles for use as a tumor contrast agent in MRI. *Magn Reson Imaging* **1996**, *14* (4), 391-402.
7. Kellar, K. E.; Fujii, D. K.; Gunther, W. H.; Briley-Saebo, K.; Bjornerud, A.; Spiller, M.; Koenig, S. H. NC100150 injection: A preparation of optimized iron oxide nanoparticles for positive-contrast MR angiography. *J Magn Reson Imaging* **2000**, *11* (5), 488-494.
8. Hifumi, H.; Yamaoka, S.; Tanimoto, A.; Citterio, D.; Suzuki, K. Gadolinium-based hybrid nanoparticles as a positive MR contrast agent. *J Am Chem Soc* **2006**, *128* (47), 15090-15091.
9. Cullity, B. D.; Graham, C. D. *Introduction to Magnetic Materials.* 2nd ed. Wiley-IEEE Press, **2008**.
10. Vuong, Q. L.; Van, D. S.; Bridot, J. L.; Argante, C.; Alejandro, G.; Hermann, R.; Disch, S.; Mattea, C.; Stapf, S.; Gossuin, Y. Paramagnetic nanoparticles as potential MRI contrast agents: Characterization, NMR relaxation, simulations and theory. *MAGMA* **2012**, *25* (6), 467-478.
11. Das, G. K.; Heng, B. C.; Ng, S. C.; White, T.; Loo, J. S.; D'Silva, L.; Padmanabhan, P.; Bhakoo, K. K.; Selvan, S. T.; Tan, T. T. Gadolinium oxide ultranarrow nanorods as multimodal contrast agents for optical and magnetic resonance imaging. *Langmuir* **2010**, *26* (11), 8959-8965.
12. Yoon, Y.; Lee, B.; Lee, K. S.; Im, G. H.; Byeon, S. H.; Lee, J. H.; Lee, I. S. Surface modification of exfoliated layered gadolinium hydroxide for the development of multimodal contrast agents for MRI and fluorescence imaging. *Adv Funct Mater* **2009**, *19* (21), 3375-3380.
13. Wu, E. X.; Tang, H.; Jensen, J. H. Applications of ultrasmall superparamagnetic iron oxide contrast agents in the MR study of animal models. *NMR Biomed* **2004**, *17* (7), 478-483.
14. Yan, G. P.; Robinson, L.; Hogg, P. Magnetic resonance imaging contrast agents: Overview and perspectives. *Radiography* **2007**, *13*, e5-e19.
15. McDonald, M. A.; Watkin, K. L. Investigations into the physicochemical properties of dextran small particulate gadolinium oxide nanoparticles. *Acad Radiol* **2006**, *13* (4), 421-427.
16. Engstrom, M.; Klasson, A.; Pedersen, H.; Vahlberg, C.; Kall, P. O.; Uvdal, K. High proton relaxivity for gadolinium oxide nanoparticles. *MAGMA* **2006**, *19* (4), 180-186.
17. Ahren, M.; Selegard, L.; Klasson, A.; Soderlind, F.; Abrikossova, N.; Skoglund, C.; Bengtsson, T.; Engstrom, M.; Kall, P. O.; Uvdal, K. Synthesis and characterization of PEGylated Gd_2O_3 nanoparticles for MRI contrast enhancement. *Langmuir* **2010**, *26* (8), 5753-5762.

18. Petoral, R. M.; Jr. Soderlind, F.; Klasson, A.; Suska, A.; Fortin, M. A.; Abrikossova, N.; Selegard, L.; Kall, P. O.; Engstrom, M.; Uvdal, K. Synthesis and characterization of Tb^{3+}-doped Gd_2O_3 nanocrystals: A bifunctional material with combined fluorescent labeling and MRI contrast agent properties. *J Phys Chem C* **2009**, *113* (17), 6913-6920.
19. Raymond, K. N.; Pierre, V. C. Next generation, high relaxivity gadolinium MRI agents. *Bioconjugate Chem* **2004**, *16* (1), 3-8.
20. Klasson, A.; Ahren, M.; Hellqvist, E.; Soderlind, F.; Rosen, A.; Kall, P. O.; Uvdal, K.; Engstrom, M. Positive MRI contrast enhancement in THP-1 cells with Gd_2O_3 nanoparticles. *Contrast Media Mol I* **2008**, *3* (3), 106-111.
21. Aime, S.; Barge, A.; Cabella, C.; Crich, S. G.; Gianolio, E. Targeting cells with MR imaging probes based on paramagnetic Gd (Ⅲ) chelates. *Curr Pharm Biotechnol* **2004**, *5* (6), 509-518.
22. Toth, E.; Helm, L.; Merbach, A. E. Relaxivity of MRI contrast agents. In: *Contrast Agents I.* Springer: Berlin Heidelberg, **2002**.
23. Merbach, A.; Helm, L.; Toth, E. *Front Matter, in the Chemistry of Contrast Agents in Medical Magnetic Resonance Imaging.* 2nd ed. John Wiley & Sons, **2013**.
24. Geraldes, C. F. G. C.; Sherry, A. D.; Cacheris, W. P.; Kuan, K. T.; Brown Ⅲ, R. D.; Koenic, S. H.; Spillers, M. Number of inner-sphere water molecules in Gd^{3+} and Eu^{3+} complexes of DTPA-amide and -ester conjugates. *Magn Reson Med* **1988**, *8* (2), 191-199.
25. Kurki, T.; Komu, M. Spin-lattice relaxation and magnetization transfer in intracranial tumors *in vivo*: Effects of Gd-DTPA on relaxation parameters. *Magn Reson Imaging* **1995**, *13* (3), 379-385.
26. Papadopoulos, A.; Vlahos, L.; Xenelis, J.; Papafragou, C.; Adamopoulos, G. Value of Gd-DTPA-enhanced MR imaging of the labyrinth in patients with sudden hearing loss. *Magn Reson Imaging* **1995**, *13* (3), 387-391.
27. Caravan, P. Strategies for increasing the sensitivity of gadolinium based MRI contrast agents. *Chem Soc Rev* **2006**, *35* (6), 512-523.
28. Yang, J. J.; Yang, J.; Wei, L.; Zurkiya, O.; Yang, W.; Li, S.; Zou, J.; Zhou, Y.; Maniccia, A. L.; Mao, H.; Zhao, F.; Malchow, R.; Zhao, S.; Johnson, J.; Hu, X.; Krogstad, E.; Liu, Z. R. Rational design of protein-based MRI contrast agents. *J Am Chem Soc* **2008**, *130* (29), 9260-9267.
29. Weinmann, H. J.; Gries, H. Tolerance of extracellular contrast agents for magnetic resonance imaging. *Invest Radiol* **1990**, *25*, S49-S50.
30. Weinmann, H. J.; Brasch, R. C.; Press, W. R.; Wesbey, G. E. Characteristics of gadolinium-DTPA complex: A potential NMR contrast agent. *Am J Roentgenol* **1984**, *142*, 619-624.
31. Schmitt-Willich, H.; Brehm, M.; Ewers, C. L.; Michl, G.; Muller-Fahrnow, A.; Petrov, O.; Platzek, J.; Radühel, B.; Sulzle, D. Synthesis and physicochemical characterization of a new gadolinium chelate: The liver-specific magnetic resonance imaging contrast agent Gd-EOB-DTPA. *Inorg Chem* **1999**, *38*, 1134-1144.
32. Uggeri, F.; Aime, S.; Anelli, P. L.; Botta, M.; Brocchetta, M.; Haeen, C. D.; Ermondi, G.; Grandi, M.; Paoli, P. Novel contrast agents for magnetic resonance

imaging. Synthesis and characterization of the ligand BOPTA and its Ln (Ⅲ) complexes (Ln = Gd, La, Lu). X-ray structure of disodium (TPS-9-145337286-C-S)-[4-Carboxy-5,8,11-tris(carboxymethyl)-1-phenyl-2-oxa- 5,8,11-triazatridecan-13-oato(5-)]gadolinate(2-) in a mixture with its enantiomer. *Inorg Chem* **1995**, *34*, 633-643.

33. Wang, Z.; Su, M.; Najafi, A.; Nalcioglu, O. Effect of vasodilator hydralazine on tumor microvascular random flow and blood volume as measured by intravoxel incoherent motion (IVIM) weighted MRI in conjunction with Gd-DTPA-albumin enhanced MRI. *Magn Reson Imaging* **2001**, *19*, 1063-1072.
34. Bligh, S. W.; Harding, C. T.; Sadler, P. J.; Bulman, R. A.; Bydder, G. M.; Pennock, J. M.; Kelly, J. D.; Latham, I. A.; Marriott, J. A. Use of paramagnetic chelated metal derivatives of polysaccharides and spin-labeled polysaccharides as contrast agents in magnetic resonance imaging. *Magn Reson Med* **1991**, *17*, 516-532.
35. Rebizak, R.; Schaefer, M.; Dellacherie, E. Polymeric conjugates of Gd^{3+}-diethylenetriaminepentaacetic acid and dextran. 1. Synthesis, characterization, and paramagnetic properties. *Bioconjugate Chem* **1997**, *8*, 605-610.
36. Rongved, P.; Klaveness, J. Water-soluble polysaccharides as carriers of paramagnetic contrast agents for magnetic resonance imaging: Synthesis and relaxation properties. *Carbohyd Res* **1991**, *214*, 315-323.
37. Corsi, D. M.; Elst, L. V.; Muller, R. N.; Bekkum, H. V.; Joop, A. Peters, J. A. Inulin as a carrier for contrast agents in magnetic resonance imaging. *Chem - Eur J* **2001**, *7*, 64-71.
38. Zhuo, R.; Fu, Y.; Liao, J. Synthesis, relaxivity and biodistribution of novel magnetic resonance imaging (MRI) contrast agents: Polylysine (Gd-DTPA/DOTA) with pendent galactose moieties as hepatocyte-targeting groups. *Chinese Chem Lett* **1997**, *8*, 157-160.
39. Margerum, L.D.; Campion, B. K.; Koo, M.; Shargill, N.; Lai, J. J.; Marumoto, A.; Sontum, P. C. Gadolinium (Ⅲ) DO3A macrocycles and polyethylene glycol coupled to dendrimers effect of molecular weight on physical and biological properties of macromolecular magnetic resonance imaging contrast agents. *J Alloy Compd* **1997**, *249*, 185-190.
40. Sun, G.; Feng, J.; Jing, F.; Pei, F.; Liu, M. Synthesis and evaluation of novel polysaccharide-Gd-DTPA compounds as contrast agent for MRI. *J Magn Magn Mater* **2003**, *265*, 123-129.
41. Chandrasekharan, P.; Yong, C. X.; Poh, Z.; He, T.; He, Z.; Liu, S.; Robins, E. G.; Chuang, K. H.; Yang, C. T. Gadolinium chelate with DO3A conjugated 2-(diphenylphosphoryl)-ethyldiphenylphosphonium cation as potential tumor-selective MRI contrast agent. *Biomaterials* **2012**, *33*, 9225-9231.
42. Ye, M.; Qian, Y.; Tang, J.; Hu, H.; Sui, M.; Shen, Y. Targeted biodegradable dendritic MRI contrast agent for enhanced tumor imaging. *J Control Release* **2013**, *169*, 239-245.
43. Han, L.; Li, J.; Huang, S.; Huang, R.; Liu, S.; Hu, X.; Yi, P.; Shan, D.; Wang, X.; Lei, H.; Jiang, C. Peptide-conjugated polyamidoamine dendrimer as a nanoscale tumor-targeted T_1 magnetic resonance imaging contrast agent. *Biomaterials* **2011**, *32*, 2989-2998.

44. Shiraishi, K.; Kawano, K.; Minowa, T.; Maitani, Y.; Yokoyama, M. Preparation and *in vivo* imaging of PEG-poly (L-lysine)-based polymeric micelle MRI contrast agents. *J Control Release* **2009**, *136*, 14-20.
45. Park, J. Y.; Baek, M. J.; Choi, E. S.; Woo, S.; Kim, J. H.; Kim, T. J.; Jung, J. C.; Chae, K. S.; Chang, Y.; Lee, G. H. Paramagnetic ultrasmall gadolinium oxide nanoparticles as advanced T_1 MRI contrast agent: Account for large longitudinal relaxivity, optimal particle diameter, and *in vivo* T_1 MR images. *ACS Nano* **2009**, *3*, 3663-3669.
46. Evanics, F.; Diamente, P. R.; Veggel, F. C. J. M.; Stanisz, G. J.; Prosser, R. S. Water-soluble GdF_3 and GdF_3/LaF_3 nanoparticles physical characterization and NMR relaxation properties. *Chem Mater* **2006**, *18*, 2499-2505.
47. Fortin, M. A.; Petoral, R. M.; Soderlind, F.; Klasson, A.; Engstrom, M.; Veres, T.; Kall, P. O.; Uvdal, K. Polyethylene glycol-covered ultra-small Gd_2O_3 nanoparticles for positive contrast at 1.5 T magnetic resonance clinical scanning. *Nanotechnol* **2007**, *18*, 395501.
48. Bridot, J. L.; Faure, A. C.; Laurent, S.; Riviere, C.; Billotey, C.; Hiba, B.; Janier, M.; Josserand, V.; Coll, J. L.; Elst, L. V.; Muller, R.; Roux, S.; Perriat, P.; Tillement, O. Hybrid gadolinium oxide nanoparticles: Multimodal contrast agents for *in vivo* imaging. *J Am Chem Soc* **2007**, *129*, 5076-5084.
49. Soderlind, F.; Pedersen, H.; Jr, R. M. P.; Kall, P. O.; Uvdal, K. Synthesis and characterisation of Gd_2O_3 nanocrystals functionalized by organic acids. *J Colloid Interf Sci* **2005**, *288*, 140-148.
50. Ma, X. H.; Gong, A.; Xiang, L. C.; Chen, T. X.; Gao, Y. X.; Liang, X. J.; Shen, Z. Y.; Wu, A. G. Biocompatible composite nanoparticles with large longitudinal relaxivity for targeted imaging and early diagnosis of cancer. *J Mater Chem B* **2013**, *1*, 3419-3428.
51. Kelkar, S. S.; Reineke, T. M. Theranostics: Combining imaging and therapy. *Bioconjugate Chem* **2011**, *22*, 1879-1903.
52. Spernyak, J. A.; White, W. H.; Ethirajan, M.; Patel, N. J.; Goswami, L.; Chen, Y.; Turowski, S.; Missert, J. R.; Batt, C.; Mazurchuk, R.; Pandey, R. K. Hexylether derivative of pyropheophorbide-a (HPPH) on conjugating with 3gadolinium (Ⅲ) aminobenzyldiethylenetriaminepentaacetic acid shows potential for *in vivo* tumor imaging (MR, fluorescence) and photodynamic therapy. *Bioconjugate Chem* **2010**, *21*, 828-835.
53. Howell, M.; Mallela, J.; Wang, C.; Ravi, S.; Dixit, S.; Garapati, U.; Mohapatra, S. Manganese-loaded lipid-micellar theranostics for simultaneous drug and gene delivery to lungs. *J Control Release* **2013**, *167*, 210-218.
54. Bae, K. H.; Lee, K.; Kim, C.; Park, T. G. Surface functionalized hollow manganese oxide nanoparticles for cancer targeted siRNA delivery and magnetic resonance imaging. *Biomaterials* **2011**, *32*, 176-184.
55. Hossann, M.; Wang, T.; Syunyaeva, Z.; Wiggenhorn, M.; Zengerle, A.; Issels, R. D.; Reiser, M.; Lindner, L. H.; Peller, M. Non-ionic Gd-based MRI contrast agents are optimal for encapsulation into phosphatidyldiglycerol-based thermosensitive liposomes. *J Control Release* **2013**, *166*, 22-29.
56. Sun, S. H.; Zeng, H.; Robinson, D. B.; Raoux, S.; Rice, P. M.; Wang, S. X.; Li, G. X. Monodisperse MFe_2O_4 (M=Fe, Co, Mn) nanoparticles. *J Am Chem Soc* **2004**, *126*, 273-279.

57. Park, J.; An, K.; Hwang, Y.; Park, J. G.; Noh, H. J.; Kim, J. Y.; Park, J. H.; Hwang, N. M.; Hyeon, T. Ultra-large-scale syntheses of monodisperse nanocrystals. *Nat Mater* **2004**, *3*, 891-895.
58. Park, J.; Lee, E.; Hwang, N. M.; Kang, M.; Kim, S. C.; Hwang, Y.; Park, J. G.; Noh, H. J.; Kim, J. Y.; Park, J. H.; Hyeon, T. One-nanometer-scale size-controlled synthesis of monodisperse magnetic iron oxide nanoparticles. *Angew Chem* **2005**, *117*, 2932-2937.
59. Wang, X.; Zhuang, J.; Peng, Q.; Li, Y. D. Ageneral strategy for nanocrystal synthesis. *Nature* **2005**, *437*, 121-124.
60. Tian, Y.; Yu, B. B.; Li, K. Facile solvothermal synthesis of monodisperse Fe_3O_4 nanocrystals with precise size control of one nanometer as potential MRI contrast agents. *J Mater Chem* **2011**, *21*, 2476-2481.
61. Hu, F. Q.; Wei, L.; Zhou, Z.; Ran, Y. L.; Li, Z.; Gao, M. Y. Preparations of biocompatible magnetite nanocrystals for *in vivo* magnetic resonance detection of cancer. *Adv Mater* **2006**, *18*, 2553-2556.
62. Liu, D. F.; Wu, W.; Ling, J. J.; Wen, S.; Gu, N.; Zhang, X. Z. Effective PEGylation of iron oxide nanoparticles for high performance *in vivo* cancer imaging. *Adv Funct Mater* **2011**, *21*, 1498-1504.
63. Hao, R.; Xing, R. J.; Xu, Z. C.; Hou, Y. L.; Gao, S.; Sun, S. H. Synthesis, functionalization and biomedical applications of multifunctional magnetic nanoparticles. *Adv Mater* **2010**, *22*, 2729-2742.
64. Lu, J.; Ma, S. L.; Sun, J. Y.; Xia, C. C.; Liu, C.; Wang, Z. Y.; Zhao, X. N.; Gao, F. B.; Gong, Q. Y.; Song, B.; Shuai, X. T.; Ai, H.; Gu, Z. W. Manganese ferrite nanoparticle micellar nanocomposites as MRI contrast agent for liver imaging. *Biomaterials* **2009**, *30*, 2919-2928.
65. Lee, J. H.; Huh, Y. M.; Jun, Y. W.; Seo, J. W.; Jang, J. T.; Song, H. T.; Kim, S.; Cho, E. J.; Yoon, H. G.; Suh, J. S.; Cheon, J. Artificially engineered magnetic nanoparticles for ultra-sensitive molecular imaging. *Nat Med* **2007**, *13*, 95-99.
66. Jun, Y. W.; Huh, Y. M.; Choi, J. S.; Lee, J. H.; Song, H. T.; Kim, S.; Yoon, S.; Kim, K. S.; Shin, J. S.; Cheon, J. Nanoscale size effect of magnetic nanocrystals and their utilization for cancer diagnosis *via* magnetic resonance imaging. *J Am Chem Soc* **2005**, *127*, 5732-5733.
67. Lee, N.; Chou, Y.; Lee, Y.; Park, M.; Moon, W. K.; Choi, S. H.; Hyeon, T. Water-dispersible ferromagnetic iron oxide nanocubes with extremely high r_2 relaxivity for highly sensitive *in vivo* MRI of tumors. *Nano Lett* **2012**, *12*, 3127-3131.
68. Xu, C.; Miranda-Nieves, D.; Ankrum, J. A.; Matthiesen, M. E.; Phillips, J. A.; Roes, I.; Wojtkiewicz, G. R.; Juneja, V.; Kultima, J. R.; Zhao, W.; Vemula, P. K.; Lin, C. P.; Nahrendorf, M.; Karp, J. M. Tracking mesenchymal stem cells with iron oxide nanoparticle loaded poly (lactide-co-glycolide) microparticles. *Nano Lett* **2012**, *12*, 4131-4139.
69. Wang, X. Q.; Wei, F.; Liu, A. J.; Wang, L.; Wang, J. C.; Ren, L.; Liu, W. M.; Tu, Q.; Li, L.; Wang, J. Y. Cancer stem cell labeling using poly (L-lysine)-modified iron oxide nanoparticles. *Biomaterials* **2012**, *33*, 3719-3732.
70. Lewin, M.; Carlesso, N.; Tung, C. H.; Tang, X. W.; Cory, D.; Scadden, D. T.; Weissleder, R. Tat peptide-derivatized magnetic nanoparticles allow *in vivo* tracking and recovery of progenitor cells. *Nat Biotechnol* **2000**, *18*, 410-414.

71. Tromsdorf, U. I.; Bruns, O. T.; Salmen, S. C.; Beisiegel, U. B.; Weller, H. A highly effective, nontoxic T_1 MR contrast agent based on ultrasmall PEGylated iron oxide nanoparticles. *Nano Lett* **2009**, *9*, 4434-4440.

72. Taboada, E.; Rodríguez, E.; Roig, A.; Oró, J.; Roch, A.; Muller, R. N. Relaxometric and magnetic characterization of ultrasmall iron oxide nanoparticles with high magnetization evaluation as potential T_1 magnetic resonance imaging contrast agents for molecular imaging. *Langmuir* **2007**, *23*, 4583-4588.

73. Zeng, L. Y.; Ren, W. Z.; Zheng, J. J.; Cui. P.; Wu, A. G. Ultrasmall water-soluble metal-iron oxide nanoparticles as T_1-weighted contrast agents for magnetic resonance imaging.*Phys Chem Chem Phys* **2012**, *14*, 2631-2636.

74. Kim, B. H.; Lee, N.; Kim, H.; An, K.; Park, Y. I.; Choi, Y.; Shin, K.; Lee, Y.; Kwon, S. K.; Na, H. B.; Park, J. G.; Ahn, T. Y.; Kim, Y. W.; Moon, W. K.; Choi, S. H.; Hyeon, T. Large-scale synthesis of uniform and extremely small-size iron oxide nanoparticles for high-resolution T_1 magnetic resonance imaging contrast agents. *J Am Chem Soc* **2011**, *133*, 12624-12631.

75. Zhou, Z. J.; Huang, D. T.; Bao, J. F.; Chen, Q. L.; Liu, G.; Chen, Z.; Chen, X. Y.; Gao, J. H. Asynergistically enhanced T_1-T_2 dual-modal contrast agent. *Adv Mater* **2012**, *24*, 6223-6228.

76. Yang, H.; Zhuang, Y. M.; Sun, Y.; Dai, A. T.; Shi, X. Y.; Wu, D. M.; Li, F. Y.; Hu, H.; Yang, S. P. Targeted dual-contrast T_1- and T_2-weighted magnetic resonance imaging of tumors using multifunctional gadolinium-labeled superparamagnetic iron oxide nanoparticles. *Biomaterials* **2011**, *32*, 4584-4593.

77. Bae, K. H.; Kim, Y. B.; Lee, Y.; Hwang, J. Y.; Park, H. W.; Park, T. G. Bioinspired synthesis and characterization of gadolinium-labled magnetic nanoparticles for dual contrast T_1- and T_2-weighted magnetic resonance imaging. *Bioconjugate Chem* **2010**, *21*, 505-512.

78. Choi, J. S.; Lee, J. H.; Shin, T, H.; Song, H. T.; Kim, E. Y.; Cheon, J. Self-confirming "AND" logic nanoparticles for fault-free MRI. *J Am Chem Soc* **2010**, *132*, 11015-11017.

79. Im, G. H.; Kim, S. M.; Lee, D. G.; Lee, W. J.; Lee, J. H.; Lee, I. S. Fe_3O_4/MnO hybrid nanocrystals as a dual contrast agent for both T_1- and T_2-weighted liver MRI. *Biomaterials* **2013**, *34*, 2069-2076.

80. Yang, F.; Li, Y. X.; Chen, Z. P.; Zhang, Y.; Wu, J. R.; Gu, N. Superparamagnetic iron oxide nanoparticle-embedded encapsulated microbubbles as dual contrast agents of magnetic resonance and ultrasound imaging. *Biomaterials* **2009**, *30*, 3882-3890.

81. Sun, Y.; Zheng, Y. Y.; Ran, H. T.; Zhou, Y.; Shen, H. X.; Chen, Y.; Chen, H. R.; Krupka, T. M.; Li, A.; Li, P.; Wang, Z. B.; Wang, Z. G. Superparamagnetic PLGA-iron oxide microcapsules for dual-modality US/MR imaging and high intensity focused US breast cancer ablation. *Biomaterials* **2012**, *33*, 5854-5864.

82. Liu, Z.; Lammers, T.; Ehling, J.; Fokong, S.; Bornemann, J.; Kiessling, F.; Gätjens, J. Iron oxide nanoparticle-containing microbubble composites as contrast agents for MR and ultrasound dual-modality imaging. *Biomaterials* **2011**, *32*, 6155-6163.

83. Zhu, X. J.; Zhou, J.; Chen, M.; Shi, M.; Feng, W.; Li, F. Y. Core-shell Fe_3O_4@$NaLuF_4$:Yb, Er/Tm nanostructure for MRI, CT and upconversion luminescence tri-modality imaging. *Biomaterials* **2012**, *33*, 4618-4627.

84. Zhu, H.; Tao, J.; Wang, W. H.; Zhou, Y. J.; Li, P. H.; Li, Z.; Yan, K.; Wu, S. L.; Yeung, K. W. K.; Xu, Z. S.; Xu, H. B.; Chu, P. K. Magnetic, fluorescent, and thermo-responsive Fe_3O_4/rare earth incorporated poly (St-NIPAM) core-shell colloidal nanoparticles in multimodal optical/magnetic resonance imaging probes. *Biomaterials* **2013**, *34*, 2296-2306.
85. Torres, R.; de Rosales, M.; Tavaré, R.; Paul, R. L.; Jauregui-Osoro, M.; Protti, A.; Glaria, A.; Varma, G.; Szanda, I.; Blower, P. J. Synthesis of $^{64}Cu^{II}$-bis (dithiocarbamatebisphosphonate) and its conjugation with superparamagnetic iron oxide nanoparticles: *In vivo* evaluation as dual-modality PET-MRI agent. *Angew Chem Int Ed* **2011**, *50*, 5509-5513.
86. Xing, H. Y.; Bu, W. B.; Zhang, S. J.; Zheng, X. P.; Li, M.; Chen, F.; He, Q. J.; Zhou, L. P.; Peng, W. J.; Hua, Y. Q.; Shi, J. L. Multifunctional nanoprobes for upconversion fluorescence, MR and CT trimodal imaging. *Biomaterials* **2012**, *33*, 1079-1089.
87. Xie, J.; Chen, K.; Huang, J.; Lee, S.; Wang, J. H.; Gao, J. H.; Li, X. G.; Chen, X. Y. PET/NIRF/MRI triple functional iron oxide nanoparticles. *Biomateirals* **2010**, *31*, 3016-3022.
88. Yang, M.; Cheng, K.; Qi, S. B.; Liu, H. G.; Jiang, Y. X.; Jiang, H.; Li, J. B.; Chen, K.; Zhang, H. M.; Cheng, Z. Affibody modified and radiolabeled gold-iron oxide hetero-nanostructures for tumor PET, optical and MR imaging. *Biomaterials* **2013**, *34*, 2796-2806.
89. Yigit, M. V.; Zhu, L. Y.; Ifediba, M. A.; Zhang, Y.; Carr, K.; Moore, A.; Medarova, Z. Noninvasive MRI-SERS imaging in living mice using an innately bimodal nanomaterial. *ACS Nano* **2011**, *5*, 1056-1066.
90. Fortin, J. P.; Wilhelm, C.; Servais, J.; Ménager, C.; Bacri, J. C.; Gazeau, F. Size-sorted anionic iron oxide nanomagnets as colloidal mediators for magnetic hyperthermia. *J Am Chem Soc* **2007**, *129*, 2628-2635.
91. Guardia, P.; Corato, R. D.; Lartigue, L.; Wilhelm, C.; Espinosa, A.; Garcia-hernandez, M.; Gazeau, F.; Manna, L.; Pellegrino, T. Water-soluble iron oxide nanocubes with high values of specific absorption rate for cancer cell hyperthermia treatment. *ACS Nano* **2012**, *6*, 3080-3091.
92. Bae, K. H.; Park, M.; Do, M. J.; Lee, N.; Ryu, J. H.; Kim, G. W.; Kim, C.; Park, T. G.; Hyeon, T. Chitosan oligosaccharide-stabilized ferromagnetic iron oxide nanocubes for magnetically modulated cancer hyperthermia. *ACS Nano* **2012**, *6*, 5266-5273.
93. Lee, J. H.; Jang, J. T.; Choi, J. S.; Moon, S. H.; Noh, S. H.; Kim, J. W.; Kim, J. G.; Kim, I. S.; Park, K. I.; Cheon, J. Exchange-coupled magnetic nanoparticles for efficient heat induction. *Nat Nanotechnol* **2011**, *6*, 418-422.
94. Hadjipanayis, C. G.; Bonder, M. J. Metallic iron nanoparticles for MRI contrast enhancement and local hyperthermia. *Small* **2008**, *4*, 1925-1929.
95. Jang, J. T.; Nah, H.; Lee, J. H.; Moon, S. H.; Kim, M. G.; Cheon, J. Critical enhancements of MRI contrast and hyperthermic effects by dopant-controlled magnetic nanoparticles. *Angew Chem Int Ed* **2009**, *48*, 1234-1238.
96. McCarthy, J. R; Jaffer, F. A.; Weissleder, R. A macrophase-targeted theranostic nanoparticle for biomedical applications. *Small* **2006**, *8-9*, 983-987.
97. Huang, P.; Li, Z. M.; Lin, J.; Yang, D. P.; Gao, G.; Xu, C.; Bao, L.; Zhang, C. L.; Wang, K.; Song, H.; Hu, H. Y.; Cui, D. X. Photosensitizer-conjugated

magnetic nanoparticles for *in vivo* simultaneous magnetofluorescent imaging and targeting therapy. *Biomaterials* **2011**, *32*, 3447-3458.

98. Gianella, A.; Jarzyna, P. A.; Mani, V.; Ramachandran, S.; Calcagno, C.; Tang, J.; Kann, B. K.; Dijk, W. J. R.; Thijssen, V. L.; Griffioen, A. W.; Storm, G.; Fayad, Z. A.; Mulder, W. J. M. Multifunctional nanoemulsion platform for imaging guided therapy evaluated in experimental cancer. *ACS Nano* **2011**, *5*, 4422-4433.

99. Park, Y.; Kim, H.; Kim, J.; Moon, K.; Yoo, B.; Lee, K.; Lee, N.; Choi, Y.; Park, W.; Ling, D.; Na, K.; Moon, W.; Choi, S.; Park, H.; Yoon, S.; Suh, Y.; Lee, S.; Hyeon, T. Theranostic probe based on lanthanide-doped nanoparticles for simultaneous *in vivo* dual-modal imaging and photodynamic therapy. *Adv Mater* **2012**, *24*, 5755-5761.

100. Qiao, X. F.; Zhou, J. C.; Xiao, J. W; Wang, Y. F.; Sun, L. D.; Yan, C. H. Triple-functional core-shell structured upconversion luminescent nanoparticles covalently grafted with photosensitizer for luminescent, magnetic resonance imaging and photodynamic therapy *in vitro*. *Nanoscale* **2012**, *4*, 4611-4623.

101. Zeng, L. Y.; Xiang, L. C.; Ren, W. Z.; Zheng, J. J.; Li, T. H.; Chen, B.; Zhang, J. C.; Mao, C. W.; Li, A. G.; Wu, A. G. Multifunctional photosensitizer-conjugated core-shell Fe_3O_4@ $NaYF_4$:Yb/Er nanocomplexes and their applications in T_2-weighted magnetic resonance /upconversion luminescence imaging and photodynamic therapy of cancer cells. *RSC Adv* **2013**, *3*, 13915-13925.

102. Rozhkova, E. A.; Ulasov, I.; Lai, B.; Dimitrijevic, N. M.; Lesniak, M. S.; Rajh, T. A high-performance nanobio photocatalyst for targeted brain cancer therapy. *Nano Lett* **2009**, *9*, 3337-3342.

103. Seo, J. W.; Chung, H.; Kim, M. Y.; Lee, J.; Choi, I. H.; Cheon, J. Development of water-soluble single crystalline TiO_2 nanoparticles for photocatalytic cancer-cell treatment. *Small* **2007**, *3*, 850-853.

104. Li, Q.; Wang, X.; Lu, X.; Tian, H.; Jiang, H.; Lv, G.; Guo, D.; Wu, C.; Chen, B. The incorporation of daunorubicin in cancer cells through the use of titanium dioxide whiskers. *Biomaterials* **2009**, *30*, 4708-4715.

105. Zeng, L. Y.; Ren, W. Z.; Xiang, L. C.; Zheng, J. J.; Chen, B.; Wu, A. G. Multifunctional Fe_3O_4-TiO_2 nanocomposites for magnetic resonance imaging and potential photodynamic therapy. *Nanoscale* **2013**, *5*, 2107-2113.

106. Kim, J.; Park ,S.; Lee, J. E.; Jin, S. M.; Lee, J. H.; Lee, I. S.; Yang, H.; Kim, J. S.; Kim, S. K.; Cho, M. H.; Hyeon, T. Designed fabrication of multifunctional magnetic gold nanoshells and their application to magnetic resonance imaging and photothermal therapy. *Angew Chem Int Ed* **2006**, *45*, 7754-7758.

107. Dong, W. J.; Li, Y. S.; Niu, D. C.; Ma, Z.; Gu, J. L.; Chen, Y.; Zhao, W. R.; Liu, X. H.; Liu, C. S.; Shi, J. L. Facile synthesis of monodisperse superparamagnetic Fe_3O_4 core @ hybrid @ Au shell nanocomposite for bimodal imaging and photothermal therapy. *Adv Mater* **2011**, *23*, 5392-5397.

108. Bardhan, R.; Chen, W.; Torres, C. P.; Bartels, M.; Huschka, R. M.; Zhao, L. L.; Morosan, E.; Pautler, R. G.; Joshi, A.; Halas, N. Nanoshells with targeted simultaneous enhancement of magnetic and optical imaging and photothermal therapeutic response. *Adv Funct Mater* **2009**, *19*, 3901-3909.

109. Cheng, L.; Yang, K.; Li, Y. G.; Zeng, X.; Shao, M. W.; Lee, S. T.; Liu, Z. Multifunctional nanoparticles for upconversion luminescence/MR multimodal

imaging and magnetically targeted photothermal therapy. *Biomaterials* **2012**, *33*, 2215-2222.

110. Chen, W. H.; Cao, Y. H.; Liu, M.; Zhao, Q. H.; Huang, J.; Zhang, H. L.; Deng, Z. W.; Dai, J. W.; Williams, D. F.; Zhang, Z. J. Rotavirus capsid surface protein VP4-coated Fe_3O_4 nanoparticles as a theranostic platform for cellular imaging and drug delivery. *Biomaterials* **2012**, *33*, 7895-7902.

111. Sanson, C.; Diou, O.; Thévenot, J.; Ibarboure, E.; Soum, A.; Brûlet, A.; Miraux, S.; Thiaudière, E.; Tan, Sisareuth.; Brisson, A.; Dupuis, V.; Sandre, O.; Lecommandoux, S. Doxorubicin loaded magnetic polymersomes: Theranostic nanocarriers for MR imaging and magneto-chemotherapy.*ACS Nano* **2011**, *5*, 1122-1140.

112. Lee, J.; Kim, H.; Kim, S.; Lee, H.; Kim, J.; Kim, N.; Park, H. J.; Choi, E. K.; Lee, J. S.; Kim, C. A multifunctional mesoporous nanocontainer with an iron oxide core and a cyclodextrin gatekeeper for an efficient theranostic platform.*J Mater Chem* **2012**, *22*, 14061-14067.

113. Quan, Q. M.; Xie, J.; Gao, H. K.; Yang, M.; Zhang, F.; Liu, G.; Lin, X.; Wang, A.; Eden, H. S.; Lee, S.; Zhang, G. X.; Chen, X. Y. HAS coated iron oxide nanoparticles as drug delivery vehicles for cancer therapy. *Mol Pharmaceut* **2011**, *8*, 1669-1676.

114. Yu, M. K.; Jeong, Y. Y.; Park, J.; Park, S.; Kim, J. W.; Min, J. J.; Kim, K.; Jon, S. Drug-loaded superparamagnetic iron oxide nanoparticles for combined cancer imaging and therapy *in vivo*. *Angew Chem Int Ed* **2008**, *47*, 5362-5365.

115. Yang, J.; Lee, C. H.; Ko, H. J.; Suh, J. S.; Yoon, H. G.; Lee, K.; Huh, Y. M.; Haam, S. Multifunctional magneto-polymeric nanohybrids for targeted detection and synergistic therapeutic effects on breast cancer. *Angew Chem Int Ed* **2007**, *46*, 8836-8839.

116. Yang, X. Q.; Grailer, J. J.; Rowland, I. J.; Javadi, A.; Hurley, S. A.; Steeber, D, A.; Gong, S. Q. Multifunctional SPIO/DOX-loaded wormlike polymer vesicles for cancer therapy and MR imaging. *Biomaterials* **2010**, *31*, 9065-9073.

117. Kim, J.; Lee, J. E.; Lee, S. H.; Yu, J. H.; Lee, J. H.; Park, T. G.; Hyeon, T. Designed fabrication of a multifunctional polymer nanomedical platform for simultaneous cancer-targeted imaging and magnetically guided drug delivery. *Adv Mater* **2008**, *20*, 478-483.

118. Yallapu, M. M.; Othman, S. F.; Curtis, E. T.; Gupta, B. K.; Jaggi, M.; Chauhan, S. C. Multi-functional magnetic nanoparticles for magnetic resonance imaging and cancer therapy. *Biomaterials* **2011**, *32*, 1890-1905.

119. Lee, G. Y.; Qian, W. P.; Wang, L.; Wang, Y. A.; Staley, C. A.; Satpathy, M.; Nie, S.; Yang, L. Theranostic nanoparticles with controlled release of gemcitabine for targeted therapy and MRI of pancreatic cancer. *ACS Nano* **2013**, *7*, 2078-2089.

120. Lee, J. E.; Lee, N.; Kim, H.; Kim, J.; Choi, S. H.; Kim, J. H.; Kim, T.; Song, I. C.; Park, S. P.; Moon, W. K.; Hyeon, T. Uniform mesoporous dye-doped silica nanoparticles decorated with multiple magnetite nanocrystals for simultaneous enhanced magnetic resonance imaging, fluorescence imaging, and drug delivery. *J Am Chem Soc* **2010**, *132*, 552-557.

121. Li. L.; Liu. C.; Zhang, L. Y.; Wang, T. T.; Yu, H.; Wang, C. G.; Su, Z. M. Multifunctional magnetic-fluorescent eccentric-(concentric-$Fe_3O_4@SiO_2$) @

polyacrylic acid core-shell nanocomposites for cell imaging and pH-responsive drug delivery. *Nanoscale* **2013**, *5*, 2249-2253.

122. Santra, S.; Kaittanis, C.; Grimm, J.; Perez, J. M. Drug/dye-loaded, multifunctional iron oxide nanoparticles for combined targeted cancer therapy and dual optical/magnetic resonance imaging. *Small* **2009**, *5*, 1862-1868.
123. Kievit, F. M.; Veiseh, O.; Bhattarai, N.; Fang, C.; Gunn, J. W.; Lee, D.; Ellenbogen, R. G.; Olson, J.; Zhang, M. PEI-PEG-chitosan copolymer coated iron oxide nanoparticles for safe gene delivery: Synthesis, complexation, and transfection. *Adv Funct Mater* **2009**, *19*, 2244-2251.
124. Veiseh, O.; Kievit, F. M.; Fang, C.; Mu, N.; Jana, S.; Leung, M. C.; Mok, H.; Ellenbogen, R. G.; Park, J. O.; Zhang, M. Chlorotoxin bound magnetic nanovector tailored for cancer cell targeting imaging, and siRNA delivery. *Biomaterials* **2010**, *31*, 8032-8042.
125. Lee, J. H.; Lee, K.; Moon, S. H.; Lee, Y.; Park, T. G.; Cheon, J. All-in-one target-cell-specific magnetic nanoparticles for simultaneous molecular imaging and siRNA delivery. *Angew Chem Int Ed* **2009**, *48*, 4174-4179.

Chapter 4

Quantum Dots (QDs) for Tumor Targeting Theranostics

*Yufei Ma, He Shen, Mengxin Zhang, and Zhijun Zhang**
Suzhou Key Laboratory of Nanobiomedicine, Division of Nanobiomedicine, Suzhou Institute of Nano-Tech and Nano-Bionics, Chinese Academy of Sciences, Suzhou 215123, China

Semiconductor quantum dots exhibit excellent optical and electronic properties such as broad absorption, narrow emission, and remarkable photostability, rendering them promising candidates for a variety of applications. This chapter introduces the synthesis, surface functionalization, physicochemical properties, and biomedical applications of different types of quantum dots, with focus on their biomedical applications, such as biological imaging, targeted drug delivery and cancer therapy. A brief discussion on the toxicity of quantum dots, as well as the future directions of quantum dots nanotheranostics, is also presented.

4.1 Introduction

Quantum dots (QDs) are nanocrystals of semiconductor materials. Their sizes are so small that, under quantum mechanics considerations, the electron energies that can exist within them are limited. These energy levels defined by the size of quantum dots consequently define the bandgaps. In addition, the nanocrystals can be technically controlled to any needed size, allowing convenient approaches to tune the bandgaps across a wide range without changing the underlying material or construction techniques.

The family of QDs has been extended to a variety of quantum-sized materials, including group Ⅱ-Ⅵ QDs (CdS, CdSe, and CdTe)[1,2], group Ⅲ-V QDs (InP, InAs, GaAs, GaN)[3], and recently developed group Ⅳ QDs (C, Si, and Ge)[4–7]. These QDs have been massively proved useful for both diagnosis and drug delivery[1,3], especially for cancer therapy. By coating QDs (QDs refers to Ⅱ-Ⅵ QDs and Ⅲ-V QDs in this chapter, unless otherwise specifically denoted) with a protective layer and functionalizing them with cancer-fighting drugs, biocompatible molecules, and targeting molecules such as antibodies, QDs are made ready for administration. Furthermore, when QDs conjugated with antibodies enter the bloodstream and localize inside cancer cells, the luminescent QDs under appropriate irradiation facilitate locating the tumor and monitoring

∗ Corresponding author: zjzhang2007@ sinano.ac.cn

the release of the anticancer drug directly into the cancer cells.[2,8]

4.1.1 *Fluorescence of QDs*

Compared to organic dyes, QDs, especially group II-VI QDs[9], possess many superior chemical and optical properties, such as excellent stability against photobleaching, size-tunable absorption and emission (Fig.4.1), extremely broad and intense absorption enabling a unique flexibility in excitation, high fluorescence quantum yields even in the NIR wavelengths and large two-photon action cross-sections. These distinct features ensure QDs great potential in bioimaging and other applications. Therefore, we will briefly discuss the origin of QDs photoluminescence in this section.

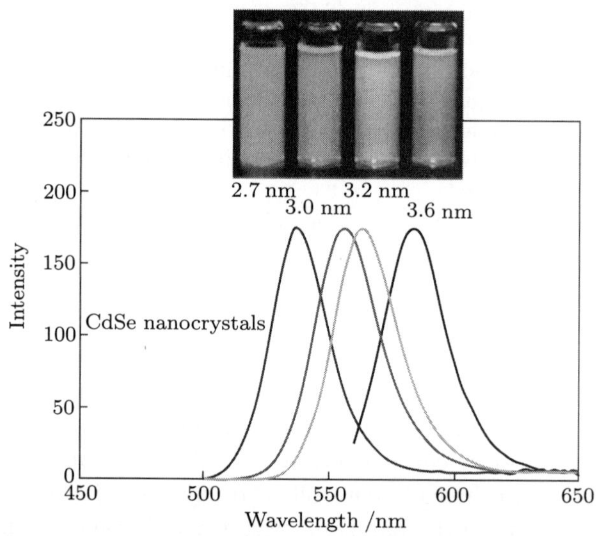

Fig. 4.1 Emission maxima shift to longer wavelength as sizes of QDs increase. All spectra are typically around 30 to 50 nm (full width at half maximum). The inset shows the photo of QDs with different size upon UV irradiation. (color figure at the end of the book)

When modeling using particle in a box, the energy levels of a quantum dot depend on its size.[10] The electrons crowding in a quantum dot with radius smaller than exciton Bohr radius (typically several nanometers for QDs; CdSe: 4.9 nm; CdS: 2.8 nm) result in the splitting of the original energy levels. Consequently, these QDs possess smaller gaps between each successive level instead of continuous levels in bulk semiconductor crystals, thus presenting so-called strong confinement regime (Fig.4.2).[11]

Due to the domination of quantum confinement effects, QDs respond to their sizes with the electronic and optical property changes and show characteristic fluorescence properties in contrast to standard fluorophores. The absorp-

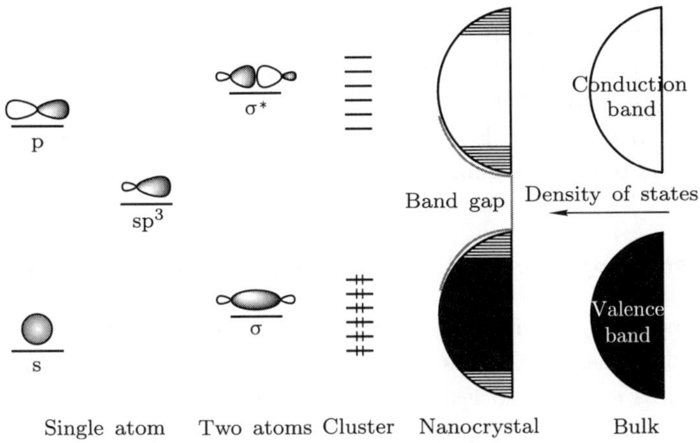

Fig. 4.2 Molecular orbital (MO) model for evolution of discrete MOs into continuous bands.

tion of a photon with energy above the semiconductor band gap energy creates an electron-hole pair, or so-called exciton. Since the absorption at higher energies has an increased probability, the absorption spectrum of a QD appears a broadband, as shown in Fig.4.3. Due to the radiative recombination of an exciton, the emission of a photon occurs in a relatively long time (>10 ns) and presents a narrow, symmetric energy band.[1,12] In a simplified model of the excitation, the energy of the emitted photon can be seen as the bond energy of the exciton. Therefore, the optical properties of QDs massively change as a function of size, according to quantum confinement effects.

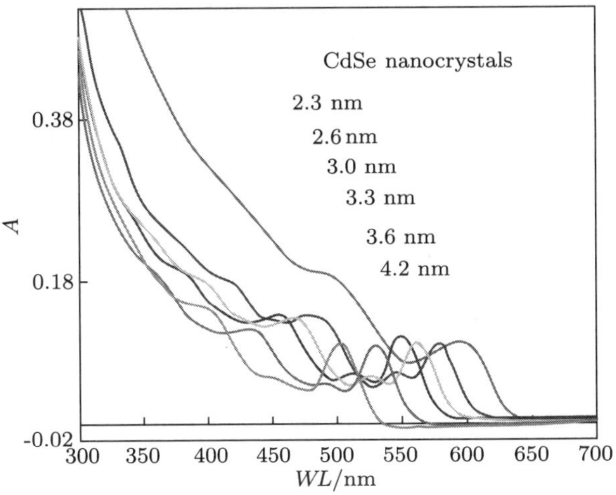

Fig. 4.3 Absorption spectra of CdSe QDs with different sizes. (color figure at the end of the book)

Arising from their distinct structures, QDs display several advantages, such as strong and broad absorption, narrow and tunable emission color, and good chemical and photostability compared to organic dyes. These characteristic features of QDs made them competitive to conventional organic fluorophores soon after their discovery. Extensive studies have been conducted in the field of life science from bioimaging to theranostics.

4.1.2 *History of QDs and Their Biomedical Applications*

QDs emerged from a number of research labs in the early 1980s, with the natural research transition from semiconductor thin-films (2D) to 0D materials (Fig.4.4). Efros *et al.*[13] and Ekimov *et al.*[14] grew semiconducting nanocrystals and microcrystals in glass matrices; Louis Brus at Bell Lab (now at Columbia University), who received the first Kavli Prize in Nanoscience in 2008, studied "quantum-sized" CdS crystallites in aqueous solution.[15] Other synthetic methods of QDs were also developed in different medias, including high-temperature organic solvents, and solid substrates.[1,3] However, the goal toward narrow size distributions of QDs was only made in 1993 by Bawendi and coworkers.[16] They introduced TOPO (trioctylphosphine oxide) to the high temperature synthesis of monodispersed cadmium-based (CdS, CdSe, and CdTe) QDs, making the synthetic chemistry of QDs at a large scale quickly advanced. Later in 1997[17], following Guyot-Sionnest's pioneering work[18], they successively improved the fluorescence efficiency of QDs from 5%–15% to 30%–50%, with the help of anoverlayer coating of ZnS on the CdSe core.

Upon the development of highly luminescent QDs, scientists started to look into the applications of QDs in life science. Alivisatos[19] and Nie[20] simultaneously reported, for the first time,the cellular labeling using CdSe QDs in 1998. They presented that QDs were superior to conventional fluorophores, due to their brighter and more stable fluorescent emission with narrow, tunable, symmetric emission spectrum. Dubertret *et al.*[21] encapsulated individual nanocrystals in phospholipid block-copolymer micelles, and demonstrated that CdSe/ZnS QDs were capable of both *in vitro* and *in vivo* imagings. Later in 2003, Webb and coworkers[22] dynamically visualized QDs through the skin of living mice, in capillaries hundreds of micrometers deep. Following these work, there are massive publications extending QDs for optimal biomedical imaging.[23,24] For example, fluorescent QDs were conjugated with bioactive moieties (*e.g.*, antibodies, receptor ligands) and employed to specifically monitor biological events and target cellular structures, such as labeling neoplastic cells[25,26], peroxisomes[27], DNA[21], and cell membrane receptors[28,29]. Bioconjugated QDs were also explored for cancer targeting and imaging in living animals.[25]

Since many of the early studies on QDs were designed by researchers of nanotechnology rather than of toxicology or health, questions concerning QDs

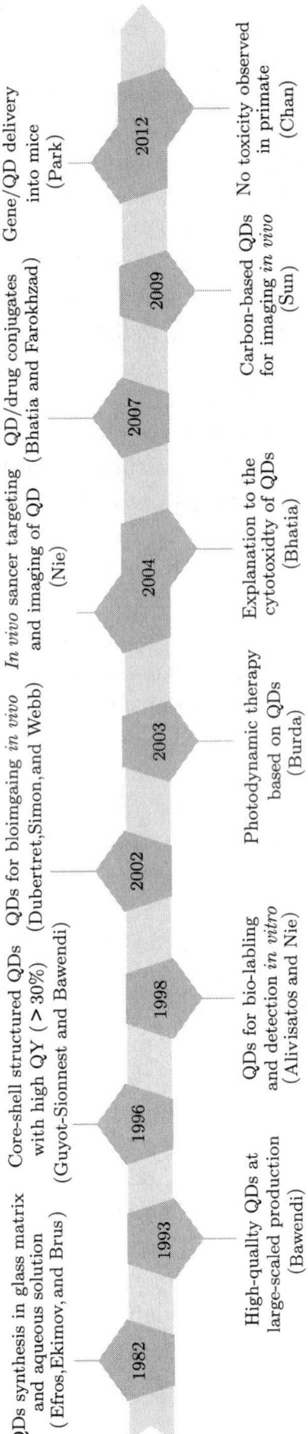

Fig. 4.4 Milestones during QDs evolution in life science.

toxicity focused on their physicochemical properties such as fluorescence, detectability, stability, and cell labeling efficacy. In addition, the diversity of QDs being synthesized complicates the safety issues of QDs.[30] Several *in vitro* and *in vivo* studies demonstrated unobservable QDs-induced cytotoxicity[21,22,31–34], while some studies did suggest QDs affecting cell growth and viability[35]. To answer this question, Bhatia and coworkers[36] showed that CdSe QDs were indeed cytotoxic under certain conditions, and they suggested that the cytotoxicity correlated with the liberation of free Cd^{2+} ions due to the deterioration of the CdSe lattice. However, as a very recent study[37] reported, CdSe-CdS-ZnS QDs slowly degraded and the free metal ions redistributed to various organs over time, but no toxic damage to the particle-accumulated organs was witnessed in the rhesus monkeys within 90 days period.

The positive results of cytotoxicity of quantum dots encouraged people to explore biomedical applications of QDs, although the long-term biological effects remained unknown. Loading drugs onto QDs is technically difficult due to their small sizes and thus little conjugating sites for therapeutics, but researchers had demonstrated the possibility of QDs as drug/gene delivery platforms as early as 2007. Bhatia and coworkers[38] demonstrated that the delivery of F3/siRNA-QDs to EGFP-transfected HeLa cells and the release from their endosomal entrapment led to significant knockdown of EGFP signal; Jon, Farokhzad and coworkers[39] utilized a QD-aptamer(Apt)-doxorubicin(DOX) conjugate as a complex system with targeted cancer imaging, therapy, and sensing. QDs were also utilized as fluorescent tags for other drug carriers even bacteria and viruses.[40–42] In another direction, Bawendi group employed QDs to model DDS (drug delivery system) behaviors depending on size and shape *in vivo*.[43–46] Their work suggested that smaller ($\sim$12 nm) nanoparticles were ideal candidates for cancer therapy due to their superior tumor penetration capability. These innovative approaches had opened up exciting opportunities in the targeted drug (DNA and RNA) delivery.

The carbon-based QDs represent new types of heavy-metal-free nanomaterials which could be more suitable for drug delivery than classical II-VI QDs. As the extensive drug delivery research on carbon nanotubes and graphene have shown[47,48], carbon-based QDs can retain large surface area for drug loading and low cytotoxicity, while producing fluorescence with fluorescence quantum yield (QY) as high as 20%[4]. Current researches on carbon-based QDs are now focusing on the synthesis and *in vitro* imaging[49,50], and only few studies go further to the *in vivo* imaging[51]. However, we expect that new types of QDs from carbonaceous materials, which are much safer, inexpensive, and capable of loading drugs more efficiently than the widely studied II-VI QDs, are developed toward final clinical applications.

4.2 Synthesis, Modification and Functionalization of QDs

4.2.1 *Synthesis of QDs*

4.2.1.1 *Synthesis of II-VI QDs*

To date, various QDs have been fabricated by different chemical or physical methods. Among them, II-VI element materials, such as CdSe, CdTe and CdS, are most intensively studied and commonly used in biotechnology and medical applications due to their bright fluorescence emission and excellent stability. The most widely and intensively employed method for II-VI QDs synthesis is the high temperature pyrolysis approach developed by Bawendi and coworkers in 1993.[16] The synthetic route usually requires organometallic compounds as the precursors, organic surfactants as the stabilizers and solvents. A typical procedure[16] of synthesizing QDs (Fig.4.5) is presented below:

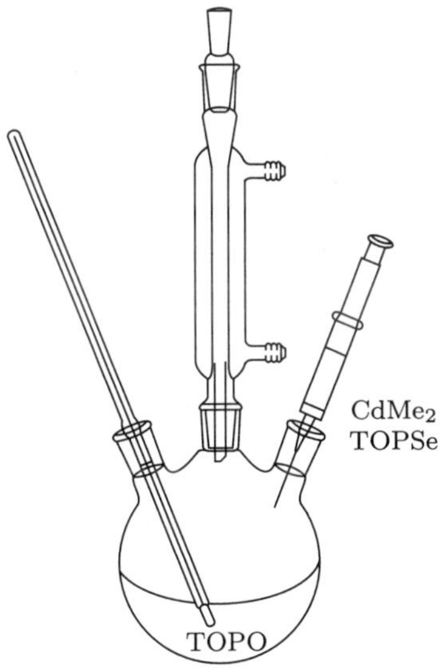

Fig. 4.5 Diagram showing a typical apparatus for preparing semiconductor nanocrystals.

Trioctylphosphine oxide (TOPO) as solvent is dried and degassed at 200 °C in argon atmosphere, and then is heated to 300 °C for the addition with syringe of $CdMe_2$ as the metal source and trioctylphosphine selenide (TOPSe) in TOPO. The rapid introduction of the reagent mixture generates a deep yellow/orange solution (absorption at 440–460 nm in the case of CdSe QDs), accompanied by suddenly decreasing temperature to 180 °C and reheating to 230–260 °C. A second injection of $CdMe_2$ may be desired sometimes to control the size and size distribution of QDs.

The process has two stages. In the first stage of nucleation, when heated to a high temperature (~300 °C), the organometallic compound began to decompose into active atoms (intermediates), and then these atoms combined to form the initial seeds. In the second stage, the seeds began to grow up to form larger particles until the temperature decreased beneath 250 °C. Since the second stage plays a critical role in the growth of QDs, the seed concentration and the reaction time are key factors affecting the size of QDs.

Since QDs emit light with size-dependent colors, a narrow size distribution during QDs preparation is highly desired. A secondary injection of precursors can be used to achieve this goal. After injecting the precursor to the hot solutions, the monomer concentration increases while the already existing size distribution of nanocrystals has not changed.[52] Anther commonly applied method to sort out a wide range of sizes for QDs is called "size selective precipitation". Due to the larger QDs having relatively strong interactions with each other, when stepwise adding a bad solvent, such as ethanol and acetone, to stable solution of QDs, the larger QDs tend to aggregate first while the small ones stay in the solution.

Other methods of QDs synthesis utilize hydrothermal decomposition[53,54], microemulsion[55], Sol-Gel[56], and microwaves[57], mainly due to the concern of toxic precursors. In addition, CdSe QDs can also be synthesized in living yeast cells by intracellular biochemical reactions[58], and the fluorescence can be simply tuned by controlling the incubation time of materials. However, these new synthetic strategies can hardly guarantee high quality QDs.

4.2.1.2 *Core/Shell Structure*

Surface defects in the crystal structure can be considered as temporary "traps" for the electron or hole, which prevent the relaxation of excited state and thus lose the fluorescence emission. The phenomena of blinking QDs at the single-molecule level result from the alternation of trapping and untrapping events, and reduce the overall quantum yield.[1] One popular way to solve these problems is to grow a shell onto the QDs' core with a few atomic layers of an inorganic material which has larger band gap than the core material. The shell not only protect surface atoms from oxidation and other chemical reactions[59], but may also transfer energies to the core to improve the overall brightness due to its larger bandgap. If this shell is carefully designed, quantum yields of QDs can be improved to as high as 90%.[60]

Several high-bandgap-energy inorganic shells (*e.g.*, CdS and ZnS) have been proved to effectively passivate the photoactive core of QDs.[17,61] Carefully choosing the shell material according to the QDs' core would minimize the lattice strain between the core and shell and maximize the QD photophysical properties. Furthermore, thicker shells (4–6 monolayers) provide more protection for the core from photooxidation and degradation, even though thin shells of 1–2 monolayers often produce the highest fluorescence yields. In addition, a thicker shell might significantly reduce QD blinking[62,63] which causes signal fluctuations in ultrasensitive detection, and thus avoid loss of distance information in single-molecule tracking and spectral jumping (change in the emission peak position).

4.2.1.3 *Synthesis of Other Types of QDs*

The major concern regarding biomedical applications of II-VI semiconductor QDs is the intrinsic toxicity of cadmium.[36] To address this issue, a variety of Cd-free QDs has been developed. The III-V dots (such as InP, InAs, GaA) have become promising candidates for imaging in biological systems[64–66], but high quality InP QDs can hardly be prepared because the precursors and surfactants are sensitive to the reaction condition. Si QDs can be achieved through solution-phase reductive process[67], microemulsion[6], supercritical fluid[5], etching[68], laser ablation[69] or electrochemical method[70], but the large scale production remains a problem. Other cadmium-free semiconductor nanocrystals, such as $CuInS_2$[71], CuInSe[72], and Ag_2S[73], have been explored for bioimaging mostly in the near-infrared region. In addition, there are some commercially supplied QDs free of heavy metals, for instance, ZnCuInS and CuInS core/shell QDs from Ocean NanoTech (*http://www.oceannanotech.com/*).

Fluorescent carbon-based nanomaterials have also aroused great research in drug delivery studies, as well as bioimaging, due to their abundant resources, large surface area, nontoxicity and biocompatibility.[4,49,50,74] The class of carbon-based QDs has been enriched with nanodiamond, carbon dots and graphene QDs (GQDs), and shows great potential in bioimaging and drug-delivery applications. Carbonaceous QDs can be produced by laser ablation of graphite powder[4], cutting graphite through oxidation[51,75], or hydrothermal approach[76], and by decomposition of carbon-rich compounds at high temperature[77] or in microwave environment[78]. Although carbon quantum dots have been applied to the *in vivo* imaging[51], the development of carbon-based QDs still struggles with the weak luminescence and broad size distribution so far.

4.2.2 *Surface Modification*

QDs synthesized by a high temperature pyrolysis are usually stabilized by native hydrophobic ligands and are only soluble in organic solutions. In order to apply QDs into a biological system, it is essential to modify the QDs surface with hydrophilic ligands, which render the QDs aqueous solubility, and at the same time remain the photophysical stability and possess reactive sites for conjugating biomolecules. Since the first report on the biological application of QDs in 1998[19,20], vast research developing suitable ligands and different methods have been reported. The typical surface modification methods include ligand exchange, surface silanization and ligand capping/encapsulation,[79,80] will be described below.

4.2.2.1 *Ligand Exchange*

In a ligand exchange approach, the local hydrophobic surface ligands (TOPO) on the QDs surface are replaced with hydrophilic groups. Usually the hydrophilic ligands contain high affinity bonding groups to the QDs surface and other hydrophilic groups to make QDs soluble in the aqueous media.

Thiol-containing ligand molecules are frequently used to replace the original ligands since they have a strong affinity to the QDs surface. Monothiol ligands including mercaptoacetic acid (MAA), mercaptopropionic acid (MPA), and mercaptoundecanoic acid (MUA) are commonly used ligands; bidentatethiols such as dihydrolipoic acid (DHLA)[81] can further stabilize QDs in water. Other functional groups which can coordinate to metals include amines, carboxylic acids, and oligomeric phosphine ligands.[82]

Ligand exchange can be easily processed as most thiol ligands are commercially available. Additionally, the as-prepared QDs have small hydrodynamic sizes, which is useful in FRET experiments. On the other hand, ligand exchange methods introduce defects to QDs surface and result in reduced fluorescence quantum yield of the QDs. Moreover, thiol groups are susceptible to oxidation and pH variation, leading to the ligand detachment from the surface and gradual aggregation of QDs.[83]

4.2.2.2 *Surface Silanization*

This procedure usually contains two steps.[8] First, the ligand exchange strategy is applied to replace the original hydrophobic groups like TOPO/TOP with thiol-derived silane ligand [such as mercaptopropy-ltris(methyloxy)silane (MPS)], which simultaneously anchors the nanoparticle and cross-linked by forming siloxane bonds. In the next step, another type of silanes is added on the outer layer to generate a different charge and to provide functional sites. This step is also called silica shell growth. Polyethylene glycol (PEG)-silane, aminopropyl-silanes (APS), and phospho-silanes are the most commonly used ligands. The silica shell can be further functionalized with biomolecules by using standard silane coupling chemistry.

The silanization improves the fluorescence of QYs and the photochemical stability, as well as the biocompatibility of QDs. However, the silica shell should be hydrolyzed eventually[63] and it is usually a difficult task to encapsulate a single QD within a silica layer. In addition, the thick silica shells result in relatively large hydrodynamic size of QDs and easy aggregation of QDs in aqueous solution.

4.2.2.3 *Ligand Capping/Encapsulation*

In a ligand capping or encapsulation method, amphiphilic ligands decorate QDs and at the same time the original surfactants of QDs remain. The hydrophobic chains from the amphiphilic ligands intercalate with the hydrophobic groups of QDs, while the hydrophilic tails exposed to the outside to ensure water solubility of QDs. Commonly employed amphiphilic ligands are phospholipids[21,84] and amphiphilic polymers[85,86]. For instance, the alkyl chain of polyacrilic acid modified with amphiphilic macromolecules is quite stable in aqueous buffers[87].

The ligand capping preserves the original surface groups of QDs and reduces the possibility of introducing more surface defects, thus avoids deterioration of the photoluminescence properties of QDs. However, particles obtained *via* this approach are several times larger than the original QDs, which could lead to a poor cellular uptake, therefore limiting their biological and medical applications (Fig.4.6).

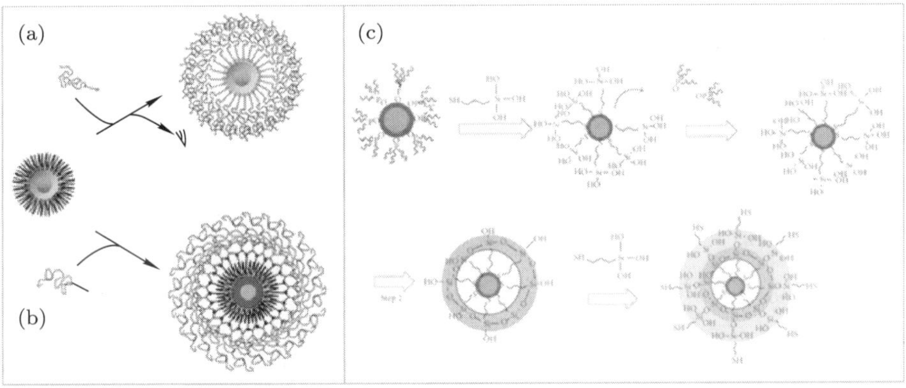

Fig. 4.6 Scheme of QD modifications. (a) Ligand exchange strategy. (b) Ligand capping strategy. Reproduced from Ref. 79 with permission of Elsevier, Copyright © 2008. (c) Surface silanization procedure.

4.2.3 *Functionalization*

In order to facilitate QDs for biomedical applications, it is necessary to conjugate QDs with particular functional biomolecules to recognize specific targets in biological environments. Several successful approaches have been developed to anchor biological molecules onto QDs, including mercapto (—SH) exchange, electrostatic interaction and covalent binding strategies.

Through mercapto exchange process, biomolecules with thiol groups can anchor to the ZnS shell of QDs and serve as the ligand for solubilization, but QDs tend to precipitate due to the weak bonding between Zn and thiol. Another strategy utilizes electrostatic interactions between charged biomolecules and surface molecules on QDs, such as polysaccharide, thiolated peptides with cysteine residues and aptamers. For example, hyaluronic acid (HA) with negative charge self-assembled onto QDs with positive charge, enabling HA-QD conjugates for targeting lymphatic vessels *in vitro* and *in vivo*.[88]

Covalent attachment of biomolecules onto QD surfaces with cross-linkers is another common and more stable method. Water-solublization of QDs introduces different kinds of functional groups, such as carboxylic acid, thiol and amino groups, which can act as reactive sites for further biofunctionalization. The most widely used cross-linkers are 1-ethyl-3-(3-dimethylaminopropyl) carbodiimide (EDC) and 4-(N-maleimidomethyl)- cyclohexanecarboxylicacid N-hydroxysuccinimideester (SMCC). EDC is usually used to link —NH_2 and —COOH groups, while SMCC can mediate —SH and —NH_2 groups. The advantage for carboxylic acid and amine reaction lies in that many proteins originally possess primary amine and carboxylic acid groups, thus the conjugation can take place without further modification and at a mild condition. The molecules to target specific parts inside cells or animal bodies so far have been extensively studied, including small molecules (folic acid, dopamine)[64,89], peptides (RGD)[90–93], nucleic acids (aptamer)[94–96], proteins (EGF, transferrin)[20,97–99] or antibodies[100–102].

4.3 QDs for Bioimaging

The fluorescence-based biological imaging is one of the most essential techniques in biology research, and has been widely used *in vivo* and *in vitro* studies.[8]Also, research based on fluorescent probes for biological sensing, imaging and tracking influences benefits from chemistry, biology and biomedicine. Organic dyes and fluorescent proteins, among a wide variety of fluorescent labels, are mostly used ones in biomedical field, but their applications for long-term *in vitro* or *in vivo* imaging are limited due to severe photobleaching. The advent of QDs, especially Ⅱ-Ⅵ QDs, offers solutions to the problems for conventional fluorescent dyes. Compared to traditional fluorescent dyes, QDs possess unique advantages, such as tunable emission color, strong absorption, broad excitation and narrow emission spectra, high fluorescence quantum yield, and remarkable photostability. Consequently, QDs have been widely employed in the biomedical field as a class of ideal fluorescent probes, particularly for *in vitro* and *in vivo* imaging.[2,3,8,23]

In 1998, two groups, led by Alivisatos and Nie independently, illustrated for the first time that Ⅱ-Ⅵ QDs are greatly promising for cell imaging.[19,20] Thereafter, these classical Ⅱ-Ⅵ QDs have been intensely investigated, in various biotechnological applications including DNA array technology, immunofluorescence assays[23], as well as *in vitro* and *in vivo* studies[1]. The early and most successful applications of QDs have spread in labeling fixed cells and tissues, staining membrane proteins[26,29,59,103,104], microtubules[26], actin[19,26] and nuclear antigens[26], and *in situ* hybridization on chromosomes[105–107] or combed DNA[63].

4.3.1 *In Vitro Imaging*

4.3.1.1 *Staining Pathology*

Fluorescence microscopy is widely employed for the assessment of phenotypes of healthy cells and the detection of target molecules of diseases. Classical techniques, such as fluorescence *in situ* hybridization (FISH) and immunohistochemistry (IHC), are developed for the detection of nucleic acids and protein biomarkers within cells and tissue specimens with high sensitivity and improved spatial resolution. QDs exhibit excellent photophysical properties, and have been further proven powerful to sensitively quantify molecular profiling of cells and tissues, providing unique identification of individual cell lineages and uncovering molecular signatures of pathological processes.[108–110] In addition, utilization of QDs for staining of fixed cells and tissue specimens does not impose strict requirements on the probe biocompatibility, toxicity, or stability in biological media. Therefore, there are massive researches on the cellular imaging using QDs.

QDs are able to transfer electrons and holes to biomolecules, and QD-dopamine conjugates have been used to visualize the redox sensitive pattern of living cells. Clarke *et al.*[89] reported the first example of a cellular biosensor based on electron transfer between a small biomolecule (dopamine, the neurotransmitter) and CdSe/ZnS QDs, and they only found fluorescent QDs in the cell periphery and lysosomes.

When used for living cell imaging, QDs may aggregate inside the cells, and thus stay in endocytotic vesicles, including endosomes and lysosomes. However, there are few successful examples for delivery of monodispersed QDs into the membrane-bound compartments of living cells, such as nucleus and mitochondria. Wu *et al.*[26] utilized bioconjugates of QD-streptavidin and QD-antibody to label both of membrane-associated HER2 receptor and a nuclear antigen simultaneously in breast cancer cells. Tholouli *et al.*[111] also employed the biotin-streptavidin interaction to prepare QD-oligonucleotide probes for FISH-based studies of mRNA. Biotinylated DNA probes pre-incubated with QD-streptavidin conjugates enable the detection of 3 mRNA targets in a 1-step FISH procedure. In another example, Chan *et al.*[112] developed a more controlled procedure for conjugating exactly one oligonucleotide probe with one QD. They blocked excessive streptavidin sites on QD-streptavidin conjugates with biocytin, and thus obtained QDs on which only a few biotinylated oligonucleotides are attached. With the help of agarose gel electrophoresis, further purification of the conjugates yields mono oligonucleotide FISH probes, which are relatively small and suitable for multiplexed mRNA detection under mild specimen permeabilization. At last, the QD-based probes were used for the comparison of cellular distribution patterns between vesicular monoamine transporter (VMAT2) mRNA and immunoreactivity of tyrosine hydroxylase in dopaminergic neurons.

It is worth noting that the multi-color imaging turns possible with QDs, owing to the broad excitation spectra and size-tunable emission color. Ghazani and coworkers[113] explored the three-color staining of lung carcinoma xenografts for epidermal growth factor receptor (EGFR), E-cadherin, and cytokeratin with 655, 605, and 565 nm QD-based assays. They also noticed a significant enhancement of 655 nm signal over 565 nm one, attributing this phenomenon to FRET transferring energy from smaller to larger QDs. In another study, Yezhelyev *et al.*[114] employed relative QD intensities to quantify three clinically significant breast cancer markers (HER2, ER, and PR) on formalin-fixed paraffin-embedded (FFPE) breast cancer cells, in addition to the multiplexed labeling. In order to account for signal improvement of red QDs and to assess the expression levels of biomarkers within one sample, they adjusted the acquired data according to the relative QD intensities (QD655:QD605:QD565 = 8:4:1 at the same QD concentrations), and obtained relative biomarker abundance consistent to that calculated with conventional techniques (Western blot, IHC, and FISH). This technology has been further validated by the detection and quantification of a panel of five biomarkers on FFPE breast cancer tissue biopsies.

Combining QDs with other contrast agents for multimodal imaging has been an active research field, where various imaging strategies can be integrated into a single nanoparticle system.[115] QDs have been conjugated with imaging techniques such as computerized tomography (CT), magnetic positron emission tomography (PET), optical imaging and especially magnetic resonance imaging (MRI), by forming core/shell structures[116], doping paramagnetic ions or attaching paramagnetic complexes. For example, Selvan *et al.* prepared hetorodimers of CdS-(CdSe) QDs and Fe_2O_3 (or Fe_3O_4) nanoparticles with tunable fluorescence emissions over 500–600 nm[116], and coated the dimers with silica shell for *in vitro* imaging of mouse breast cancer cells (4T1). Wang *et al.*[117] doped Mn^{2+} ions to the shell of CdSe/ZnS QDs while preserving the strong luminescence and maximizing the relaxivity for MRI, and verified both optical and MR imaging of mouse macrophages. Mulder *et al.*[118] have conjugated CdSe/ZnS QDs with Gd-DTPA and RGD peptides and illustrated the application of multimodal imaging probes for *in vitro* fluorescent and MR imaging of human umbilical vein endothelial cells (HUVEC).

4.3.1.2 *Monitoring Dynamic Molecular Processes*

Real-time imaging of live cells enables the study of highly complex and dynamic biological event at molecular level.[3] Since the biomarkers expressed on the cell membrane is usually readily achievable for QDs, the reported applications focus on dynamics of membrane proteins (*e.g.*, receptor diffusion) and membrane-associated processes (*e.g.*, endocytosis and intracellular trafficking) rather than monitoring of intracellular targets. For instance, Lidke *et al.* coated QDs with epidermal growth factor (EGF, a ligand for erbB/HER transmembrane receptors) to study the early steps of receptor-mediated signal transduction.[29] And a later report by the same group presented the antigen uptake and processing by dendritic cells using QDs functionalized with pathogen-specific ligands.[119]

The fluorescence of QDs is resistant to photobleaching and degradation, enabling the long-term probe monitoring up to several days. For example, Jaiswal *et al.* have utilized this property for the visualization of QDs endocytic uptake and specific cell-surface labeling of P-glycoprotein transporters over the course of 14 h, acquiring images at a rate of 1 frame per minute.[33] Furthermore, localization of particles within the endosomes of live HeLa cells and D. discoideum amoebae could be monitored over the course of more than a week with minimal loss of QD fluorescence.

Much success in the study of cell receptor diffusion and interaction has been achieved. In a single-molecule imaging study, Dahan *et al.*[120] used commercially available QDs for labeling individual glycine receptors on the surface of cultured spinal neurons and tracking the receptor diffusion in and out of the synaptic cleft. In addition, the results showed that differential 2D diffusion coefficients of receptors can be measured over time spans 240 times longer than previously achieved using organic dye tags, with 4 to 8-fold better spatial resolution, and with a signal-to-noise ratio almost being an order of magnitude higher. In another work, Choquet and coworkers[121] took advantages of QDs to reveal a previously unknown receptor diffusion mechanism for recovery from synaptic depression in neurons. Tracking of the rapid lateral diffusion of QD-labeled AMPA glutamate receptors has shown diffusion behavior comparable to that of organic dye-labeled receptors, while providing a robust fluorescence signal for the duration of experiment.

Due to the small size, individual QD can be taken up by the cells *via* endocytosis, incorporated within the endosomes, and transports like any other endocytosed cargo without interfering with the mechanism of the process. Thus QDs represent a useful model system for the detailed study of intracellular processes such as endocytosis and intracellular trafficking. For example, Cui *et al.*[122] studied the dynamics of axonal internalization and neuronal retrograde transport of the nerve growth factor (NGF) by tagging native NGF with QDs. As they showed, the real-time monitoring of individual QD-NGF-containing endosomes revealed a "stop-and-go" behavior and occasional anterograde movement, thus providing insight into the diversity in transport mechanisms. Zhang *et al.*[123] also utilized QDs with pH-dependent fluorescence to study the dynamics of synaptic vesicles during multiple rounds of neuronal transmission without perturbing the vesicle cycling. Monitoring of individual QD-loaded synaptic vesicles enabled to characterize complete vesicle fusion (full-collapse fusion) and transient fusion (so-called kiss-and-run behavior) in terms of time and frequency of impulse firing, and also uncovered new aspects of the neurotransmitter release and replenishment mechanisms. Highly stable ligand-QD conjugates mimicking viruses and pathogenic microorganisms can be modeled to characterize the immune response mechanisms in details. For example, Roullier *et al.*[124]functionalized QDs with a chelator, tris-nitrilotriacetic acid (tris-NTA), pre-loaded with Ni for labeling of biomolecules with ubiquitous HIS tags.

4.3.1.3 *Labeling Intracellular Targets*

Labeling intracellular targets in live cells enables the study of molecular processes occurring within living systems. A variety of techniques, such as microinjection, electroporation, chemical transfection, and ligand-mediated uptake, have been developed for the delivery of macromolecular cargoes within cells.[125]

Mechanical techniques represent the most straightforward approach to QD intracellular delivery. For example, peptide-functionalized QD probes were directly delivered into the cytoplasm *via* microinjection, and successfully revealed active peptide-specific transport mechanisms to reach target compartments, nucleus and mitochondria, within several hours after delivery. In another example, Yum *et al.*[126] employed gold-coated boron nitride nanotubes (with a diameter of 50 nm) to send QDs into the cytoplasm or nucleus of live HeLa cells, and consequently monitored QD diffusion within those compartments for 30 min. Furthermore, Park *et al.*[127] engineered arrays of vertically aligned carbon nanosyringes to provide cytosolic access for injection of unmodified QDs, obtaining high-throughput intracellular delivery of QDs.

Non-mechanical approaches are increasingly popular, as they offer QDs intracellular delivery with minimal intrusion to cell physiology. Functionalization of QDs with engineered peptides or small versatile biomolecules provides great flexibility in tuning the QD interaction with cell components[128]. Delehanty *et al.*[129] modified QDs with HIS-tagged cell penetrating peptide based on the HIV-1 Tat protein motif, and presented efficient internalization of QDs *via* endocytosis. Rozenzhak *et al.*[130] attached the nuclear localization sequence to QDs for nuclear targeting and apoptotic GH3 domain for triggering cell death. Other examples achieving cell entry of QDs utilize cationic peptides, such as polyarginine[131], polylysine[132], and other engineered peptides or small versatile biomolecules[128]. Despite the versatility of QD-peptide conjugates for labeling intracellular targets, this approach still suffers from the uncontrolled probe aggregation and lysosomal sequestration inside cells.

Recent work employs endocytosis/pinocytosis as a universal delivery mechanism and endosome destabilization/lysis as a cytosolic access route.[133] Courty *et al.*[134] reported a strategy involving QD cell-loading using osmotic lysis of pinocytic vesicles to load QD-tagged kinesin motors on living HeLa cells and monitor single-motor movement within the cytoplasm. Similarly, Kim *et al.*[135] used 100-nm biodegradable vesicles made of poly(lactic-co-glycolic acid) (PLGA) polymer for intracellular delivery of unmodified QD probes. Duan and Nie[136] coated QDs with hybrid poly(ethylene glycol)/polyethylenimine (PEG/PEI) polymers to contain multifunctionalities within single QDs, and they prepared nanoparticles with a small HD (15–22 nm), endosome-disrupting capacity, good stability and biocompatibility. Yezhelyev *et al.*[137] decorated negatively-charged polymer-coated QDs with tertiary amines, thus produced proton-absorbing QDs that efficiently achieve intracellular endosomal release with bright fluorescence and good colloidal stability.

McGrath *et al.*[138] has taken advantage of FRET between QD-transferrin probes and dye-transferrin conjugates. Dimerization of transferrin receptors results in FRET during receptor mediated endocytosis. To further minimize the accidental non-FRET excitation of acceptor dyes, Clapp *et al.*[139] utilized a two-photon QD excitation route to significantly enhance the signal-to-noise ratio of intracellular FRET (Fig.4.7).

4.3.2 *In Vivo Imaging*

Compared to *in vitro* study which relies on the living cells, tissue thickness becomes a major concern while imaging a multicellular organism, because of the poor transmission of visible light through biological tissues. Using of near infrared (NIR) window is considered to be the key to optical imaging in deep tissues.[140] Few organic dyes emit brightly in the NIR region without severe photobleaching.[141] One of the exciting advantages of QDs for *in vivo* imaging is their tunable emission wavelengths throughout the NIR spectrum affected by their composition and size, leading to photostable fluorophores that sustain in biological buffers.[142]

4.3.2.1 *Vascular Imaging*

Properly modified QDs are particularly suitable for vascular imaging because of their small size, photostability and fluorescence for real-time visualization with existing inexpensive technology.[143] A very useful and possible application of QDs is the *in situ* detection of sentinel lymph nodes (SLN) during surgery, with superior effectiveness to Lymphoscintigraphy and deeper penetration than

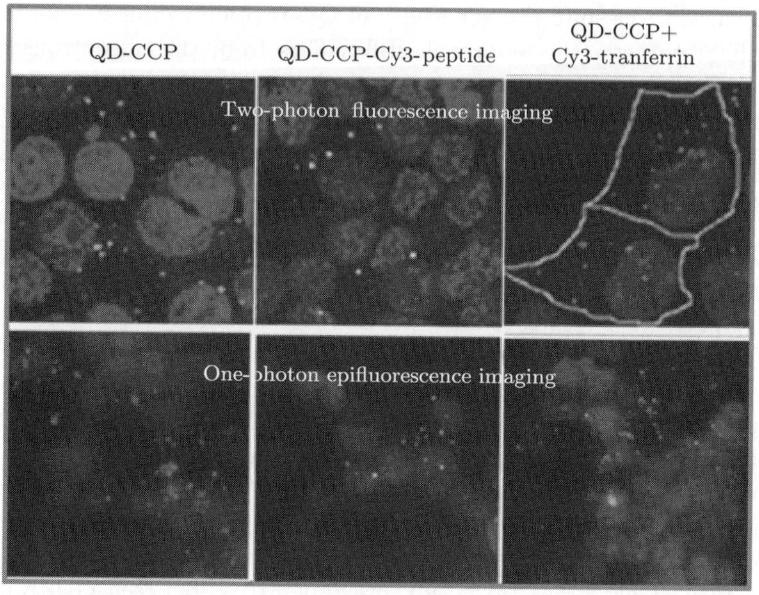

Fig. 4.7 Two-and one-photon excitation of QD-Cy3 FRET system. QDs are efficiently excited by both methods, enabling fluorescence of conjugated Cy3 molecules *via* FRET. QD is located outside the cell nuclei. However, only two-photon excitation precludes non-FRET excitation of Cy3 dye, reducing significant background produced by conventional one-photon fluorescence. Reproduced from Ref. 139 with permission of Wiley-VCH Verlag GmbH & Co. KGaA, Copyright © 2007. (color figure at the end of the book)

Blue isosulfan dye.[144]

Pioneering work was done by Larson and collaborators[22] in 2003, which demonstrated the possibility of QDs for dynamic imaging of the blood vasculature of skin and adipose tissue in live mice. In this work, polymer-encapsulated QDs provided bright and persistent fluorescence contrast after intravenous injection, enabling the investigators to assess blood flow velocities by performing line scans across capillaries and monitoring the propagation of QD fluorescence. Furthermore, the large two-photon excitation action cross-section of QDs presented nearly background-free vasculature imaging at tissue depths of several hundred microns with two-photon fluorescence intravital microscopy. Stroh *et al.*[145] combined two photon intravital microscopy, blue-emitting QDs encapsulated in PEG-phospholipid micelles, and a transgenic mice model of GFP-expressing perivascular cells to study the morphology of the tumor vasculature. In this study, QDs highlighted the vessel boundary and provided a clear picture of tumor vessel morphology for at least 30 min, while GFP fluorescence indicated the distribution of perivascular cells. Kim *et al.*[146] took advantage of poor extravasation of QDs for studying the patho-physiology of viral infection of the central nervous system in mice. Using intravital two-photon microscopy, QD extravasation from brain microvasculature has been monitored as a measure of disease-associated vascular injury and blood-brain barrier breakdown.

SLN mapping with QDs was first explored in 2003 by intradermally injecting NIR QDs (16–19 nm) in mice and pigs.[142] Following intradermal injection, QDs migrated rapidly to the lymphatics and colocalized with isosulfan blue in the nodes. Fluorescence contrast of these nodes beneath the skin surface could be distinguished up to 1 cm. They also revealed that smaller QDs (HD ~9 nm) could move further into the lymphatic system with fluorescence signals from up to 5 nodes.[147] The later study demonstrated the migration of QDs to the SLN following injection directly into tumors, further supporting their potential for SLN mapping.[148] This technique also helps scientists to identify lymph nodes downstream from the lungs[149,150], esophagus[151], and from subcutaneous tumors[148].

In experimental settings, multicolor imaging has been exploited to simultaneously image 5 different lymphatic basins in mice.[152] Following intracutaneous injection of 5 types of polymer-coated carboxy-QDs (emission wavelength 565–800 nm, HD 15–19 nm) into the paws, ears and chin of mice, transportation of QDs through lymphatic networks and accumulation in SLNs were monitored. Spectral imaging using the Maestro In-Vivo imaging system (CRi, Waltham, MA, USA) enabled the visualization of each QD and thus the lymphatic drainage to the primary lymph nodes of each injection site. By employing visible QDs, the real-time five color lymphatic imaging achieved with much less expensive installment setting than the spectral imager.[153] Most of QDs were taken up by macrophages and dendritic cells in the lymphatic sinuses. Imaging was possible seven days after injection with little signal loss after the first day, a feature unique to QDs.

In other work, NIR QDs have been explored to image the coronary vasculature of a rat heart[154], and the blood vessels of chicken embryos[155]. Jayagopal *et al.*[156] also illustrated that QDs could serve as molecular agents for vascular imaging. QDs with distinct emission colors were functionalized with three different cell adhesion molecules (CAMs), and injected intravenously in rats under a diabetic rat model. These QDs freely spread in the vessels, stayed in the endothelium, and underwent leukostasis, allowing researchers to visualize the inflammation-specific behavior of individual leukocytes due to the fluorescence enhancement of CAMs, as fluorescence angiography of the retinal vasculature showed. Thanks to the unique spectral properties of QDs, up to four spectrally distinct QD tags were allowed to be simultaneously imaged.

Santra *et al.*[157] synthesized CdS:Mn/ZnS QDs with fluorescent, radioopacity, and paramagnetic properties in a single unit for brain blood vessels imaging in mice. QDs surface was aminated to completely prevent particle agglomeration and further conjugate with a TAT peptide. After endovascularly administering QDs suspension to a rat, the labeling efficacy of these QDs was demonstrated on the basis of the histological analysis of the microtome sliced brain tissue, clearly showing that TAT-conjugated QDs stained brain blood vessels.

4.3.2.2 *Cell Tracking*

Cells loaded with QDs can be administered to an organism, enabling the identification of the original cells and their progeny within the organism. By microinjecting QDs into the cytoplasms of single frog embryos, Dubertret *et al.*[21] demonstrated the first example of QDs-based cell tracking on a small organism scale. While the embryos grew and the cells divided, each cell retained a portion of the fluorescent cytoplasm from the original labeled cells, thus enabling continuous cell lineage tracing in real time. In the work of Hoshino *et al.*[32] and Voura *et al.*[34], tumor cells loaded with QDs were intravenously injected into mice, and the QDs distributions in animals were revealed through tissue dissection by fluorescence imaging. Similarly, Gao *et al.*[25] subcutaneously injected QDs-loaded human cancer cells in an immune-compromised mouse. The solid tumor divided from the cancer cells could be determined without sacrificing the mouse by using a spectral imaging system (CRI). Rosen *et al.*[158] also incorporated human mesenchymal stem cells (hMSC) loaded with QDs into a regenerative implant, which was for canine hearts with a surgically-induced defect. QDs were found to remain fluorescent within the cells even after eight weeks from the implantation, and helped the authors to track the behaviors (locations and fates) of the hMSC cells. In addition, QDs-labeled stem cells were directly injected into the canine myocardium, and were employed to elaborately locate these cells in the heart, based on the fluorescence signals in cardiac tissue sections.

In another example, Stroh *et al.*[145] utilized two-photon intravital microscopy to track the transport of bone marrow-derived progenitor cells through tumor vessels in real time. Cells loaded with TAT-functionalized orange-emitting QDs were injected into the carotid artery of live mice along with blue-emitting QDs for vessel contrast. As two spectrally distinct QD types can both be excited by 800 nm two-photon light, simultaneous highlighting of tumor vasculature and continuous cell tracking at B1 frame/s is possible. A 3-dimensional tracking of individual transplanted haematopoietic stem cells has been recently achieved by Celso *et al.*[159], while fluorescent dyes have been used for cell labeling and QDs have been only utilized for microvasculature imaging. This technique can be readily expanded to QD tagging of cells and multiplexed high-resolution 3D cell tracking, thus providing access to real-time dynamic studies of cell interaction within physiological niches.

4.3.3 *Cancer Imaging*

The QD-based fluorescence molecular imaging represents an attractive technique for the detection of specific biomarker expressing cells *in vivo*. As previously discussed, QDs have been successfully applied to cellular and vascular imaging and cell tracking as organ targeting. However, the QD-based targeted molecular imaging *in vivo* remains a significant challenge due to the lack of specificity in heterogeneous physiological conditions and pervasive foreign-particle clearance mechanisms.[3] Thus designing specific functionalities of QDs for efficiently passive and active targeting pathways (Fig.4.8) represents the major focus of current research.

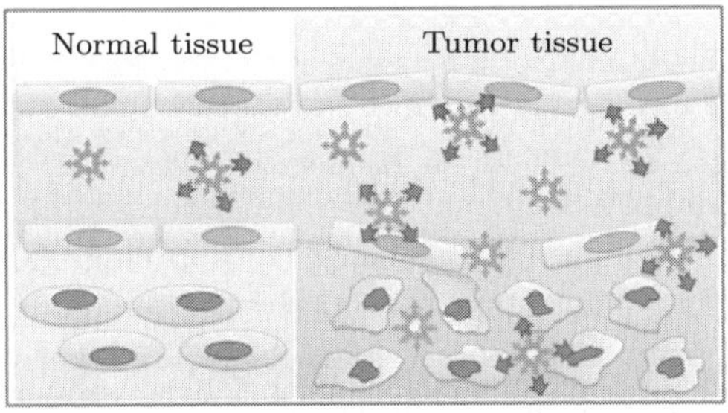

Fig. 4.8 Routes for *in vivo* QD targeting. Abnormal highly permeable tumor blood vasculature and poor tumor lymphatic drainage enable passive QD targeting and accumulation within the tumor site *via* enhanced permeability and retention effect. Similar mechanism can be employed for labeling the regions with damaged or abnormal vasculature. Intact endothelium represents a significant barrier for QD extravasation. Labeling of biomarkers specifically expressed on the vasculature of a particular organ or tissue (*e.g.*, tumor site) provides an efficient way of targeting *in vivo* imaging with bulky QD probes. When QD extravasation is possible, active targeting of biomarkers expressed on the cell surface enables specific labeling deep within the tissues of interest, while reducing non-specific QD accumulation in non-targeted areas.[3]

The progress towards targeted imaging of cancer was first made in 2002, when Akerman and coworkers coupled green and red MAA-coated QDs with peptides which had affinity for lung endothelium and tumor vasculature, and injected the conjugates into live mice for *in vivo* imaging.[103] While whole organs could not yet be visualized by QDs *in vivo*, the idea of using targeting peptides to target specific antigens with fluorescent QD markers holds a great potential. For example, Cai *et al.*[90] later functionalized NIR CdTe/ZnS QDs with cyclic RGD (arginine-glycine-aspartic acid) for targeting integrin $\alpha_v\beta_3$, which is an up-regulated biomarker in cancerous tissue during proliferation, metastasis, and angiogenesis.

Another step towards molecular imaging of cancer with QDs was achieved soon thereafter, in 2004, with the imaging of tumor xenografts in live animals.[25]

In this study, an amphiphilictriblock copolymer was employed to protect and disperse QDs, PEG molecules to improve biocompatibility and circulation time, and/or an anti-PSMA-monoclonal antibody to target the cancer. After intravenously injecting QDs into mice bearing subcutaneous human prostate cancers, the authors demonstrated that both passive and active targeting of tumor xenografts expressed the prostate specific membrane antigen (PSMA) in live animal models. Furthermore, the results proved that QDs with surface PEG groups can accumulated passively though the EPR effect due to longer circulation times; and they also showed that the QD-antibody conjugates accumulated and were retained by the PSMA expressing tumors, generating intense signals in whole animal fluorescence imaging. Yu *et al.*[160] also coated QDs with an antibody against alpha-fetoprotein to rationally target and image mice with human liver cancer.

QDs are powerful imaging tools for the study of the complicated anatomy and pathophysiology of cancers in animal models. Stroh *et al.*[145] reported that QDs largely improved current intravital microscopic techniques for the visualization of tumor microenvironment. By applying QDs for blood-vessels visualization with two-photon excitation, the authors captured images of extracellular matrix from autofluorescent collagen and perivascular cells from fluorescent protein expression simultaneously. In addition, the authors employed QD-tagged beads with a variety of size ranges to simulate the size-dependent distribution of nanotherapeutics in tumors. Single-particle tracking in live animals has also been reported by Tada and coworkers[100] through combining sensitive QDs with high resolution 3D confocal real-time intravital microscopy techniques. Following injection through the tail vein, QDs conjugated to trastuzumab (a monoclonal antibody against the HER2/neu receptor) could be visualized within HER2-overexpressing human breast cancer tumors grown in mice. With the help of intravital fluorescence microscopy, the authors observed how individual QDs traveled in the bloodstream, spread into the tumor and extracellular matrix, attached onto tumor cells *via* their receptors, and finally translocated into the perinuclear region of the cells.

Fluorescence imaging in NIR-II (1.0–1.4 μm) is more powerful than imaging in visible (450–750 nm) and traditional NIR-I (750–900 nm) region, due to reduced light scattering, deeper tissue penetration, and less autofluorescence.[161] In order to avoid the cytotoxicity concern of heavy metals[162], Wang and coworkers developed silver sulfide (Ag_2S) QDs as new NIR-II fluorophores[73], which had high fluorescence quantum yield (15.5 %) of NIR-II emission at approximately 1 200 nm, and utilized them for imaging of xenograft tumors[163]. The results showed that Ag_2S QDs can locate tumors *via* passive tumor targeting with a high signal-to-background ratio.

The most distinct features of QDs are their superior fluorescence properties, thus bioimaging using QDs has been extensively studies from cell staining, vascular imaging, to targets tracking and biological process monitoring inside cells or tissues. Following the trend of the QDs evolution, we can see that current research trend has been deepened to multimodal imaging and multifunctional systems with high sensitivity and resolution, and we believe that the future direction will still rely on imaging-based study due to the intrinsic fluorescence advantages of QDs.

4.4 QDs for Theranostics

Due to their unique advantages, QDs are widely used as fluorescence probes for sensitive assessment of phenotypes of healthy cells and tissues, detection of molecular makers of diseases, and monitoring various biological events. Thus, combining QDs with therapeutic agents (such as small chemotherapeutic drugs, genes, and photosensitizers) will provide QDs a potential multifunctional platform for biomedicine.

4.4.1 *QDs for Gene Delivery*

Gene therapy is a promising strategy for the treatment of various diseases due to genetic disorders, such as Parkinson's disease and cancer.[164] Genes, including plasmid DNA (pDNA), small interfering RNA (siRNA), and micro RNA (mRNA), can be employed as potential therapeutic agents through protein expressing or gene silencing.[165,166] However, clinical applications of genes are limited, mainly due to low cellular uptake efficiency and fast degradation under biological conditions[167], which are generally overcome by employing a gene vector to protect and deliver the genes[168].

QDs were initially used as a potential photostable fluorescence imaging agent to track siRNA-delivery process *in vitro* by combining QDs with siRNA in cationic liposomes.[40,169] Chen *et al.*[40] firstly reported that co-transfection of different siRNA molecules with multiple colored QDs allowed multi-channeled tracking of gene expression. In another example, Burgess and coworkers[170] attached pDNA onto QDs through specific interactions between pDNA and peptide nucleic acids (PNA) on QDs, and thus developed an optical tag for intracellular tracking of pDNA with a delivery system based on cationic liposomes.

Based on the researches of QDs in a drug delivery system, QDs were exploited as both fluorescence beacons and gene vectors by Sangeeta N. Bhatia for the first time in 2007.[38] They functionalized PEGylated QDs with tumor-homing peptides (F3) to develop a gene delivery platform, by conjugating siRNA *via* chemical cross-linker onto the surface of the nanoparticles. Delivery of F3/siRNA-QDs to cancer cells and the consequent release from endosomes caused remarkable knockdown signal of siRNA. This strategy was adapted to simultaneously targeted gene treatment and to image metastatic cancers. QD-gene complexes could serve as dual-modality probes of both optical and fluorescence responses, allowing real-time tracking with ultrasensitive localization of QDs through the whole delivery and transfection process. Furthermore, the complexes of QDs and genes could be further optimized to improve cancer cells targeting, tumor fluorescence probing and drug tracing, making them promising candidates for gene therapy.

In addition, pDNA could be also transfected by QD-based vehicles *via* chemical cross-linker onto the surface of QDs. Srinivasan *et al.*[170] attached pDNA (PNA) with phospholipid-coated QDs through a pDNA-N- succinimidyl-3-(2-pyridylthio) propionate (PNA-SPDP) linker. These QD-DNA conjugates could be internalized and then expressed the reporter protein. However, the huge cost of the SPDP-PNA linker would limit its application. So that, to avert this problem, MPA-coated CdTe/CdS core/shell QDs were reported for delivering a suicide gene, particularly herpes simplex virus thymidine kinase (HSV-TK) gene. In this strategy, pDNA were also chemically conjugated to the surface of QDs.[171] QDs-TK conjugates could transfect with high efficiency, kill tumor with high efficiency and kill tumor cells comparable to the parental TK gene after adding GCV, whereas QDs do not show obvious cytotoxicity, and distinctly monitor gene intracellular trafficking. Oligonucleotides, regardless of their base length and sequence, could be effectively internalized into cells by thioglycolic acid (TGA)-coated CdTe QDs delivery. Anti-survivin antisense oligonucleotides (ASON) were covalently linked to CdTe QD surface. CdTe-ASON were found to effectively induce cell apoptosis, and consequently visualize the intracellular tracking of ASON, which were in cytoplasm and tended to accumulate around the nucleus, according to biological activity of ASON.[172]

Although constructing the transfection nanomaterials *via* covalent bonds was expected to be useful to visualize the genes within cells, there still remained some problems. As siRNA was conjugated onto the surface of QDs *via* chemical reactions rather than the electrostatic interactions, siRNA/QDs showed different silencing efficiency depending on the cross-linkers, and the number of attachment sites on particles was limited. To overcome the influence of chemical cross linker between QDs and siRNA, a strategy of loading siRNA onto QDs *via* physical adsorption is introduced by Gao's group. They explored polymer to encapsulate QDs so as to develop a multifunctional gene nanovector for efficiency RNA interference and imaging with less toxicity (Fig.4.9).[137] In their approach, siRNA was loaded onto proton-sponge polymeric coated QDs *via* electrostatic interactions. Positive charges coated QD explored as gene nanocarrier had several advantages, such as protecting nucleic acids from enzymatic degradation[173], and promoting cell surface binding and intracellular transport[174], as well as endosomal release. More importantly, the efficiency of transfection and gene silence was extremely high, and the obtained silencing efficiency was tens of times higher than three most commonly used transfection reagents, which were lipofectamine, TransIT-TKO and Jet-PEI. The QD-siRNA composites were also dual-modality probes with optical and electron microscopy responses, facilitating *in situ* tracking and precise localization of QDs during the process of transfection. Another study of using positively charged polymer coated QDs to deliver siRNA was carried out by Zhang's group. In this strategy, a natural polymer, chitosan, and HER2 antibody were used to modify QDs for HER2/neu siRNA targeted gene therapy.[41] SiO_2 nanoparticles doped with CdSe/ZnS QDs with both imaging and gene transfection capabilities can be internalized by primary cortical neural cells with good biocompatibility *in vitro* and *in vivo*.[175]

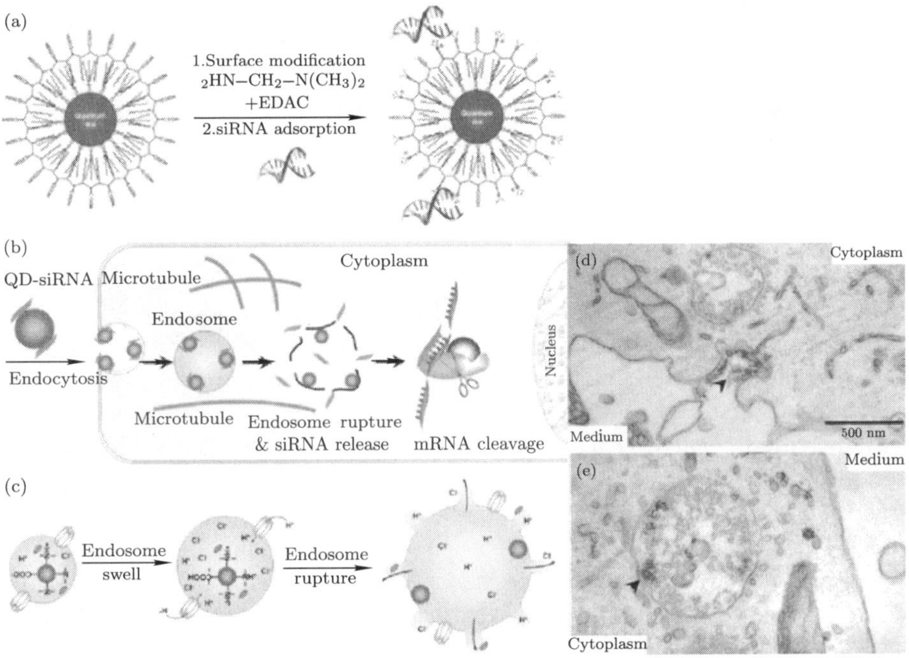

Fig. 4.9 Use of cationic polymer-coated QDs as a multifunctional nanoplatform for gene transfection and intracellular imaging. (a) Functionalization of QDs loaded with siRNA. (b) Routes of siRNA-QD in membrane binding, cellular uptake, endosomal escape, and consequently gene silence effect. (c) Illustration of the proton-sponge effect of the involvement of the membrane proton pump, and endosome rupture. (d,e) TEM images of localization of QDs inside cells: (d) endocytosis process; and (e) QD attachment to the inner vesicle membrane. Image is adapted with permission from Ref. 137. Copyright (2008) American Chemical Society.

The modification of QDs will obviously influence the efficiency of gene delivery. Li *et al.*[176] synthesized two L-arginine (L-Arg)-modified CdSe/ZnSe QDs in the absence or presence of β-cyclodextrin (β-CD), as gene carriers to overcome barriers, including siRNA protection, cellular penetration, endosomal unpacking, carrier releasing, intracellular transporting and efficiently gene silencing. It was found that β-CD-L-Arg-QDs could much more efficiently transfer siRNA and gene silence *in vitro* than L-Arg-QDs, because β-CD improved the aqueous solubility and biocompatibility, which would help the complexes to enter targeted cells. Despite applying QDs to study gene delivery has shown its promising futures, challenges of the long-term investigation still hinder the biological applications of QDs at animal level. To develop bio-stable and non-cytotoxic QDs, a polypeptide copolymer named dcBSA-PEO-TA, which was derived from denatured and cationized bovine serum albumin,was used to coat CdSe/CdZnS QDs for pDNA transporting and imaging.[177] QDs incorporated with the branched hCT-derived carrier peptide can efficiently deliver genes into different cell lines, exhibiting high stability and low toxicity.[178] Besides, Gao's group also combined QDs with amphipols for efficient siRNA delivery which worked in both serum-free and complete cell culture medium and real-time imaging.[179]

Recently, Park's group[180] studied QDs transfected pDNA to endothelial progenitor cells (EPCs) *in vivo*, for the first time. EPCs, which induced angiogenesis and vascularization[181], differentiated into endothelial cells (ECs) after transfection with specific genes, both *in vitro* and *in vivo*.[182] In angiogenesis, vascular endothelial growth factor (VEGF) mediated vascular permeability and participated in the multistep process of EC formation.[183] Transfection of EPCs with VEGF gene by PEI modified QDs led to the expression of specific genes and proteins for mature ECs. The yield of cellular uptake by EPCs was extremely high, nearly 52% at 1 nmol of QDs. A model of hind limb ischemia was introduced in nude mice, and then VEGF gene-transfected EPCs were intramuscularly transplanted into the ischemic limbs (Fig.4.10). After that, the VEGF gene-transfected EPCs largely enhanced gene expression, the number of differentiated ECs, and blood perfusion levels. Furthermore, the expression levels of specific gene and protein of mature EC markers and angiogenic factors increased after the transplantation of VEGF gene-transfected EPCs in the animal model. Besides, QDs could serve as effective carriers and sensitive probes for EPCs differentiating into ECs and for real-time tracking when the cells were transplanted into an animal model of severe ischemia.

4.4.2 *QDs for Drug Delivery*

Inspired by that many nanomaterial-based drug delivery systems always required an additional fluorescence labeling, Farokhzad and coworkers[39] developed for the first time QD-aptamer-doxorubicin conjugates that integrated localized cancer imaging, therapy, and diagnosis in 2007, which represented a big advance in the nanomedicine area. In the approach, CdSe/ZnS core-shelled QDs were first modified with A10 RNA aptamer (Apt), which could specifi-

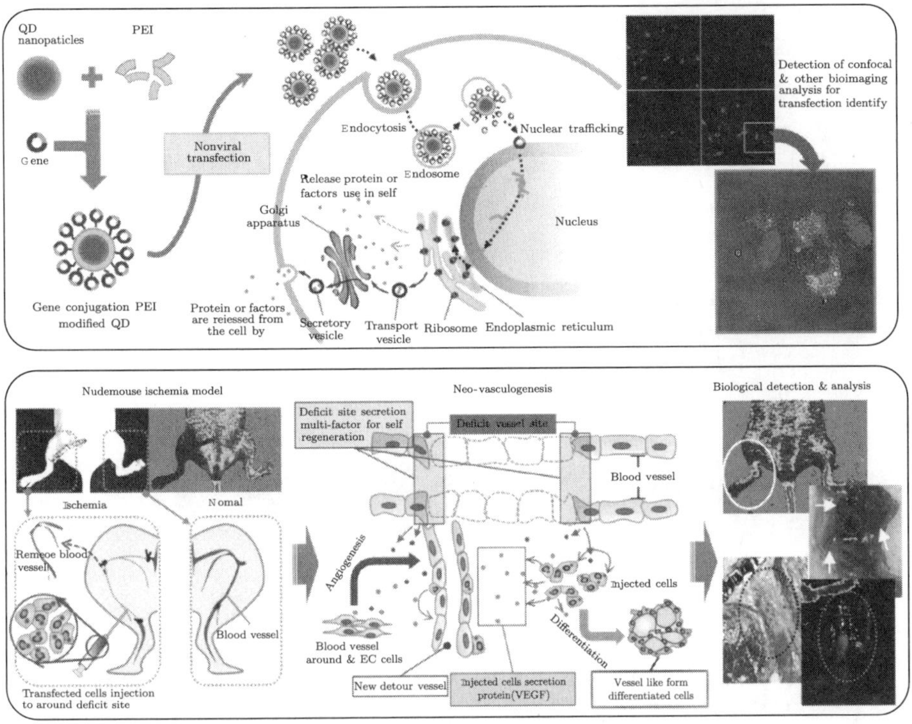

Fig. 4.10 Diagram of the VEGF gene delivery into endothelial progenitor cells (EPCs) using PEI modified QD and bioimaging of transplanted EPCs into hind limb ischemic model. Reproduced from Ref. 180 with permission of Elsevier. Copyright © 2012.

cally recognize membrane antigen of prostate cancer cells. Then doxorubicin (DOX) which has been widely used as an anthracycline drug was introduced to intercalate within the double-stranded CG sequence of Apt. The drug loading/release process can be monitored by the fluorescence change due to the fluorescence resonance energy transfer (FRET) between QDs and DOX, *via* the change of fluorescence intensity. QD-Apt-DOX exhibited high cytotoxicity for the targeted cancer cells owing to the drug release after the endocytic cellular uptake (Fig.4.11). This QD-Apt drug nanovector has several advantages: ① targeted drug delivery, ② monitoring the delivery process, ③ *in situ* monitoring the drug delivery process, ④ imaging the cancer cells, ⑤ loading efficiency of DOX would be controllable by incorporation of multiple CG sequences within the stem of the aptamers. This QD-drug conjugate acted as a specific and sensitive multifunctional theranostic system.

Since then, several groups started to explore QD as a drug nanocarrier for theranostic. For instance, a QD-pseudopolyrotaxane supramolecule system with high stability, targeted drug transportation and pH-sensitive release characteristics was presented for early diagnosis and therapy.[184] Another work of

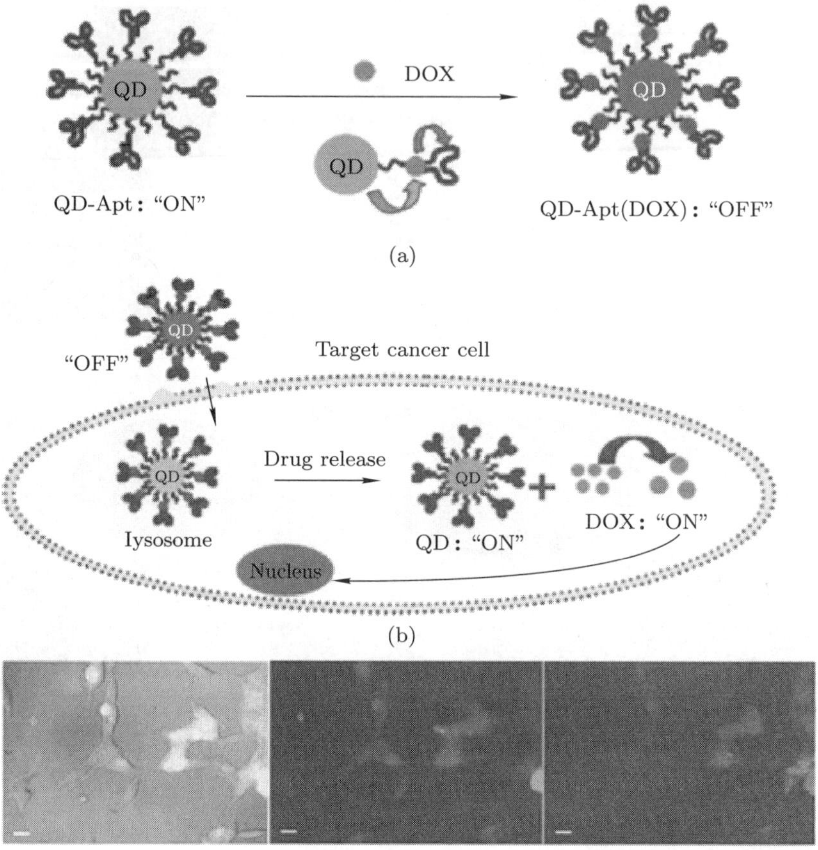

Fig. 4.11 Schematic images of (a) QD-Apt-DOX FRET system. QD was functionalized with Apt and then loaded DOX, leading to the fluorescence quench, (b) tumor-targeted delivery of QD-Apt-DOX. The release of DOX from the QD-Apt-DOX conjugates induces fluorescence recovery, thereby detecting the intracellular delivery of drug and QD localization, as well as killing of cancer cells. (c) Images of target cells after incubation with QD-Apt-DOX. DOX and QD are shown in red and green, respectively. (scale bar: 20 μm). Image is adapted with permission from Ref. 39. Copyright (2007) American Chemical Society. (color figure at the end of the book)

QD-based nanomaterial for pH-regulated drug delivery has been reported.[185] The *in situ* immobilization of hydroxypropylcellulose-poly(acrylic acid)-CdSe hybrid nanogels (HPC-PAA-CdSe) was utilized as pH-sensitive drug delivery systems based on its reversible swelling/shrinking transition owing to the pH-sensitive PAA chains. However, the complicated synthetic route of the supramolecules and low drug loading efficiency restricted the usage of this composite. So that, the utilization of small molecules to cap QDs for drug delivery was investigated. CdTe QDs modified with negatively charged mercap to propionic acid (MPA), forming MPA-QD composites, can facilitate the cellular internalization of anticancer agent daunorubicin (DNR) on drug-sensitive and drug-resistant leukemia cells.[186] It has been demonstrated that MPA-QDs could serve as an effective drug nanocarrier to inhibit the growth of these leukemia cells even at low drug concentrations, while acting as a bio-labeling agent. Modification of QDs with MUC1 aptamer which is a tumor targeting moiety/penetration enhancer was also employed as multi functional nanocarrier.[187] DOX was attached to QDs *via* a pH-sensitive hydrazone bond to improve the stability of the complex in blood circulation and facilitate drug release in acidic environment like tumors. According to rough calculation, the authors found that each QD contains 46 DOX molecules. QD-MUC1-DOX could target to ovarian cancer cells inducing higher toxicity compared with free DOX *in vitro* and preferentially accumulated in ovarian tumor *in vivo.*

It should be mentioned that QDs could deliver drugs to either cancers or tissues such as lung. Knight's group[188] demonstrated the first example of the feasibility of CdSe/CdS/ZnS QDs loading with drug for *in vivo* targeted immune-inflammatory therapy in 2011. CdSe/CdS/ZnS QDs could enhance drug uptake by the alveolar macrophages (aMØs) and further induced cells apoptosis without inducing a concurrent proinflammatory cytokine response. aMØs is known as a vital factor in the pathogenesis of inflammatory lung injuries. An accumulation of DOX in nucleus and QDs in the cytoplasm was observed *in vitro*. More importantly, QD-DOX complexes were transported into the pulmonary airspaces of rats through the oropharyngeal aspiration. Several parameters assessing inflammatory injury, such as proinflammatory cytokines, albumin leakage and neutrophil infiltration, were monitored after the administration of QD-DOX or DOX, and the results showed that QD-DOX treatment was comparable to DOX. Therefore, CdSe/CdS/ZnS QDs were proved to be useful for the targeted macrophage-selective therapy and diagnosis of pulmonary disease.

In order to reduce the cytotoxicity of CdSe QD-based drug nanovehicles, highly photoluminescent and water dispersible ZnO QDs functionalized with folic acid (FA) were introduced as a platform for pH responsive and targeted delivery of DOX.[189] DOX loaded on ZnO-FA QDs was stable in the physiological condition, but readily disintegrated in the weak acidic intracellular environment of cancer cells and endosomes/lysosomes. FA also acted as a targeting ligand to specific cancer cells. Thus, ZnO QDs rapidly dissoluted in cancer cells, and consequently released DOX and induced apoptosis. It proved that ZnO QDs-based nanovectors could be used for simultaneous diagnosis and therapy of targeting cancer cells with excellent biocompatibility. Another research of using ZnO QDs as a nanocarrier for drug delivery was developed by Misra's group.[190] Biodegradable molecules (chitosan) and tumor targeted agents (FA) were used to modify ZnO QDs to improve their stability, biocompatibility, tumor-targeting and cellular uptaken. DOX was loaded on to ZnO-chitosan-FA surface with high loading efficiency (~80%) *via* physical and chemical interactions. The drug release behavior of the DOX-loaded ZnO-chitosan-FA nanocarrier was a pH-controlled release response. Recently, another FA and chitosan capped zinc-based QDs with functions of active targeting, efficient drug delivery and sensitive optical imaging of cancer cells were developed by Jayakumar's group.[191] An anticancer drug 5-Fluorouracil (5-FU) loaded onto the manganese doped zinc sulphide QDs coordinated to FA linked carboxymethyl chitosan (CMC) (FA-CMC-ZnS:Mn) with high drug encapsulation efficiency of nearly 92%. FA-CMC-ZnS:Mn encapsulated with 5-FU could induced breast cancer cells death, while FA-CMC-ZnS:Mn was found to be nontoxic.

Chang's group[192] presented a multifunctional platform which combined the magnetic resonance imaging (MRI) and optical imaging with drug delivery. $CuInS_2$/ZnS QDs and magnetitc nanoparticles, Fe_3O_4, were encapsulated with SiO_2 and polyethylene glycol (PEG) molecules to provide better biocompatibility, subsequent conjugation and efficient cellular internalization (Fig.4.12). Pt(IV)-anticancer drugs were chemically coupled to the surface of the silica nanohybrids, resulting in Pt(IV)-conjugated SiO_2 nanohybrids and higher cytotoxicity than the free Pt(IV) anticancer drug *in vitro*. This work demonstrated the potential of SiO_2 matrix with $CuInS_2$/ZnS QDs and Fe_3O_4 NPs acting simultaneously as highly effective dual-modality imaging probes and drug delivery system for cancer diagnosis and chemotherapy.

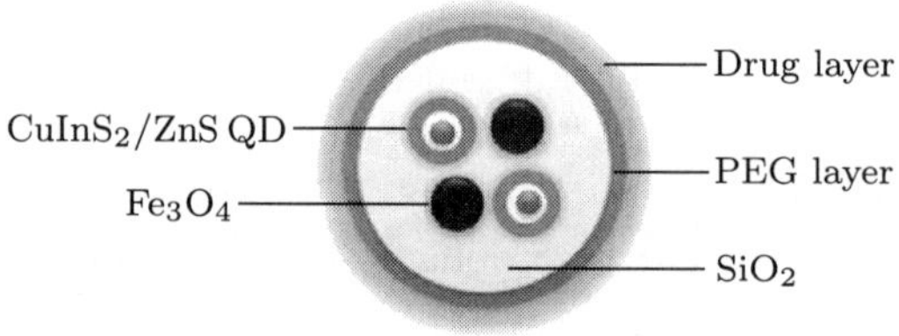

Fig. 4.12 Schematic illustration of SiO_2 nanohybrids incorporated with QDs and Fe_3O_4, and subsequently conjugated with Pt(IV)-anticancer drugs.[192]

4.4.3 *QDs for Photodynamic Therapy*

Photodynamic therapy (PDT) is a promising strategy for cancer treatment since it selectively destroys the diseased tissues. It utilizes the interaction between light and a photosensitizing compound to initiate cell death.[193] The photosensitizing compound is firstly irradiated by light, and then produces reactive singlet oxygen species (ROS) and thus induces *in vitro* toxicity.[194]

QDs offer great promise in PDT applications, which can be classified as agents for simultaneous imaging and therapy. Burda's group[195] has demonstrated that the combination of CdSe QDs with PDT photosensitizers enables applying an excitation wavelength beyond the absorption of photosensitizer alone. They utilized QD-based FRET for the first time to enable the excitation of photosensitizer for photodynamic cancer therapy. In this work, QDs were chemically linked with a silicon Pc photosensitizer (Pc4), so that QDs can be used to sensitize either a Pc4 *via* FRET or molecular oxygen undergoing triplet energy transfer (TET), as shown in Fig.4.13. Since QDs possess broad absorption, QD-photosensitizer conjugation could adapt to different excitation wavelengths and then activate the photosensitizer. Soon after that, another group also evaluated the potential of exploring CdSe QDs for targeting PDT.[196] Anti-CD antibody was modified on the surface of QDs which specifically targets to leukemia cells. Bio-targeted QDs improved the sensitivity of leukemia cells to UV irradiation and promoted the effect of different photosensitizers, such as plasma membrane-specific photosensitizer (SALPC) and phenothiazine photosensitizer (TFPZ). In addition, some investigations focused on the photosensitization by QDs in biological conditions[197,198], exhibiting singlet oxygen generation by QD-mesotetra (4-sulfonatophenyl) porphine (TSPP) and QD-chlorine e6 composites.

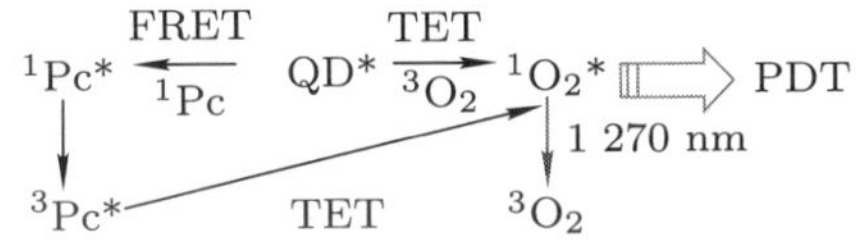

Fig. 4.13 Schematics of the $^1O_2$ generation in QD-based PDT systems. Image is adapted with permission from Ref. 195. Copyright (2003) American Chemical Society.

4.5 Biosafety of QDs

Semiconductor QDs have strong potential for biological applications such as drug delivery and bioimaging, as we described above. Although they provide remarkable biomedical benefits[1,23,25,199], it might potentially cause adverse human health effects due to exposure to QDs. Therefore, further investigation on the fate and toxicity of QDs, such as the cellular uptake, distribution, circulation, and role of QDs in biological systems should be addressed. In this section, we mainly discuss the cellular uptake mechanism, biodistribution, and *in vitro* and *in vivo* toxicity of semiconductor QDs.[200] More importantly, the toxicity of QDs relies on multiple factors, including physicochemical properties, such as concentration, size, charge, and modification, as well as oxidative, photolytic, and stability of QDs. We summarized studies both *in vitro* and *in vivo* to imply the biosafety considerations of QDs.

4.5.1 *Cellular Uptake Mechanisms of QDs*

The pathway of cellular uptake of QDs is closely related to the mechanism of cytotoxicity *in vitro*. Therefore, a profound understanding of the QDs' cellular uptake mechanism is required.

The size of QDs is one of the crucial factors determining the cellular internalization of QDs. Unlike larger nanoparticles, smaller CdSe/ZnS core/shell QDs (~4 nm) modified by D-penicillamine (DPA) first accumulated at the plasma membrane before internalization. Interestingly, uptake efficiency did not scale nonlinearly with the concentration, indicating that triggering the cellular internalization of QDs requires a proper concentration of QDs during incubation. DPA-QDs were mainly internalized through clathrin-mediated endocytosis and the minority by macropinocytosis. It has been demonstrated that the dynamin-dependent pathway explained the internalization of 70% of DPA-QDs. Aggregates of DPA-QDs were observed in endosomes, due to active transportation along microtubules toward the perinuclear region by microtubule-based dyneins. After that, a large portion of endocytosed DPA-QDs was detected in lysosomes, while the others were actively transported to the cell periphery and exocytosed very soon (21 min of half-life). It is the first work on clearance QDs *in vitro* (Fig.4.14).[201]

The cellular uptake is also controlled by the surface chemistry of QDs. QDs with various surface coatings had similar kinetics and mechanism of their cellular uptake in two human mammary cells in general but different in details. A large amount of QD655-COOH (hydrodynamic diameter 20–30 nm) was internalized by both MCF-7 and MCF-10A (non-tumori-genic) cells and then localized in lysosomes, but the percentage of QD-loaded cells in MCF-7 cell line was slightly higher than in MCF-10A cell line. A possible pathway for QD uptake into cells involves three major stages: endocytosis, sequestration in early endosomes, and translocation to later endosomes or lysosomes. Besides, QDs containing COOH groups were internalized quickly with the majority by both cancerous and non-cancerous cells, while QDs coated with PEG significantly reduced non-specific internalization of QDs into cells.[202] In addition, the positively charged polymer coated QDs moved along microtubules inside living cells and accumulated outside the cell nucleus forming large multi-vesicular structures. QDs with positive charges were not found in the cytoplasm but aggregated and stayed on the organelle membranes, because of electrostatic interactions between the positive charges of QDs and the negative charges of cell membranes.[137] In addition, it is worthy of noting the mechanism of improved efficiencies of cellular uptake of oligonucleotide-coated CdTe QDs with negative charges. The cellular internalization of negatively charged CdTe QDs was through an energy- and lipid rafts-dependent endocytosis. The anionic CdTe QDs also followed certain endocytic pathways, due to the smaller nano-scaled protrusion and depressions of membrane compared with that of QDs, which minimized the repulsive forces of negatively charged QDs to cell membrane, and enhanced the cellular uptake of the anionic QDs, too.[172] It has also been reported that non-functionalized QDs were located in cells, and that the cellular internalization depended on the size of QDs in the case of human blood monocyte-derived macrophages.[203]

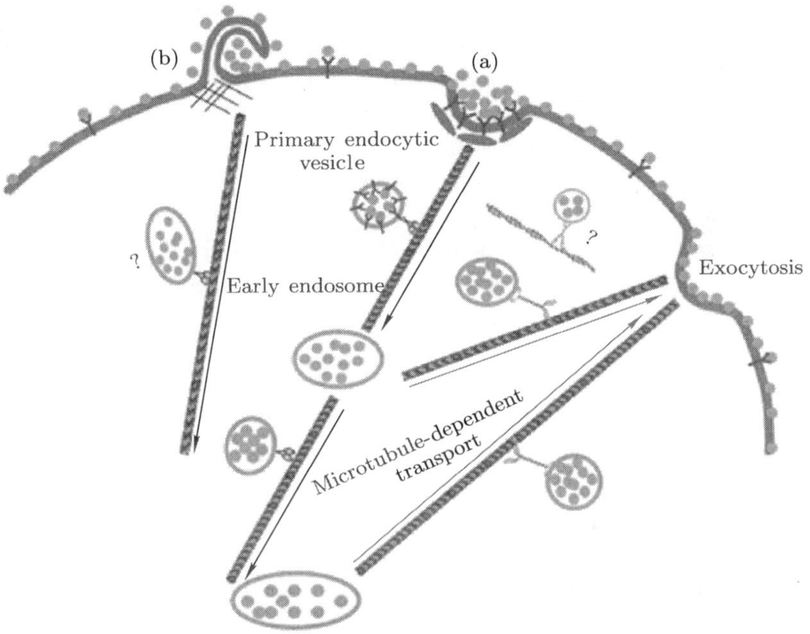

Fig. 4.14 Diagram of routes involved in DPA-QD intracelluar uptake and fate. DPA-QDs first binded to the plasma membrane, and then started internalization process mainly through (a) the clathrin-mediated endocytosis and (b) to a lesser part *via* macropinocytosis. Transportation of DPA-QDs resided in early endosomes, and then developed into late endosomes and finally lysosomes. About 50% of the internalized DPA-QDs were then exocytosed. Image is adapted with permission from Ref. 201. Copyright (2010) American Chemical Society.

Additionally, different cell lines can affect the cell uptake mechanism of QDs. For example, monocytes internalized QD655-COOH, but lymphocytes did not. Besides, monocyte-derived dendritic cells (DCs) employed receptors for clathrin and scavenger and regulated F-actin and phospholipase C taking up QD655-COOH *via* endocytic pathways, while the results were different in the human epidermal keratinocytes (HEK)[204] cellular uptake of QDs. In HEK cells, microtubules, not F-actin, influenced cellular endocytosis of QD655-COOH. In addition, lipid rafts, rather than clathrin, caveolae or macropinocytosis, were responsible for the uptake of QD655-COOH. As DCs favored the clathrin-regulated endocytic pathways of QDs, there are some other specific receptors (SR, PLC) that may influence the cellular uptake in DCs, unlike other cell lines.[205]

4.5.2 *Cytotoxicity of QDs*

The cytotoxicity is a vital factor to be considered before the biomedical applications of QDs. The unique physicochemical properties of QDs, such as particle size, composition, dosage, exposure route, surface charge, and surface coating, influence the *in vitro* and *in vivo* toxicity of QDs.[206]

In 2004, Derfus and coworkers[36] illustrated the first example of the cytotoxicity and its mechanism of CdSe QDs *in vitro*. They utilized a variety of strategies to oxidize the surface of QDs, and found that the surface modification led to reduced Cd concentration on the QD surface and thus free cadmium ions in the solution, which was responsible for the cell death. The cytotoxicity of CdSe QDs exposed to UV light would increase as a progressive release of toxic metals. CdSe QDs with modified surface of ZnS or BSA were demonstrated to largely reduce, but not eliminate, the QDs toxicity to cells by preventing surface oxidation. Therefore, it is important to point out that the surface coating of CdSe QDs surface not only determined the biological stability of QDs, but also decreased the dissolution and leakage of Cd^{2+} from CdSe cores.[36] The oxidation of QDs' surfaces also can occur through ROS generation upon UV irradiation.[207] As Cd^{2+} is the major response of the toxic effect contributed to QDs, the surface coating preventing Cd leakage is one of the most important physicochemical factors of QD toxicity. Several studies have reported that the cytotoxicity of QDs is dependent on the properties of the coating ligands. Yamamoto and coworkers[35] showed that different surface modifications to QDs with similar nanoparticle size, for example, CdSe/ZnS QDs modified with cysteamine (QD-NH_2), mercaptoundecanoic acid (MUA, QD-COOH), and thioglycerol (QD-OH) possess different cytotoxicity. The surfaces of QDs were highly negatively charged in the form of QD-COOH and QD-COOH/OH, or were positively charged in QD-NH_2 and QD-NH_2/OH. It should be noted that QD-OH was less negatively charged than QD-COOH. QD-COOH induced cell death, while QD-COOH/OH, QD-NH_2, QD-NH_2/OH and QD-OH, which are less negatively charged or positively charged, showed little DNA damage, demonstrating that the cytotoxicity of QDs is influenced by the surface coatings. Kirchner *et al.*[208] also confirmed that the cytotoxicity depended on Cd^{2+} releasing and surface-capping. They coated CdSe and CdSe/ZnS QDs, either with MPA, integrated in a silica shell or in an amphiphilic polymer shell. After a size-dependent cellular internalization, either modified QDs were found in vesicles around nucleus.

Another mechanism of QD cytotoxicity is that the photoactivation of QDs resulted in a free radical generation, such as reactive oxygen intermediates

(ROI), and consequently causing damage to DNA. It is reported that DNA was nicked after exposure to CdSe/ZnS QDs.[209] In addition, Lu *et al.*[210] also reported that CdSe QDs induced human osteoblasts apoptosis through ROI generation, inducing the activation of certain enzymes to trigger apoptotic death.

Generally, cellular lines, size, composition, surface coating and dosage of QDs affect cytotoxicity. A summary of the influence of these factors on QDs cytotoxicity is listed in Table 4.1.[30]

4.5.3 *In Vivo Fate and Toxicity of QDs*

Before exploring QDs in clinic, their possible undesirable effects should be avoided. Thus, it is essential to understand the *in vivo* toxicity of QDs, as well as long-term toxicological and pharmacokinetic effects, including the degradation, excretion, persistence and immune response of QDs.

Mancini *et al.*[211] proposed that CdSe QDs toxicity originated from hypochlorous acid and hydrogen peroxide, both produced by phagocytic cells. In their oxidative mechanism, hypochlorous acid and hydrogen peroxide diffused across polymeric secondary coatings, and subsequently dissolved the core. Also, *in vivo* oxidative mechanisms may explain how QDs degraded in animals in the presence or absence of secondary coatings.

As well, the *in vivo* toxicity of QDs is affected by their chemical composition, size, shape, aggregation, and surface coating. Interactions with blood components, accumulation and excretion in main organs, and immune responses can induce *in vivo* toxic effects of QDs. For the amphiphilic polymer-coated QDs on mice after i.v. injection, there was no necrosis observed in the tissues after 133 days.[31] Moreover, high dosages (3 600 pmol/mouse) of carboxyl-CdSe QDs led to pulmonary vascular thrombosis after i.v. injection due to that negatively charged QDs effectively activated the coagulation cascade.[212]

Biodistribution of QDs is also essential for evaluating *in vivo* effects caused by nanoparticles. Some researchers have focused on this issue to draw a more detail and clear understanding of biological effectiveness of QDs.

It is well known that the size of nanoparticles determines their circulation time. Amphiphilic poly (acrylic acid) coated QDs, methoxy PEG (750Da)-, or carboxylic-acid-functionalized PEG (3400Da)-coated QDs were cleared from blood stream after injection for 1 h, while methoxyPEG (5000Da)-coated QDs of the largest size, still remained in the circulation for more than 3 h, and then deposited in the liver, skin and bone marrow.[31] In addition, the hydrodynamic size of QDs was a significant factor for the excretion.[213] Small QDs (< 5.5 nm) would be cleared out very quickly through efficient urinary excretion, while large QDs (>15 nm) skipped the renal excretion.

If Cd-based QDs were intravenously injected into small animals, the Cd concentration in the livers and kidneys of the animals would increase in four

Table 4.1 Details of QDs cytotoxicity. Table modified from Ref. 30.

QD	Model	Administration	Concentration	Exposure duration	Cytotoxicity	Study
CdSe/ZnS-SSA	EL-4 cells	1×10^6 cells/ well	0.1–0.4 mg/mL	0–24 h	Altered cell growth at low concentration; most cells nonviable at 0.4 mg/mL	Hoshino *et al.*, 2004
		200 μL cell suspension injected (i.v.) into the mice	0.1 mg/mL QDs per 5×10^7 cells	2 h to 7 days	No toxicity in mice *in vivo*	Hoshino *et al.*, 2004 (*in vivo*)
CdSe/ZnS conjugates: NH_2, OH, OH/COOH, H_2/OH, MUA, COOH	WTK1 cells	5×10^4 cells/ mL	2 μM	12 h	DNA damage at 2 h	Hoshino *et al.*, 2004
CdSe/ZnS-MUA	Vero, HeLa, primary human hepatocytes	100 μL QDs/ 3×10^4 cells	0–0.4 mg/mL	24 h	0.1–0.2 mg/mL cytotoxic	Shiohara *et al.*, 2004
CdTe	Rat pheochromocytoma cells, murine, microglial cells	1×10^5 cells/cm^2	0.01–100 μg/mL	2–24 h	10 μg/mL cytotoxic	Lovric *et al.*, 2005
CdSe-MAA, TOPO	Primary rat hepatocytes		62.5–1 000 μg/mL	1–8 h	62.5 μg/mL cytotoxic under oxidative/photolytic conditions	Derfus, 2004

Table 4.1 (Continued).

QD	Model	Administration	Concentration	Exposure duration	Cytotoxicity	Study
QD micelles: CdSe/ZnS QDs in (PEG-PE) and phosphatydil-choline	*Xenopus* blastomeres	5×10^9 QDs/cell (~ 0.23 pmol/cell)	1.5–3 nL of 2.3 μM QDs injected, $\sim 2.1 \times 10^9$–4.2×10^9 injected QDs/cell	Days	5×10^9 QDs/cell: cell abnormalities, altered viability and motility; No toxicity at 2×10^9 QDs/cell	Dubertret *et al.*, 2002
CdSe/ZnS-DHLA	*Dictyostelium-discoideum* and HeLa cells		400–600 nM	45–60 min	No effects	Jaiswal *et al.*, 2003
Avidin-conjugated CdSe/ZnS	HeLa cells		0.5–1.0 μM	15 min	No effect on cell growth	Jaiswal *et al.*, 2003
CdSe/ZnS-DHLA	Mice, B16F10 cells	5×10^4 B16F10 cells with 10 μL QDs (~10 pmol); injection (i.v.)	100 μL, $\sim 2 \times 10^5$– 4×10^5 cells used for injection	4–6 h cell incubation Mice sacrificed at 1–6 h	No observed toxicity	Voura *et al.*, 2004
CdSe/ZnS-MUA QD-SSA complexes	Vero cells	0.4 mg/mL	0.24 mg/mL	2 h	0.4 mg/mL MUA/SSA-QD complexes did not affect cell viability	Hanaki *et al.*, 2003
CdSe/ZnS	HeLa cells	1×10^6 cells	10 pmol QDs/ 1×10^5 cells (~ 10 nM)	10 days (cell culture)	10 nM QD had minimal cytotoxic	Chen and Gerion, 2004
CdTe	HEK293 cells	1×10^5 cells	300 or 600 nM	3 days	Inhibited cell growth from the beginning	Nan Chen *et al.*, 2012

Table 4.1 (Continued).

QD	Model	Administration	Concentration	Exposure duration	Cytotoxicity	Study
CdTe-gelatinized/ nongelatinized	PC12 cells	1×10^5 cells/cm^2	1–100 nM	72 h	At 1 nM, did not initiate any detrimental effects; At 100 nM, resulted in the death of all cells	Babu R Prasad *et al.*, 2010
CdTe, CdTe/CdS, CdTe/CdS/ZnS	K562 and HEK293T human cell lines	1×10^5 cells	0.2–3.0 μM	0–48 h	CdTe and CdTe/CdS QDs induced cell death	Su *et al.*, 2009
CdSe/ZnS-PEG (EviTag T1 490 QD)	Caco-2 cells lines	10^6 cells/mL, 0.2 mL/well	0.84–105 μM	0–24 h	Low cytotoxicity; Induced cell detachment	Wang *et al.*, 2008
CdSe	Primary rat hippocampal neuron cells in culture	10^4–10^5 cells/mL	1 nM, 10 nM, and 20 nM	24 h	1 nM QD showed no effect on cells; Cells treated with 10 and 20 nM QD for 24 h showed 20% and 30% decrease in cell viability	Tang *et al.*, 2008

weeks. According to the Cd signals in major organs, kidneys contains reached 10% of the injected dose while livers had 40%, and QDs *in vivo* could degrade over time.[2] Chen *et al.*[214] also reported that silica-coated CdSe QDs showed severe accumulation in the livers, kidneys, spleens, and lungs after i.v. injection. Akerman *et al.*[103] found peptide-capped CdSe QDs were mainly in livers and spleens of mice. It is interesting to figure out that the accumulation of QDs in tissues would be regulated by specific targeting peptide, which was coated on the QD surface.

To further study the dosing effect of QDs through the short- and long-term observation in small animals, different concentrations (2.5–15.0 nmol) of CdSe-ZnS core-shell QDs were injected to healthy Sprague–Dawley rats and tracked for short-term (<7 days) study and long-term (>80 days) investigation. The QDs did not induce obvious toxicity at high doses (15 nmol), even after their degradation *in vivo* across this long-term study.[215] Another long-term study of PEG-ZnS/CdTe showed that livers and spleens were the main accumulative organs after a single i.v. injection into ICR mice. Although the plasma half-life of QDs in mice was short, QDs could stay in the spleens, livers, and kidneys for 28 days, which raised serious consideration on the toxicity of QDs.[216] Moreover, a 6-month long investigation on the disposition and pharmacokinetics of QDs in male ICR mice was carried out by Lin and coworkers.[217]After a single i.v. injection of QDs, they revealed that QDs accumulated and stayed in the spleens, livers and kidneys for more than 28 days with little degradation, and were gradually cleared in 6 months. In addition, they found mitochondrial changes in renal tubular epithelial cells on day 28 and 180 of post-exposure. Similarly, PEG-coated CdTe QDs can be eliminated from the injection sites and persisted in the regional lymph nodes of the livers, spleens and kidneys, after s.c. injection into female BALB/c mice.[218]

Wang's group[37] reported a pilot study on the biological effect of CdSe/CdS/ZnS QDs within phospholipid micelles through rhesus macaques to mimic humans. Rhesus macaques lead to both chimpanzees and humans, and thus are closer to humans, and more abundant and easily maintained. In this work, a 90-days study of QDs based on male rhesus macaques were carried out, during which 25 mg/kg QDs encapsulated in phospholipid micelles were intravenously administrated. They found that body weight and temperature of the animals had no obvious fluctuation. In addition, eating, drinking, grooming, exploratory behavior, activity, physical features, urination or neurological status of these rhesus macaques seemed normal. However, the results also showed that QDs had mainly stayed in the kidneys (84 mg/g of Ca concentration), spleens (63 mg/g) and livers (18 mg/g), which is slightly different with small animals. Selenium uptake was also very high in the livers, spleens and kidney. The ICP-MS analysis showed that QDs slowly degraded in the liver and spleen through releasing free cadmium ions which then accumulated in the kidney. No

Table 4.2 Biodistribution and *in vivo* toxicity of QDs. Table modified from Ref. 200.

QD	Model	Dose	Exposure duration	Biodistibution	Toxicity	Study
CdSe coated with: GFE, F3, LyP-1, PEG	Mice, BALB/c	100–200 μg in 0.1–0.2 mL	i.v., once	Tissue targeting, accumulated in liver and spleen	Did not observe any acute toxicity	Akerman *et al.*, 2002
CdSe coated with: poly(acrylic acid) polymer or mPEG-750, 5000, COOH-PEG-3400	Mice, BALB/c	50–500 pmol in 50–200 μL saline	i.v., once	In liver, spleen, and bone marrow; mPEG-750CdSe remained in liver, lymph nodes, and bone marrow for a month	mPEG-5000 have longer circulation time and lower accumulation	Ballou *et al.*, 2004
CdSe-MUA, CdSe- BSA	Rats, male, SD	5 mmol in 0.2 mL	i.v., once	Liver took up 40% of the QD-MUA while 99% of QD-BSA were sequestered	—	Fischer *et al.*, 2006
CdSe coated with silica	Mice, male, ICR	5 nmol/mouse in 100 μL	i.v., once	In liver, kidneys, spleen, and lungs; 33.3% and 23.8% of QDs *via* feces and urine	—	Chen *et al.*, 2008
CdSe-fetoprotein monoclonal anti-body	Mice, male, BALB/c	1.44–3 600 pmol/mouse	i.v., once	In liver, spleen, and kidneys, at 24 h	No animal died in the test group and no difference in body weight	Chen *et al.*, 2008
Carboxyl- CdSe, Amine- CdSe	Mice, male, BALB/c	—	i.v., once	In lungs, liver, and blood	Marked vascular thrombosis for carboxyl-QDs at high dose	Geys *et al.*, 2008

Table 4.2 (Continued).

QD	Model	Dose	Exposure duration	Biodistibution	Toxicity	Study
CdSe-fat	Rats, Wistar	—	orally	QDs were degraded in the digestive tract; QDs with fat were more stable and degraded more slowly than bare ones	—	Karabanovas *et al.*, 2008
CdSe-carboxylic acid	Mice, female, Fox	40 pmol	s.c., once	In axillary lymph nodes; Most of the injected QDs remained at the injection site	—	Robe *et al.*, 2008
CdTe	Rat, male	2 μg/kg	i.v., once	—	Few signs of functional toxicity or clinical changes	Zhang *et al.*, 2007
CdTe-mPEG-5000	—	40 pmol	i.v., once	In spleen, liver, and kidneys for at least 28 days; Partially eliminated after 6 months	Mitochondrial alterations in renal tubular epithelial cells	Lin *et al.*, 2008
CdTe	Mice, female, ApoE	137.5 μg	i.t., once	—	Inflammation, genotoxicity	Jacobsen *et al.*, 2009
CdTe	Mice, ICR	1 mmol/L, 10 μL	nasal, once	Located in the olfactory bulb after 30 min	—	Yin *et al.*, 2010
CdSe/ZnS-amphiphilic micelle	—	60 μM QD/g animal weight	i.v., injection	—	Mice showed no noticeable ill effects	Larson *et al.*, 2003

pathological changes were discovered in the heart, lung, kidney, liver or spleen. Besides, there were no signs of inflammatory response. Continued evaluation of QDs on two treated monkeys for one year showed no evidence of abnormality. This study suggests that the QDs encapsulated in phospholipid micelles have low toxicity in primates.

It is concluded that multiple factors, including the stability, size, surface charge, and surface coatings of QDs, determine the toxicity,*in vivo* half lifetime and biodistribution of QDs. Table 4.2 presents a summary of the fate and toxicity of QDs. To overcome the challenge of the metal iron toxicity of QDs, some research groups[219,220] are currently developing Cd-free QDs to replace traditional QDs for long term *in vitro* and *in vivo* biological applications.

4.6 Outlook

In this chapter, different types of QDs are described for their potentially biomedical applications such as imaging, sensing, targeted delivery and therapy, both *in vitro* and *in vivo*. Techniques and strategies of QDs being applied in the life sciences are also introduced, ranging from multiplexing sensing, multimodal imaging, microarray technology, and fluorescence *in situ* hybridization to targeted drug delivery and therapy. However, most of the developed solutions for using QDs are individual ones and far from markets. Meanwhile, QDs act more like nanocolloids rather than small molecules, which complicate their behaviors in biological environments and raise more safety concerns in addition to their instinct cytotoxicity. Therefore, there will be increasing interests from researchers to develop novel QDs with the bright luminescence emission and low toxicity for the applications in clinic. The research on Cd-free QDs based on $CuInS_2$, CuInSe and Ag_2S, and carbonaceous materials is a relatively new direction and deserves a special attention, though the quantum yield of these new types of QDs aforementioned is still low compared to that of II-IV QDs. The control over size and size distribution of carbon-based QDs, especially GQDs, is still a big challenge facing us. Furthermore, more advanced QDs-based theranostic systems will be developed, in which diagnosis, drug delivery and therapy will be integrated. However, no matter how studies on quantum dots are advanced, the applications of QDs in life science will still mainly lie on their distinct characteristics of excellent photoluminescence properties superior to the organic dyes, such as strong and broad absorbance, tunable and bright emission for multiplexing imaging, photostability for long-term monitoring, non-blinking properties for luminescence quantification and two-photo excitation for the clear *in vivo* imaging.

References

1. Michalet, X.; Pinaud, F. F.; Bentolila, L. A.; Tsay, J. M.; Doose, S.; Li, J. J.; Sundaresan, G.; Wu, A. M.; Gambhir, S. S.; Weiss, S. Quantum dots for live cells, *in vivo* imaging, and diagnostics. *Science* **2005**, *307* (5709), 538-544.
2. Smith, A. M.; Duan, H.; Mohs, A. M.; Nie, S. Bioconjugated quantum dots for *in vivo* molecular and cellular imaging. *Adv Drug Delivery Rev* **2008**, *60* (11), 1226-1240.
3. Zrazhevskiy, P.; Sena, M.; Gao, X. Designing multifunctional quantum dots for bioimaging, detection, and drug delivery. *Chem Soc Rev* **2010**, *39* (11), 4326-4354.
4. Sun, Y.-P.; Zhou, B.; Lin, Y.; Wang, W.; Fernando, K. A. S.; Pathak, P.; Meziani, M. J.; Harruff, B. A.; Wang, X.; Wang, H.; Luo, P. G.; Yang, H.; Kose, M. E.; Chen, B.; Veca, L. M.; Xie, S.-Y. Quantum-sized carbon dots for bright and colorful photoluminescence. *J Am Chem Soc* **2006**, *128* (24), 7756-7757.
5. Holmes, J. D.; Ziegler, K. J.; Doty, R. C.; Pell, L. E.; Johnston, K. P.; Korgel, B. A. Highly luminescent silicon nanocrystals with discrete optical transitions. *J Am Chem Soc* **2001**, *123* (16), 3743-3748.
6. Warner, J. H.; Hoshino, A.; Yamamoto, K.; Tilley, R. D. Water-soluble photoluminescent silicon quantum dots. *Angew Chem Int Ed* **2005**, *117* (29), 4626-4630.
7. Ross, F.; Tersoff, J.; Tromp, R. Coarsening of self-assembled Ge quantum dots on Si (001). *Phys Rev Lett* **1998**, *80* (5), 984-987.
8. Alivisatos, A. P.; Gu, W.; Larabell, C. Quantum dots as cellular probes. *Annu Rev Biomed Eng* **2005**, *7*, 55-76.
9. Resch-Genger, U.; Grabolle, M.; Cavaliere-Jaricot, S.; Nitschke, R.; Nann, T. Quantum dots versus organic dyes as fluorescent labels. *Nat Methods* **2008**, *5* (9), 763-775.
10. Brus, L. E. A simple model for the ionization potential, electron affinity, and aqueous redox potentials of small semiconductor crystallites. *J Chem Phys* **1983**, *79* (11), 5566-5571.
11. Bawendi, M. G.; Steigerwald, M. L.; Brus, L. E. The quantum mechanics of larger semiconductor clusters ("quantum dots"). *Annu Rev Phys Chem* **1990**, *41* (1), 477-496.
12. Efros, A. L.; Rosen, M. The electronic structure of semiconductor nanocrystals. *Annu Rev Mater Sci* **2000**, *30* (1), 475-521.
13. Efros, A. L.; Efros, A. Interband absorption of light in a semiconductor sphere. *SPIE Milestone Series* **2005**, *180*, 71-74.
14. Ekimov, A.; Onushchenko, A. Quantum size effect in the optical spectra of semiconductor microcrystals. *Sov Phys Semicond* **1982**, *16* (7), 775-778.
15. Rossetti, R.; Nakahara, S.; Brus, L. E. Quantum size effects in the redox potentials, resonance Raman spectra, and electronic spectra of CdS crystallites in aqueous solution. *J Chem Phys* **1983**, *79* (2), 1086-1088.
16. Murray, C.; Norris, D.; Bawendi, M. G. Synthesis and characterization of nearly monodisperse CdE (E= sulfur, selenium, tellurium) semiconductor nanocrystallites. *J Am Chem Soc* **1993**, *115* (19), 8706-8715.
17. Dabbousi, B. O.; Rodriguez-Viejo, J.; Mikulec, F. V.; Heine, J. R.; Mattoussi, H.; Ober, R.; Jensen, K. F.; Bawendi, M. G. (CdSe)ZnS core-shell quantum dots:

Synthesis and characterization of a size series of highly luminescent nanocrystallites. *J Phys Chem B* **1997**, *101* (46), 9463-9475.

18. Hines, M. A.; Guyot-Sionnest, P. Synthesis and characterization of strongly luminescing ZnS-capped CdSe nanocrystals. *J Phys Chem* **1996**, *100* (2), 468-471.
19. Bruchez, M.; Moronne, M.; Gin, P.; Weiss, S.; Alivisatos, A. P. Semiconductor nanocrystals as fluorescent biological labels. *Science* **1998**, *281* (5385), 2013-2016.
20. Chan, W. C.; Nie, S. Quantum dot bioconjugates for ultrasensitive nonisotopic detection. *Science* **1998**, *281* (5385), 2016-2018.
21. Dubertret, B.; Skourides, P.; Norris, D. J.; Noireaux, V.; Brivanlou, A. H.; Libchaber, A. *In vivo* imaging of quantum dots encapsulated in phospholipid micelles. *Science* **2002**, *298* (5599), 1759-1762.
22. Larson, D. R.; Zipfel, W. R.; Williams, R. M.; Clark, S. W.; Bruchez, M. P.; Wise, F. W.; Webb, W. W. Water-soluble quantum dots for multiphoton fluorescence imaging *in vivo*. *Science* **2003**, *300* (5624), 1434-1436.
23. Alivisatos, P. The use of nanocrystals in biological detection. *Nat Biotechnol* **2003**, *22* (1), 47-52.
24. Chan, W. C. W.; Maxwell, D. J.; Gao, X.; Bailey, R. E.; Han, M.; Nie, S. Luminescent quantum dots for multiplexed biological detection and imaging. *Curr Opin Biotechnol* **2002**, *13* (1), 40-46.
25. Gao, X.; Cui, Y.; Levenson, R. M.; Chung, L. W.; Nie, S. *In vivo* cancer targeting and imaging with semiconductor quantum dots. *Nat Biotechnol* **2004**, *22* (8), 969-976.
26. Wu, X.; Liu, H.; Liu, J.; Haley, K. N.; Treadway, J. A.; Larson, J. P.; Ge, N.; Peale, F.; Bruchez, M. P. Immunofluorescent labeling of cancer marker HER2 and other cellular targets with semiconductor quantum dots. *Nat Biotechnol* **2003**, *21* (1), 41-46.
27. Colton, H. M.; Falls, J. G.; Ni, H.; Kwanyuen, P.; Creech, D.; McNeil, E.; Casey, W. M.; Hamilton, G.; Cariello, N. F. Visualization and quantitation of peroxisomes using fluorescent nanocrystals: Treatment of rats and monkeys with fibrates and detection in the liver. *Toxicol Sci* **2004**, *80* (1), 183-192.
28. Beaurepaire, E.; Buissette, V.; Sauviat, M.-P.; Giaume, D.; Lahlil, K.; Mercuri, A.; Casanova, D.; Huignard, A.; Martin, J.-L.; Gacoin, T.; Boilot, J.-P.; Alexandrou, A. Functionalized fluorescent oxide nanoparticles: Artificial toxins for sodium channel targeting and imaging at the single-molecule level. *Nano Lett* **2004**, *4* (11), 2079-2083.
29. Lidke, D. S.; Nagy, P.; Heintzmann, R.; Arndt-Jovin, D. J.; Post, J. N.; Grecco, H. E.; Jares-Erijman, E. A.; Jovin, T. M. Quantum dot ligands provide new insights into erbB/HER receptor-mediated signal transduction. *Nat Biotechnol* **2004**, *22* (2), 198-203.
30. Hardman, R. A toxicologic review of quantum dots: Toxicity depends on physicochemical and environmental factors. *Environ Health Perspect* **2006**, *114* (2), 165-172.
31. Ballou, B.; Lagerholm, B. C.; Ernst, L. A.; Bruchez, M. P.; Waggoner, A. S. Noninvasive imaging of quantum dots in mice. *Bioconjugate Chem* **2004**, *15* (1), 79-86.

32. Hoshino, A.; Hanaki, K.; Suzuki, K.; Yamamoto, K. Applications of T-lymphoma labeled with fluorescent quantum dots to cell tracing markers in mouse body. *Biochem Biophys Res Commun* **2004**, *314* (1), 46-53.
33. Jaiswal, J. K.; Mattoussi, H.; Mauro, J. M.; Simon, S. M. Long-term multiple color imaging of live cells using quantum dot bioconjugates. *Nat Biotechnol* **2002**, *21* (1), 47-51.
34. Voura, E. B.; Jaiswal, J. K.; Mattoussi, H.; Simon, S. M. Tracking metastatic tumor cell extravasation with quantum dot nanocrystals and fluorescence emission-scanning microscopy. *Nat Med* **2004**, *10* (9), 993-998.
35. Hoshino, A.; Fujioka, K.; Oku, T.; Suga, M.; Sasaki, Y. F.; Ohta, T.; Yasuhara, M.; Suzuki, K.; Yamamoto, K. Physicochemical properties and cellular toxicity of nanocrystal quantum dots depend on their surface modification. *Nano Lett* **2004**, *4* (11), 2163-2169.
36. Derfus, A. M.; Chan, W. C.; Bhatia, S. N. Probing the cytotoxicity of semiconductor quantum dots. *Nano Lett* **2004**, *4* (1), 11-18.
37. Ye, L.; Yong, K.-T.; Liu, L.; Roy, I.; Hu, R.; Zhu, J.; Cai, H.; Law, W.-C.; Liu, J.; Wang, K. A pilot study in non-human primates shows no adverse response to intravenous injection of quantum dots. *Nat Nanotechnol* **2012**, *7* (7), 453-458.
38. Derfus, A. M.; Chen, A. A.; Min, D.-H.; Ruoslahti, E.; Bhatia, S. N. Targeted quantum dot conjugates for siRNA delivery. *Bioconjugate Chem* **2007**, *18* (5), 1391-1396.
39. Bagalkot, V.; Zhang, L.; Levy-Nissenbaum, E.; Jon, S.; Kantoff, P. W.; Langer, R.; Farokhzad, O. C. Quantum dot-aptamer conjugates for synchronous cancer imaging, therapy, and sensing of drug delivery based on bi-fluorescence resonance energy transfer. *Nano Lett* **2007**, *7* (10), 3065-3070.
40. Chen, A. A.; Derfus, A. M.; Khetani, S. R.; Bhatia, S. N. Quantum dots to monitor RNAi delivery and improve gene silencing. *Nucleic Acids Res* **2005**, *33* (22), e190.
41. Tan, W. B.; Jiang, S.; Zhang, Y. Quantum-dot based nanoparticles for targeted silencing of HER2/neu gene *via* RNA interference. *Biomaterials* **2007**, *28* (8), 1565-1571.
42. Akin, D.; Sturgis, J.; Ragheb, K.; Sherman, D.; Burkholder, K.; Robinson, J. P.; Bhunia, A. K.; Mohammed, S.; Bashir, R. Bacteria-mediated delivery of nanoparticles and cargo into cells. *Nat Nanotechnol* **2007**, *2* (7), 441-449.
43. Popovic, Z.; Liu, W.; Chauhan, V. P.; Lee, J.; Wong, C.; Greytak, A. B.; Insin, N.; Nocera, D. G.; Fukumura, D.; Jain, R. K.; Bawendi, M. G. A nanoparticle size series for *in vivo* fluorescence imaging. *Angew Chem Int Ed* **2010**, *49* (46), 8649-8652.
44. Wong, C.; Stylianopoulos, T.; Cui, J.; Martin, J.; Chauhan, V. P.; Jiang, W.; Popovi, Z.; Jain, R. K.; Bawendi, M. G.; Fukumura, D. Multistage nanoparticle delivery system for deep penetration into tumor tissue. *Proc Natl Acad Sci* **2011**, *108* (6), 2426-2431.
45. Chauhan, V. P.; Popovic, Z.; Chen, O.; Cui, J.; Fukumura, D.; Bawendi, M. G.; Jain, R. K. Fluorescent nanorods and nanospheres for real-time *in vivo* probing of nanoparticle shape-dependent tumor penetration. *Angew Chem Int Ed* **2011**, *50* (48), 11417-11420.

46. Chauhan, V. P.; Stylianopoulos, T.; Martin, J. D.; Popović, Z.; Chen, O.; Kamoun, W. S.; Bawendi, M. G.; Fukumura, D.; Jain, R. K. Normalization of tumour blood vessels improves the delivery of nanomedicines in a size-dependent manner. *Nat Nanotechnol* **2012**, *7* (6), 383-388.
47. Shen, H.; Zhang, L.; Liu, M.; Zhang, Z. Biomedical applications of graphene. *Theranostics* **2012**, *2* (3), 283-294.
48. Yang, K.; Feng, L.; Shi, X.; Liu, Z. Nano-graphene in biomedicine: Theranostic applications. *Chem Soc Rev* **2013**, *42* (2), 530-547.
49. Li, H.; Kang, Z.; Liu, Y.; Lee, S.-T. Carbon nanodots: Synthesis, properties and applications. *J Mater Chem* **2012**, *22* (46), 24230-24253.
50. Zhang, Z.; Zhang, J.; Chen, N.; Qu, L. Graphene quantum dots: An emerging material for energy-related applications and beyond. *Energy Environ Sci* **2012**, *5* (10), 8869-8890.
51. Yang, S.-T.; Cao, L.; Luo, P. G.; Lu, F.; Wang, X.; Wang, H.; Meziani, M. J.; Liu, Y.; Qi, G.; Sun, Y.-P. Carbon dots for optical imaging *in vivo*. *J Am Chem Soc* **2009**, *131* (32), 11308-11309.
52. Yin, Y.; Alivisatos, A. P. Colloidal nanocrystal synthesis and the organic-inorganic interface. *Nature* **2005**, *437* (7059), 664-670.
53. Zhang, H.; Wang, L.; Xiong, H.; Hu, L.; Yang, B.; Li, W. Hydrothermal synthesis for high-quality CdTe nanocrystals. *Adv Mater* **2003**, *15* (20), 1712-1715.
54. Gao, X.; Liu, X.; Lin, Z.; Liu, S.; Su, X. $CuInS_2$ quantum dots as a near-infrared fluorescent probe for detecting thrombin in human serum. *Analyst* **2012**, *137* (23), 5620-5624.
55. Karanikolos, G. N.; Alexandridis, P.; Itskos, G.; Petrou, A.; Mountziaris, T. J. Synthesis and size control of luminescent ZnSe nanocrystals by a microemulsion-gas contacting technique. *Langmuir* **2004**, *20* (3), 550-553.
56. Bera, D.; Qian, L.; Holloway, P. H. Time-evolution of photoluminescence properties of ZnO/MgO core/shell quantum dots. *J Phys D: Appl Phys* **2008**, *41* (18), 182002.
57. Zhu, J.; Palchik, O.; Chen, S.; Gedanken, A. Microwave assisted preparation of CdSe, PbSe, and $Cu_{2-x}Se$ nanoparticles. *J Phys Chem B* **2000**, *104* (31), 7344-7347.
58. Cui, R.; Liu, H.-H.; Xie, H.-Y.; Zhang, Z.-L.; Yang, Y.-R.; Pang, D.-W.; Xie, Z.-X.; Chen, B.-B.; Hu, B.; Shen, P. Living yeast cells as a controllable biosynthesizer for fluorescent quantum dots. *Adv Funct Mater* **2009**, *19* (15), 2359-2364.
59. Sukhanova, A.; Devy, J.; Venteo, L.; Kaplan, H.; Artemyev, M.; Oleinikov, V.; Klinov, D.; Pluot, M.; Cohen, J. H.; Nabiev, I. Biocompatible fluorescent nanocrystals for immunolabeling of membrane proteins and cells. *Anal Biochem* **2004**, *324* (1), 60-67.
60. Reiss, P.; Bleuse, J.; Pron, A. Highly luminescent CdSe/ZnSe core/shell nanocrystals of low size dispersion. *Nano Lett* **2002**, *2* (7), 781-784.
61. Peng, X.; Schlamp, M. C.; Kadavanich, A. V.; Alivisatos, A. P. Epitaxial growth of highly luminescent CdSe/CdS core/shell nanocrystals with photostability and electronic accessibility. *J Am Chem Soc* **1997**, *119* (30), 7019-7029.

62. Nirmal, M.; Dabbousi, B. O.; Bawendi, M. G.; Macklin, J. J.; Trautman, J. K.; Harris, T. D.; Brus, L. E. Fluorescence intermittency in single cadmium selenide nanocrystals. *Nature* **1996**, *383* (6603), 802-804.
63. Michalet, X.; Pinaud, F.; Lacoste, T. D.; Dahan, M.; Bruchez, M. P.; Alivisatos, A. P.; Weiss, S. Properties of fluorescent semiconductor nanocrystals and their application to biological labeling. *Single Mol* **2001**, *2* (4), 261-276.
64. Bharali, D. J.; Lucey, D. W.; Jayakumar, H.; Pudavar, H. E.; Prasad, P. N. Folate-receptor-mediated delivery of InP quantum dots for bioimaging using confocal and two-photon microscopy. *J Am Chem Soc* **2005**, *127* (32), 11364-11371.
65. Gao, J.; Chen, K.; Luong, R.; Bouley, D. M.; Mao, H.; Qiao, T.; Gambhir, S. S.; Cheng, Z. A novel clinically translatable fluorescent nanoparticle for targeted molecular imaging of tumors in living subjects. *Nano Lett* **2011**, *12* (1), 281-286.
66. Gao, J.; Chen, K.; Xie, R.; Xie, J.; Lee, S.; Cheng, Z.; Peng, X.; Chen, X. Ultrasmall near-infrared non-cadmium quantum dots for *in vivo* tumor imaging. *Small* **2009**, *6* (2), 256-261.
67. Yang, C.-S.; Bley, R. A.; Kauzlarich, S. M.; Lee, H. W.; Delgado, G. R. Synthesis of alkyl-terminated silicon nanoclusters by a solution route. *J Am Chem Soc* **1999**, *121* (22), 5191-5195.
68. Ahire, J. H.; Wang, Q.; Coxon, P. R.; Malhotra, G.; Brydson, R.; Chen, R.; Chao, Y. Highly luminescent and nontoxic amine-capped nanoparticles from porous silicon: Synthesis and their use in biomedical imaging. *ACS Appl Mater Interfaces* **2012**, *4* (6), 3285-3292.
69. Riabinina, D.; Durand, C.; Chaker, M.; Rosei, F. Photoluminescent silicon nanocrystals synthesized by reactive laser ablation. *Appl Phys Lett* **2006**, *88* (7), 073105-073105-3.
70. Kang, Z.; Tsang, C. H. A.; Zhang, Z.; Zhang, M.; Wong, N. B.; Zapien, J. A.; Shan, Y.; Lee, S. T. A polyoxometalate-assisted electrochemical method for silicon nanostructures preparation: From quantum dots to nanowires. *J Am Chem Soc* **2007**, *129* (17), 5326-5327.
71. Pons, T.; Pic, E.; Lequeux, N.; Cassette, E.; Bezdetnaya, L.; Guillemin, F.; Marchal, F.; Dubertret, B. Cadmium-free $CuInS_2$/ZnS quantum dots for sentinel lymph node imaging with reduced toxicity. *ACS Nano* **2010**, *4* (5), 2531-2538.
72. Cassette, E.; Pons, T.; Bouet, C.; Helle, M.; Bezdetnaya, L.; Marchal, F.; Dubertret, B. Synthesis and characterization of near-infrared Cu-In-Se/ZnS core/shell quantum dots for *in vivo* imaging. *Chem Mater* **2010**, *22* (22), 6117-6124.
73. Du, Y.; Xu, B.; Fu, T.; Cai, M.; Li, F.; Zhang, Y.; Wang, Q. Near-infrared photoluminescent Ag_2S quantum dots from a single source precursor. *J Am Chem Soc* **2010**, *132* (5), 1470-1471.
74. Mochalin, V. N.; Shenderova, O.; Ho, D.; Gogotsi, Y. The properties and applications of nanodiamonds. *Nat Nanotechnol* **2011**, *7* (1), 11-23.
75. Peng, J.; Gao, W.; Gupta, B. K.; Liu, Z.; Romero-Aburto, R.; Ge, L.; Song, L.; Alemany, L. B.; Zhan, X.; Gao, G.; Vithayathil, S. A.; Kaipparettu, B. A.; Marti, A. A.; Hayashi, T.; Zhu, J. J.; Ajayan, P. M. Graphene quantum dots derived from carbon fibers. *Nano Lett* **2012**, *12* (2), 844-849.

76. Pan, D.; Zhang, J.; Li, Z.; Wu, M. Hydrothermal route for cutting graphene sheets into blue-luminescent graphene quantum dots. *Adv Mater* **2009**, *22* (6), 734-738.
77. Han, B.; Wang, W.; Wu, H.; Fang, F.; Wang, N.; Zhang, X.; Xu, S. Polyethyleneimine modified fluorescent carbon dots and their application in cell labeling. *Colloids Surf B* **2012**, *100*, 209-214.
78. Zhu, H.; Wang, X.; Li, Y.; Wang, Z.; Yang, F.; Yang, X. Microwave synthesis of fluorescent carbon nanoparticles with electrochemiluminescence properties. *Chem Commun* **2009**, (34), 5118-5120.
79. Hezinger, A. F.; Tessmar, J.; Gopferich, A. Polymer coating of quantum dots: A powerful tool toward diagnostics and sensorics. *Eur J Pharm Biopharm* **2008**, *68* (1), 138-152.
80. Wang, J.; Han, S.; Ke, D.; Wang, R. Semiconductor quantum dots surface modification for potential cancer diagnostic and therapeutic applications. *J Nanomater* **2012**, *2012*, 1-8.
81. Clapp, A. R.; Goldman, E. R.; Mattoussi, H. Capping of CdSe-ZnS quantum dots with DHLA and subsequent conjugation with proteins. *Nat Protoc* **2006**, *1* (3), 1258-1266.
82. Kim, S.; Bawendi, M. G. Oligomeric ligands for luminescent and stable nanocrystal quantum dots. *J Am Chem Soc* **2003**, *125* (48), 14652-14653.
83. Aldana, J.; Lavelle, N.; Wang, Y.; Peng, X. Size-dependent dissociation pH of thiolate ligands from cadmium chalcogenide nanocrystals. *J Am Chem Soc* **2005**, *127* (8), 2496-2504.
84. Fan, H.; Leve, E. W.; Scullin, C.; Gabaldon, J.; Tallant, D.; Bunge, S.; Boyle, T.; Wilson, M. C.; Brinker, C. J. Surfactant-assisted synthesis of water-soluble and biocompatible semiconductor quantum dot micelles. *Nano Lett* **2005**, *5* (4), 645-648.
85. Pellegrino, T.; Manna, L.; Kudera, S.; Liedl, T.; Koktysh, D.; Rogach, A. L.; Keller, S.; Rädler, J.; Natile, G.; Parak, W. J. Hydrophobic nanocrystals coated with an amphiphilic polymer shell: A general route to water soluble nanocrystals. *Nano Lett* **2004**, *4* (4), 703-707.
86. Yu, W. W.; Chang, E.; Falkner, J. C.; Zhang, J.; Al-Somali, A. M.; Sayes, C. M.; Johns, J.; Drezek, R.; Colvin, V. L. Forming biocompatible and nonaggregated nanocrystals in water using amphiphilic polymers. *J Am Chem Soc* **2007**, *129* (10), 2871-2879.
87. Luccardini, C.; Tribet, C.; Vial, F.; Marchi-Artzner, V.; Dahan, M. Size, charge, and interactions with giant lipid vesicles of quantum dots coated with an amphiphilic macromolecule. *Langmuir* **2006**, *22* (5), 2304-2310.
88. Bhang, S. H.; Won, N.; Lee, T.-J.; Jin, H.; Nam, J.; Park, J.; Chung, H.; Park, H.-S.; Sung, Y.-E.; Hahn, S. K. Hyaluronic acid-quantum dot conjugates for *in vivo* lymphatic vessel imaging. *ACS Nano* **2009**, *3* (6), 1389-1398.
89. Clarke, S. J.; Hollmann, C. A.; Zhang, Z.; Suffern, D.; Bradforth, S. E.; Dimitrijevic, N. M.; Minarik, W. G.; Nadeau, J. L. Photophysics of dopamine-modified quantum dots and effects on biological systems. *Nat Mater* **2006**, *5* (5), 409-417.
90. Cai, W.; Shin, D.-W.; Chen, K.; Gheysens, O.; Cao, Q.; Wang, S. X.; Gambhir, S. S.; Chen, X. Peptide-labeled near-infrared quantum dots for imaging tumor vasculature in living subjects. *Nano Lett* **2006**, *6* (4), 669-676.

91. Cai, W.; Chen, X. Preparation of peptide-conjugated quantum dots for tumor vasculature-targeted imaging. *Nat Protoc* **2008**, *3* (1), 89-96.
92. Whaley, S. R.; English, D. S.; Hu, E. L.; Barbara, P. F.; Belcher, A. M. Selection of peptides with semiconductor binding specificity for directed nanocrystal assembly. *Nature* **2000**, *405* (6787), 665-668.
93. Xu, J.; Ruchala, P.; Ebenstain, Y.; Li, J. J.; Weiss, S. Stable, compact, bright biofunctional quantum dots with improved peptide coating. *J Phys Chem B* **2012**, *116* (36), 11370-11378.
94. Choi, Y.; Kim, H. P.; Hong, S. M.; Ryu, J. Y.; Han, S. J.; Song, R. *In situ* visualization of gene expression using polymer-coated quantum-dot-DNA conjugates. *Small* **2009**, *5* (18), 2085-2091.
95. Medintz, I. L.; Berti, L.; Pons, T.; Grimes, A. F.; Douglas, S.; Alessandrini, A.; Facci, P.; Mattoussi, H. A reactive peptidic linker for self-assembling hybrid quantum dot-DNA bioconjugates. *Nano Lett* **2007**, *7* (6), 1741-1748.
96. Li, Z.; Huang, P.; He, R.; Lin, J.; Yang, S.; Zhang, X.; Ren, Q.; Cui, D. Aptamer-conjugated dendrimer-modified quantum dots for cancer cell targeting and imaging. *Mater Lett* **2010**, *64* (3), 375-378.
97. Hanaki, K.; Momo, A.; Oku, T.; Komoto, A.; Maenosono, S.; Yamaguchi, Y.; Yamamoto, K. Semiconductor quantum dot/albumin complex is a long-life and highly photostable endosome marker. *Biochem Biophys Res Commun* **2003**, *302* (3), 496-501.
98. Mattoussi, H.; Mauro, J. M.; Goldman, E. R.; Anderson, G. P.; Sundar, V. C.; Mikulec, F. V.; Bawendi, M. G. Self-assembly of CdSe-ZnS quantum dot bioconjugates using an engineered recombinant protein. *J Am Chem Soc* **2000**, *122* (49), 12142-12150.
99. Xing, Y.; Chaudry, Q.; Shen, C.; Kong, K. Y.; Zhau, H. E.; Chung, L. W.; Petros, J. A.; O'Regan, R. M.; Yezhelyev, M. V.; Simons, J. W. Bioconjugated quantum dots for multiplexed and quantitative immunohistochemistry. *Nat Protoc* **2007**, *2* (5), 1152-1165.
100. Tada, H.; Higuchi, H.; Wanatabe, T. M.; Ohuchi, N. *In vivo* real-time tracking of single quantum dots conjugated with monoclonal anti-HER2 antibody in tumors of mice. *Cancer Res* **2007**, *67* (3), 1138-1144.
101. Goldman, E. R.; Anderson, G. P.; Tran, P. T.; Mattoussi, H.; Charles, P. T.; Mauro, J. M. Conjugation of luminescent quantum dots with antibodies using an engineered adaptor protein to provide new reagents for fluoroimmunoassays. *Anal Chem* **2002**, *74* (4), 841-847.
102. Kuo, Y.-C.; Wang, Q.; Ruengruglikit, C.; Yu, H.; Huang, Q. Antibody-conjugated CdTe quantum dots for Escherichia coli detection. *J Phys Chem C* **2008**, *112* (13), 4818-4824.
103. Åkerman, M. E.; Chan, W. C.; Laakkonen, P.; Bhatia, S. N.; Ruoslahti, E. Nanocrystal targeting *in vivo*. *Natl Acad Sci* **2002**, *99* (20), 12617-12621.
104. Tokumasu, F.; Dvorak, J. Development and application of quantum dots for immunocytochemistry of human erythrocytes. *J Microsc* **2003**, *211* (3), 256-261.
105. Pathak, S.; Choi, S.-K.; Arnheim, N.; Thompson, M. E. Hydroxylated quantum dots as luminescent probes for *in situ* hybridization. *J Am Chem Soc* **2001**, *123* (17), 4103-4104.

106. Xiao, Y.; Barker, P. E. Semiconductor nanocrystal probes for human metaphase chromosomes. *Nucleic Acids Res* **2004**, *32* (3), e28.
107. Bentolila, L. A.; Weiss, S. Single-step multicolor fluorescence *in situ* hybridization using semiconductor quantum dot-DNA conjugates. *Cell Biochem Biophys* **2006**, *45* (1), 59-70.
108. Byers, R. J.; Di Vizio, D.; O'Connell, F.; Tholouli, E.; Levenson, R. M.; Gossard, K.; Twomey, D.; Yang, Y.; Benedettini, E.; Rose, J.; Ligon, K. L.; Finn, S. P.; Golub, T. R.; Loda, M. Semiautomated multiplexed quantum dot-based *in situ* hybridization and spectral deconvolution. *J Mol Diagn* **2007**, *9* (1), 20-29.
109. True, L. D.; Gao, X. Quantum dots for molecular pathology: Their time has arrived. *J Mol Diagn* **2007**, *9* (1), 7-11.
110. Jain, K. K. Personalised medicine for cancer: From drug development into clinical practice. *Expert Opin Pharmacother* **2005**, *6* (9), 1463-1476.
111. Tholouli, E.; Hoyland, J. A.; Di Vizio, D.; O'Connell, F.; MacDermott, S. A.; Twomey, D.; Levenson, R.; Yin, J. A. L.; Golub, T. R.; Loda, M.; Byers, R. Imaging of multiple mRNA targets using quantum dot based *in situ* hybridization and spectral deconvolution in clinical biopsies. *Biophys Res Commun* **2006**, *348* (2), 628-636.
112. Chan, P.; Yuen, T.; Ruf, F.; Gonzalez-Maeso, J.; Sealfon, S. C. Method for multiplex cellular detection of mRNAs using quantum dot fluorescent *in situ* hybridization. *Nucleic Acids Res* **2005**, *33* (18), e161.
113. Ghazani, A. A.; Lee, J. A.; Klostranec, J.; Xiang, Q.; Dacosta, R. S.; Wilson, B. C.; Tsao, M. S.; Chan, W. C. High throughput quantification of protein expression of cancer antigens in tissue microarray using quantum dot nanocrystals. *Nano Lett* **2006**, *6* (12), 2881-2886.
114. Yezhelyev, M. V.; Al-Hajj, A.; Morris, C.; Marcus, A. I.; Liu, T.; Lewis, M.; Cohen, C.; Zrazhevskiy, P.; Simons, J. W.; Rogatko, A.; Nie, S.; Gao, X.; O'Regan, R. M. *In situ* molecular profiling of breast cancer biomarkers with multicolor quantum dots. *Adv Mater* **2007**, *19* (20), 3146-3151.
115. Louie, A. Multimodality imaging probes: Design and challenges. *Chem Rev* **2010**, *110* (5), 3146-3195.
116. Selvan, S. T.; Patra, P. K.; Ang, C. Y.; Ying, J. Y. Synthesis of silica-coated semiconductor and magnetic quantum dots and their use in the imaging of live cells. *Angew Chem Int Ed* **2007**, *46* (14), 2448-2452.
117. Wang, S.; Jarrett, B. R.; Kauzlarich, S. M.; Louie, A. Y. Core/shell quantum dots with high relaxivity and photoluminescence for multimodality imaging. *J Am Chem Soc* **2007**, *129* (13), 3848-3856.
118. Mulder, W. J. M.; Koole, R.; Brandwijk, R. J.; Storm, G.; Chin, P. T. K.; Strijkers, G. J.; de Mello Donegá, C.; Nicolay, K.; Griffioen, A. W. Quantum dots with a paramagnetic coating as a bimodal molecular imaging probe. *Nano Lett* **2005**, *6* (1), 1-6.
119. Cambi, A.; Lidke, D. S.; Arndt-Jovin, D. J.; Figdor, C. G.; Jovin, T. M. Ligand-conjugated quantum dots monitor antigen uptake and processing by dendritic cells. *Nano Lett* **2007**, *7* (4), 970-977.
120. Dahan, M.; Levi, S.; Luccardini, C.; Rostaing, P.; Riveau, B.; Triller, A. Diffusion dynamics of glycine receptors revealed by single-quantum dot tracking. *Sci Signal* **2003**, *302* (5644), 442-445.

121. Heine, M.; Groc, L.; Frischknecht, R.; Béïque, J.-C.; Lounis, B.; Rumbaugh, G.; Huganir, R. L.; Cognet, L.; Choquet, D. Surface mobility of postsynaptic AMPARs tunes synaptic transmission. *Sci Signal* **2008**, *320* (5873), 201-205.
122. Cui, B.; Wu, C.; Chen, L.; Ramirez, A.; Bearer, E. L.; Li, W.-P.; Mobley, W. C.; Chu, S. One at a time, live tracking of NGF axonal transport using quantum dots. *Proc Natl Acad Sci* **2007**, *104* (34), 13666-13671.
123. Zhang, Q.; Li, Y.; Tsien, R. W. The dynamic control of kiss-and-run and vesicular reuse probed with single nanoparticles. *Science* **2009**, *323* (5920), 1448-1453.
124. Roullier, V.; Clarke, S.; You, C.; Pinaud, F.; Gouzer, G.; Schaible, D.; Marchi-Artzner, V.; Piehler, J.; Dahan, M. High-affinity labeling and tracking of individual histidine-tagged proteins in live cells using Ni^{2+} tris-nitrilotriacetic acid quantum dot conjugates. *Nano Lett* **2009**, *9* (3), 1228-1234.
125. Delehanty, J.; Mattoussi, H.; Medintz, I. Delivering quantum dots into cells: Strategies, progress and remaining issues. *Anal Bioanal Chem* **2009**, *393* (4), 1091-1105.
126. Yum, K.; Na, S.; Xiang, Y.; Wang, N.; Yu, M.-F. Mechanochemical delivery and dynamic tracking of fluorescent quantum dots in the cytoplasm and nucleus of living cells. *Nano Lett* **2009**, *9* (5), 2193-2198.
127. Park, S.; Kim, Y.-S.; Kim, W. B.; Jon, S. Carbon nanosyringe array as a platform for intracellular delivery. *Nano Lett* **2009**, *9* (4), 1325-1329.
128. Zhou, M.; Ghosh, I. Quantum dots and peptides: A bright future together. *Pept Sci* **2007**, *88* (3), 325-339.
129. Delehanty, J. B.; Medintz, I. L.; Pons, T.; Brunel, F. M.; Dawson, P. E.; Mattoussi, H. Self-assembled quantum dot-peptide bioconjugates for selective intracellular delivery. *Bioconjugate Chem* **2006**, *17* (4), 920-927.
130. Rozenzhak, S. M.; Kadakia, M. P.; Caserta, T. M.; Westbrook, T. R.; Stone, M. O.; Naik, R. R. Cellular internalization and targeting of semiconductor quantum dots. *Chem Commun* **2005**, (17), 2217-2219.
131. Lagerholm, B. C.; Wang, M.; Ernst, L. A.; Ly, D. H.; Liu, H.; Bruchez, M. P.; Waggoner, A. S. Multicolor coding of cells with cationic peptide coated quantum dots. *Nano Lett* **2004**, *4* (10), 2019-2022.
132. Silver, J.; Ou, W. Photoactivation of quantum dot fluorescence following endocytosis. *Nano Lett* **2005**, *5* (7), 1445-1449.
133. Qi, L.; Gao, X. Emerging application of quantum dots for drug delivery and therapy. *Expert Opin Drug Del* **2008**, *5* (3), 263-267.
134. Courty, S.; Luccardini, C.; Bellaiche, Y.; Cappello, G.; Dahan, M. Tracking individual kinesin motors in living cells using single quantum-dot imaging. *Nano Lett* **2006**, *6* (7), 1491-1495.
135. Kim, B. Y. S.; Jiang, W.; Oreopoulos, J.; Yip, C. M.; Rutka, J. T.; Chan, W. C. W. Biodegradable quantum dot nanocomposites enable live cell labeling and imaging of cytoplasmic targets. *Nano Lett* **2008**, *8* (11), 3887-3892.
136. Duan, H.; Nie, S. Cell-penetrating quantum dots based on multivalent and endosome-disrupting surface coatings. *J Am Chem Soc* **2007**, *129* (11), 3333-3338.
137. Yezhelyev, M. V.; Qi, L.; O'Regan, R. M.; Nie, S.; Gao, X. Proton-sponge coated quantum dots for siRNA delivery and intracellular imaging. *J Am Chem Soc* **2008**, *130* (28), 9006-9012.

138. McGrath, N.; Barroso, M. Quantum dots as fluorescence resonance energy transfer donors in cells. *J Biomed Opt* **2008**, *13* (3), 031210.

139. Clapp, A. R.; Pons, T.; Medintz, I. L.; Delehanty, J. B.; Melinger, J. S.; Tiefenbrunn, T.; Dawson, P. E.; Fisher, B. R.; O'Rourke, B.; Mattoussi, H. Two-photon excitation of quantum-dot-based fluorescence resonance energy transfer and its applications. *Adv Mater* **2007**, *19* (15), 1921-1926.

140. Weissleder, R. A clearer vision for *in vivo* imaging. *Nat Biotechnol* **2001**, *19* (4), 316-317.

141. Frangioni, J. V. *In vivo* near-infrared fluorescence imaging. *Curr Opin Chem Biol* **2003**, *7* (5), 626-634.

142. Kim, S.; Lim, Y. T.; Soltesz, E. G.; De Grand, A. M.; Lee, J.; Nakayama, A.; Parker, J. A.; Mihaljevic, T.; Laurence, R. G.; Dor, D. M. Near-infrared fluorescent type Ⅱ quantum dots for sentinel lymph node mapping. *Nat Biotechnol* **2003**, *22* (1), 93-97.

143. Rosenblum, L. T.; Kosaka, N.; Mitsunaga, M.; Choyke, P. L.; Kobayashi, H. *In vivo* molecular imaging using nanomaterials: General *in vivo* characteristics of nano-sized reagents and applications for cancer diagnosis. *Mol Membr Biol* **2010**, *27* (7), 274-285.

144. Jain, R.; Dandekar, P.; Patravale, V. Diagnostic nanocarriers for sentinel lymph node imaging. *J Controlled Release* **2009**, *138* (2), 90-102.

145. Stroh, M.; Zimmer, J. P.; Duda, D. G.; Levchenko, T. S.; Cohen, K. S.; Brown, E. B.; Scadden, D. T.; Torchilin, V. P.; Bawendi, M. G.; Fukumura, D. Quantum dots spectrally distinguish multiple species within the tumor milieu *in vivo*. *Nat Med* **2005**, *11* (6), 678-682.

146. Kim, J. V.; Kang, S. S.; Dustin, M. L.; McGavern, D. B. Myelomonocytic cell recruitment causes fatal CNS vascular injury during acute viral meningitis. *Nature* **2009**, *457* (7226), 191-195.

147. Zimmer, J. P.; Kim, S.-W.; Ohnishi, S.; Tanaka, E.; Frangioni, J. V.; Bawendi, M. G. Size series of small indium arsenide-zinc selenide core-shell nanocrystals and their application to *in vivo* imaging. *J Am Chem Soc* **2006**, *128* (8), 2526-2527.

148. Ballou, B.; Ernst, L. A.; Andreko, S.; Harper, T.; Fitzpatrick, J. A.; Waggoner, A. S.; Bruchez, M. P. Sentinel lymph node imaging using quantum dots in mouse tumor models. *Bioconjugate Chem* **2007**, *18* (2), 389-396.

149. Soltesz, E. G.; Kim, S.; Laurence, R. G.; DeGrand, A. M.; Parungo, C. P.; Dor, D. M.; Cohn, L. H.; Bawendi, M. G.; Frangioni, J. V.; Mihaljevic, T. Intraoperative sentinel lymph node mapping of the lung using near-infrared fluorescent quantum dots. *Ann Thorac Surg* **2005**, *79* (1), 269-277.

150. Parungo, C. P.; Colson, Y. L.; Kim, S.-W.; Kim, S.; Cohn, L. H.; Bawendi, M. G.; Frangioni, J. V. Sentinel lymph node mapping of the pleural space. *CHEST Journal* **2005**, *127* (5), 1799-1804.

151. Parungo, C. P.; Ohnishi, S.; Kim, S.-W.; Kim, S.; Laurence, R. G.; Soltesz, E. G.; Chen, F. Y.; Colson, Y. L.; Cohn, L. H.; Bawendi, M. G. Intraoperative identification of esophageal sentinel lymph nodes with near-infrared fluorescence imaging. *J Thorac Cardiov Sur* **2005**, *129* (4), 844-850.

152. Kobayashi, H.; Hama, Y.; Koyama, Y.; Barrett, T.; Regino, C. A.; Urano, Y.;

Choyke, P. L. Simultaneous multicolor imaging of five different lymphatic basins using quantum dots. *Nano Lett* **2007**, *7* (6), 1711-1716.

153. Kosaka, N.; Ogawa, M.; Sato, N.; Choyke, P. L.; Kobayashi, H. *In vivo* real-time, multicolor, quantum dot lymphatic imaging. *J Invest Dermatol* **2009**, *129* (12), 2818-2822.
154. Lim, Y. T.; Kim, S.; Nakayama, A.; Stott, N. E.; Bawendi, M. G.; Frangioni, J. V. Selection of quantum dot wavelengths for biomedical assays and imaging. *Mol Imaging* **2003**, *2* (1), 50-64.
155. Smith, J. D.; Fisher, G. W.; Waggoner, A. S.; Campbell, P. G. The use of quantum dots for analysis of chick CAM vasculature. *Microvasc Res* **2007**, *73* (2), 75-83.
156. Jayagopal, A.; Russ, P. K.; Haselton, F. R. Surface engineering of quantum dots for *in vivo* vascular imaging. *Bioconjugate Chem* **2007**, *18* (5), 1424-1433.
157. Santra, S.; Yang, H.; Holloway, P. H.; Stanley, J. T.; Mericle, R. A. Synthesis of water-dispersible fluorescent, radio-opaque, and paramagnetic CdS:Mn/ZnS quantum dots: A multifunctional probe for bioimaging. *J Am Chem Soc* **2005**, *127* (6), 1656-1657.
158. Rosen, A. B.; Kelly, D. J.; Schuldt, A. J.; Lu, J.; Potapova, I. A.; Doronin, S. V.; Robichaud, K. J.; Robinson, R. B.; Rosen, M. R.; Brink, P. R. Finding fluorescent needles in the cardiac haystack: Tracking human mesenchymal stem cells labeled with quantum dots for quantitative *in vivo* three-dimensional fluorescence analysis. *Stem Cells* **2007**, *25* (8), 2128-2138.
159. Celso, C. L.; Fleming, H. E.; Wu, J. W.; Zhao, C. X.; Miake-Lye, S.; Fujisaki, J.; Côté, D.; Rowe, D. W.; Lin, C. P.; Scadden, D. T. Live-animal tracking of individual haematopoietic stem/progenitor cells in their niche. *Nature* **2008**, *457* (7225), 92-96.
160. Yu, X.; Chen, L.; Li, K.; Li, Y.; Xiao, S.; Luo, X.; Liu, J.; Zhou, L.; Deng, Y.; Pang, D. Immunofluorescence detection with quantum dot bioconjugates for hepatoma *in vivo*. *J Biomed Opt* **2007**, *12* (1), 014008-014008-5.
161. Smith, A. M.; Mancini, M. C.; Nie, S. Bioimaging: Second window for *in vivo* imaging. *Nat Nanotechnol* **2009**, *4* (11), 710-711.
162. Bakueva, L.; Gorelikov, I.; Musikhin, S.; Zhao, X. S.; Sargent, E. H.; Kumacheva, E. PbS quantum dots with stable efficient luminescence in the near-IR spectral range. *Adv Mater* **2004**, *16* (11), 926-929.
163. Hong, G.; Robinson, J. T.; Zhang, Y.; Diao, S.; Antaris, A. L.; Wang, Q.; Dai, H. *In vivo* fluorescence imaging with Ag_2S quantum dots in the second near-infrared region. *Angew Chem Int Ed* **2012**, *124* (39), 9956-9959.
164. Yang, Z.; Wang, H.; Zhao, J.; Peng, Y.; Wang, J.; Guinn, B.; Huang, L. Recent developments in the use of adenoviruses and immunotoxins in cancer gene therapy. *Cancer Gene Ther* **2007**, *14* (7), 599-615.
165. Nuhn, L.; Hirsch, M.; Krieg, B.; Koynov, K.; Fischer, K.; Schmidt, M.; Helm, M.; Zentel, R. Cationic nanohydrogel particles as potential siRNA carriers for cellular delivery. *ACS Nano* **2012**, *6* (3), 2198-2214.
166. Kim, D. H.; Rossi, J. J. Strategies for silencing human disease using RNA interference. *Nat Rev Genet* **2007**, *8* (3), 173-184.
167. Novina, C. D.; Sharp, P. A. The RNAi revolution. *Nature* **2004**, *430* (6996), 161-164.

168. Naldini, L.; Blömer, U.; Gallay, P.; Ory, D.; Mulligan, R.; Gage, F. H.; Verma, I. M.; Trono, D. *In vivo* gene delivery and stable transduction of nondividing cells by a lentiviral vector. *Science* **1996**, *272* (5259), 263-267.
169. Derfus, A. M.; Chan, W. C.; Bhatia, S. N. Intracellular delivery of quantum dots for live cell labeling and organelle tracking. *Adv Mater* **2004**, *16* (12), 961-966.
170. Srinivasan, C.; Lee, J.; Papadimitrakopoulos, F.; Silbart, L. K.; Zhao, M.; Burgess, D. J. Labeling and intracellular tracking of functionally active plasmid DNA with semiconductor quantum dots. *Mol Ther* **2006**, *14* (2), 192-201.
171. Shao, D.; Zeng, Q.; Fan, Z.; Li, J.; Zhang, M.; Zhang, Y.; Li, O.; Chen, L.; Kong, X.; Zhang, H. Monitoring HSV-TK/ganciclovir cancer suicide gene therapy using CdTe/CdS core/shell quantum dots. *Biomaterials* **2012**, *33* (17), 4336-4344.
172. Li, Y.; Duan, X.; Jing, L.; Yang, C.; Qiao, R.; Gao, M. Quantum dot-antisense oligonucleotide conjugates for multifunctional gene transfection, mRNA regulation, and tracking of biological processes. *Biomaterials* **2011**, *32* (7), 1923-1931.
173. He, X.; Wang, K.; Tan, W.; Liu, B.; Lin, X.; He, C.; Li, D.; Huang, S.; Li, J. Bioconjugated nanoparticles for DNA protection from cleavage. *J Am Chem Soc* **2003**, *125* (24), 7168-7169.
174. Conner, S. D.; Schmid, S. L. Regulated portals of entry into the cell. *Nature* **2003**, *422* (6927), 37-44.
175. Bardi, G.; Malvindi, M. A.; Gherardini, L.; Costa, M.; Pompa, P. P.; Cingolani, R.; Pizzorusso, T. The biocompatibility of amino functionalized CdSe/ZnS quantum-dot-doped SiO_2 nanoparticles with primary neural cells and their gene carrying performance. *Biomaterials* **2010**, *31* (25), 6555-6566.
176. Li, J. M.; Zhao, M. X.; Su, H.; Wang, Y. Y.; Tan, C. P.; Ji, L. N.; Mao, Z. W. Multifunctional quantum-dot-based siRNA delivery for HPV18 E6 gene silence and intracellular imaging. *Biomaterials* **2011**, *32* (31), 7978-7987.
177. Wu, Y.; Eisele, K.; Doroshenko, M.; Algara-Siller, G.; Kaiser, U.; Koynov, K.; Weil, T. A quantum dot photoswitch for DNA detection, gene transfection, and live-cell imaging. *Small* **2012**, *8* (22), 3465-3475.
178. Walther, C.; Meyer, K.; Rennert, R.; Neundorf, I. Quantum dot-carrier peptide conjugates suitable for imaging and delivery applications. *Bioconjugate Chem* **2008**, *19* (12), 2346-2356.
179. Qi, L.; Gao, X. Quantum dot-amphipol nanocomplex for intracellular delivery and real-time imaging of siRNA. *ACS Nano* **2008**, *2* (7), 1403-1410.
180. Yang, H. N.; Park, J. S.; Woo, D. G.; Jeon, S. Y.; Park, K. H. Transfection of VEGF(165) genes into endothelial progenitor cells and *in vivo* imaging using quantum dots in an ischemia hind limb model. *Biomaterials* **2012**, *33* (33), 8670-8684.
181. Fadini, G. P.; Losordo, D.; Dimmeler, S. Critical reevaluation of endothelial progenitor cell phenotypes for therapeutic and diagnostic use. *Circ Res* **2012**, *110* (4), 624-637.
182. Lockard, M. M.; Witkowski, S.; Jenkins, N. T.; Spangenburg, E. E.; Obisesan, T. O.; Hagberg, J. M. Thrombin and exercise similarly influence expression of cell cycle genes in cultured putative endothelial progenitor cells. *J Appl Physiol* **2010**, *108* (6), 1682-1690.
183. Petrova, T. V.; Makinen, T.; Alitalo, K. Signaling *via* vascular endothelial growth factor receptors. *Exp Cell Res* **1999**, *253* (1), 117-130.

184. Adeli, M.; Hakimpoor, F.; Parsamanesh, M.; Kalantari, M.; Sobhani, Z.; Attyabi, F. Quantum dot-pseudopolyrotaxane supramolecules as anticancer drug delivery systems. *Polymer* **2011**, *52* (11), 2401-2413.
185. Wu, W.; Aiello, M.; Zhou, T.; Berliner, A.; Banerjee, P.; Zhou, S. *In-situ* immobilization of quantum dots in polysaccharide-based nanogels for integration of optical pH-sensing, tumor cell imaging, and drug delivery. *Biomaterials* **2010**, *31* (11), 3023-3031.
186. Zhou, Y.; Shi, L.; Li, Q.; Jiang, H.; Lv, G.; Zhao, J.; Wu, C.; Selke, M.; Wang, X. Imaging and inhibition of multi-drug resistance in cancer cells *via* specific association with negatively charged CdTe quantum dots. *Biomaterials* **2010**, *31* (18), 4958-4963.
187. Savla, R.; Taratula, O.; Garbuzenko, O.; Minko, T. Tumor targeted quantum dot-mucin 1 aptamer-doxorubicin conjugate for imaging and treatment of cancer. *J Controlled Release* **2011**, *153* (1), 16-22.
188. Chakravarthy, K. V.; Davidson, B. A.; Helinski, J. D.; Ding, H.; Law, W.-C.; Yong, K.-T.; Prasad, P. N.; Knight, P. R. Doxorubicin-conjugated quantum dots to target alveolar macrophages and inflammation. *Nanomed Nanotechnol Biol Med* **2011**, *7* (1), 88-96.
189. Muhammad, F.; Guo, M.; Guo, Y.; Qi, W.; Qu, F.; Sun, F.; Zhao, H.; Zhu, G. Acid degradable ZnO quantum dots as a platform for targeted delivery of an anticancer drug. *J Mater Chem* **2011**, *21* (35), 13406-13412.
190. Yuan, Q.; Hein, S.; Misra, R. D. K. New generation of chitosan-encapsulated ZnO quantum dots loaded with drug: Synthesis, characterization and *in vitro* drug delivery response. *Acta Biomater* **2010**, *6* (7), 2732-2739.
191. Mathew, M. E.; Mohan, J. C.; Manzoor, K.; Nair, S.; Tamura, H.; Jayakumar, R. Folate conjugated carboxymethyl chitosan-manganese doped zinc sulphide nanoparticles for targeted drug delivery and imaging of cancer cells. *Carbohydr Polym* **2010**, *80* (2), 442-448.
192. Hsu, J.-C.; Huang, C.-C.; Ou, K.-L.; Lu, N.; Mai, F.-D.; Chen, J.-K.; Chang, J.-Y. Silica nanohybrids integrated with $CuInS_2$/ZnS quantum dots and magnetite nanocrystals: Multifunctional agents for dual-modality imaging and drug delivery. *J Mater Chem* **2011**, *21* (48), 19257-19266.
193. Morris, R. L.; Azizuddin, K.; Lam, M.; Berlin, J.; Nieminen, A. L.; Kenney, M. E.; Samia, A. C.; Burda, C.; Oleinick, N. L. Fluorescence resonance energy transfer reveals a binding site of a photosensitizer for photodynamic therapy. *Cancer Res* **2003**, *63* (17), 5194-5197.
194. Allen, C. M.; Sharman, W. M.; Van Lier, J. E. Current status of phthalocyanines in the photodynamic therapy of cancer. *J Porphyrins Phthalocyanines* **2001**, *05* (02), 161-169.
195. Samia, A. C. S.; Chen, X.; Burda, C. Semiconductor quantum dots for photodynamic therapy. *J Am Chem Soc* **2003**, *125* (51), 15736-15737.
196. Bakalova, R.; Ohba, H.; Zhelev, Z.; Nagase, T.; Jose, R.; Ishikawa, M.; Baba, Y. Quantum dot anti-CD conjugates: Are they potential photosensitizers or potentiators of classical photosensitizing agents in photodynamic therapy of cancer? *Nano Lett* **2004**, *4* (9), 1567-1573.
197. Shi, L.; Hernandez, B.; Selke, M. Singlet oxygen generation from water-soluble quantum dot-organic dye nanocomposites. *J Am Chem Soc* **2006**, *128* (19),

6278-6279.

198. Tsay, J. M.; Trzoss, M.; Shi, L.; Kong, X.; Selke, M.; Jung, M. E.; Weiss, S. Singlet oxygen production by peptide-coated quantum dot-photosensitizer conjugates. *J Am Chem Soc* **2007**, *129* (21), 6865-6871.
199. Roco, M. C. Nanotechnology: Convergence with modern biology and medicine. *Curr Opin Biotechnol* **2003**, *14* (3), 337-346.
200. Li, Y. F.; Chen, C. Fate and toxicity of metallic and metal-containing nanoparticles for biomedical applications. *Small* **2011**, *7* (21), 2965-2980.
201. Jiang, X.; Röcker, C.; Hafner, M.; Brandholt, S.; Dörlich, R. M.; Nienhaus, G. U. Endo-and exocytosis of zwitterionic quantum dot nanoparticles by live HeLa cells. *ACS Nano* **2010**, *4* (11), 6787-6797.
202. Yan, X.; Samuel, F.; Xiugong, G.; David, H. R.; William, T.; Alessandro, T. Dynamics and mechanisms of quantum dot nanoparticle cellular uptake. *J Nanobiotechnology* **2010**, *8*, 13-13-9.
203. Nabiev, I.; Mitchell, S.; Davies, A.; Williams, Y.; Kelleher, D.; Moore, R.; Byrne, S.; Rakovich, Y. P.; Donegan, J. F.; Sukhanova, A. Nonfunctionalized nanocrystals can exploit a cell's active transport machinery delivering them to specific nuclear and cytoplasmic compartments. *Nano Lett* **2007**, *7* (11), 3452-3461.
204. Zhang, L. W.; Monteiro-Riviere, N. A. Mechanisms of quantum dot nanoparticle cellular uptake. *Toxicol Sci* **2009**, *110* (1), 138-155.
205. Zhang, L. W.; Bäumer, W.; Monteiro-Riviere, N. A. Cellular uptake mechanisms and toxicity of quantum dots in dendritic cells. *Nanomed Nanotechnol Biol Med* **2011**, *6* (5), 777-791.
206. Yong, K. T.; Wang, Y.; Roy, I.; Rui, H.; Swihart, M. T.; Law, W. C.; Kwak, S. K.; Ye, L.; Liu, J.; Mahajan, S. D.; Reynolds, J. L. Preparation of quantum dot/drug nanoparticle formulations for traceable targeted delivery and therapy. *Theranostics* **2012**, *2* (7), 681-694.
207. Cho, S. J.; Maysinger, D.; Jain, M.; Röder, B.; Hackbarth, S.; Winnik, F. M. Long-term exposure to CdTe quantum dots causes functional impairments in live cells. *Langmuir* **2007**, *23* (4), 1974-1980.
208. Kirchner, C.; Liedl, T.; Kudera, S.; Pellegrino, T.; Muñoz Javier, A.; Gaub, H. E.; Stölzle, S.; Fertig, N.; Parak, W. J. Cytotoxicity of colloidal CdSe and CdSe/ZnS nanoparticles. *Nano Lett* **2004**, *5* (2), 331-338.
209. Green, M.; Howman, E. Semiconductor quantum dots and free radical induced DNA nicking. *Chem Commun* **2005**, (1), 121-123.
210. Lu, H.; Shiao, N.; Chan, W. CdSe quantum dots induce apoptosis *via* activation of JNK and PAK2 in a human osteoblast cell line. *J Med Biol Eng* **2006**, *26* (2), 89-96.
211. Mancini, M. C.; Kairdolf, B. A.; Smith, A. M.; Nie, S. Oxidative quenching and degradation of polymer-encapsulated quantum dots: New insights into the long-term fate and toxicity of nanocrystals *in vivo*. *J Am Chem Soc* **2008**, *130* (33), 10836-10837.
212. Geys, J.; Nemmar, A.; Verbeken, E.; Smolders, E.; Ratoi, M.; Hoylaerts, M. F.; Nemery, B.; Hoet, P. H. Acute toxicity and prothrombotic effects of quantum dots: Impact of surface charge. *Environ Health Perspect* **2008**, *116* (12), 1607-1613.

213. Soo Choi, H.; Liu, W.; Misra, P.; Tanaka, E.; Zimmer, J. P.; Itty Ipe, B.; Bawendi, M. G.; Frangioni, J. V. Renal clearance of quantum dots. *Nat Biotechnol* **2007**, *25* (10), 1165-1170.

214. Chen, Z.; Chen, H.; Meng, H.; Xing, G.; Gao, X.; Sun, B.; Shi, X.; Yuan, H.; Zhang, C.; Liu, R. Bio-distribution and metabolic paths of silica coated CdSeS quantum dots. *Toxicol Appl Pharmacol* **2008**, *230* (3), 364-371.

215. Hauck, T. S.; Anderson, R. E.; Fischer, H. C.; Newbigging, S.; Chan, W. C. *In vivo* quantum-dot toxicity assessment. *Small* **2010**, *6* (1), 138-144.

216. Yang, R. S.; Chang, L. W.; Wu, J.-P.; Tsai, M.-H.; Wang, H.-J.; Kuo, Y.-C.; Yeh, T.-K.; Yang, C. S.; Lin, P. Persistent tissue kinetics and redistribution of nanoparticles, quantum dot 705, in mice: ICP-MS quantitative assessment. *Environ Health Perspect* **2007**, *115* (9), 1339-1343.

217. Lin, P.; Chen, J.-W.; Chang, L. W.; Wu, J.-P.; Redding, L.; Chang, H.; Yeh, T.-K.; Yang, C. S.; Tsai, M.-H.; Wang, H.-J. Computational and ultrastructural toxicology of a nanoparticle, quantum dot 705, in mice. *Environ Sci Technol* **2008**, *42* (16), 6264-6270.

218. Pic, E.; Pons, T.; Bezdetnaya, L.; Leroux, A.; Guillemin, F.; Dubertret, B.; Marchal, F. Fluorescence imaging and whole-body biodistribution of near-infrared-emitting quantum dots after subcutaneous injection for regional lymph node mapping in mice. *Mol Imaging Biol* **2010**, *12* (4), 394-405.

219. Zhang, L.; Xing, Y.; He, N.; Zhang, Y.; Lu, Z.; Zhang, J.; Zhang, Z. Preparation of graphene quantum dots for bioimaging application. *J Nanosci Nanotechnol* **2012**, *12* (3), 2924-2928.

220. Chen, M.-L.; He, Y.-J.; Chen, X.-W.; Wang, J.-H. Quantum-dot-conjugated graphene as a probe for simultaneous cancer-targeted fluorescent imaging, tracking, and monitoring drug delivery. *Bioconjugate Chem* **2013**, *24* (3), 387-397.

Chapter 5

Mesoporous Silica Nanoparticles (MSNs) for Cancer Theranostics

*Yu Hsia, Maharajan Sivasubramanian, Nai-Tzu Chen, and Leu-Wei Lo**
Institute of Biomedical Engineering and Nanomedicine, National Health Research Institutes, Zhunan, Miaoli 350, Taiwan, China

Optimizing the balance between therapeutic efficacy and side effects is a major challenge in developing drugs. To circumvent unwanted side effects, nanoformulation of drugs for targeted delivery and controlled release is being researched and developed actively in translational nanomedicine. Among the various organic, inorganic, and hybrid nanomaterials used in formulating drugs, mesoporous silica nanoparticles (MSNs) have been drawing increasing attention because they provide a solid framework with a mesoporous structure and a high surface area that can be modified with diverse functional groups to control morphology. MSNs possess 3 topological domains, framework, nanochannels, and outermost surface, each of which can be functionalized sequentially with distinct moieties, making MSN a unique theranostic nanoplatform. The modified functional groups on the surface allow MSNs to be conjugated with diverse targeting molecules, peptides, and aptamers for efficient uptake and accumulation by specific target sites. Furthermore, to release the incorporated drug payload, nanochannels in MSNs can be controlled by varying external and/or internal stimuli, for example by photoirradiating or by altering the pH of subcellular domains and tumor microenvironments. Multifunctionalized MSNs have been proven to be a key platform for 3-in-1 nanotheranostics that orchestrates traceable imaging, targeted delivery, and selective therapeutics for cancer treatments.

5.1 Traditional Nanocarriers and Current Challenges

Nanocarriers, synthesized using a variety of materials, provide specific benefits over traditional drug delivery agents. In designing ideal nanocarriers, substantial effort is devoted to enhancing drug delivery to cells, increasing the circulation of nanocarriers *in vivo*, and achieving tissue-specific delivery. Among various nanocarriers, organic nanocarriers have drawn considerable attention because they possess favorable features, such as biodegradability and biocompatibility. Most of the organic nanocarriers are prepared using conventional biomaterials, such as lipids and polymers, but a major drawback with these nanocarriers is that they release cargo prematurely. Recently, inorganic

* Corresponding author: lwlo@nhri.org.tw

nanocarriers composed of gold or silver have also been used as drug carriers, and substantial advances have been made in the development of inorganic materials to improve drug delivery by inorganic nanocarriers.

The most popular examples of organic nanocarriers are liposomes. Liposomes are lipid-based carriers consisting of a lipid bilayer that forms amphiphilic, spherical structures. Because liposomes are composed of phospholipids, they are biocompatible and can be degraded easily without causing toxicity, and since they are amphiphilic, through hydrophobic interactions liposomes imbibe drugs that are poorly soluble in water. The therapeutic efficacy of liposome-based drug delivery depends mainly on the physiochemical properties of liposomes such as size, stability, charge, and interactions with blood-serum proteins. Liposomal nanocarriers have been demonstrated to accumulate drugs and release them at target sites, which is achieved by chemically modifying the surface of liposomes with targeting moieties or ligands that enhance the localization and accumulation of drugs loaded in the liposomes at specific disease sites. For instance, liposomes functionalized with anti-HER2 scFv F5 antibody accumulate specifically in breast tumors.[1]

Like liposomes, polymer-based nanocarriers are synthesized from biocompatible and biodegradable polymers, and poly(lactic-co-glycolic acid) is one of the polymer-based nanocarriers that has been approved by the Food and Drug Administration (FDA, USA) for clinical applications.[2,3] Although traditional nanocarriers have obtained FDA approval or reached clinical trials, many of these materials suffer serious drawbacks: the blood environment can interfere with the molecular interactions of phospholipids to destabilize liposomes and make them leak their cargo, and few techniques are available for controlling drug-release rates effectively and for enabling the nanocarriers to pass biological barriers.

5.2 Design of MSNs

5.2.1 *Why Use MSNs?*

Mesoporous silica nanoparticles (MSNs) have been originated when a co-hydrolysis method was used to synthesize mobile crystalline material-41 (MCM-41).[4] Over the past decade, MSN-based bioapplications have became highly popular in the fields of nanobiotechnology and nanomedicine for diagnosing or treating diseases such as cancer, diabetes, inflammation, and infectious diseases.[5] MSNs were developed as multifunctional nanocarriers to transport diagnostic or therapeutic agents for use with a variety of cell types and animal models,[6,7] and low toxicity differentiates MSNs from other inorganic materials: MSNs showed little or no cytotoxicity *in vitro* on various normal and cancer cell lines[8], and MSNs were biodegradable and biocompatible in a mouse model.[9,10]

MSNs were found to be biologically safe nanocarriers because >90% of the injected MSNs were excreted in feces/urine, and no differences were detected in blood analysis and histological examinations of control mice and mice injected with MSNs.[11] MSNs offer several advantages over traditional and other inorganic nanocarriers. Liposomes, for example, can be loaded with drugs readily and taken up by specific cells, but liposomes destabilize *in vitro* or lose their structural integrity when diluted, which results in premature drug release, a reduction in the efficiency of drug delivery, and toxic side effects. With inorganic nanomaterials like gold, the nanocarriers have a firm structure, but the cargo or drug has to be conjugated covalently to these particles. In contrast, MSNs provide a solid structure, they absorb cargo through particle pores without covalent bonding, and the MSN surface can be tailored for distinct purposes. Moreover, with "gatekeeping" molecules designed to cover MSN pores, cargo release can be controlled using external or internal stimuli. Lastly, major advances in surfactant-templated synthesis and sol-gel chemistry have made producing MSNs easy. Using these new synthetic processes and the numerous silica precursors available, MSNs can be manufactured in large amounts inexpensively.

5.2.2 *Synthesis*

In general, mesoporous materials are mainly synthesized by the combination of sol-gel process and supramolecular self-assembly involving surfactant molecules. In this process, the inorganic components self-assemble around the well defined surfactant templates. After synthesis, the surfactants are removed by extraction with solvent or calcination to get hollow pores. A large amount of different mesoporous materials have been synthesized such as two-dimensional/three-dimensional hexagonal, three-dimensional cubic and bicontinuous cubic formation. For example, the most studied mesoporous silica materials is MCM-41 which consists of hexagonally ordered rod-shaped micelles surrounded by silica network.

To classify the structure of mesoporous silica nanoparticles is fundamental for comprehending and predicting their properties. For instance, the pore size and the connectivity of channels influence their cargo loading capacity. The pore size can be modified by altering the surfactant or by changing the synthesis parameters. The architecture of pores can be verified and measured by different instruments, such as N_2 adsorption-desorption measurements or high-resolution transmission electron microscopy (TEM).

5.2.3 *Surface Modification*

An interesting feature of MSNs is that their outer and inner surfaces can be modified. MSNs were modified mainly with methyl phosphonate to increase their solubility in aqueous solution and to reduce their aggregation.[12] Unmodified MSNs tend to aggregate because of hydrogen bonding between amine groups and the silanols, but the strong electrostatic repulsion between the inserted methyl phosphonate functional groups increases the negative potential of thereby reducing aggregation.[13] Active targeting of nanocarriers can maximize their uptake and reduce the required dose of the drug-distribution vehicle. Similarly, conjugating the surface of MSNs with specific targeting moieties can enhance the efficacy of drug delivery. For example, MSNs conjugated with targeting moieties such as folate[14], transferrin (Tf), mannose[15], and Arg-Gly-Asp (RGD) peptides have been developed.[16] Folate-linked MSNs can be delivered specifically to cancer cells because they are taken up robustly through interactions with specific folate receptors that are overexpressed in many cancer cells; this interaction enhances therapeutic delivery of the modified MSNs. Tf is another molecule whose receptors are selectively overexpressed in many cancer cells, and conjugating Tf to MSNs can actively target the MSNs to cancer cells. The RGD peptide binds specifically to integrin $\alpha_v\beta_3$, which is overexpressed in metastatic cancer cells and primary tumors. When MSNs were conjugated with a 5-amino-acid analog of the RGD peptide, cyclo(Arg-Gly-Asp-D-Phe-Cys) (c[RGDfC]), they were targeted to cancer cells by the binding of the RGD-peptide analog to receptors on cancer cells.[17] The c[RGDfC] peptide has a free thiol group in the cysteine residue that allows the peptide to be conjugated by disulfide bonding to MSNs modified with 3-mercaptopropyltrimethoxysilane.

Cancer cells have a higher metabolic rate than normal cells and continuously require nutrients. Thus, MSNs uptake by cancer cells can be increased by linking sugars such as mannose to the MSN surface.[15] MSNs can be conjugated with mannose by functionalizing the MSN surface with aminopropyltriethoxysilane, which reacts with ethyl squarate-functionalized mannose. Mannose modification increases the uptake of MSNs in a colorectal HCT-116 and breast cancer cell lines (MCF7 and MDA-MB231) that are known to overexpress mannose-specific receptors.[18]

5.2.4 *Cellular Uptake/Exocytosis/Biocompatibility/ Biodistribution*

5.2.4.1 *Cellular Uptake and Exocytosis*

MSNs are used in nanomedicine because these nanoparticles are taken up rapidly by cells. MSNs' uptake was energy dependent[19] and likely mediated by active endocytosis, because MSNs' uptake was inhibited at 4 °C or when cells were incubated with molecules that degrade microtubules. When cells that had endocytosed fluorescent MSNs were labeled with antibodies against LAMP-I (a lysosome marker) and examined by fluorescence microscopy, MSNs and lysosomes were found to colocalize. However, the cellular distribution of MSNs can vary, and in cells such as mesenchymal stem cells, MSNs were detected in mitochondria and not in lysosomes.[20] The above mentioned studies also showed that MSNs with diameters ranging from 50 nm to 300 nm were not cytotoxic at concentrations <100 μg/mL, which makes them favorable drug-delivery candidates.[4] The above mentioned studies further revealed that the cellular uptake of MSNs depended on their size, because MSNs with diameters >150 nm were not endocytosed as efficiently as MSNs with diameters <110 nm.[21] Moreover, MSNs shape also affected uptake,[22] with spherical and short rod-shaped MSNs being endocytosed more inefficiently than long rod-shaped MSNs. Although understanding how MSNs deliver cargo to cells is critical, determining whether MSNs endocytosed by cells accumulate inside cells or leave the cells is also essential. MSNs' exocytosis was observed recently where the surface of MSNs being exocytosed was found to be coated by proteins,[23] and more than 80% of MSNs taken up by a lung cancer cell line were shown to be exocytosed within 24 h after uptake. MSNs have different efficacy of exocytosis in different cancer cell lines but the mechanism is unclear and requires further investigation.

5.2.4.2 *MSNs Biocompatibility*

Safety and toxicity, which are the major concerns in designing drug-delivery agents, have been tested *in vitro* for MSNs using cultured cells. Surface-modified MSNs in the 100–200 nm range are usually tolerated well by several cell types,[24–27] but unmodified MSNs show higher toxicity. The effects of MSN size, shape, charge, and chemical modifications on cellular toxicity have been reviewed recently.[28] How do particle characteristics affect cellular toxicity? Non-functionalized MSNs were found to be lethal when injected intraperitoneally and intravenously, but not lethal if injected subcutaneously;[29] postmortem analyses showed that thrombosis might have caused death, and that although subcutaneous injection was non-lethal, it triggered systemic toxicity. Functionalized MSNs injected*in vivo* showed no acute toxicity in animal models.[30] MSNs in the size range of 50–100 nm could be targeted specifically to the liver, and non-porous silica particles of 50–200 nm in size were found to be cleared by the liver and kidney in a size-dependent manner.[31] Larger particles were found to accumulate in the spleen and liver and become trapped by macrophages, whereas smaller particles were cleared preferentially. However, the long-term toxicity of the accumulated particles in liver or others organs has not been investigated, and how size, charge, or surface modifications of MSNs affect the *in vivo* biodistribution and toxicity of MSN remains unknown. However, MSNs show sufficient biocompatibility to be suitable for biomedical applications.

5.2.4.3 *MSNs Biodistribution*

MSNs' biodistribution *in vivo* is a critical concern that must be addressed. When administered subcutaneously, unmodified MSNs accumulate in tumors through enhanced permeability and retention (EPR) effect. MSNs also accumulate in kidneys and lungs, but when targeted actively, MSNs can be directed to cancer cells specifically. Conversely, positively charged MSNs are secreted by the liver into the GI tract and excreted with feces by animals, whereas negatively charged MSNs accumulate in the liver.[32] Thus, surface charge may affect MSNs' biodistribution. Moreover, morphology also alters the biodistribution patterns of MSNs. For example, long rod-shaped MSNs accumulate in the spleen, but short rod-shaped MSNs accumulate in the liver.[33] MSNs are excreted with urine and feces[34,35]: mice excreted almost 94% of phosphonate-modified MSNs 4 days after injection (73% in urine, 21% in feces),[36] and TEM revealed that the excreted MSNs retained their structure. Further research is required to determine, provided MSNs in the urine were intact, how MSNs passed through the glomerular basement membrane of the kidney, which filters blood and prevents large molecules from leaving the circulating blood.

5.3 MSNs as a Drug Carrier

5.3.1 *Anticancer Drugs Delivery*

Drug delivery agents that show promising features such as the ability to target specific cells and remain in circulation for long periods are tested in tumor-bearing animal models. The ultimate aims of a drug delivery system are to increase the aqueous solubility of drugs that dissolve poorly in water and to prolong their circulation in blood. Hydrophobic drugs such as doxorubicin (DOX) can be targeted more effectively by loading them in MSNs. DOX can be delivered to cells efficiently using MSNs coated with a polyethyleneimine-poly (ethyleneglycol) copolymer[11]: PEI-PEG-coated MSNs with a diameter of 110 nm were used, and 120 mg/kg of nanoparticles was administered weekly to a KB-31 xenograft model for 3 weeks. The DOX-loaded MSN/PEI-PEG inhibited tumor by 85% compared with 70% by the free drug. Previously, mannose-functionalized MSNs were loaded with a photosensitizer for photodynamic therapy (PDT) in a mouse model bearing HCT-116 xenografts. The results showed that tumor growth was inhibited, and moreover that after PDT, the mice did not show any tumor metastases to the colon and liver compared to thecontrol mice.[18]

5.3.2 *Cargo Release*

5.3.2.1 *Simple Diffusion*

Small molecular drugs trapped within MSN pores are released by diffusion obeying first-order kinetics. Drug-release rates are affected by MSN pore size and polarity, and by the drug's solubility and reactive state inside the pore. Moreover, the rates and mechanism of drug release are affected by the strength of the drug-MSN interaction. When an anionic dye was used that interacted weakly with MSNs, the dye was released by simple diffusion. By contrast, when a cationic dye was used, most of the dye was adsorbed on the negatively charged pore walls of MSNs, mostly near the surface.[37] These results indicate that dye release was affected by adsorbent-adsorbate interaction. Furthermore, MSN pore numbers and the size of adsorption sites for drugs have been shown recently to influence adsorbent-adsorbate interactions and drug-release kinetics. However, even in the absence of strong adsorbent-adsorbate interactions, unexpectedly slow drug release not obeying first-order release kinetics has also been observed and it was suggested to result from distinct adsorbent-adsorbate interactions in inner and outer sections of the MSN surface or from MSN pores being blocked partially by the precipitate of hydroxyapatite crystals on the MSN surface.[38]

Drug release by simple diffusion often displays an initial "burst" in drug release, which is reduced in MSN-based delivery systems. Ideally, the MSNs should block ineffective drug release before reaching target site. Below, we discuss the stimuli used to control the release of drugs.

5.3.2.2 *Controlled Drug Release Using Internal Stimuli: pH/Enzyme/Redox/Temperature*

1. Controlled Release Using Internal Stimuli: pH-operated Machines

Drugs are delivered into the body mainly by oral and intravenous routes. To prevent premature drug release in the gastrointestinal tract, MSNs are designed to release drugs in neutral but not acidic pH through the actions of gatekeepers that control drug release. For example, in pH-sensitive MSNs, covalently linked charged polyamines, in the form of silanes, functioned as gatekeepers by swelling and keeping the MSNs closed to prevent drug release.[39] In another case, the release of vitamin B required neutral pH conditions, and the release relied on the ionic composition of the medium.[40] In both the cases, pH-dependent, controlled breaking of hydrogen-bond interactions with amines and the formation of amine-anion complexes are considered to be responsible for the observed drug release patterns. Another example is ibuprofen encapsulated by MSNs coated with a polymer-based layer that also showed pH-dependent release.[41] In these pH-dependent MSNs, the silane was linked to the silica covalently, and polymer swelling was controlled by pH-dependent protonation of carboxylic groups in the methacrylate.

MSNs endocytosed by cells subsequently colocalize with lysosomes. Lysosomes are degradative compartments containing hydrolases that function at low pH. When the goal is to have MSNs release drugs inside cells after being taken but not in the extracellular milieu, pH-dependent controlled release is designed with gatekeepers that close at neutral pH but open at low pH. For example, in our study to design pH-dependent release of DOX, a pH-sensitive linker was conjugated on the surface of MSN channels by hydrazone bonds (MSN-hydrazone-DOX). The MSN channels were modified first with aldehyde groups by post-modification of the MSN inner surface with triethoxysilylbutyraldehyde. Next, the aldehyde and one hydrazide group of adipic acid dihydrazide were reacted to produce a hydrazone bond. The other hydrazide group of adipic acid was reacted with the ketone group of DOX to create extra hydrazone bonds for DOX conjugation. These pH-dependent linkers conjugated with DOX provided 2 hydrazone bonds that were cleaved at endosomal pH (Fig.5.1),[42] and this pH-dependent drug delivery agent also produced fewer systemic side effects than traditional delivery agents.

Another example of a pH-dependent system is of MSNs capped by covalently linked gold nanoparticles on surface through acid-labile acetal linkers.[43] The cargo was released only at pH of 4.0 and zero-release was detected at neutral pH.

2. Controlled Release Using Internal Stimuli: Enzyme Activation

Enzyme-responsive nanocontainers based on MSNs have been developed. In the first enzyme-triggered drug-release MSN system, a 2-rotaxane-conjugated MSN and an ester-linked α-cyclodextrin stopper were present.[44] The loaded dye was released from MSN pores when the ester bond was cleaved by porcine liver esterase. In another case, cyclodextrin served as the gatekeeper and its cleavage by α-amylase triggered rapid drug release.[45]

In another interesting study, a matrix metalloproteinase (MMP)-responsive system was developed;[46] MMPs are overexpressed in tumor microenvironments and can degrade the extracellular matrix. MSNs were loaded with DOX and coated with an MMP-substrate polypeptide and polyethylene glycol diacrylate. When added to 3T3-J2 cells, MSNs coated with the degradable MMP-substrate peptide induced faster apoptosis than PEG-coated MSNs, which supported the conclusion that higher chemotoxicity resulted from efficient DOX release triggered by activity of MMPs in cancer cells.

3. Controlled Release Using Internal Stimuli: Redox Activation

A redox-sensitive rotaxane model developed was similar to the enzyme-triggered

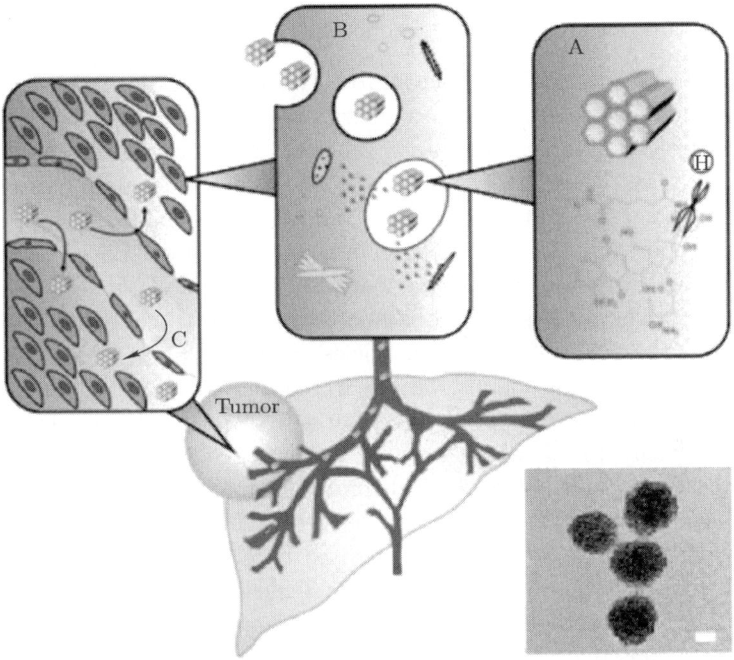

Fig. 5.1 The pH-dependent MSNs encapsulated with DOX to a tumor by EPR effect, and nonspecific uptake by tumor cells through endocytosis. The pH-dependent linker was hydrolysis of hydrazone bond in endosomes / lysosomes because of their low pH value (pH 5–6), then release the DOX form the channel of MSN to kill the tumor cells. There is a TEM images of the structure of MSN-hydrazone-DOX. Scale bars: 20 nm. Reproduced from Ref. 42 with permission of Wiley-VCH.

drug-release system designed by Ambrogio *et al.*[47] Rotaxane contains disulfide bonds, which are sensitive to glutathione, a thiol-containing tripeptide that can reduce disulfide bonds. Glutathione is abundant in the cytoplasm, whereas it is rarely found in blood plasma. Cargo or payloads diffused out of pores when gatekeepers bearing disulfide bonds were reduced by the action of glutathione. In the extracellular environment, glutathione concentrations were low and the disulfide bonds in gatekeeper molecules were mostly oxidized, but inside cells the disulfide bonds were reduced because of the high concentrations of glutathione.[48] The first redox-sensitive MSNs developed contained CdS nanoparticles as gatekeepers, which were anchored to the surface by silanes possessing S—S bonds.[49] In the presence of reducing agents such as 2-mercaptoethanol, the disulfide bonds were reduced to SH groups, which allowed cargo release. Other redox-sensitive MSNs have been developed using gold[50] and iron oxide[51] as gatekeepers. There was also a report of MSNs assembled using oppositely charged polyelectrolytes conjugated layer-by-layer by S—S crosslinking also showed redox-cleavable gatekeeper property.[52]

4. Temperature-responsive Gatekeepers

Tumor environments have been found to be warmer than normal tissues, a feature that could be used to develop temperature-sensitive drug-delivery systems. Temperature-sensitive systems based on poly (N-isopropylacrylamide) (PNIPAM) have been investigated extensively. The chain conformation of PNIPAM is sensitive to changes in temperatures,[53] and PNIPAM exhibits a lower critical solution temperature (LCST) of 32 °C in aqueous solution. Based on this idea, MSNs were developed to deliver payloads to cancer cells in a stimuli-sensitive manner.[54–56] The MSN surface was functionalized with PNIPAM to trap the drugs inside the pores of the nanoparticles. When LCST of PNIPAM was below 32 °C, its molecular chains were hydrated and they blocked the release of cargo from MSN channels. At temperatures higher than LCST, the polymer chains were dehydrated, which resulted in cargo release from MSN channels. Increasing LCST under physiological conditions might be an effective method for delivering drugs to treat tumors. Besides PNIPAM, acrylamide[57] and N-isopropylmethacrylamide[58] can also be used by copolymerization. In another temperature-sensitive system, designed by Schlossbauer *et al.*, MSN channels were blocked by double-stranded (ds) DNA sequences with specific melting points; cargo was released when the dsDNA separated into single strands because of increasing temperatures.[59]

5.3.2.3 *Controlled Drug Release Using External Stimuli: Light Activation/Magnetic Field Activation*

1. Controlled Release Using External Stimuli: Light Activation

Controlled drug release by MSN-based delivery agents is activated not just by internal factors such as enzymes, reducing agents, and low pH, but also by external factors. Fujiwara and coworkers.[60] developed the uptake and release of molecules by MCM-41 nanoparticles and could control the release of drugs using light; the MSN pores were covered by the intermolecular dimerization of coumarin. In 2008, Lu *et al.*[61] developed MSNs in which azobenzene connected to the nanopores and functioned as impellers. This structure is unlike previous designs because the pores are without gatekeepers, which are replaced by conjugated azobenzene derivatives that are linked to the surface of channels and control drug release through their reversible cis-trans photoisomerization. In the steady state, the azobenzene derivative functions like a gatekeeper to block cargo diffusion from MSNs. When photoactivated, azobenzene undergoes a cis-trans conformational change by a rotation-inversion mechanism, which mixes cargo in the MSN pores and accelerates their release in a light intensity-dependent manner. The external stimulus required for this drug-delivery system is harmless light, and only the cells that take up drug-loaded MSNs are killed when stimulated by light. Another light-stimulated MSNs contained azobenzene derivatives and β-CD, but the design was unlike the one used before: the azobenzene formed a part of the stalk and the β-CD acted as the cap for the pore, and drug release was controlled by the binding affinity of the β-CD-azobenzene interaction.[62] In aqueous solutions, β-CD interacted strongly with trans-azobenzene and weakly with cis-azobenzene.[63] When irradiated with 351 nm light, azobenzene isomerized form trans to cis, which resulted in the dissociation of β-CD from azobenzene and therefore the release of cargo. For the same purpose, an alternative design used a photocleavable o-nitrobenzyl ester bond as the stalk that was covalently linked to β-CD on the MSN surface.[64]

Another drug-release design based on LCST depended on the incorporation of hydrophilic and hydrophobic monomers into PNIPAM.[65] The polymers were light-responsive when the PNIPAM backbone incorporated light-responsive moieties such as 2-nitrobenzyl and azobenzene polymers. The LCST of the polymers could be regulated using UV radiation. Pores opened to release cargo when the polymers changed their conformation from a collapsed structure to a coil after UV irradiation. Recently, a cargo (sulforhodamine 101) was encapsulated inside pores using $Ru(bpy)_2(PPh_3)$-moieties as mercaptopropyl functional groups.[66] The coordination bonds formed by the mercaptopropyl groups with the inner surface blocked the nanopores; cargo release was triggered when the Ru—S coordination bond was cleaved by irradiating with visible light (455 nm).

2. Controlled Release Using External Stimuli: Magnetic Field Activation

Magnetic field has been used as an external stimulus to control the release of cargo from MSNs bearing Fe_3O_4 nanoparticles as gatekeepers.[67] The Fe_3O_4 gatekeepers were removed by using magnetic fields to break their chemical bonds and release cargo.

Substantial interest has developed in combining MSNs and super paramagnetic (SPM) metals because of the unique behavior of MSNs in the presence of magnetic fields. A magnetically controlled release platform was developed by combining zinc-coated iron oxide nanocrystals (ZnNCs) and MSN particles[68]. Magnetic fields heated the nanocrystals internally through molecular disassembly, and this combined material induced apoptosis in a breast cancer cell line (MDA-MB 231) in a magnetic-field-dependent manner.

Ruiz-Hernández *et al.*[69] designed a magnetic-DNA gatekeeper using MSNs with SPM iron oxide to develop an on-off drug delivery platform. The MSN surface and iron oxide nanocrystals were linked by single-stranded (ss) DNA, and the MSN was capped with iron oxide particles by the hybridization of the ssDNA strands. In the presence of a magnetic field, internal heating denatured the dsDNA, which resulted in the dissociation of iron oxide gatekeepers and the release of encapsulated cargo. The magnetic-DNA gatekeepers prepared in this study worked reversibly. The cargo release was inhibited at room temperature, whereas increasing the temperature released the cargo. This controllable, on-off drug-release system holds promise for the treatment of cancer.

5.3.2.4 *Dual Stimuli Responsive Gatekeepers*

Sophisticated nanomaterials with multiple molecular triggers that respond to sequential or concurrent chemical or physiological stimulation have studied widely. For example, a multi-responsive MSN was developed by chemically conjugating β-CD-bearing polymers to the MSN surface to form a polymeric network bearing, cross-linked disulfide groups, as gatekeepers.[70] From these MSNs, cargo could be released in response to distinct stimuli: ① UV light, which affected the azobenzene groups' isomerization; ② α-CD, which competed with β-CD; and ③ disulfide-reducing agents, which cleaved the disulfide bonds between polymer chains and β-CD.

Thermo- and pH-responsive core-shell particles have been prepared recently.[71] The system comprises a magnetic MSN as the core and a cross-linked outer shell of poly (N-isopropylacrylamide-co-methacrylic acid) [P (NIPAM-co-MAA)] polymer. The thermo-responsiveness of volume phase transition temperature (VPTT) was regulated by the concentration of NaCl and molar ratios of MAA/NIPAM. More drugs were released above the VPTT and less below VPTT, showing distinct responses to thermo- and pH-stimuli. In another example, MSN surface was conjugated with ds cDNAs as gatekeepers, which could be denatured by increasing temperatures and also by cleaving with endonucleases, resulting in cargo release from pores.[72]

5.4 MSNs as Theranostic Agents

5.4.1 *Diagnostic Applications of MSNs*

Positron emission tomography (PET) and magnetic resonance imaging (MRI) are major diagnostic applications in medical imaging. PET detects biological activity, whereas MRI diagnoses ultrastructural transforms sensitively. For cancerous growths, early detection improves the chances of successfully preventing cancers, and effectively treating and recovering from cancers, but early tumors are difficult to detect because monitoring instruments lack the required level of sensitivity. Tumors can be identified better by enhancing their detection specificity and increasing signals from them by transporting imaging reagents to tumors or lesions. MSNs can facilitate MRI and PET imaging when imaging reagents are attached chemically to the MSNs. Another method used commonly is fluorescence imaging, which can be quantified easily *in vitro*. However, most fluorescent dyes are not yet suitable for clinical applications, and the other major disadvantages of fluorescence microscopy include poor tissue penetration of fluorescence photons and signal interference by auto fluorescence from endogenous molecules.

5.4.1.1 *MRI*

MRI is a non-invasive diagnostic tool for imaging brain, central nervous system and for detecting tumors. MRI is one of the most sensitive tools for molecular and cellular imaging. Since MRI can produce images of soft tissues with high resolution, it may be very useful to detect early lesions. Despite these merits, MRI needs contrast agents to enhance the accuracy of the images. MRI agents excite the relaxation state of water protons to generate image contrast. The different contrast mechanisms are processed by using specific contrast agents.[73] The T_1 contrast agents are mainly Gd(Ⅲ) complexes, which increases the longitudinal relaxation rates of water to generate a positive image contrast. On the contrary, the T_2 contrast agents increase the transverse relaxation rates of surrounding water protons to generate the negative image contrast. The most common T_2 contrast agents are superparamagnetic nanoparticles which can be magnetized in the presence of external magnetic field. Among T_1/T_2 contrast agents, T_2 agents provide higher sensitivity of detection by producing stronger magnetic susceptibility to influence a larger number of water protons.

The development of diagnostic imaging tool requires high sensitive and targeted contrast agents. The core-shell particles assembled by a magnetic core and mesoporous silica shell are the most common type for MRI imaging. Mesoporous silica shell permits water freely mobile of the channels by their porosity, the rigidity of frame also blocks the rotation movement of the metal core.

For example, Mou and coworkers[74] developed a potential MRI contrast agent with MSN as the metal carrier. Gadolinium loaded MSN (Gd-MSN) was synthesized using a long chain cationic surfactant as the template. The synthesized Gd-MSN has a pore size around 2 nm with a large surface area. All Gd-MSN samples showed better longitudinal and transverse relaxivity (r_1 and r_2). On the one hand, the r_1 values of various Gd-MSN decreased with the increasing loading percentage of Gd. Authors explained that, this might be due to the predominant dipole-dipole interaction among the metals. Moreover, r_2 was insensitive to the loading percentage of Gd. The *in vitro* phantom imaging results demonstrated that, in the T_1 weighed image sequence, as the amount of Gd-MSN increased, the T_1 images lighten up faster than Magnevist (a commercial Gd based contrast agent). With T_2 weighed image sequence, the image darkens as the concentration is increased compared to Magnevist. These results clearly indicate that, Gd-MSN prepared in this study could be a better relaxivity enhancer.

Highly efficient MRI contrast agent based on MSN was developed by Lin and coworkers.[75] MSN was synthesized by a base catalyzed condensation reaction using surfactant template. The surfactant extracted MSN was coated with gadolinium complex (Gd-Si-DTTA) through siloxane linkage and for the *in vitro* experiments, rhodamine was also coated with MSN. The Gd coated MSN (MSN-Gd) exhibited very large longitudinal (r_1) and transverse (r_2) releaxivities. The confocal observations showed that, MSN-Gd was readily taken up monocyte cells with strong fluorescence signals. For MR imaging, the monocyte cells incubated with MSN-Gd showed that positive signal was significantly enhanced in T_1 weighed image and negative signal was enhanced in T_2 weighed image. To realize the concept *in vivo*, DBA/IJ mouse was given tail injection of MSN-Gd (2.1 μmol/kg), T_1 weighed image was significantly enhanced with a dose level lower than current contrast agents. This demonstrates the utility of MSN-Gd as intravascular MR contrast agent. Authors also showed that, MSN-Gd could be an efficient T_2 weighed contrast agent at high dose.[75] Potential T_1 positive contrast agent based on luminescent rattle type mesoporous silica microspheres was developed by Lin and coworkers.[76] The whole architecture is a multifunctional composite microsphere possessing the advantages of luminescent and temperature sensitiveness. To realize this idea, authors synthesized hollow mesoporous silica capsules by a sacrificial template method, followed by the loading of Gd_2O_3: Eu^{3+} in the internal cavity by an incipient wetness impregnation method. To impart the temperature sensitiveness to the system, the inner cavity of rattle type mesoporous silica capsules were filled with temperature sensitive hydrogel by photopolymerization of PNIPAM-based hydrogel. From the *in vitro* cell viability tests, the nanocomposite material was found to be biocompatible and was readily taken up SKOV3 cells and the uptake was found to increase with time. From the *in vitro* MR imaging test, the composite material showed promising spin lattice relaxation time (T_1) weighed effect. In addition, the nanocomposite exhibited temperature dependent release of Indomethacin.

5.4.1.2 *Biosensing*

The large surface area of MSNs can be used for conjugating sensor molecules for detecting biological activities. Numerous MSNs for *in vitro* biosensing have been developed that can detect glucose.[77], H_2O_2[78], and NO_2.[79] The NO_2-sensing MSN, for example, was conjugated with the protein myoglobin (Mb) in the channels, whose structure allowed Mb to maintain its electrocatalytic activity. The conjugated Mb showed stable and reproducible electrocatalytic activity and high affinity for reducing H_2O_2 and nitrite. Thus, these MSNs not only provide a potential diagnostic application by sensing H_2O_2 and nitrite, but also offer a new platform for studying electron transfer by proteins.

To cite another example of biosensing, the inside and outside surfaces of MSNs were used for differentiating between neurotransmitters with similar structures but dissimilar functions[80]: the diffusion rates of neurotransmitters functionalized distinctly were distinguished by detecting the change in particles charges. Despite these advances, biosensing systems based on MSNs are not yet utilized for clinical diagnostics and are used mainly for developing drug delivery agents. Combining biosensing molecules on MSNs with antibodies that target specific biomarkers on cancer may provide a system suitable for detecting cancers early as well as facilitating cancer therapeutics.

5.4.2 *Therapeutic Applications of MSNs*

Cancer is the leading cause of death in most of the developed and developing countries worldwide. Success in treating cancers depends on detecting cancers at early stages. Major developments in nanotechnology have provided powerful tools that help in detecting, preventing, and curing cancers. The ultimate goal of cancer therapy is to achieve a high local concentration of drugs at tumor sites without affecting the surrounding healthy tissues. The efficiency of nanoscale drug-delivery platforms depends on parameters such as structural integrity, biodistribution, and the ability to avoid rapid clearance by macrophages. Various nanomaterial architectures have shown distinct abilities to penetrate biological barriers, escape macrophage clearance, and accumulate in tumors. Below, we will discuss MSN-based drug-delivery strategies for cancer therapeutics.

5.4.2.1 *Efficient Drug Delivery*

Conventional chemotherapy is associated with life-threatening side effects produced by the non-specific uptake of chemotherapeutic agents by normal cells. Moreover, the vast therapeutic potential of anticancer drugs is offset by factors such as poor solubility, susceptibility to proteolytic enzymes, and inefficient cellular uptake of the drugs. To overcome these limitations, substantial effort has been invested in developing drug-delivery systems. An ideal drug-delivery agent should encapsulate drugs at high concentrations, increase the blood-circulation time of the drugs and protect them from proteolytic degradation, and target the drugs to specific cells and enhance their cellular uptake. Recently, MSNs have been shown to be excellent intracellular delivery agents for both hydrophobic and hydrophilic anticancer drugs.[81] For example, methotrexate-conjugated MSNs induced more apoptosis in cancers cells than the same concentration of the free drug, a concentration at which normal cell lines were unaffected. Because the MSN surface can be modified easily and since MSNs have a high drug-loading capacity, these nanoparticles are potential multidrug delivery agents. This could be helpful because diverse signaling pathways crosstalk in tumors, which often become resistant to the drugs used for treating them. Recently, the possibility of multidrug delivery by MSNs has been suggested by the co-delivery of 2 hydrophobic fluorescent dyes, $1,1'$-dioctadecyl-$3,3,3',3'$-tetramethyl-indocarbocyanine perchlorate and

3, 3′-dioctadecyloxacarbocyanine perchlorate, which were encapsulated in MSN pores at the same time.[82]

While the potential for using MSNs as drug-delivery agents *in vitro* has been demonstrated by numerous studies, few studies have discussed the therapeutic efficiency of drug-loaded MSNs *in vivo*. Dai *et al.* applied MSN powder directly on wounds in rabbits to show that MSNs coated with silver and calcium induced the coagulation cascade by the intrinsic pathway to promote blood clotting in liver and after femoral artery damage.[83] In another study, Mellaerts *et al.* encapsulated the insoluble drug itraconazole in MSN to achieve higher availability after oral delivery.[84] However, MSNs are still not popular in clinical use.

5.4.2.2 *Targeted Therapy*

The major challenge in cancer therapy is to increase the concentration of anticancer drugs at the tumor site, while sparing the surrounding healthy tissues. Hence, the tumor targeting becomes an integral part of cancer therapy. The purpose of tumor targeting could be achieved *via* passive targeting and active targeting strategies. The drug delivery systems reach the tumor location *via* blood stream. After circulating to the tumor site, nanoparticles need translocation from the vasculature to the solid tumor by extravasation. Tumors have pathological characteristics that are not observed in normal tissues such as impaired lymphatic drainage system, extensive angiogenesis and defective vasculature. Nanoparticles accumulate in the tumor site *via* the leaky vasculature by passive targeting, also called as the EPR effect.[85] But experimental data revealed that, the passive targeting efficiency is lower than 10% in animal model.[86] Hence many researchers have focused on the active targeting strategy to enhance the antitumor efficacy. Active targeting strategy involves the use of targeting ligands, whose receptors are overexpressed in the cancer cells.

Based on the active targeting strategy, MSN was conjugated with various specific targeting ligands, such as folic acid (FA) for targeting folate receptor[87], cyclic-RGD for targeting $\alpha_v\beta_3$ integrin, DNA aptamer sgc8 for targeting human protein tyrosine kinase-7[52] and monoclonal antibody for targeting the human epidermal growth factor receptor 2 (HER2).[88] The significant design of active targeting drug delivery systems were based on the property and grafting density of targeting ligands. The choice of targeting ligands should have the high specific affinity to receptors over expressed in tumor tissues.[89] The conjugation density of targeting ligands onto MSN must be optimized, because lower density of targeting ligand leads to poor efficiency and the multivalent ligands may enhance the immune response.[90]

Wang and coworkers[52] reported an efficient cell targeting and intracellular controlled release drug delivery system. The system is based on MSNs that possess several advantages of mesopores, cell targeting and controlled drug release. MSN was synthesized using cetyltrimethyl ammonium bromide (CTAB) as a template followed by the loading of fluoroisothiocynate (FITC) and anti-cancer drug DOX. Then, cargo loaded MSN was sequentially deposited with poly (vinylpyrrolidone) and thiolated poly (methacrylic acid). After the formation of multilayer, thiolated aptamer was incubated with this nanoparticle for the formation of disulfide bond. Aptamer conjugated MSN were readily taken up by HeLa cells *via* receptor mediated endocytosis with a strong fluorescent signal, while bare MSN showed low fluorescent signals of FITC. Cell viability studies revealed that, bare MSN was biocompatible and viability significantly decreased for DOX loaded MSN aptamer conjugate. Authors explained that, this high toxicity of DOX loaded MSN aptamer conjugate might be due to the enhanced uptake by HeLa cells. In addition, the MSN exhibited remarkable reduction sensitive release of FITC and DOX.[52] In a more recent study, Tamanoi *et al.*[91] prepared folate modified MSN based drug delivery system for the enhanced tumor suppression. MSN was synthesized by a template method using CTAB. Then, MSN was successfully conjugated with FA followed by loading of an anticancer drug, camptothecin. The effect of folate conjugation on antitumor efficacy was studied using two different human pancreatic cancer xenografts on different mouse species. The folate conjugated MSN exhibited antitumor efficacy with a significant reduction in the tumor volume. Authors observed that, the folate conjugate exhibited dose dependent tumor suppression and found 0.5 mg of camptothecin loaded FITC conjugate as a minimum dose required to inhibit complete tumor growth.

Lang and coworkers[92] developed silica nanorattle based tumor targeted delivery using mesenchymal stem cells (MSC) as a targeting ligand. A study reported that, MSC have tumor tropic property towards malignant cells and hence the attachment of drug loaded nanoparticles could be an efficient strategy to directly target the tumor. To achieve this, DOX loaded silica nanorattle was attached to the MSC by antigen-antibody interactions. Nanorattles loaded MSC (up to 1 500 nanoparticles) demonstrated high cell viability and tumor tropic ability. Nanorattles loaded MSC could successfully target U251 glioma tumor cells and release DOX with long retention time in tumor tissues, which resulted in the enhanced tumor cell apoptosis. Even though numerous strategies have been developed to enhance the active targeting efficacy, more attention is needed to understand the connection between tumor targeting efficiency and the materials properties. This understanding will help the researchers to fine tune and improve the targeted cancer therapy.

5.4.2.3 *Anti-multidrug Resistance*

Multidrug resistance (MDR) is a serious setback in cancer therapy, which affects the efficacy of cancer treatment even with targeted drugs or combination therapy. In the conventional chemotherapy, drug sensitive cells were killed, however, a proportion of the drug resistant cells manage to survive, which leads to tumor metastases. Increased drug efflux is one of the major mechanisms responsible for MDR mediated by ATP-dependent drug transpoters typically of ATP-binding cassette superfamily. The ATP-dependent transporters eject the delivery of anticancer into tumor cells and decrease the intracellular accumulation.[93] A recent study has discovered that the existence of cancer stem cells (CSCs) also have a profound effect on the function of drug resistance.[94] Notch signaling is a specific CSCs signaling for regulating stem cells property.[95] The strategy to inhibit the Notch pathway by delivering γ-secretase inhibitors (GSIs) encapsulated into folate bioconjugated MSN showed enhanced tumor therapy efficacy.[35]

MSN has several advantages than conventional nanomaterials to overcome MDR. To mention a few, the high drug loading capacity and easily tunable surface, which can enhance the cellular uptake and increase the intracellular drug concentration. MSN has more than 1 000 times higher drug loading capacity than liposomes. The DOX encapsulated MSN coated with lipid bilayer showed 100 times toxicity to MDR cancer cells than DOX loaded liposomes.[96] He and coworkers developed a co-delivery strategy to overcome the multi drug resistance in cancer cells using MSN. The strategy uses siRNA to silence mRNA encoding Bcl-2 protein, which activates cellular anti-apoptotic defence. The MSN surface was modified with second generation amine terminated polyamidoamine (G2-PAMAM) dendrimers to electrostatically bind siRNA without pronounced toxicity. With the outer surface covered by siRNA, the inner nanopore is available for the loading of anticancer drug DOX. The following experimental results suggest that MSN could be used to co-deliver siRNA together with DOX. Due to the simultaneous delivery of siRNA with DOX, the anticancer efficacy of DOX increased 132 times compared to free DOX. The authors suggest that the co-delivery significantly suppressed the Bcl-2 expression.[97]

5.4.2.4 *PDT*

PDT is a new therapeutic modality for treating cancer, which uses a targeted photosensitizer (PS) and a light of appropriate wavelength. These two components are combined to induce cellular or tissue effects in an oxygen-dependent manner.[15] In PDT, light induces cell damage or death by producing free radicals and reactive oxygen (*ex.* $^1O_2$). Large amounts of PS can be loaded in MSN channels and the outer surface of MSNs can be modified chemically for targeting MSNs or enhancing their cellular uptake.[98] The short-lived $^1O_2$ has to react directly with molecules in intracellular pathways to produce maximum cytotoxicity, and the unique architecture of MSNs can protect PS from degradative environments. One well-studied PS, protoporphyrin IX, was taken up by cells efficiently and showed phototoxicity that depended on light intensity and irradiation time. The key benefit of PDT in cancer treatment is that cytotoxicity occurs only in irradiated regions, and cells outside these regions are not damaged (Fig.5.2).

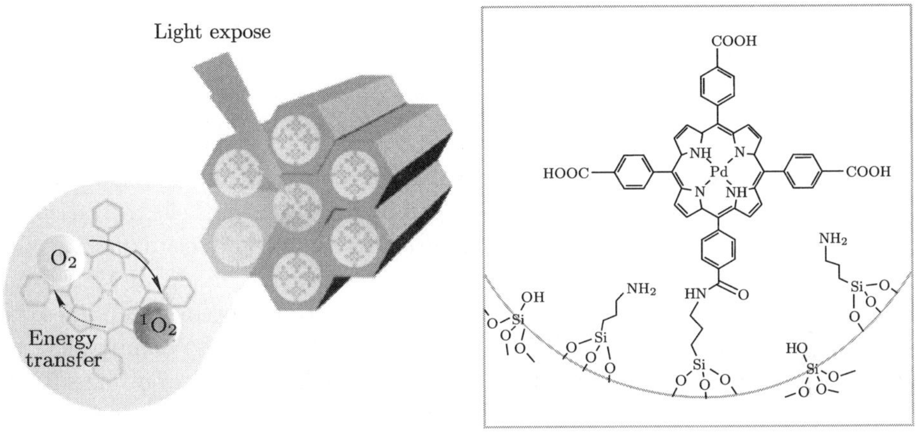

Fig. 5.2 A photodynamic model of MSN and the covalent modification of PdTPP onto the nanochannel surface of MSN. Reproduced from Ref. 98 with permission of Royal Society of Chemistry.

In 2011, we demonstrated that 2-photon activated-photodynamic therapy (TPA-PDT) allowed spatially selective treatment of cancers and also enhanced the depth to which tissues were penetrated by incident light. MSNs co-encapsulating 2-photon-absorption dyes and PS enhance the high-energy transfer rate for TPA-PDT. The energy transfer rate was increased to 93% because of the well-ordered mesoporous structure of MSN. The cytotoxicity induced by the $^1O_2$, which was produced by intra-cellular energy transfer, was also highly

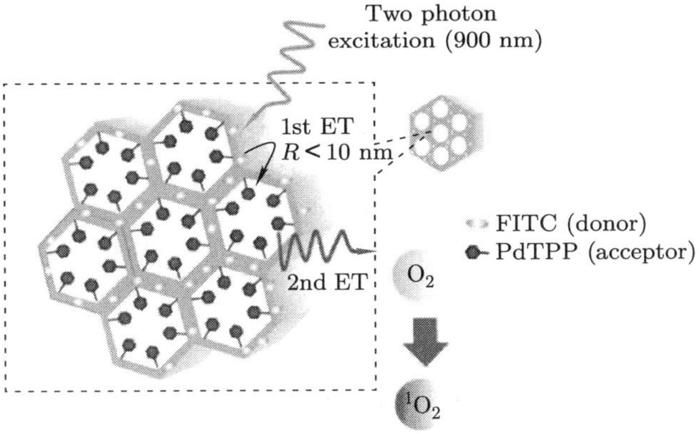

Fig. 5.3 Two step intra-MSN energy transfer form FITC within the framework to PdTPP onto the nanochannels to generate singlet oxygen by two photon excitation. Reproduced from Ref. 99 with permission of Elsevier.

efficient *in vitro* and *in vivo* in breast cancer models (Fig.5.3).[99] These results suggest that MSNs modified for TPA-PDT can potentially be used in clinical applications.

5.4.3 *Gene Delivery System*

Successful clinical application of gene therapy relies on delivering an appropriate amount of therapeutic genes to target tissues without causing substantial toxicity. Thus, efficient and safe gene-delivery agents must be designed and developed. Cationic polymers or materials can bind nucleic acids electrostatically, condense them into particles up to approximately 100 nm in diameter, and protect them effectively against degradation. Cationically modified MSNs can condense plasmid DNA and transfect them into CHO and HeLa cell lines.[100] Based on this idea, Chen *et al.* designed a double-delivery MSN encapsulating DOX, and used the G2-PAMAM dendrimers as gatekeepers in targeting siRNA complexes to knockdown the mRNA expression of Bcl-2 and thereby suppress MDR in cancer cells.[97] To increase gene transfection efficiency, the MSN surface was decorated with mannose[101] and non-modified PEI.[102] PEI is ideal for forming complexes with nucleic acids because PEI is a cationic polymer with a highly positive charge density. PEI also acts as a buffer under low pH conditions that can destabilize the lysosome membrane. Similarly, MSNs modified with G2-PAMAM dendrimers were taken up more efficiently by gliomas than esterified MSNs with lower surface charge.[103]

5.5 Multifunctional Design of MSNs

5.5.1 *Hybrid Materials*

Organic-inorganic hybrid materials based on MSNs were developed by modifying the MSN surface chemically with lipids, polymers[104], or metals to achieve nanoparticles with unique features. For example, lipid modification of MSNs results in diminished non-specific protein binding, and MSNs coated with PEGylated phospholipids could be suspended effectively in solution.[105] To increase the efficiency of cellular uptake, in hybrid systems the phospholipids are functionalized and folate is added as a targeting agent. MSNs can also be coated with polyelectrolyte multilayers by continually overlaying thiolated poly (methacrylic acid) and poly (vinylpyrrolidone).[52] The release of cargo can be controlled because the polyelectrolyte multilayer's disulfide groups are reduced using dithiothreitol, and the experimental results showed controlled release of DOX by these MSNs in HeLa cells. MSN-polymer hybrid materials were also demonstrated to enhance circulation time and increase accumulation in tumors. For example, PEG-coated MSN exhibited a better biodistribution with substantially lower accumulation in the liver.[11] In this study, the accumulation of PEI/PEG co-polymer-coated MSN in tumor increased from 3% to 12% at 72 h post injection, and the tumor growth was also inhibited strongly.

5.5.2 *Core-shell Particles*

By combining mesoporous silica particles and other materials to develop core-shell nanostructures that are more uniform than the all-silica structures, MSNs have increased their functionality and applicability in biomedical imaging and are used widely in the biomaterials field. In several cases, solid cores were made using diverse materials such as silver[106], gold[107,108], platinum[109], and quantum dots[110] were encapsulated by mesoporous silica shells successfully. Mesoporous, silica-coated magnetic materials have received considerable attention because ferro- or superparamagnetic iron oxide (SPIO) can be used for magnetic-response controlled drug delivery, and because they provide sufficient material for MRI. Several methods for encapsulating magnetic materials using mesoporous silica have been described.[111] The most common method for synthesizing core-shell magnetic MSN particles is by coating a silica shell surrounding the magnetic, iron-oxide nanocrystal core. The mesoporous silica layer can be deposited by diverse synthetic processes such as directly overlaying onto the magnetic cores[10,112], and pre-coating the magnetic crystal material with a dense silica layer[113,114] by Stöber process[115] (which is a method used commonly to synthesize metal-oxide crystal cores with silica-shell particles). The silica layer protects the iron-oxide core against the acidic environment and helps the cationic surfactant template to complex with the negatively charged surface through self assembly.[116] The development of these magnetic cores with silica-shell particles has challenges to surmount, such as gaining separable particles after a single crystal-coating process[112] and obtaining uniform magnetic cores with distinct magnetizations. Nevertheless, the core-shell particles have considerable potential for biomedical applications.

5.5.3 *Multifunctional MSNs*

MSNs possess unique features such as high surface area ($\sim$1 000 m^2/g), uniform pore volumes ($\sim$1.0 cm^3/g) and well defined surface properties. Owing to these features, MSN are considered as the building blocks for the construction of sophisticated nanomaterials for drug delivery. In addition, the three topologically different domains namely silica framework, nanochannels/pores, and the outermost surface can be tailor made for different biomedical applications. Tunable and versatile nature of MSN allows the accommodation of variety of guest molecules. The nature of guest molecules ranges from hydrophobic/hydrophilic small molecule drugs to macromolecular therapeutics such as peptides, proteins and nucleic acids. Much of the attention was paid towards MSN for the delivery of small molecule drugs such as camptothecin, DOX, ibuprofen and cisplatin. Only a few studies have utilized MSN for the delivery of macromolecular therapeutics. Proteins and genes hold great promise in treating wide range of diseases due to their high specificity. Unfortunately, their clinical applications have been severely hampered due to their intrinsic nature such as instability, high molecular weight and fragile structures. Hence, an ideal system is required to deliver this therapeutics unscathed to the target site of action. MSN with tunable surface area and nanopores are considered to be the promising candidates for the delivery of genes. The genes encapsulated inside the mesopores are highly stable and can escape from proteolytic digestion. Additionally, the loading efficiency could be significantly enhanced by decorating the surface of MSN with polycations. The negatively charged genes can bind electrostatically to the polycations, moreover, the presence of positive charge can also facilitate endosomal escape by the well known "proton sponge effect".

Nel and coworkers[117] developed a co-delivery strategy that uses siRNA to silence the P-glycoprotein expression and an anticancer drug DOX to kill drug resistant KB-V1cells in a synergistic fashion. For this purpose, authors modified the surface of MSN with phosphonate group, which can electrostatically bind DOX to porous interior, and allows the coating of external surface with polyethyleneimine to complex with siRNA. The loaded DOX could be released in the presence of acidic medium and complexed siRNA could safely escape endosomes through "proton sponge effect". The results demonstrated that, the dual delivery of DOX and siRNA in drug resistant cancer cell increased the inter or intra cellular drug concentration to exceeding levels compared to free DOX or DOX delivered without siRNA. In conclusion, MSN based carriers for siRNA could be more effective in less P-glycoprotein expressing cancer cells.[117]

Lin and coworkers[118] developed a multimodal imaging agent based on MSN and gadolinium (Gd) chelates. To realize this concept, two diethylenetriaminepentaacetic acid based Gd chelating ligands were synthesized. The first ligand, 3-aminopropyl (trimethoxysilyl)- diethylenetriamine acetic acid contains one trialkoxysilane functional group for attachment to the MSN. The second ligand, bis (3-aminopropyltriethoxysilyl)-diethylenetriaminepentaacetic acid contains two trialkoxysilane group to allow incorporation into the MSN. For the *in vitro* experiments, MSN-Gd (30 wt%) was labeled with rhodamine and to enable cell specific targeting, RGD peptide sequence was chemically grafted on to the surface. By varying the amounts of two ligands, different MSN-Gd contrast agents were developed by co-condensation method. The synthesized MSN particle size ranged from 75 nm to few hundreds of nm and the particle size increased with increase in Gd wt%. RGD embedded MSN-Gd (30 wt%) particle was successfully taken up by RGD receptor expressing cancer cell *via* the receptor mediated endocytosis. Cells incubated with targeted MSN showed enhanced T_2-weighed contrast compared to cells incubated with non-targeted MSN. Imaging plays an important role in biomedicine because we can visually monitor the cargo delivery to the target site. Despite these advantages, the *in vivo* utility of this system is restricted due to the large sizes and non-biodegardable nature.[118]

Cheng and coworkers[119] developed a novel theranostic platform using mesoporous silica coated gold nanorods. The outer silica layer serves as an effective drug carrier by allowing the loading of DOX, while the inner gold nanorod core functions as two-photon imaging agent. To realize this concept, MSN was successfully coated over gold nanorods with a thickness and pore diameter of 30 nm and 5 nm respectively. DOX was loaded into the MSN through electrostatic interaction and the release behavior of loaded DOX was investigated in the presence of pH and laser radiation. More DOX was released under low pH, due to the protonation of silanol groups. Interestingly, in the presence of 250 mW NIR laser radiation, the DOX release becomes much faster compared to non- radiation treatment. Authors explained that, generated heat dissociated the strong interaction between DOX and silica, thus more DOX was released. Cellular uptake studies in the presence of A549 cells demonstrated that, 5–10 times DOX could be internalized when they were loaded in MSN. Using two-photon confocal microscope, most of the nanoparticles were observed to remain in the endosome/lysosomes and only a few entered mitochondria, suggesting an opportunity for image guided therapy. Both free DOX and DOX loaded MSN showed dose and time dependent toxicity, while bare MSN exhibited little or no toxicity. To demonstrate the hyperthermic potential, two groups of MSN-DOX internalized A549 cells were irradiated with 24 and 48 mW/cm^2 laser light. Strong disruption in the lysosomal membrane was observed immediately after 3 min of 48 mW/cm^2, which lead to the fast cell death. In the case of 24 mW/cm^2 laser radiation, the cells remained unchanged. Chemotherapeutic potential of MSN-DOX were also evaluated using low power radiation (20 mW/cm^2) to avoid killing cells directly through radiation. No acute cell death was immediately observed after 3 min of laser irradiation, MSN-DOX treated cells lost viability with increasing time with a maximum loss of cell viability at 24 h. This result supports that, it is chemotherapy responsible for the curative effect of the treatment. This type of novel theranostic systems could be useful for the treatment of cancer.[120] In our group, we investigated multifunctional MSN particles that consisted of a palladium-porphyrin based theranostic photosensitizer for photodynamic therapy, near-infrared fluorescent agents ATTO647N for optical tracking, and cell specific homing RGD peptide (Fig.5.4).[121] To demonstrate targeting specificity and cell uptake of the nanoparticles, two different cell lines were used. One was a breast cancer cell line (MCF-7), which express little $\alpha_v\beta_3$ integrin; and the other was $\alpha_v\beta_3$ over expressing glioblastoma cells (U87-MG). RGD bearing nanoparticles were significantly taken up by U87-MG cells by receptor mediated endocytosis, while bare nanoparticles without RGD did not show any significant uptake. But, neither of the particles was sensitive to MCF-

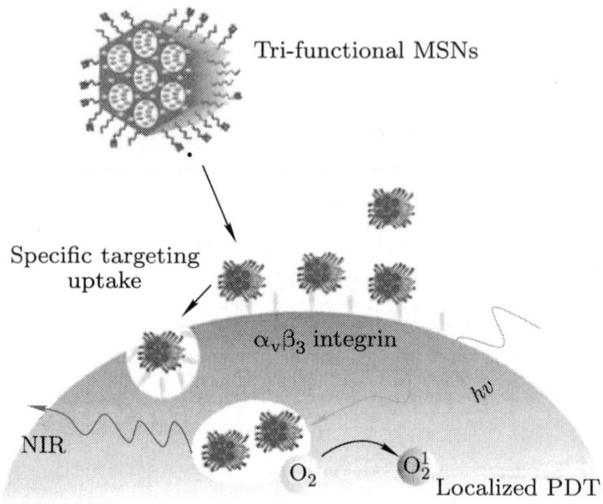

Fig. 5.4 The illustration of the tri-functionalized MSN for use as theranostic application that possess separate domains for traceable optical imaging, localized photodynamic therapy, and efficient targeted delivery. Reproduced from Ref. 121 with permission of Royal Society of Chemistry.

7 cells, which clearly indicated the absence of receptor mediated endocytosis. To demonstrate therapeutic potential of photodynamic therapy, U87-MG cells treated with RGD bearing formulation showed dose dependent cytotoxicity on irradiation, while formulations in the absence of RGD peptide was insensitive to therapy. The multifunctional MSN delivery system developed in this study could be useful for a highly efficient and selective photodynamic therapy.

5.6 Future Perspective

Currently, a variety of approaches are available to deliver drugs to the target site of action. The desired drugs can be either directly loaded into the nanoparticles that have the ability to reach the target site, or indirectly designed in the form of prodrugs that are not active until they meet specific molecules or enzymes. For realization of a clinically translatable nanoplatform, the efficient performances in all three facets of traceable imaging, targeted delivery and selective therapy are highly demanded. For the preparation of such a versatile system, MSN has become an excellent candidate due to possession of three topologically different domains for individual functionalities. In spite of all these architectural advantage for exploiting MSN on drug delivery, there are still several challenges need to be surpassed in the immediate future for the success of translation in clinics. Firstly, to increase loading efficiency of larger size therapeutic moiety from genes to proteins by enlargement of MSN pores without compromising architectural integrity; secondly, to advance the selectivity and specificity of controlled release using external physical stimuli and/or endogenous biological/chemical reactions; thirdly, to improve the biosafety by active control of MSN biodegradation and bioelimination with delicate manipulation of surface chemistry in response to a variety of biological microenvironments. All these challenges could be coped with necessary physicochemical modifications of MSN following the understanding of targeted diseases through molecular to physiological levels.

Towards the ultimate goal of disease cure, the prognosis strategy is the most critical. To function as facilitator, it is either to design the nanoparticle possessing requested multiple functionalities all-in-one using hybrid materials approach, or to simply apply different nanoparticles metronomically and each works with high efficiency in its respective phase of treatment. No matter with the all-in-one or the metronomic strategy, in consideration of the versatility of utilization and potential improvement of biosafety, MSN serves as one of the most efficient enabling nanoplatforms to date and is worth ensuing diligence in translational development.

References

1. Drummond, D. C.; Noble, C. O.; Guo, Z.; Hayes, M. E.; Connolly-Ingram, C.; Gabriel, B. S.; Hann, B.; Liu, B.; Park, J. W.; Hong, K.; Benz, C. C.; Marks, J. D.; Kirpotin, D. B. Development of a highly stable and targetable nanoliposomal formulation of topotecan. *J Control Release* **2010**, *141* (1), 13-21.
2. Maurer, N.; Fenske, D. B.; Cullis, P. R. Developments in liposomal drug delivery systems. *Expert Opin Biol Th* **2001**, *1* (6), 923-947.
3. Kamaly, N.; Xiao, Z.; Valencia, P. M.; Radovic-Moreno, A. F.; Farokhzad, O. C. Targeted polymeric therapeutic nanoparticles: Design, development and clinical translation. *Chem Soc Rev* **2012**, *41* (7), 2971-3010.
4. Slowing, I. I.; Vivero-Escoto, J. L.; Wu, C.-W.; Lin, V. S. Y. Mesoporous silica nanoparticles as controlled release drug delivery and gene transfection carriers. *Adv Drug Deliver Rev* **2008**, *60* (11), 1278-1288.
5. Tang, F.; Li, L.; Chen, D. Mesoporous silica nanoparticles: Synthesis, biocompatibility and drug delivery. *Adv Mater* **2012**, *24* (12), 1504-1534.
6. Coti, K. K.; Belowich, M. E.; Liong, M.; Ambrogio, M. W.; Lau, Y. A.; Khatib, H. A.; Zink, J. I.; Khashab, N. M.; Stoddart, J. F. Mechanised nanoparticles for drug delivery. *Nanoscale* **2009**, *1* (1), 16-39.
7. Lee, J. E.; Lee, N.; Kim, T.; Kim, J.; Hyeon, T. Multifunctional mesoporous silica nanocomposite nanoparticles for theranostic applications. *Accounts Chem Res* **2011**, *44* (10), 893-902.
8. Li, Z.; Barnes, J. C.; Bosoy, A.; Stoddart, J. F.; Zink, J. I. Mesoporous silica nanoparticles in biomedical applications. *Chem Soc Rev* **2012**, *41* (7), 2590-2605.
9. He, X.; Nie, H.; Wang, K.; Tan, W.; Wu, X.; Zhang, P. *In vivo* study of biodistribution and urinary excretion of surface-modified silica nanoparticles. *Anal Chem* **2008**, *80* (24), 9597-9603.

10. Lin, Y.-S.; Haynes, C. L. Synthesis and characterization of biocompatible and size-tunable multifunctional porous silica nanoparticles. *Chem Mater* **2009**, *21* (17), 3979-3986.
11. Meng, H.; Xue, M.; Xia, T.; Ji, Z.; Tarn, D. Y.; Zink, J. I.; Nel, A. E. Use of size and a copolymer design feature to improve the biodistribution and the enhanced permeability and retention effect of doxorubicin-loaded mesoporous silica nanoparticles in a murine xenograft tumor model. *ACS Nano* **2011**, *5* (5), 4131-4144.
12. Lu, J.; Liong, M.; Zink, J. I.; Tamanoi, F. Mesoporous silica nanoparticles as a delivery system for hydrophobic anticancer drugs. *Small* **2007**, *3* (8), 1341-1346.
13. Bagwe, R. P.; Hilliard, L. R.; Tan, W. Surface modification of silica nanoparticles to reduce aggregation and nonspecific binding. *Langmuir* **2006**, *22* (9), 4357-4362.
14. Slowing, I.; Trewyn, B. G.; Lin, V. S. Y. Effect of surface functionalization of MCM-41-type mesoporous silica nanoparticles on the endocytosis by human cancer cells. *J Am Chem Soc* **2006**, *128* (46), 14792-14793.
15. Brevet, D.; Gary-Bobo, M.; Raehm, L.; Richeter, S.; Hocine, O.; Amro, K.; Loock, B.; Couleaud, P.; Frochot, C.; Morere, A.; Maillard, P.; Garcia, M.; Durand, J.-O. Mannose-targeted mesoporous silica nanoparticles for photodynamic therapy. *Chem Commun* **2009**, (12), 1475-1477.
16. Ferris, D. P.; Lu, J.; Gothard, C.; Yanes, R.; Thomas, C. R.; Olsen, J.-C.; Stoddart, J. F.; Tamanoi, F.; Zink, J. I. Synthesis of biomolecule-modified mesoporous silica nanoparticles for targeted hydrophobic drug delivery to cancer cells. *Small* **2011**, *7* (13), 1816-1826.
17. Aumailley, M.; Gurrath, M.; Müller, G.; Calvete, J.; Timpl, R.; Kessler, H. Arg-Gly-Asp constrained within cyclic pentapoptides: Strong and selective inhibitors of cell adhesion to vitronectin and laminin fragment P1. *FEBS Lett* **1991**, *291* (1), 50-54.
18. Gary-Bobo, M.; Mir, Y.; Rouxel, C.; Brevet, D.; Basile, I.; Maynadier, M.; Vaillant, O.; Mongin, O.; Blanchard-Desce, M.; Morère, A.; Garcia, M.; Durand, J.-O.; Raehm, L. Mannose-functionalized mesoporous silica nanoparticles for efficient two-photon photodynamic therapy of solid tumors. *Angew Chem Inter Ed* **2011**, *50* (48), 11425-11429.
19. Lu, J.; Liong, M.; Sherman, S.; Xia, T.; Kovochich, M.; Nel, A. E.; Zink, J. I.; Tamanoi, F. Mesoporous silica nanoparticles for cancer therapy: Energy-dependent cellular uptake and delivery of paclitaxel to cancer cells. *Nanobiotechnol* **2007**, *3* (2), 89-95.
20. Huang, D.-M.; Hung, Y.; Ko, B.-S.; Hsu, S.-C.; Chen, W.-H.; Chien, C.-L.; Tsai, C.-P.; Kuo, C.-T.; Kang, J.-C.; Yang, C.-S.; Mou, C.-Y.; Chen, Y.-C. Highly efficient cellular labeling of mesoporous nanoparticles in human mesenchymal stem cells: Implication for stem cell tracking. *FASEB J* **2005**, *19* (14), 2014-2016.
21. Lu, F.; Wu, S.-H.; Hung, Y.; Mou, C.-Y. Size effect on cell uptake in well-suspended, uniform mesoporous silica nanoparticles. *Small* **2009**, *5* (12), 1408-1413.
22. Huang, X.; Teng, X.; Chen, D.; Tang, F.; He, J. The effect of the shape of mesoporous silica nanoparticles on cellular uptake and cell function. *Biomaterials*

2010, *31* (3), 438-448.

23. Slowing, I. I.; Vivero-Escoto, J. L.; Zhao, Y.; Kandel, K.; Peeraphatdit, C.; Trewyn, B. G.; Lin, V. S. Y. Exocytosis of mesoporous silica nanoparticles from mammalian cells: From asymmetric cell-to-cell transfer to protein harvesting. *Small* **2011**, *7* (11), 1526-1532.
24. Di Pasqua, A. J.; Sharma, K. K.; Shi, Y.-L.; Toms, B. B.; Ouellette, W.; Dabrowiak, J. C.; Asefa, T. Cytotoxicity of mesoporous silica nanomaterials. *J Inorg Biochem* **2008**, *102* (7), 1416-1423.
25. Fisichella, M.; Dabboue, H.; Bhattacharyya, S.; Saboungi, M.-L.; Salvetat, J.-P.; Hevor, T.; Guerin, M. Mesoporous silica nanoparticles enhance MTT formazan exocytosis in HeLa cells and astrocytes. *Toxicology in Vitro* **2009**, *23* (4), 697-703.
26. Trewyn, B. G.; Slowing, I. I.; Giri, S.; Chen, H.-T.; Lin, V. S. Y. Synthesis and functionalization of a mesoporous silica nanoparticle based on the sol-gel process and applications in controlled release. *Accounts Chem Res* **2007**, *40* (9), 846-853.
27. Vallhov, H.; Gabrielsson, S.; Strømme, M.; Scheynius, A.; Garcia-Bennett, A. E. Mesoporous silica particles induce size dependent effects on human dendritic cells. *Nano Lett* **2007**, *7* (12), 3576-3582.
28. Rosenholm, J. M.; Sahlgren, C.; Linden, M. Towards multifunctional, targeted drug delivery systems using mesoporous silica nanoparticles: Opportunities & challenges. *Nanoscale* **2010**, *2* (10), 1870-1883.
29. Hudson, S. P.; Padera, R. F.; Langer, R.; Kohane, D. S. The biocompatibility of mesoporous silicates. *Biomaterials* **2008**, *29* (30), 4045-4055.
30. Ferrari, M. Cancer nanotechnology: Opportunities and challenges. *Nat Rev Cancer* **2005**, *5* (3), 161-171.
31. Cho, M.; Cho, W.-S.; Choi, M.; Kim, S. J.; Han, B. S.; Kim, S. H.; Kim, H. O.; Sheen, Y. Y.; Jeong, J. The impact of size on tissue distribution and elimination by single intravenous injection of silica nanoparticles. *Toxicology Lett* **2009**, *189* (3), 177-183.
32. Souris, J. S.; Lee, C.-H.; Cheng, S.-H.; Chen, C.-T.; Yang, C.-S.; Ho, J.-a. A.; Mou, C.-Y.; Lo, L.-W. Surface charge-mediated rapid hepatobiliary excretion of mesoporous silica nanoparticles. *Biomaterials* **2010**, *31* (21), 5564-5574.
33. Huang, X.; Li, L.; Liu, T.; Hao, N.; Liu, H.; Chen, D.; Tang, F. The shape effect of mesoporous silica nanoparticles on biodistribution, clearance, and biocompatibility *in vivo*. *ACS Nano* **2011**, *5* (7), 5390-5399.
34. He, Q.; Zhang, Z.; Gao, F.; Li, Y.; Shi, J. *In vivo* biodistribution and urinary excretion of mesoporous silica nanoparticles: Effects of particle size and pegylation. *Small* **2011**, *7* (2), 271-280.
35. Mamaeva, V.; Rosenholm, J. M.; Bate-Eya, L. T.; Bergman, L.; Peuhu, E.; Duchanoy, A.; Fortelius, L. E.; Landor, S.; Toivola, D. M.; Linden, M.; Sahlgren, C. Mesoporous silica nanoparticles as drug delivery systems for targeted inhibition of notch signaling in cancer. *Mol Ther* **2011**, *19* (8), 1538-1546.
36. Lu, J.; Liong, M.; Li, Z.; Zink, J. I.; Tamanoi, F. Biocompatibility, biodistribution, and drug-delivery efficiency of mesoporous silica nanoparticles for cancer therapy in animals. *Small* **2010**, *6* (16), 1794-1805.

37. Cheng, K.; Landry, C. C. Diffusion-based deprotection in mesoporous materials: A strategy for differential functionalization of porous silica particles. *J Am Chem Soc* **2007**, *129* (31), 9674-9685.
38. Mortera, R.; Fiorilli, S.; Garrone, E.; Verné, E.; Onida, B. Pores occlusion in MCM-41 spheres immersed in SBF and the effect on ibuprofen delivery kinetics: A quantitative model. *Chem Eng J* **2010**, *156* (1), 184-192.
39. Casasús, R.; Marcos, M. D.; Martínez-Máñez, R.; Ros-Lis, J. V.; Soto, J.; Villaescusa, L. A.; Amorós, P.; Beltrán, D.; Guillem, C.; Latorre, J. Toward the development of ionically controlled nanoscopic molecular gates. *J Am Chem Soc* **2004**, *126* (28), 8612-8613.
40. Bernardos, A.; Aznar, E.; Coll, C.; Martínez-Mañez, R.; Barat, J. M.; Marcos, M. D.; Sancenón, F.; Benito, A.; Soto, J. Controlled release of vitamin B2 using mesoporous materials functionalized with amine-bearing gate-like scaffoldings. *J Control Release* **2008**, *131* (3), 181-189.
41. Gao, Q.; Xu, Y.; Wu, D.; Sun, Y.; Li, X. pH-responsive drug release from polymer-coated mesoporous silica spheres. *J Phys Chem C* **2009**, *113* (29), 12753-12758.
42. Lee, C.-H.; Cheng, S.-H.; Huang, I. P.; Souris, J. S.; Yang, C.-S.; Mou, C.-Y.; Lo, L.-W. Intracellular pH-responsive mesoporous silica nanoparticles for the controlled release of anticancer chemotherapeutics. *Angew Chem Int Edit* **2010**, *49* (44), 8214-8219.
43. Liu, R.; Zhang, Y.; Zhao, X.; Agarwal, A.; Mueller, L. J.; Feng, P. pH-responsive nanogated ensemble based on gold-capped mesoporous silica through an acid-labile acetal linker. *J Am Chem Soc* **2010**, *132* (5), 1500-1501.
44. Patel, K.; Angelos, S.; Dichtel, W. R.; Coskun, A.; Yang, Y.-W.; Zink, J. I.; Stoddart, J. F. Enzyme-responsive snap-top covered silica nanocontainers. *J Am Chem Soc* **2008**, *130* (8), 2382-2383.
45. Park, C.; Kim, H.; Kim, S.; Kim, C. Enzyme responsive nanocontainers with cyclodextrin gatekeepers and synergistic effects in release of guests. *J Am Chem Soc* **2009**, *131* (46), 16614-16615.
46. Singh, N.; Karambelkar, A.; Gu, L.; Lin, K.; Miller, J. S.; Chen, C. S.; Sailor, M. J.; Bhatia, S. N. Bioresponsive mesoporous silica nanoparticles for triggered drug release. *J Am Chem Soc* **2011**, *133* (49), 19582-19585.
47. Ambrogio, M. W.; Pecorelli, T. A.; Patel, K.; Khashab, N. M.; Trabolsi, A.; Khatib, H. A.; Botros, Y. Y.; Zink, J. I.; Stoddart, J. F. Snap-top nanocarriers. *Org Lett* **2010**, *12* (15), 3304-3307.
48. Meister, A.; Anderson, M. E. Glutathione. *Annual Rev Biochem* **1983**, *52* (1), 711-760.
49. Lai, C.-Y.; Trewyn, B. G.; Jeftinija, D. M.; Jeftinija, K.; Xu, S.; Jeftinija, S.; Lin, V. S. Y. A mesoporous silica nanosphere-based carrier system with chemically removable CdS nanoparticle caps for stimuli-responsive controlled release of neurotransmitters and drug molecules. *J Am Chem Soc* **2003**, *125* (15), 4451-4459.
50. Torney, F.; Trewyn, B. G.; Lin, V. S. Y.; Wang, K. Mesoporous silica nanoparticles deliver DNA and chemicals into plants. *Nat Nano* **2007**, *2* (5), 295-300.
51. Giri, S.; Trewyn, B. G.; Stellmaker, M. P.; Lin, V. S. Y. Stimuli-responsive

controlled-release delivery system based on mesoporous silica nanorods capped with magnetic nanoparticles. *Angew Chem Inter Edit* **2005**, *44* (32), 5038-5044.

52. Zhu, C.-L.; Song, X.-Y.; Zhou, W.-H.; Yang, H.-H.; Wen, Y.-H.; Wang, X.-R. An efficient cell-targeting and intracellular controlled-release drug delivery system based on MSN-PEM-aptamer conjugates. *J Mater Chem* **2009**, *19* (41), 7765-7770.
53. Fu, Q.; Rama Rao, G. V.; Ward, T. L.; Lu, Y.; Lopez, G. P. Thermoresponsive transport through ordered mesoporous silica/PNIPAAm copolymer membranes and microspheres. *Langmuir* **2006**, *23* (1), 170-174.
54. Park, J.-H.; Lee, Y.-H.; Oh, S.-G. Preparation of thermosensitive PNIPAM-grafted mesoporous silica particles. *Macromol Chem Phy* **2007**, *208* (22), 2419-2427.
55. You, Y.-Z.; Kalebaila, K. K.; Brock, S. L.; Oupický, D. Temperature-controlled uptake and release in PNIPAM-modified porous silica nanoparticles. *Chem Mater* **2008**, *20* (10), 3354-3359.
56. Zhu, Y.; Kaskel, S.; Ikoma, T.; Hanagata, N. Magnetic SBA-15/poly(N-isopropylacrylamide) composite: Preparation, characterization and temperature-responsive drug release property. *Micropor Mesopor Mater* **2009**, *123* (1-3), 107-112.
57. Zintchenko, A.; Ogris, M.; Wagner, E. Temperature dependent gene expression induced by PNIPAM-based copolymers: Potential of hyperthermia in gene transfer. *Bioconjugate Chem* **2006**, *17* (3), 766-772.
58. Keerl, M.; Smirnovas, V.; Winter, R.; Richtering, W. Copolymer microgels from mono- and disubstituted acrylamides: Phase behavior and hydrogen bonds. *Macromolecules* **2008**, *41* (18), 6830-6836.
59. Schlossbauer, A.; Warncke, S.; Gramlich, P. M. E.; Kecht, J.; Manetto, A.; Carell, T.; Bein, T. A programmable DNA-based molecular valve for colloidal mesoporous silica. *Angew Chem Int Edit* **2010**, *49* (28), 4734-4737.
60. Mal, N. K.; Fujiwara, M.; Tanaka, Y. Photocontrolled reversible release of guest molecules from coumarin-modified mesoporous silica. *Nature* **2003**, *421* (6921), 350-353.
61. Lu, J.; Choi, E.; Tamanoi, F.; Zink, J. I. Light-activated nanoimpeller-controlled drug release in cancer cells. *Small* **2008**, *4* (4), 421-426.
62. Ferris, D. P.; Zhao, Y.-L.; Khashab, N. M.; Khatib, H. A.; Stoddart, J. F.; Zink, J. I. Light-operated mechanized nanoparticles. *J Am Cheml Soc* **2009**, *131* (5), 1686-1688.
63. Sanchez, A. M.; de Rossi, R. H. Effect of β-cyclodextrin on the thermal cis-trans isomerization of azobenzenes. *J Org Chem* **1996**, *61* (10), 3446-3451.
64. Park, C.; Lee, K.; Kim, C. Photoresponsive cyclodextrin-covered nanocontainers and their sol-gel transition induced by molecular recognition. *Angew Chem Int Edit* **2009**, *48* (7), 1275-1278.
65. Lai, J.; Mu, X.; Xu, Y.; Wu, X.; Wu, C.; Li, C.; Chen, J.; Zhao, Y. Light-responsive nanogated ensemble based on polymer grafted mesoporous silica hybrid nanoparticles. *Chem Commun* **2010**, *46* (39), 7370-7372.
66. Knezevic, N. Z.; Trewyn, B. G.; Lin, V. S. Y. Functionalized mesoporous silica nanoparticle-based visible light responsive controlled release delivery system. *Chem Commun* **2011**, *47* (10), 2817-2819.

67. Chen, P.-J.; Hu, S.-H.; Hsiao, C.-S.; Chen, Y.-Y.; Liu, D.-M.; Chen, S.-Y. Multifunctional magnetically removable nanogated lids of Fe_3O_4-capped mesoporous silica nanoparticles for intracellular controlled release and MR imaging. *J Mater Chem* **2011**, *21* (8), 2535-2543.
68. Thomas, C. R.; Ferris, D. P.; Lee, J.-H.; Choi, E.; Cho, M. H.; Kim, E. S.; Stoddart, J. F.; Shin, J.-S.; Cheon, J.; Zink, J. I. Noninvasive remote-controlled release of drug molecules *in vitro* using magnetic actuation of mechanized nanoparticles. *J Am Chem Soc* **2010**, *132* (31), 10623-10625.
69. Ruiz-Hernández, E.; Baeza, A.; Vallet-Regí, M. Smart drug delivery through DNA/magnetic nanoparticle gates. *ACS Nano* **2011**, *5* (2), 1259-1266.
70. Liu, R.; Zhang, Y.; Feng, P. Multiresponsive supramolecular nanogated ensembles. *J Am Chem Soc* **2009**, *131* (42), 15128-15129.
71. Chang, B.; Sha, X.; Guo, J.; Jiao, Y.; Wang, C.; Yang, W. Thermo and pH dual responsive, polymer shell coated, magnetic mesoporous silica nanoparticles for controlled drug release. *J Mater Chem* **2011**, *21* (25), 9239-9247.
72. Chen, C.; Geng, J.; Pu, F.; Yang, X.; Ren, J.; Qu, X. Polyvalent nucleic acid/mesoporous silica nanoparticle conjugates: Dual stimuli-responsive vehicles for intracellular drug delivery. *Angew Chem Int Edit* **2011**, *50* (4), 882-886.
73. Khemtong, C.; Kessinger, C. W.; Gao, J. Polymeric nanomedicine for cancer MR imaging and drug delivery. *Chem Commun* **2009**, (24), 3497-3510.
74. Lin, Y.-S.; Hung, Y.; Su, J.-K.; Lee, R.; Chang, C.; Lin, M.-L.; Mou, C.-Y. Gadolinium(Ⅲ)-incorporated nanosized mesoporous silica as potential magnetic resonance imaging contrast agents. *J Phys Chem B* **2004**, *108* (40), 15608-15611.
75. Taylor, K. M. L.; Kim, J. S.; Rieter, W. J.; An, H.; Lin, W.; Lin, W. Mesoporous silica nanospheres as highly efficient MRI contrast agents. *J Am Chem Soc* **2008**, *130* (7), 2154-2155.
76. Kang, X.; Cheng, Z.; Yang, D.; Ma, P. A.; Shang, M.; Peng, C.; Dai, Y.; Lin, J. Drug delivery: Design and synthesis of multifunctional drug carriers based on luminescent rattle-type mesoporous silica microspheres with a thermosensitive hydrogel as a controlled switch. *Adv Function Mater* **2012**, *22* (7), 1539.
77. Wei, Y.; Dong, H.; Xu, J.; Feng, Q. Simultaneous immobilization of horseradish peroxidase and glucose oxidase in mesoporous sol-gel host materials. *ChemPhysChem* **2002**, *3* (9), 802-808.
78. Dai, Z.; Liu, S.; Ju, H.; Chen, H. Direct electron transfer and enzymatic activity of hemoglobin in a hexagonal mesoporous silica matrix. *Biosens Bioelectron* **2004**, *19* (8), 861-867.
79. Dai, Z.; Xu, X.; Ju, H. Direct electrochemistry and electrocatalysis of myoglobin immobilized on a hexagonal mesoporous silica matrix. *Anal Biochem* **2004**, *332* (1), 23-31.
80. Radu, D. R.; Lai, C.-Y.; Wiench, J. W.; Pruski, M.; Lin, V. S. Y. Gatekeeping layer effect: A poly(lactic acid)-coated mesoporous silica nanosphere-based fluorescence probe for detection of amino-containing neurotransmitters. *J Am Chem Soc* **2004**, *126* (6), 1640-1641.
81. Slowing, I. I.; Trewyn, B. G.; Lin, V. S. Y. Mesoporous silica nanoparticles for intracellular delivery of membrane-impermeable proteins. *J Am Chem Soc* **2007**, *129* (28), 8845-8849.

82. Rosenholm, J. M.; Peuhu, E.; Eriksson, J. E.; Sahlgren, C.; Lindén, M. Targeted intracellular delivery of hydrophobic agents using mesoporous hybrid silica nanoparticles as carrier systems. *Nano Lett* **2009**, *9* (9), 3308-3311.

83. Dai, C.; Yuan, Y.; Liu, C.; Wei, J.; Hong, H.; Li, X.; Pan, X. Degradable, antibacterial silver exchanged mesoporous silica spheres for hemorrhage control. *Biomaterials* **2009**, *30* (29), 5364-5375.

84. Mellaerts, R.; Mols, R.; Jammaer, J. A. G.; Aerts, C. A.; Annaert, P.; Van Humbeeck, J.; Van den Mooter, G.; Augustijns, P.; Martens, J. A. Increasing the oral bioavailability of the poorly water soluble drug itraconazole with ordered mesoporous silica. *Eur J Pharmaceut Biopharmaceut* **2008**, *69* (1), 223-230.

85. Rosenholm, J.; Sahlgren, C.; Linden, M. Cancer-cell targeting and cell-specific delivery by mesoporous silica nanoparticles. *J Mater Chem* **2010**, *20* (14), 2707-2713.

86. De Wolf, H. K.; Snel, C. J.; Verbaan, F. J.; Schiffelers, R. M.; Hennink, W. E.; Storm, G. Effect of cationic carriers on the pharmacokinetics and tumor localization of nucleic acids after intravenous administration. *Inter J Pharmaceut* **2007**, *331* (2), 167-175.

87. Rosenholm, J. M.; Peuhu, E.; Bate-Eya, L. T.; Eriksson, J. E.; Sahlgren, C.; Lindén, M. Cancer-cell-specific induction of apoptosis using mesoporous silica nanoparticles as drug-delivery vectors. *Small* **2010**, *6* (11), 1234-1241.

88. Tsai, C.-P.; Chen, C.-Y.; Hung, Y.; Chang, F.-H.; Mou, C.-Y. Monoclonal antibody-functionalized mesoporous silica nanoparticles (MSN) for selective targeting breast cancer cells. *J Mater Chem* **2009**, *19* (32), 5737-5743.

89. Huang, X.; Peng, X.; Wang, Y.; Wang, Y.; Shin, D. M.; El-Sayed, M. A.; Nie, S. A reexamination of active and passive tumor targeting by using rod-shaped gold nanocrystals and covalently conjugated peptide ligands. *ACS Nano* **2011**, *5* (8), 6765.

90. Ruoslahti, E.; Bhatia, S. N.; Sailor, M. J. Targeting of drugs and nanoparticles to tumors. *J Cell Biology* **2010**, *188* (6), 759-768.

91. Lu, J.; Liong, M.; Zink, Jeffrey I.; Tamanoi, F. Mesoporous silica nanoparticles as a delivery system for hydrophobic anticancer drugs. *Small* **2007**, *3* (8), 1341-1346.

92. Nakamizo, A.; Marini, F.; Amano, T.; Khan, A.; Studeny, M.; Gumin, J.; Chen, J.; Hentschel, S.; Vecil, G.; Dembinski, J.; Andreeff, M.; Lang, F. F. Human bone marrow-derived mesenchymal stem cells in the treatment of gliomas. *Cancer Research* **2005**, *65* (8), 3307-3318.

93. Gottesman, M. M.; Fojo, T.; Bates, S. E. Multidrug resistance in cancer: Role of Atp-dependent transporters. *Nat Rev Cancer* **2002**, *2* (1), 48-58.

94. Hanahan, D.; Weinberg, Robert A. Hallmarks of cancer: The next generation. *Cell* **2011**, *144* (5), 646-674.

95. Ranganathan, P.; Weaver, K. L.; Capobianco, A. J. Notch signalling in solid tumours: A little bit of everything but not all the time. *Nat Rev Cancer* **2011**, *11* (5), 338-351.

96. Ashley, C. E.; Carnes, E. C.; Phillips, G. K.; Padilla, D.; Durfee, P. N.; Brown, P. A.; Hanna, T. N.; Liu, J.; Phillips, B.; Carter, M. B.; Carroll, N. J.; Jiang, X.; Dunphy, D. R.; Willman, C. L.; Petsev, D. N.; Evans, D. G.; Parikh, A. N.;

Chackerian, B.; Wharton, W.; Peabody, D. S.; Brinker, C. J. The targeted delivery of multicomponent cargos to cancer cells by nanoporous particle-supported lipid bilayers. *Nat Mater* **2011**, *10* (6), 476.

97. Chen, A. M.; Zhang, M.; Wei, D.; Stueber, D.; Taratula, O.; Minko, T.; He, H. Co-delivery of doxorubicin and Bcl-2 siRNA by mesoporous silica nanoparticles enhances the efficacy of chemotherapy in multidrug-resistant cancer cells. *Small* **2009**, *5* (23), 2673-2677.
98. Cheng, S.-H.; Lee, C.-H.; Yang, C.-S.; Tseng, F.-G.; Mou, C.-Y.; Lo, L.-W. Mesoporous silica nanoparticles functionalized with an oxygen-sensing probe for cell photodynamic therapy: Potential cancer theranostics. *J Mater Chem* **2009**, *19* (9), 1252-1257.
99. Cheng, S.-H.; Hsieh, C.-C.; Chen, N.-T.; Chu, C.-H.; Huang, C.-M.; Chou, P.-T.; Tseng, F.-G.; Yang, C.-S.; Mou, C.-Y.; Lo, L.-W. Well-defined mesoporous nanostructure modulates three-dimensional interface energy transfer for two-photon activated photodynamic therapy. *Nano Today* **2011**, *6* (6), 552-563.
100. Radu, D. R.; Lai, C.-Y.; Jeftinija, K.; Rowe, E. W.; Jeftinija, S.; Lin, V. S. Y. A polyamidoamine dendrimer-capped mesoporous silica nanosphere-based gene transfection reagent. *J Am Chem Soc* **2004**, *126* (41), 13216-13217.
101. Park, I. Y.; Kim, I. Y.; Yoo, M. K.; Choi, Y. J.; Cho, M.-H.; Cho, C. S. Mannosylated polyethylenimine coupled mesoporous silica nanoparticles for receptor-mediated gene delivery. *Inter J Pharmaceut* **2008**, *359* (1-2), 280-287.
102. Hom, C.; Lu, J.; Liong, M.; Luo, H.; Li, Z.; Zink, J. I.; Tamanoi, F. Mesoporous silica nanoparticles facilitate delivery of siRNA to shutdown signaling pathways in mammalian cells. *Small* **2010**, *6* (11), 1185-1190.
103. Yu, J.; Zhao, H.; Ye, L.; Yang, H.; Ku, S.; Yang, N.; Xiao, N. Effect of surface functionality of magnetic silica nanoparticles on the cellular uptake by glioma cells *in vitro*. *J Mater Chem* **2009**, *19* (9), 1265-1270.
104. Liu, R.; Zhao, X.; Wu, T.; Feng, P. Tunable redox-responsive hybrid nanogated ensembles. *J Am Chem Soc* **2008**, *130* (44), 14418-14419.
105. Wang, L.-S.; Wu, L.-C.; Lu, S.-Y.; Chang, L.-L.; Teng, I. T.; Yang, C.-M.; Ho, J.-a. A. Biofunctionalized phospholipid-capped mesoporous silica nanoshuttles for targeted drug delivery: Improved water suspensibility and decreased non-specific protein binding. *ACS Nano* **2010**, *4* (8), 4371-4379.
106. Liong, M.; France, B.; Bradley, K. A.; Zink, J. I. Antimicrobial activity of silver nanocrystals encapsulated in mesoporous silica nanoparticles. *Adv Mater* **2009**, *21* (17), 1684-1689.
107. Nooney, R. I.; Dhanasekaran, T.; Chen, Y.; Josephs, R.; Ostafin, A. E. Self-assembled highly ordered spherical mesoporous silica/gold nanocomposites. *Adv Mater* **2002**, *14* (7), 529-532.
108. Nooney, R. I.; Thirunavukkarasu, D.; Chen, Y.; Josephs, R.; Ostafin, A. E. Self-assembly of mesoporous nanoscale silica/gold composites. *Langmuir* **2003**, *19* (18), 7628-7637.
109. Lin, K. J.; Chen, L. J.; Prasad, M. R.; Cheng, C. Y. Core-shell synthesis of a novel, spherical, mesoporous silica/platinum nanocomposite: Pt/PVP@MCM-41. *Adv Mater* **2004**, *16* (20), 1845-1849.
110. Gorelikov, I.; Matsuura, N. Single-step coating of mesoporous silica on

cetyltrimethyl ammonium bromide-capped nanoparticles. *Nano Lett* **2007**, *8* (1), 369-373.

111. Wang, Y.; Ren, J.; Liu, X.; Wang, Y.; Guo, Y.; Guo, Y.; Lu, G. Facile synthesis of ordered magnetic mesoporous γ-Fe_2O_3/SiO_2 nanocomposites with diverse mesostructures. *J Colloid Int Sci* **2008**, *326* (1), 158-165.
112. Kim, J.; Kim, H. S.; Lee, N.; Kim, T.; Kim, H.; Yu, T.; Song, I. C.; Moon, W. K.; Hyeon, T. Multifunctional uniform nanoparticles composed of a magnetite nanocrystal core and a mesoporous silica shell for magnetic resonance and fluorescence imaging and for drug delivery. *Angew Chem Int Edit* **2008**, *120* (44), 8566-8569.
113. Deng, Y.; Qi, D.; Deng, C.; Zhang, X.; Zhao, D. Superparamagnetic high-magnetization microspheres with an Fe_3O_4@SiO_2 core and perpendicularly aligned mesoporous SiO_2 shell for removal of microcystins. *J Am Chem Soc* **2007**, *130* (1), 28-29.
114. Zhao, W.; Shi, J.; Chen, H.; Zhang, L. Particle size, uniformity, and mesostructure control of magnetic core/mesoporous silica shell nanocomposite spheres. *J Mater Res* **2006**, *21* (12), 3080-3089.
115. Stöber, W.; Fink, A.; Bohn, E. Controlled growth of monodisperse silica spheres in the micron size range. *J Colloid Interf Sci* **1968**, *26* (1), 62-69.
116. Wu, P.; Zhu, J.; Xu, Z. Template-assisted synthesis of mesoporous magnetic nanocomposite particles. *Adv Function Mater* **2004**, *14* (4), 345-351.
117. Meng, H.; Liong, M.; Xia, T.; Li, Z.; Ji, Z.; Zink, J. I.; Nel, A. E. Engineered design of mesoporous silica nanoparticles to deliver doxorubicin and P-glycoprotein siRNA to overcome drug resistance in a cancer cell line. *ACS Nano* **2010**, *4* (8), 4539-4550.
118. Taylor-Pashow, K. M. L.; Rocca, J. D.; Lin, W. Mesoporous silica nanoparticles with co-condensed gadolinium chelates for multimodal imaging. *Nanomaterials* **2011**, *2* (1), 1-14.
119. Tong, L.; Wei, Q.; Wei, A.; Cheng, J.-X. Gold nanorods as contrast agents for biological imaging: Optical properties, surface conjugation and photothermal effects. *Photochem Photobiol* **2009**, *85* (1), 21-32.
120. Zhang, Z.; Wang, L.; Wang, J.; Jiang, X.; Li, X.; Hu, Z.; Ji, Y.; Wu, X.; Chen, C. Mesoporous silica-coated gold nanorods as a light-mediated multifunctional theranostic platform for cancer treatment. *Adv Mater* **2012**, *24* (11), 1418-1423.
121. Cheng, S.-H.; Lee, C.-H.; Chen, M.-C.; Souris, J. S.; Tseng, F.-G.; Yang, C.-S.; Mou, C.-Y.; Chen, C.-T.; Lo, L.-W. Tri-functionalization of mesoporous silica nanoparticles for comprehensive cancer theranostics: The trio of imaging, targeting and therapy. *J Mater Chem* **2010**, *20* (29), 6149-6157.

Chapter 6

Surface Enhanced Raman Scattering (SERS) Nanoprobes as Cancer Theranostics

*Meikun Fan**

Faculty of Geoscience and Environmental Engineering, Southwest Jiaotong University, Chengdu 611756, China

Surface enhanced Raman scattering (SERS) is an analytical method that is well known for its high sensitivity, specificity, flexibility (almost any molecules, under any excitation wavelength), and capability of providing multiplicity.[1] Today, SERS not only serves as one of the most powerful tools for exploring material properties at single molecule level, but also plays very important role in applied sciences. In environmental analysis, it has been used not only for the determination of chemical species[2], both inorganic and organic, but also live objects such as bacteria[2,3]. In biomedical sciences, SERS based methods were reported to be able to monitor the chemical reactions inside a single cell[4], or probe different cancer markers[5] and cells[6]. In this chapter, the fundamentals of Raman scattering and SERS, the development of SERS nanoparticles (NPs), and the development and current state of SERS theranostics will be discussed.

6.1 Introduction

6.1.1 *Raman Scattering*

When shining light onto molecules, it is believed that 1 in 1 000 incident photons will be scattered.[7] Once there is an energy exchange between the scattered photons and the molecule, it is called Raman scattering, and the probability is roughly 10^6–10^8 excitation photons to have one Raman scattered photon.[8] This was first experimentally observed in 1921 by C. V. Raman and named after him.[1] Incredibly, Mr. Raman only used the sun as light source and some color filters, and observed what he called as "a new type of secondary radiation" that later led to his Nobel prize in physics in 1930.

This process can be described in energy diagram as shown in Fig.6.1. When a photon with energy $h\nu_L$ hits a molecule that sits on its ground vibrational state (V_0), the molecule will get excited to a virtual state, which describes a short-lived state that depends on the energy of the laser rather than solely on its molecular structure. When the photon is re-emitted, some energy can be

* Corresponding author: meikunfan@gmail.com

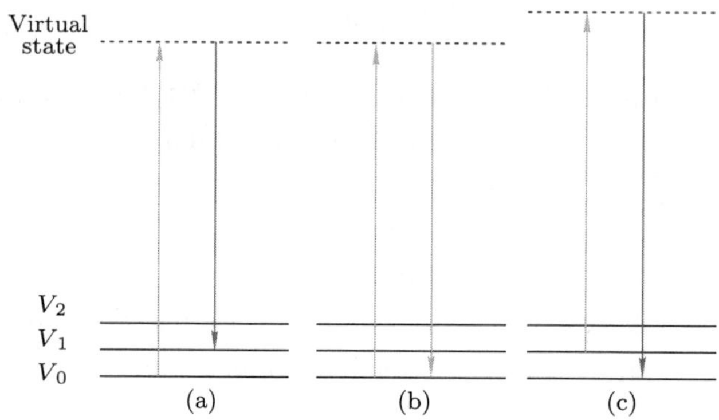

Fig. 6.1 Schematic energy diagram describing the light scattering process. (a) Stokes' Raman scattering. (b) Rayleigh scattering. (c) Anti-Stokes' Raman scattering.

transferred to the molecule. In this case, it is called Stokes' Raman scattering [Fig.6.1(a)]. If, on the contrary, energy is transferred from molecule to photon, it is called anti-Stokes' Raman scattering [Fig.6.1(c)]. If there is no energy exchange, as shown in Fig.6.1(b), then it is Rayleigh scattering. What C. V. Raman saw in the 1920s is Stokes' Raman scattering, the wavelength of which is longer than the incident light and much stronger than that of anti-Stokes' Raman scattering. The Raman scattering frequency is the difference between the incident and scattered frequencies, and sometimes it is called Raman shift. That is

$$\nu_{\text{Raman}} = \nu_{\text{L}} - \nu_{\text{S}} \tag{6.1}$$

Because of this, Raman shift in general is not dependent on the excitation laser wavelength [although the (relative) intensities of the spectrum may vary a lot due to the scattering law and possible matching with absorption band].

Both Raman and IR spectra describe the vibrational information of the molecule examined, share the same unit and roughly close band positions for same functional groups. However, they are fundamentally different in the mechanism. For Raman scattering, there has to be a change in the polarizability in the specific normal mode vibration, while in IR there needs to be a dipole moment change. Thus, the symmetric stretch of carbon dioxide is IR inactive but Raman active.

6.1.2 *Resonance Raman Scattering*

Normal Raman scattering is very inefficient. Very few molecules can be excited to their virtual state and then emit Raman scattered photons [Fig.6.2(a)]. This is because the virtual state is too short-lived. One way to improve the probability of Raman scattering is to carefully choose laser frequency to match the absorption of the molecule so that the molecule will be excited to its real electronic state [Fig.6.2(b)] instead of the virtual ones [Fig.6.2(a)]. This is called resonance Raman scattering (RRS) and the probability of Raman scattering can be greatly increased. There are reports estimated that up to 10^6 times enhancement can be achieved compared with normal Raman scattering,[9] though 2–3 orders of magnitude of enhancement is more common.

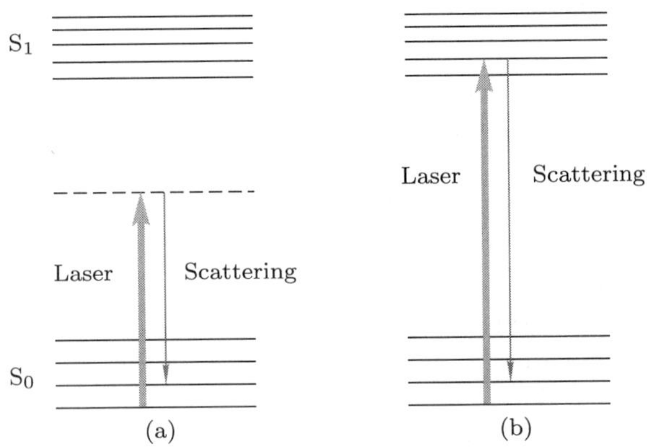

Fig. 6.2 Schematic diagram of normal Raman scattering (a) and resonance Raman scattering (RRS) (b).

6.1.3 *Localized Surface Plasmons and SERS*

Normal Raman scattering is really not a sensitive analytical method. It can mostly be used for pure materials analysis. On the other hand, the condition to trigger resonance Raman scattering is dependent on each molecule. It is not practical to tune laser wavelength to match all molecules. It was not until 1970s, when abnormal intense Raman scattering was found for pyridine adsorbed on electrochemically roughened silver electrode,[10–12] that Raman technique finally showed its promising future.

Soon after the discovery of SERS, Moskovits proposed that the gigantic enhancement is due to the pre-resonant or resonant excitation of the conducting electrons in the "metal bumps", the surface plasmons.[13] Later, Aroca pointed out that charge transfer might also play a role in the observed SERS effect.[8] Today, we know that up to 12 orders of magnitude of enhancement relative to the normal Raman response can be found for species at "hot spots" of the SERS substrate, though normally an average of 10^6 is observed.[14] Single molecule SERS detection is widely reported[15,16], even with the electrochemically roughed electrode as substrate[17]. SERS has become one of the most important analytical methods, the one that can give not only the quantity information, but also the fingerprint of the analyte of minute amount of sample, and have very wide applications, from environmental analysis to homeland security.[18] In the following part, we will discuss the currently widely accepted mechanism of SERS.

In SERS community, it is believed that there are mainly two kinds of theories to describe the origin of SERS.[13,19–22] The first one accounts for most of the observed SERS and is called the electromagnetic (EM) mechanism.[13] The second and less important one is the charge transfer (CT) mechanism.[21]

Let us start with the EM theory, which involves the excitation of surface plasmons (SPs). SPs are the collective oscillation of free electrons in the metals. For metallic NPs made of gold, silver, and copper, when the size of NPs decreases to some extent (from a few to ~100 nm), the color looks very different from their bulk materials. This is due to the excitation of SPs on the NPs.[23] When light is shining on these NPs, since light is electromagnetic field, once the frequency of the incident light matches the inherent frequency of the NPs, the free electrons in NPs will oscillate collectively with incident light and in turn causes the absorption of the electromagnetic wave, which is called (localized) surface plasmons [(L)SPs] and gives the NPs different colors (Fig.6.3). It should be noted that for gold, silver, and copper, the absorption frequency falls into visible range. It is for this reason that most widely examined metals in SERS community are these three metals. It does not necessarily mean other materials do not support SPs.

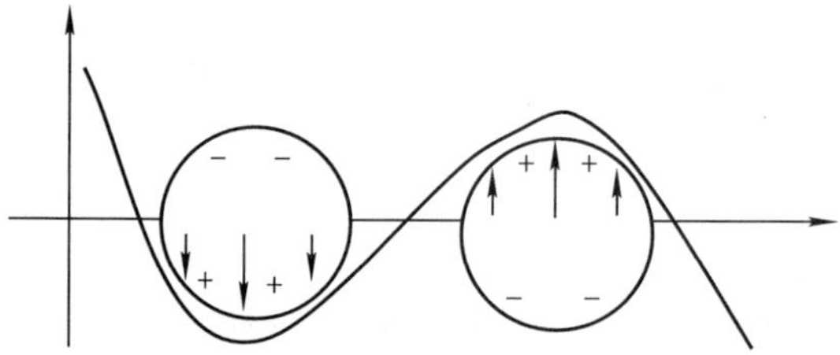

Fig. 6.3 Localized SPs on metallic NP excited by free-space light, adapted from Ref. 23.

One of the most important aspects of SPs is that the electromagnetic field can be localized by the excitation of SPs (or, in other words, "enhanced"). Calculation indicates that up to 10^4 of local electric field $|E|^2$ can be found for single NPs with specific morphology.[24] Furthermore, when two or more NPs aggregate together with proper distance (*e.g.*, 2 nm), at the junction the $|E|^2$ can get further one order of magnitude enhancement.[24,25] Some low probability optical phenomena, such as Raman scattering, can be greatly enhanced due to the localization of electromagnetic field (SERS).

We consider a single sphere gold NP,[13] experiencing an electromagnetic field E_0 from the incident laser. Due to the excitation of LSPs, the magnitude of the average field radiated by the NP E_s will be: $E_s = gE_0$, where g is the field enhancement factor. Therefore, the field strength of Raman scattering light from a molecule that adsorbed on the surface of the NP will be: $E_R \propto \alpha_R E_s \propto \alpha_R g E_0$, where α_R is the Raman tensor. But this is not the ultimate scattered field. The NP can also scatter light at the Raman-shifted frequency enhanced by a factor, say, g'. Hence, for SERS, the amplitude of the scattered field can be expressed as: $E_{\mathrm{SERS}} \propto \alpha_{\mathrm{R}} g g' E_0$. Since the intensity is just the square of the amplitude of the field, the SERS intensity now can be described as

$$I_{\mathrm{SERS}} \propto |\alpha_{\mathrm{R}}|^2 |gg'|^2 I_0 \tag{6.2}$$

where I_0 is the intensity of the incident light.

For low frequency (small shift) Raman scattering, it is very close to the wavelength of the incident light, such that we have $g \cong g'$. Therefore, Eq. (6.2) can be re-written as

$$I_{\mathrm{SERS}} \propto |\alpha_{\mathrm{R}}|^2 |g|^4 I_0 \tag{6.3}$$

It is clear that the SERS intensity would be proportional to the fourth power of the enhancement of the local field.

One more frequently talked phrase is SERS hot spot. As discussed above, when NPs are close enough, at the junction of NPs, the local field can be one order of magnitude higher than single NP, due to the coupling of induced dipole.[24] It is also true for some single NPs with sharp edge, the local field at the edge will be much higher than other places due to lighting rod effect. As a result, molecules located in this junction or the sharp edge will experience much higher field and hence contribute most to the observed SERS signal. This is both theoretically[24] and experimentally[25] proved in literature. Because of this, it is called the hot spot and the basis for SERS single molecule measurement.[1,3,26] It is also interesting to note that due to the coupling, the peak position of the LSPs will shift to longer wavelength, which can be rationalized by taking the analogy with the case of atomic orbital. The forming of "bonding orbital" of the LSP from the two NPs will make the extinction spectrum of the NP red shift.[1] This also explains when NPs aggregate, the optimum excitation laser wavelength does not necessarily match the absorption of single NP to get maximum SERS enhancement.

Another very important aspect deduced from the EM enhancement is that SERS is strongly dependent on the distance between the target molecule and the metal surface. The decay of field intensity in respect to distance from NPs' surface to SPs field is exponential.[1,27] The closer the molecules are, the larger field it will experience, and hence stronger SERS intensity is induced. This difficulty may explain why few researches characterize the adsorbates of multiple layers with SERS.[28]

Although SPs play main role in most of the SERS phenomena observed, some experiments indicate that other mechanism may apply.[19–21] For example, when the potential in electrochemical environment is changed, the SERS intensity will vary as well. This can be explained by CT mechanism. When molecules adsorbed onto metal surfaces with atomic level of roughness, they may react with adatoms and form complexes. Due to the complex formation, the energy gap between metal's Fermi level and the molecular frontier orbitals may match the energy of incident laser photon. If this happens, it is like a resonance Raman scattering process, the odds of emitting Raman photons will be much higher. By applying potential on the electrode, the Fermi level is perturbed. This can be used to tune the system to match the laser energy and achieve resonant Raman scattering effect as well. However, the enhancement from CT process is rather weak compared with that from SPs field. Some people[1] believe that CT process rarely provide enhancement more than 10 times. Further, it is worth mentioning that CT effect is a strict "first layer effect", *i.e.*, only the molecules in touch with the NPs' surface will benefit from this effect.[1]

6.2 Development of NPs Based SERS Probe

Electrochemically roughed Ag electrode was the very first SERS substrate discovered in 1974.[11] Today, there are many types of SERS substrates.[22,29] We can roughly divide them into two groups: chemically (wet) made, and physically fabricated substrates. The latter can be further divided into directly physical vapor deposition made, such as the so called silver film over nanosphere (Ag-FON)[30], and nanofabricated SERS substrates[31]. For the chemically made SERS substrates, we can also arbitrarily separate them into two categories: solution prepared "free" (dispersed) NPs/NPs aggregates[32] and nanostructures on some sort of solid support[28].

Here, we will mainly focus on dispersed metallic NPs since they are the ones that used in SERS diagnostics. Metallic NPs have been widely used in SERS experiment since 1979.[26,33,34] The major advantage of using metallic NPs is that they are easy to be synthesized through regular wet chemistry, and their sizes and shapes are controllable. As a result, the surface plasmons can be tuned to match different laser excitations. Currently, the active research in this area is aiming to improve both the sensitivity and the reproducibility. The former involves the synthesis of metallic NPs with different structures and shapes, even materials; the latter is achieved mainly from the attempts in controlling the size distributions.

6.2.1 *SERS NPs Made of Single Pure Plasmon Supporting Metals*

Gold and silver are the most common materials for preparing SERS active NPs. The most widely used SERS active NPs are nanospheres. Both Au and Ag NPs can be prepared by the using of sodium citrate as reducing reagent with heating, and $HAuCl_4$ and $AgNO_3$ as metal precursors, respectively.[22,35] For example, by adding 50 mL of 38.8 mM of sodium citrate into boiling 500 mM of $HAuCl_4$ solution, and keeping boiling while stirring for 15 min, 13 nm Au nanospheres can be prepared. The size distribution of Au nanospheres prepared in this way is very narrow, and can be stored in fridge stably for more than one year. Furthermore, the size of the nanospheres can be adjusted by tuning the amount of sodium citrate added.[36] There is even report[37] indicating by using seeded growth, gold spheres from nano (~50 nm) to micro (~1 μm) size can be prepared using $NaBH_4$ and ascorbic acid as reducing reagents in the presence of CTAB and KI. Ag NPs can also be made by reducing silver salt with citrate, the most commonly used one is of ~ 60 nm in size.[1,36,38,39] However, Ag NPs produced by sodium citrate, in general, has large size distribution and does not always turn spherical in shape. Further, it is less stable compared with Au NPs.

Gold nanorod is probably one of the most promising SERS NPs that could also be used in cancer photothermal treatment.[40–45] This is due to its special property. Its SPs excitation wavelength could be tuned into near infrared (NIR), which makes *in vivo* SERS diagnostic possible. Meanwhile, it is more efficient in absorbing the NIR laser.[46] Generally, the synthesis of gold nanorod involves surfactants directed growth of small nanosphere seeds.[47] CTAB, ascorbic acid, and $AgNO_3$ are the typically involved reagents.[48] Reports indicate that gold nanorod with an aspect ratio of ~25 can be prepared with high yielding.[47] Photochemical synthesis of gold nanorods has also been reported.[49] Interested readers are encouraged to review a recent book chapter about the preparation and characterization of gold nanorods.[50] The application of Au NPs as imaging and therapeutic element has also been reviewed recently.[45,51–53] The preparation of silver nanorods and nanowires has also been widely reported.[54]

Ag nanoprisms can be prepared through a photo-induced conversion of silver nanospheres.[55] First, the Ag nanosphere was prepared by using both $NaBH_4$ and sodium citrate as reducing reagent. Then, the as prepared Ag spheres were illuminated with conventional 40 W fluorescent lamp for 70 hours, during which almost all the nanospheres could be transferred into nanoprisms. The edge length of the as prepared nanoprism is about 100 nm. Ag nanoprisms can also be prepared through a thermal route by optimizing the concentrations of citrate, poly(vinylpyrrolidone), H_2O_2, and $NaBH_4$[56], or microwave assisted method[57]. For gold nanoprisms, reports showed that a seeded growth with the addition of CTAB and halide ions was a good approach.[58] Mirkin and coworkers also found that adding NaOH together with CTAB in a seeded growth procedure could produce gold nanoprism as well.[59]

SERS NPs with sharp edges, such as nanocubes[60,61], nanostars[62,63], nanostarfruit[64], nanocrescent[65], and even nanocages[61,66] have also been reported. Due to lighting rod effect, the SPs at the edge or corner of the NPs will be much stronger.[67] Hence, it is believed that these kinds of NPs can have higher SERS efficiency, up to a few orders of magnitude,[62] and hence will have wider application in biomedical research.

6.2.2 *Bimetallic SERS NPs*

Composition of the nanomaterials is one of the most important parameters when pursuing for optimized SERS substrates. Au and Ag are the two most used plasmonic metals. It is widely accepted that Ag is a more effective plasmonic material than Au. However, Au provides advantages in terms of chemical stability. Bimetallic Au:Ag materials have the potential to combine the advantages of both metals in terms of chemical and plasmonic properties. In fact, alloy and core-shell metallic NPs can show quite different SERS performance.[68–71] Some reports show that NPs made of both gold and silver metals can provide higher SERS activity compared with NPs made from either pure gold or silver.[69,70] Particularly, if the bimetallic NPs are core shell structures, pin-holes on the shell can bring extra 3–4 times of enhancement.[70,72] On top of that, recent work[73] indicates that, for small alloy NPs made of Au and Ag, there is preferential enhancement to specific type of reporter molecules for a given composition. That is, if working with bimetallic NPs, one needs to consider not only the property of the NPs, but also the Raman reporters, to achieve best SERS efficiency. Other than gold and silver, copper and gold bimetallic NPs have also been examined as SERS substrates.[74] It is believed that for Au core copper shell NPs, CT mechanism plays important role in the observed SERS enhancement.

There are also other approaches to have high sensitivity SERS NPs. One brilliant approach was to sandwich Raman reporter between the core and shell of bimetallic NPs,[75] which provided much higher SERS enhancement.

Regarding the fabrication of bimetallic NPs, it is relatively straight forward. In most scenarios,[71,76] a seed and a growth solution, which contains metal salt, and a mild reducing reagent, are good enough for the synthesis of bimetallic NPs. The co-reduction of both metal salts in solution is also possible.[73,77]

6.2.3 *Dielectric-metal Complex SERS NPs*

There are a few motivations for hunting dielectric-metal complex as SERS probes. For example, a multilayer dielectric-metal-dielectric-metal structure is believed to be superior in SERS efficiency compared with single pure metal nanosphere as well as dielectric-metal structure.[78] It is also believed that through such a protocol, the structures of the NPs can be tailored to match specific laser frequency.[79,80] This is very important in bioimaging, where NIR laser is preferred (to penetrate into the tissue). Thirdly, by carefully choosing the dielectric material, extra dimension of the SERS NPs can be explored. For instance, using Fe_2O_3 as core material, and coating silver or gold as shell, the as prepared SERS NPs can have magnetic property and hence can be separated using magnets. This could be used in detection of cancer cell.[81,82] Finally, it is also possible to coat metallic NPs, such as gold or silver, with a dielectric layer.

The benefits are two folds. On one hand, this can be a protecting layer. For example, thin layer of SiO_2 or Al_2O_3 have been proposed to cover the SERS NPs[83], and hence to separate possible chemical interactions between metals and samples. In biomedical application, this protocol may reduce the possible toxic effect of metallic NPs. On the other hand, the modification of bare metallic NPs generally is limited on thiol chemistry. Silica protecting layer offers versatile choices of chemical routes for modifying with different functional groups. The last practice can also be extended to encapsulate Raman reporters between metal and dielectric protecting layer[84,85], which, according to some researcher[84], is safer to load into living subjects. Derived from this structure, Wang *et al.*[86] further coated it with mesoporous Titania, and realized simultaneous fluorescence-SERS bimodal imaging as well as drug delivery.

6.2.4 *NPs Decorated Carbon Material as SERS Probe*

It is interesting to note that the fast developing carbon nanotubes (CNTs) and graphenes/graphene oxides (GO) can be decorated with metallic NPs and used as SERS probes. Ray and coworkers[87] reported that by modifying SWCNT with carboxyl group followed by reacting with aminothiophenol, gold nanopopcorn could be attached on the SWCNT through —SH linkage. The SERS nanoprobe was then further modified with tumor targeting aptamer. It was found that the decorating of gold nanopopcorn only gave one order of magnitude enhancement of the native D and G band of SWCNT. However, after incubating with tumor cell, the distinctive D and G bands of the SWCNT were enhanced 3 orders of magnitude. Cancer cell of 10 cells/mL was successfully detected. SWCNT can also be modified with DNAs, and Au and Ag NPs can be synthesized *in situ* on the SWCNT,[88] which then serves as an imaging tool as well as photothermal reagent.

In another work, GO was mixed with $HAuCl_4$ and heated to 80 °C for an hour.[89] The resulting material was GO flake of ~400 nm decorated with ~20 nm gold spheres. It was found that the characteristic G and D bands of GO were enhanced almost 4 times, and cancer cell imaging with the GO-Au NP complex was realized. Au NPs on GO can be also synthesized in cell and used for the SERS imaging of cell as well.[90] GO modified with Ag NPs[91], and reduced GO modified with Ag NPs[91], have also been reported as effective SERS probe and used for cell imaging.

6.2.5 *Other SERS Probes*

In SERS community it is widely believed that aggregated metallic NPs is the hot spot for giant enhancement.[1,26,27] In regular SERS analysis, one can simply add salt into the metallic NPs solution to trigger aggregation and achieve better sensitivity.[5] But this is not possible in most of biomedical application, since there is hardly any reproducibility, and almost impossible to realize selectivity. Composite-organic-inorganic-nanoparticles (COINs)[92–95] is imitative to inherit the giant enhancement of metallic NPs aggregates without its drawbacks (Fig.6.4). In the COINs approach, metallic NPs, usually silver, were first synthesized by regular wet chemical method. Later, a Raman reporter was carefully introduced into the colloidal solution. Depends on the nature of the Raman reporter, extra aggregating reagents, such as NaCl, may or may not need for initiating NPs aggregation. The aggregation is in a controlled fashion, such that few-NPs-cluster (COINs) assembly is preferred. Later, proteins, such as bovine serum albumin (BSA), are introduced with glutaraldehyde to complete the encapsulation of the COINs. The introduction of BSA not only protects the NPs from aggregating, but also provides functional groups for further modifying with antibodies and so on.

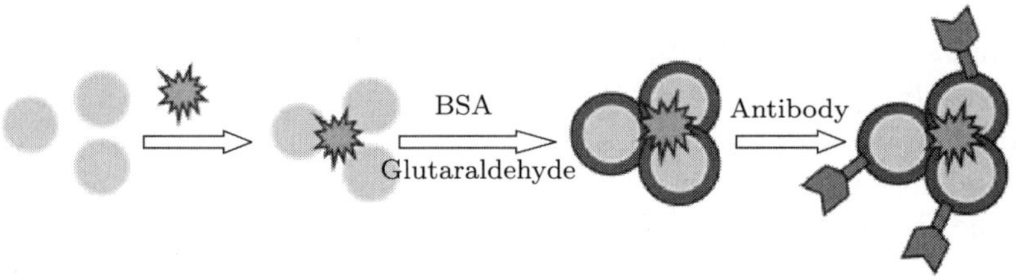

Fig. 6.4 The fabrication of antibody modified COINs.[95]

Another type of SERS substrate, which is not free NPs but has great potency in biomedical imaging, is the fiber optic based SERS probe. Since the invention of low loss optical fiber in 1970s, optical fiber sensors have attracted tremendous interests from science community.[96] Optical fibers offers many advantages,[96] some of them are irreplaceable with other techniques, including but not limited to: multiplexed detection with minimum optical alignment requirement, capable of remote sensing in a cramped or hazardous environment. By adding SERS substrate onto the distal end of an optical fiber, or combining a SERS active fiber with collection fiber(s), SERS optrode can be built. It has a great potential to be used in SERS endoscopy, for measurement in environment as small as a single cell.[97] To fabricate SERS optrode, different strategies can be applied. For example, self-assembly through electrostatic attraction between freshly cut fiber and surfactant stabilized NPs,[98] or through chemically modified optical fiber and NPs,[52,99] physical vapor deposition of silver,[100]nano-fabrication[31] and finally *in situ* chemical grown[101] have been reported. Some of them are capable of detecting Raman reporters at near single molecule sensitivity.[99]

6.3 SERS Theranostics

As discussed above, SERS has emerged as an ultrasensitive analytical method that is capable of providing analytes' quantitative and qualitative information.[22] On the other hand, the material basis for SERS, *i.e.*, gold or silver NPs, are known to have photothermal effect due to the excitation of SPs.[66] Furthermore, metallic NPs in general is a good platform for delivery chemicals into living organisms.[102] In this section, the theranostic applications of SERS nanoprobes will be discussed.

6.3.1 *Advantages of SERS Theranostics*

According to Gambhir and coworker[103], the ideal theranostic NPs shall include these features: ① capable of accumulation at diseased tissue selectively and rapidly; ② must be able to report both biochemical and morphological characteristics of the targeted tissue; ③ capable of delivering effective therapeutic; and ④ safe. Amazingly, SERS nanoprobes almost match all the criteria.

Due to the excitation of SPs, photons can be confined within gold or silver NPs, and hence dramatic light scattering (elastic and inelastic) and absorption can happen. The former leads to diagnostic applications of noble metallic NPs, such as surface enhanced Raman scattering spectroscopy. There are quite a few advantages of SERS techniques in diagnostic applications.[104] First, SERS technique is capable of performing multiplexing detection, which makes it suitable for precisely tracking of cancer cells. As shown in Fig.6.5, it is clear that the Raman bands are very narrow. In fact, in the case of Rh6G, the whole useful Raman spectrum, from ~300 cm^{-1} to 1 600 cm^{-1}, spread only a bit more than 60 nm. The full width at half maximum (FWHM) of a typical SERS band is only 1–2 nm. However, that of normal fluorescent label is 80 nm, while a quantum dot gives about 30 nm of FWHM.[105] Moreover, the material of SERS nanoprobe is gold or silica coated gold and/or silver. It is well known that both gold and silica are relatively inert in chemistry, which presumably will not react with living cell and cause toxicity. Furthermore, the modification chemistry on both gold and silica surface are mature. For example, SERS nanoprobes can be coated with PEG, which makes them exhibit excellent *in vivo* biodistribution and pharmacokinetic properties. Finally, the ultra-high sensitivity of SERS technique makes it suitable for tasks like tracking low abundance biomarkers as well as cells like circulating tumor cells (CTCs), which is only 1 to 10 cells per mL of whole blood.

The strong light absorption of gold or silver nanomaterials also makes them outstanding candidates to be used in therapeutic applications, based on photothermal ablation effect.[106] For example, gold nanorod can produce heat when shining with NIR lasers. It is reported that local temperature can be heated up to 43 °C by radiation with NIR laser, and in turn destroy the tumor cell

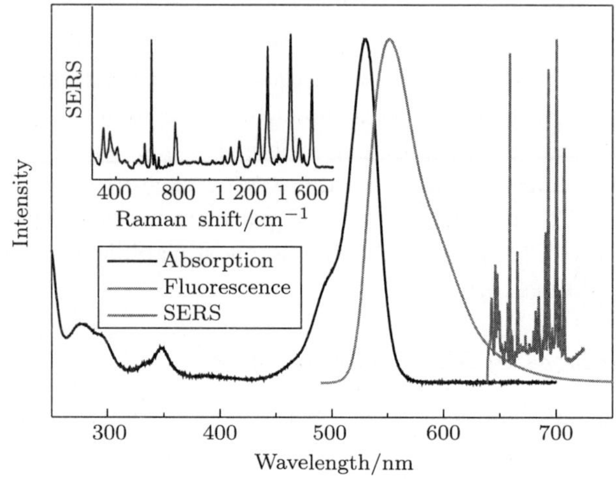

Fig. 6.5 Absorption, fluorescence, and SERS of Raman reporter Rhodamine 6G(Rh6G). Inset shows the "normal" plot of SERS of Rh6G. (color figure at the end of the book)

without causing damage to the surrounding healthy tissues.[107]

Another important property of noble metallic NPs is its capability of carrying molecules (drugs, *etc.*) around the surface of the nanoparticles. Conventional chemotherapeutic drugs are distributed in the body with almost no specificity. The outcome is two folds: there is not enough drug concentration at the tumor sites yet side effect has already happened to healthy tissues. This is one of the most serious obstacles that (chem-)anticancer therapy is facing.[107] It is reported that noble metallic NPs can be conjugated with up to a density of 10^{14} of ligands/cm^2, which is 3 orders of magnitude higher than that of polymeric NPs.[106] By modifying with several functionalities simultaneously, for example, therapeutic functioning drugs, and nucleic acids/proteins that works as functional moiety to recognize a tumor cell, intracellular drug concentration in tumor tissue can greatly be increased while minimizing the side effect to normal cells (a smart missile). NPs also offer great potential when dealing with drug resistance in cancer treatment. This is because NPs can bypass the P-glycoprotein efflux pump, one of the main drug resistance mechanisms.

6.3.2 *Encoding of SERS Nanoprobes and Modification for Targeting Bio-analytes*

Regarding the reporter molecules that attach on the SERS NPs, the rule of thumb is that it has to be close to the NPs' surface, and the closer the better (to achieve best SERS signal). However, not all reporters can stick to the NPs' surface directly.[83] Approaches to attach reporter molecules on the surface include directly chemical binding (through thiol functional group[108,109], for example), electrostatic attraction[93,110], and so on. Recently, some authors[111] are also making efforts to control the surface charge of the metallic NPs so that specific Raman reporter can reach the NPs' surface close enough, and in turn with better SERS signal. Co-aggregation with few-NPs does raise the reporters' SERS signal, but possible reporter leaching issue is needed to be addressed.[93,95] The reporter molecules can also be encapsulated with the MNP together by a thin film of silica (30–60 nm thick), which gives the nanoprobe ultrahigh sensitivity.[84,86,112]

NPs only encoded with SERS label cannot be directly used in bioimaging or carrying anticancer drugs. It has to be cross linked with antibodies, aptamers, and other molecules to give it the needed specificity. Co-adsorption of reporter molecules with targeting ligands is the first generation of encoding SERS nanoprobes for bioimaging applications.[83] For example, Raman reporters can be adsorbed on metallic NPs right before the addition of thiolated PEG and thiolated PEG with carboxyl group.[113–115] The carboxyl group is the anchoring point for future antibody attachment. Targeting ligands can also be non-covalently adsorbed on polyelectrolyte coated metallic NPs where the polyelectrolyte also acts as Raman reporter.[42] BSA is also used for encapsulating the reporters carrying NPs as well as a platform for further chemical modification.[93,95] However, several drawbacks are associated with these protocols. For example, non-wanted aggregation will cause large variation in SERS intensity. The leaching of SERS reporters will generate other problems. Synthesis of Raman reporters with functional moiety ready for biomolecules immobilization is thought to be an improvement in preparing SERS nanoprobes for bioimaging.[116] For example, Porter *et al.*[108,117] proposed a new type of Raman reporter that contains disulfides and succinimides, which make it ready to be attached on metallic NPs as well as coupled to bio-selective species. Raman active PEG has also been reported.[85] There are also reports using SWCNT and GO as both Raman reporter and platform for future labeling with bio-selective species.[87,118] In general, silica coating of SERS NPs is probably superior.[109,119] The silica layer is generally rigid and can keep the Raman reporters from leaching. On the other hand, it also provides a versatile platform for future chemical modifications.

Modification of SERS nanoprobes for bimodal diagnostic has also been proposed.[120,121] Lee *et al.*[121] suggested that, to prevent aggregation and quench unwanted fluorescence, NPs can be first modified with Raman reporter and then encapsulated with silica. The fluorescent dye can be conjugated onto the silica layer using aminosilane as linker. Hsieh *et al.*[120] on the other hand, used oxygen plasma to etch the commercially fluorescent polystyrene beads into hemisphere, followed by gold vapor deposition. The oxygen plasma also provides carboxylic group on the polystyrene beads and served as anchors for biomolecules.

6.3.3 *SERS Theranostics: Current State*

SERS in cancer related diagnostic is not uncommon. Back in 1996, SERS has already been used to study cytotoxicity of antitumor DNA intercalators.[122] By SERS technique, Chourpa *et al.*[122] examined the interaction of a potent antitumor drug amsacrine within living K562 cancer cell, and proposed that specific interaction of the drug side chain with the toposiomerase is the key mechanism for its inhibitory action. The authors also pointed out that there are two main advantages of SERS as the diagnostic tool: the non-perturbing characteristic of the adsorbing of amsacrine on Ag NPs, and the ability of Ag NPs for penetrating the cell membrane. Recently, SERS is even proved to be a promising candidate technique for the detection of CTCs.[123] SERS has been successfully used to detect CTCs in the peripheral blood of 19 patients with squamous cell carcinoma of the head and neck (SCCHN).[6]

On the other hand, the first plasmonic NPs based photothermal therapy was accomplished by directly injecting gold nanoshells into the tumor site.[106] A group in Georgia Institute of Technology[106] found gold nanorod, PEGylated gold nanorod and peptide modified PEGylated gold nanorod to be very efficient in selective tumor cell ablation and resorption *in vivo*. Up to 20 °C could be raised at the tumor site for just 10 min of laser shining. The photothermal destruction and drug delivery effect of gold nanocage was reported by Xia group.[66,124]

Ideal SERS theranostics could be defined as following. In the diagnostic part, it shall be able to monitor biological processes similar to biopsy, but non-invasive, and in real time. It shall give the information of whether there is a disease, the stage, and the response to a therapy. On the therapeutic side, applications involve drug delivery and thermo-ablative therapy.[103]

One important issue that rises when applying SERS NPs for theranostics is their potential toxicity to live organism. Thakor *et al.*[84] addressed this issue by the using of 60 nm gold NPs coated with 30 nm silica shell as an example. It is to be noted that the Raman reporter in this case was sandwiched between the gold core and silica shell. Furthermore, the NPs were wrapped in PEG molecules. The dose that administered to mice is 1 000 times higher than the minimum concentration that can currently be detected *in vivo* with Raman microscope. Temporary inflammation and some apoptosis in the liver was observed 24 hours later by intravenous injection. But they are cleared eventually through reticulo-endothelial system. Comprehensive survey of the mice did not find any ill effects of the NPs to the general health of the mice. Further, if the NPs were administered through colon, they even did not get into the circulation system.

Multifunctional magnetic plasmonic nanomaterials have also been developed in a work published in 2013.[82] The nanomaterial contains a magnetic core, $MnFe_2O_4$, silica shell that encapsulate the core, and finally a silver shell that gives it SERS activity. It is proved that heat can be produced by either AC magnetic field or laser radiation. At the same time, the NPs exhibit high sensitivity to Raman reporter Rh6G. However, the authors did not seek cell targeting capability of the nanomaterials.

Vo-Dinh and coworkers[125] developed a type of SERS NPs that can be used both in imaging of cancer cell and performing photodynamic therapy. They firstly synthesized gold nanostar and tuned it into maximum absorption at NIR frequency. Later, the nanostar was protected by using PEG-SH. Raman reporter $3,3'$-diethylthiatricarbocyanine (DTTC) was added into the PEGylated nanostar overnight. Finally, photosensitizer methylene blue was encapsulated onto the modified nanostar by adding methylene blue with tetraethyl orthosilicate (TEOS) to forming the silica layer. The authors claimed that the nanostars presented cytotoxicity to BT549 breast cancer cell upon laser irradiation. However, it is worth noting that no antibody was conjugated with the NPs and the experiment was done *in vitro*.

Tseng *et al.*[120] reported a trifunctional SERS nanoprobes aiming to recognize cancer cell, delivery drug, and monitor biological response in real time. In their work, fluorescent polystyrene beads were engineered with oxygen plasma into corrugated upper hemisphere and simultaneously modified with carboxylic groups. Gold was deposited onto the corrugated hemisphere and labeled with Raman reporter as well as sulfo-NHS-SS-biotin. Tumor targeting antibody was linked through the carboxylic group. Streptavidin was used as a model anticancer drug and linked through biotin. The authors found that the concentration of trifunctional NPs in cancer cell is 12 times higher than that in a normal cell.

A bimodal imaging and resection guiding SERS nanoprobes were reported by Gambhir*et al.*[40] In their work, gold nanorods with aspect ratio of 3.5 was synthesized and labeled with Raman reporters before wrapped with PEG-SH. No active targeting antibody was linked onto the nanoprobe. Instead, a passive targeting mode, using enhanced permeability and retention (EPR) effect, was utilized. It was found that photoacoustic signal reached at maximum in 3 hours but persisted for at least two days after nanoprobe administration in three cell lines tested. SERS imaging was utilized for resection of the tumor tissue in living mice. In another work, SERS spectraPen technique was developed.[126] Briefly, a portable Raman spectrometer with optical fiber detection head was used during the resection of tumor. The passive labeling efficiency of SERS nanoprobes were compared with fluorescent method. Another bimodal SERS and fluorescence imaging nanoprobe was reported by Wang *et al.*[86] which also bears drug delivery functionality. During the work, silver NPs were first synthesized and tagged with 4-mercpatopyridine as Raman reporter and wrapped in silica shell. After coating with mesoporous TiO_2, fluorescence dye flavin mononucleotide and an anticancer drug doxorubicin were co-adsorbed onto it.

They found that the as prepared nanoprobe is low cytotoxic to MCF-7 cells. However, drug loading efficiency was found to be only comparable to free drugs.

Au:Ag bimetallic NPs have also been used in SERS imaging and photothermal therapy.[127] In the work, aptamer s2.2 was mixed with silver salt and underwent UV irradiation to form Ag nanostructures on the aptamer. After 30 min, gold salt was added and the solution was exposed to UV light for another 30 min to form the bimetallic NPs of ~100 nm on aptamer. Then, Rh6G was allowed to adsorb on the nanomaterials. It is proved that nanoprobes made in this way can target human breast cancer MCF-7 with high affinity and specificity. Meanwhile, it can be used to destroy the MCF-7 cells photothermally at very low irradiation power density (0.25 W/cm^2, 808 nm).

The multiplexing imaging property, photothermal therapeutic efficiency, and the *in vitro* cytotoxicity of gold nanorod has been examined by a group in MIT.[41] Since Au nanorods are generally synthesized using CTAB as protecting reagent, procedure must be carefully chosen to replace them with thiol-PEG as well as Raman reporters. By directly injecting intratumorally with the nanorods based SERS nanoprobe, they found that multiplexed imaging with three labels could be utilized. Further, irradiation with 810 nm laser resulted temperature raise to 70 °C with nanoprobe labeled tissue, while minimum change of temperature was observed in control samples. Finally, they tested the cytotoxicity of the SERS nanoprobe. It was found that, at the concentration of 1 million times higher than what is needed for SERS imaging showed >70% cell viability, which is encouraging.

SWCNT-Au/Ag-PEG nanocomposites recently have gained some attention.[88] SWCNTs were first functionalized with DNA, and gold or silver were grown *in situ* to form a shell. Later, the nanocomposite was wrapped with PEG and linked with folic acid to target cancer cell. It was found that the imaging process needs significantly less time compared with SWCNT due to the SERS effect. On the other hand, the nanocomposite showed remarkable improvement in photothermal killing of cancer cell compared with gold nanorod-PEG composite. In a similar work, gold nanopopcorn was attached to SWCNT, and labeled with an aptamer for targeting breast cancer.[87] It is believed that there is about 3 orders of enhancement to the D and G bands of SWCNT, and hence as low as 10 cancer cells per mL can be detected, and 10 min of 785 nm laser irradiation at 1.5 W/cm^2 is enough to kill all the cancerous cells. The authors believed that the high efficiency of photothermal therapy is due to the fact that both SWCNT and gold nanoshells can absorb laser energy.

One brilliant approach of using gold NPs as SERS imaging and traceable intracellular drug delivery was reported by a group in Singapore.[128] In their work, gold NPs of 14 nm size was modified with PEG and a pH-sensitive polymer brush, as well as Raman reporter. Since the NPs have distinctly different hydrophobicity, they tend to form a vesicle and can trap anticancer drug inside.

After modifying with cancer targeting antibody, the plasmonic vesicles will attach on the cancer cell and pass through the membrane by endocytosis. Since the microenvironment inside the cell is different and the pH-sensitive polymer brush will respond to that. Thus, the vesicles will decompose into NPs and release the drug. Most importantly, since the gold NPs are only 14 nm in size, it is not efficient SERS probe in the form of single NPs, while in vesicle it is in an aggregated state and very SERS active. Thus, the whole drug delivery process could be monitored by SERS: one could observe SERS signal accumulating at the tumor sites and finally diminishing after the successful delivery of anticancer drugs.

6.4 Limitation and Future Development

SERS is booming, especially after its single molecule detection sensitivity was displayed in 1997.[129,130] The applications now are ranging from environmental monitoring[2,23], homeland security[18] to bio-analysis and imaging[35,131]. Recently, the combination of the imaging property with its drug delivery capability and/or photothermal therapeutic capacity of SERS nanoprobe, namely SERS theranostics[35], have gained wide attention from biomedical community as well as material chemists[130].

The advantages of SERS are prominent as they are highly sensitive, comparable to fluorescence. Yet it has many benefits that fluorescence technique cannot offer, including multiplexing detection, detecting 10 probes simultaneously has been experimentally demonstrated;[119] generally stable, has almost no or minimum photobleaching problem; working with NIR laser is not a problem, which make it possible for *in vivo* applications; probes are in low toxicity and easy to be transported into cell, and can have high drug loading capacity.[106] Even more, the SERS nanoprobe itself could be an efficient photothermal therapeutic reagent due to the excitation of SPs.[46,132] With all the pros, it is natural to think that the future of SERS theranostic will be even brighter. Sure it is, if the following cons could be addressed.

The biggest issue is associated with the penetration depth of SERS technique. When applied in biomedical exams, the absorption and scattering of both the excitation and Raman scattered photons have to be taken into consideration. Furthermore, most of the SERS imaging experiments are carried out by using a microscope in a reflection configuration. Thus, the working distance of the lens is another important factor. Currently, the deepest "regular" SERS imaging experiment indicates that, tissues as deep as 5.5 mm could be imaged.[133]

Recently, Stone *et al.*[134,135] proposed a technique called surface enhanced spatially offset Raman spectroscopy (SESORS), and obtained multiplexing mapping of porcine tissue block as deep as 50 mm, 10 times deeper than that of regular SERS technique. Most importantly, the authors claimed that at a depth of 25 mm into the tissue, 1.1×10^9 SERS NPs can be detected, two orders of magnitude smaller than what has been achieved at 5 mm by regular SERS. One of the drawbacks of the technique is loss of signal at higher wave numbers at ~50 mm deep due to interference from water and myoglobin absorption. Meanwhile, the experiment set up is in transmission configuration. Thus, it may not still be possible to examine deep organs in human. Nevertheless, the technique is a major progress in SERS diagnostic, if combined with therapeutic functionality, it may receive more attention in the near future.

The second problem involves the long term toxicity. To avoid that, SERS NPs must be cleaned out from the body in a reasonable time frame. It is reported that in quantum dots, for a rigid spherical NP, if the diameter falls in the criteria of ~6 nm and covered with zwitterionic surface, it can be cleared in 4 hours after injection.[136,137] This may propose a serious question mark to SERS based theranostics, since the optimum NP size for SERS is in the range of 10–100 nm. It can be argued that Au or Ag NPs are less toxic compared with QDs and other counterparts, but the retention of such NPs in the body could be a problem for further examinations, such as MRI, CT, PET, *etc.* are needed. Moreover, the blockage of capillaries and glomerular space by these relative large NPs is also a potential threat. Thus, Biju *et al.*[136] proposed that if the SERS reporter is built from Au or Ag NPs of small (<5 nm) size, in other words, aggregates from smaller NPs, might be a good choice for future development.

Although SERS NPs with sizes less than 5 nm is hard to achieve, report[128] shows that smaller NPs at 14 nm can be utilized in SERS theranostics, as discussed in the above section. Of course, there is plenty of room to optimize in order to improve the SERS efficiency of single NPs. We believe both the shape control and material variation, such as bimetallic NPs, could be something to start with. There are reports[71,73] indicating that for bimetallic NPs, the SERS efficiency is not only dependent on the metals composition, but also on the Raman reporter. Most importantly, in most cases, bimetallic NPs have better SERS efficiency.[73]

Regarding the selectivity of SERS theranostics, a few recent works involving plasmonic nanobubbles (PNB) have shown great promises. In the PNB approach, NPs are labeled with targeting antibodies, and are allowed to bind on the cell. Through the endocytosis process, NPs get into the cell and form clusters. When shining with pulsed laser, the NPs cluster will produce heat because of excitation of SPs and evaporates its nano-environment, forming nanobubbles (and then kill the target cell). The main advantage of the PNB approach is to further improve the selectivity of NPs theranostics, only the cancer tissue will accumulate the most NPs and that's where the therapeutic effect happens.[138] The NPs cluster can have different shapes of nanomaterials, such that the formation of PNB can be obtained through a combination of lasers, which in turn reduce the energy requirement of single laser (and cause less damage). The utilization of multiple shapes of nanomaterials also makes it possible to label with different antibodies that target the same cancer cell, and hence improve the selectivity. This approach can be easily adapted into SERS theranostics by simply labeling each shape of NPs with a separate Raman reporter, multiplexing SERS imaging with high sensitivity can be realized, while no therapeutic property will be sacrificed.

Finally, one direction that still does not gain enough attention yet and may have great potential in SERS theranostics is SERS optrode technique.[99,139] By engineering optical fibers into a needle like ultrafine probe followed by coating with SERS active materials, fiber optic based nanosensors can be developed. Since its dimension is even smaller than a cell, this type of probe can be used to monitor micro-environment in a single cell at real time. Moreover, it has been used to monitor apoptosis in cells treated with anticancer drugs, and local pH in cell.[140] When used in combination with regular SERS nanoprobes (theranostics), we think it is possible to overcome the inherent penetration depth issue of SERS technique, by using SERS optrode as a SERS endoscope. Recently, an SERS acupuncture needle[141] has been proposed. However, the chemical information versus depth of pseudo tissue is needed to be obtained.

References

1. Le Ru, E. C.; Etchegoin, P. G. *Principles of Surface-enhanced Raman Spectroscopy: And Related Plasmonic Effects.* Elsevier, **2008**.
2. Halvorson, R. A.; Vikesland, P. J. Surface-enhanced Raman spectroscopy (SERS) for environmental analyses. *Environ Sci Technol* **2010**, *44* (20), 7749-7755.
3. Ryder, A. G. Surface enhanced Raman scattering for narcotic detection and applications to chemical biology. *Curr Opin Chem Biol* **2005**, *9* (5), 489-493.
4. Ruan, C.; Wang, W.; Gu, B. Detection of alkaline phosphatase using surface-enhanced Raman spectroscopy. *Anal Chem* **2006**, *78* (10), 3379-3384.
5. Qian, X. M.; Nie, S. M. Single-molecule and single-nanoparticle SERS: From fundamental mechanisms to biomedical applications. *Chem Soc Rev* **2008**, *37* (5), 912-920.
6. Wang, X.; Qian, X.; Beitler, J. J.; Chen, Z. G.; Khuri, F. R.; Lewis, M. M.; Shin, H. J. C.; Nie, S.; Shin, D. M. Detection of circulating tumor cells in human peripheral blood using surface-enhanced Raman scattering nanoparticles. *Cancer Res* **2011**, *71* (5), 1526-1532.
7. Harris, D. C.; Bertolucci, M. D. *Symmetry and Spectroscopy: An Introduction to Vibrational and Electronic Spectroscopy.* Courier Dover Publications, **1978**.
8. Aroca, R. F. Plasmon enhanced spectroscopy. *Phys Chem Chem Phys* **2013**, *15* (15), 5355-5363.
9. Petry, R.; Schmitt, M.; Popp, J. Raman spectroscopy: A prospective tool in the life sciences. *ChemPhysChem* **2003**, *4* (1), 14-30.

10. Albrecht, M. G.; Creighton, J. A. Anomalously intense Raman spectra of pyridine at a silver electrode. *J Am Chem Soc* **1977**, *99* (15), 5215-5217.
11. Fleischmann, M.; Hendra, P.; McQuillan, A. Raman spectra of pyridine adsorbed at a silver electrode. *Chem Phys Lett* **1974**, *26* (2), 163-166.
12. Jeanmaire, D. L.; Van Duyne, R. P. Surface Raman spectroelectrochemistry: Part I. Heterocyclic, aromatic, and aliphatic amines adsorbed on the anodized silver electrode. *J Electroanal Chem* **1977**, *84* (1), 1-20.
13. Moskovits, M. Surface-enhanced Raman spectroscopy: A brief perspective. In: *Surface-enhanced Raman Scattering*. Springer, **2006**, 1-17.
14. Guieu, V.; Talaga, D.; Servant, L.; Sojic, N. O.; Lagugné-Labarthet, F. Multitip-localized enhanced Raman scattering from a nanostructured optical fiber array. *J Phys Chem C* **2008**, *113* (3), 874-881.
15. Kneipp, K.; Wang, Y.; Kneipp, H.; Perelman, L. T.; Itzkan, I.; Dasari, R. R.; Feld, M. S. Single molecule detection using surface-enhanced Raman scattering (SERS). *Phys Rev Lett* **1997**, *78* (9), 1667.
16. Delfino, I.; Bizzarri, A. R.; Cannistraro, S. Time-dependent study of single-molecule SERS signal from yeast cytochrome. *Chem Phys* **2006**, *326* (2), 356-362.
17. Santos, D. P. D.; Andrade, G. F.; Temperini, M. L.; Brolo, A. G. Electrochemical control of the time-dependent intensity fluctuations in surface-enhanced Raman scattering (SERS). *J Phys Chem C* **2009**, *113* (41), 17737-17744.
18. Golightly, R. S.; Doering, W. E.; Natan, M. J. Surface-enhanced Raman spectroscopy and homeland security: A perfect match? *ACS Nano* **2009**, *3* (10), 2859-2869.
19. Campion, A.; Kambhampati, P. Surface-enhanced Raman scattering. *Chem Soc Rev* **1998**, *27* (4), 241-250.
20. Brolo, A. G.; Irish, D. E.; Smith, B. D. Applications of surface enhanced Raman scattering to the study of metal-adsorbate interactions. *J Mol Struct* **1997**, *405* (1), 29-44.
21. Otto, A.; Futamata, M. Electronic mechanisms of SERS. In: *Surface-enhanced Raman Scattering*. Springer, **2006**, 147-182.
22. Fan, M. K.; Andrade, G. F. S.; Brolo, A. G. A review on the fabrication of substrates for surface enhanced Raman spectroscopy and their applications in analytical chemistry. *Anal Chim Acta* **2011**, *693* (1-2), 7-25.
23. Hutter, E.; Fendler, J. H. Exploitation of localized surface plasmon resonance. *Adv Mater* **2004**, *16* (19), 1685-1706.
24. Hao, E.; Schatz, G. C. Electromagnetic fields around silver nanoparticles and dimers. *J Chem Phys* **2004**, *120* (1), 357-366.
25. Qin, L.; Zou, S.; Xue, C.; Atkinson, A.; Schatz, G. C.; Mirkin, C. A. Designing, fabricating, and imaging Raman hot spots. *P Natl Acad Sci USA* **2006**, *103* (36), 13300-13303.
26. Aroca, R. *Surface Enhanced Vibrational Spectroscopy*. John Wiley & Sons, **2006**.
27. Baker, G. A.; Moore, D. S. Progress in plasmonic engineering of surface-enhanced Raman-scattering substrates toward ultra-trace analysis. *Anal Bioanal Chem* **2005**, *382* (8), 1751-1770.

28. Fan, M.; Maréchal, M.; Finn, A.; Harrington, D. A.; Brolo, A. G. Layer-by-layer characterization of a model biofuel cell anode by (*in situ*) vibrational spectroscopy. *J Phys Chem C* **2010**, *115* (1), 310-316.
29. Lin, X. M.; Cui, Y.; Xu, Y. H.; Ren, B.; Tian, Z. Q. Surface-enhanced Raman spectroscopy: Substrate-related issues. *Anal Bioanal Chem* **2009**, *394* (7), 1729-1745.
30. Yuen, J. M.; Shah, N. C.; Walsh Jr, J. T.; Glucksberg, M. R.; Van Duyne, R. P. Transcutaneous glucose sensing by surface-enhanced spatially offset Raman spectroscopy in a rat model. *Anal Chem* **2010**, *82* (20), 8382-8385.
31. Smythe, E. J.; Dickey, M. D.; Bao, J.; Whitesides, G. M.; Capasso, F. Optical antenna arrays on a fiber facet for *in situ* surface-enhanced Raman scattering detection. *Nano Lett* **2009**, *9* (3), 1132-1138.
32. Patete, J. M.; Peng, X.; Koenigsmann, C.; Xu, Y.; Karn, B.; Wong, S. S. Viable methodologies for the synthesis of high-quality nanostructures. *Green Chem* **2011**, *13* (3), 482-519.
33. Kerker, M.; Thompson, B. *Selected Papers on Surface-enhanced Raman Scattering*. SPIE Optical Engineering Press Bellingham, **1990**.
34. Moskovits, M. Surface-enhanced Raman spectroscopy: A brief retrospective. *J Raman Spectrosc* **2005**, *36* (6-7), 485-496.
35. Wang, Y.; Yan, B.; Chen, L. SERS tags: Novel optical nanoprobes for bioanalysis. *Chem Rev* **2013**, *113* (3), 1391-1428.
36. Bauer, G.; Hassmann, J.; Walter, H.; Haglmüller, J.; Mayer, C.; Schalkhammer, T. Resonant nanocluster technology: From optical coding and high quality security features to biochips. *Nanotechnology* **2003**, *14* (12), 1289.
37. Pazos-Perez, N.; Garcia de Abajo, F. J.; Fery, A.; Alvarez-Puebla, R. A. From nano to micro: Synthesis and optical properties of homogeneous spheroidal gold particles and their superlattices. *Langmuir* **2012**, *28* (24), 8909-8914.
38. Schwartzberg, A. M.; Zhang, J. Z. Novel optical properties and emerging applications of metal nanostructures. *J Phys Chem C* **2008**, *112* (28), 10323-10337.
39. Lu, X. M.; Rycenga, M.; Skrabalak, S. E.; Wiley, B.; Xia, Y. N. Chemical synthesis of novel plasmonic nanoparticles. *Annu Rev Phys Chem* **2009**, *60*, 167-192.
40. Jokerst, J. V.; Cole, A. J.; Van de Sompel, D.; Gambhir, S. S. Gold nanorods for ovarian cancer detection with photoacoustic imaging and resection guidance *via* Raman imaging in living mice. *ACS Nano* **2012**, *6* (11), 10366-10377.
41. von Maltzahn, G.; Centrone, A.; Park, J. H.; Ramanathan, R.; Sailor, M. J.; Hatton, T. A.; Bhatia, S. N. SERS-coded gold nanorods as a multifunctional platform for densely multiplexed near-infrared imaging and photothermal heating. *Adv Mater* **2009**, *21* (31), 3175-3180.
42. Huang, X.; El-Sayed, I. H.; Qian, W.; El-Sayed, M. A. Cancer cells assemble and align gold nanorods conjugated to antibodies to produce highly enhanced, sharp, and polarized surface Raman spectra: A potential cancer diagnostic marker. *Nano Lett* **2007**, *7* (6), 1591-1597.
43. Huff, T. B.; Tong, L.; Zhao, Y.; Hansen, M. N.; Cheng, J. X.; Wei, A. Hyperthermic effects of gold nanorods on tumor cells. *Nanomed-Nanotechnol* **2007**, *2* (1), 125-132.

44. Cai, W.; Gao, T.; Hong, H.; Sun, J. Applications of gold nanoparticles in cancer nanotechnology. *Nanotech Sci Appl* **2008**, *1* (1), 17-32.
45. Zhang, Z.; Wang, J.; Chen, C. Gold nanorods based platforms for light-mediated theranostics. *Theranostics* **2013**, *3* (3), 223-238.
46. Kennedy, L. C.; Bickford, L. R.; Lewinski, N. A.; Coughlin, A. J.; Hu, Y.; Day, E. S.; West, J. L.; Drezek, R. A. A new era for cancer treatment: Gold-nanoparticle-mediated thermal therapies. *Small* **2011**, *7* (2), 169-183.
47. Busbee, B. D.; Obare, S. O.; Murphy, C. J. An improved synthesis of high-aspect-ratio gold nanorods. *Adv Mater* **2003**, *15* (5), 414-416.
48. Gao, J.; Bender, C. M.; Murphy, C. J. Dependence of the gold nanorod aspect ratio on the nature of the directing surfactant in aqueous solution. *Langmuir* **2003**, *19* (21), 9065-9070.
49. Kim, F.; Song, J. H.; Yang, P. Photochemical synthesis of gold nanorods. *J Am Chem Soc* **2002**, *124* (48), 14316-14317.
50. Li, Q.; Cao, Y. Preparation and characterization of gold nanorods. In: Yalçın, O. Ed. *Nanorods.* Intech, **2012**.
51. Panchapakesan, B.; Book-Newell, B.; Sethu, P.; Rao, M.; Irudayaraj, J. Gold nanoprobes for theranostics. *Nanomedicine (London, England)* **2011**, *6* (10), 1787-1811.
52. Akhter, S.; Ahmad, M. Z.; Ahmad, F. J.; Storm, G.; Kok, R. J. Gold nanoparticles in theranostic oncology: Current state-of-the-art. *Expert Opin Drug Del* **2012**, *9* (10), 1225-1243.
53. Arvizo, R. R.; Bhattacharyya, S.; Kudgus, R. A.; Giri, K.; Bhattacharya, R.; Mukherjee, P. Intrinsic therapeutic applications of noble metal nanoparticles: Past, present and future. *Chem Soc Rev* **2012**, *41* (7), 2943-2970.
54. Sun, Y.; Mayers, B.; Herricks, T.; Xia, Y. Polyol synthesis of uniform silver nanowires: A plausible growth mechanism and the supporting evidence. *Nano Lett* **2003**, *3* (7), 955-960.
55. Jin, R.; Cao, Y.; Mirkin, C. A.; Kelly, K. L.; Schatz, G. C.; Zheng, J. G. Photoinduced conversion of silver nanospheres to nanoprisms. *Science* **2001**, *294* (5548), 1901-1903.
56. Métraux, G. S.; Mirkin, C. A. Rapid thermal synthesis of silver nanoprisms with chemically tailorable thickness. *Adv Mater* **2005**, *17* (4), 412-415.
57. Darmanin, T.; Nativo, P.; Gilliland, D.; Ceccone, G.; Pascual, C.; De Berardis, B.; Guittard, F.; Rossi, F. Microwave-assisted synthesis of silver nanoprisms/nanoplates using a "modified polyol process". *Colloid Surface A* **2012**, *395*, 145-151.
58. Ha, T. H.; Koo, H.-J.; Chung, B. H. Shape-controlled syntheses of gold nanoprisms and nanorods influenced by specific adsorption of halide ions. *J Phys Chem C* **2006**, *111* (3), 1123-1130.
59. Millstone, J. E.; Park, S.; Shuford, K. L.; Qin, L.; Schatz, G. C.; Mirkin, C. A. Observation of a quadrupole plasmon mode for a colloidal solution of gold nanoprisms. *J Am Chem Soc* **2005**, *127* (15), 5312-5313.
60. Siekkinen, A. R.; McLellan, J. M.; Chen, J.; Xia, Y. Rapid synthesis of small silver nanocubes by mediating polyol reduction with a trace amount of sodium sulfide or sodium hydrosulfide. *Chem Phys Lett* **2006**, *432* (4-6), 491-496.

61. Skrabalak, S. E.; Au, L.; Li, X.; Xia, Y. Facile synthesis of Ag nanocubes and Au nanocages. *Nat Protoc* **2007**, *2* (9), 2182-2190.
62. Yuan, H.; Liu, Y.; Fales, A. M.; Li, Y. L.; Liu, J.; Vo-Dinh, T. Quantitative surface-enhanced resonant Raman scattering multiplexing of biocompatible gold nanostars for *in vitro* and *ex vivo* detection. *Anal Chem* **2012**, *85* (1), 208-212.
63. Yuan, H. K.; Khoury, C. G.; Hwang, H.; Wilson, C. M.; Grant, G. A.; Vo-Dinh, T. Gold nanostars: Surfactant-free synthesis, 3D modelling, and two-photon photoluminescence imaging. *Nanotechnology* **2012**, *23* (7), 075102.
64. Vigderman, L.; Zubarev, E. R. Starfruit-shaped gold nanorods and nanowires: Synthesis and SERS characterization. *Langmuir* **2012**, *28* (24), 9034-9040.
65. Lu, Y.; Liu, G. L.; Kim, J.; Mejia, Y. X.; Lee, L. P. Nanophotonic crescent moon structures with sharp edge for ultrasensitive biomolecular detection by local electromagnetic field enhancement effect. *Nano Lett* **2004**, *5* (1), 119-124.
66. Xia, Y.; Li, W.; Cobley, C. M.; Chen, J.; Xia, X.; Zhang, Q.; Yang, M.; Cho, E. C.; Brown, P. K. Gold nanocages: From synthesis to theranostic applications. *Accounts Chem Res* **2011**, *44* (10), 914-924.
67. Tiwari, V. S.; Oleg, T.; Darbha, G. K.; Hardy, W.; Singh, J. P.; Ray, P. C. Non-resonance SERS effects of silver colloids with different shapes. *Chem Phys Lett* **2007**, *446* (1-3), 77-82.
68. Freeman, R. G.; Hommer, M. B.; Grabar, K. C.; Jackson, M. A.; Natan, M. J. Ag-clad Au nanoparticles: Novel aggregation, optical, and surface-enhanced Raman scattering properties. *J Phys Chem* **1996**, *100* (2), 718-724.
69. Jana, N. R. Silver coated gold nanoparticles as new surface enhanced Raman substrate at low analyte concentration. *Analyst* **2003**, *128* (7), 954-956.
70. Cui, Y.; Ren, B.; Yao, J.-L.; Gu, R.-A.; Tian, Z.-Q. Synthesis of AgcoreAushell bimetallic nanoparticles for immunoassay based on surface-enhanced Raman spectroscopy. *J Phys Chem B* **2006**, *110* (9), 4002-4006.
71. Kumar, G. V. P.; Shruthi, S.; Vibha, B.; Reddy, B. A. A.; Kundu, T. K.; Narayana, C. Hot spots in Ag core-Au shell nanoparticles potent for surface-enhanced Raman scattering studies of biomolecules. *J Phys Chem C* **2007**, *111* (11), 4388-4392.
72. Hao, E.; Li, S.; Bailey, R. C.; Zou, S.; Schatz, G. C.; Hupp, J. T. Optical properties of metal nanoshells. *J Phys Chem B* **2004**, *108* (4), 1224-1229.
73. Fan, M.; Lai, F.-J.; Chou, H.-L.; Lu, W.-T.; Hwang, B.-J.; Brolo, A. G. Surface-enhanced Raman scattering (SERS) from Au:Ag bimetallic nanoparticles: The effect of the molecular probe. *Chem Sci* **2013**, *4* (1), 509-515.
74. Cao, L.; Diao, P.; Tong, L.; Zhu, T.; Liu, Z. Surface-enhanced Raman scattering of P-aminothiophenol on a Au (core)/Cu (shell) nanoparticle assembly. *Chem Phys Chem* **2005**, *6* (5), 913-918.
75. Lim, D. K.; Jeon, K. S.; Hwang, J. H.; Kim, H.; Kwon, S.; Suh, Y. D.; Nam, J. M. Highly uniform and reproducible surface-enhanced Raman scattering from DNA-tailorable nanoparticles with 1-nm interior gap. *Nat Nanotechnol* **2011**, *6* (7), 452-460.
76. Cui, Y.; Gu, R. Preparation of AgcoreAushell composite bimetallic nanoparticles and its surface-enhanced Raman spectroscopy. *Chem J Chinese U* **2005**, *26* (11), 2090-2092.

77. Mallin, M. P.; Murphy, C. J. Solution-phase synthesis of sub-10 nm Au-Ag alloy nanoparticles. *Nano Lett* **2002**, *2* (11), 1235-1237.

78. Shen, J.; Zhu, Y.; Yang, X.; Zong, J.; Li, C. Multifunctional Fe_3O_4@Ag/SiO_2/Au core-shell microspheres as a novel SERS-activity label *via* long-range plasmon coupling. *Langmuir* **2012**, *29* (2), 690-695.

79. Wang, H.; Brandl, D. W.; Le, F.; Nordlander, P.; Halas, N. J. Nanorice: A hybrid plasmonic nanostructure. *Nano Lett* **2006**, *6* (4), 827-832.

80. Xu, X.; Cortie, M. B. Precious metal core-shell spindles. *J Phys Chem C* **2007**, *111* (49), 18135-18142.

81. Jun, B. H.; Noh, M. S.; Kim, J.; Kim, G.; Kang, H.; Kim, M. S.; Seo, Y. T.; Baek, J.; Kim, J. H.; Park, J.; Kim, S.; Kim, Y. K.; Hyeon, T.; Cho, M. H.; Jeong, D. H.; Lee, Y. S. Multifunctional silver-embedded magnetic nanoparticles as SERS nanoprobes and their applications. *Small* **2010**, *6* (1), 119-125.

82. Lai, J. J.; Lai, W. R.; Chen, C. Y.; Chen, S. W.; Chiang, C. L. Multifunctional magnetic plasmonic nanoparticles for applications of magnetic/photo-thermal hyperthermia and surface enhanced Raman spectroscopy. *J Magnetism Magnetic Mater* **2013**, *331*, 204-207.

83. Abalde-Cela, S.; Aldeanueva-Potel, P.; Mateo-Mateo, C.; Rodriguez-Lorenzo, L.; Alvarez-Puebla, R. A.; Liz-Marzan, L. M. Surface-enhanced Raman scattering biomedical applications of plasmonic colloidal particles. *J R Soc Interface* **2010**, *7*, 435-450.

84. Thakor, A. S.; Luong, R.; Paulmurugan, R.; Lin, F. I.; Kempen, P.; Zavaleta, C.; Chu, P.; Massoud, T. F.; Sinclair, R.; Gambhir, S. S. The fate and toxicity of Raman-active silica-gold nanoparticles in mice. *Sci Transl Med* **2011**, *3* (79), 79ra33.

85. Huang, Y.; Swarup, V. P.; Bishnoi, S. W. Rapid Raman imaging of stable, functionalized nanoshells in mammalian cell cultures. *Nano Lett* **2009**, *9* (8), 2914-2920.

86. Wang, Y. Q.; Chen, L. X.; Liu, P. Biocompatible triplex Ag@SiO_2@mTiO_2 core-shell nanoparticles for simultaneous fluorescence: SERS bimodal imaging and drug delivery. *Chem Eur J* **2012**, *18* (19), 5935-5943.

87. Beqa, L.; Fan, Z.; Singh, A. K.; Senapati, D.; Ray, P. C. Gold nano-popcorn attached SWCNT hybrid nanomaterial for targeted diagnosis and photothermal therapy of human breast cancer cells. *ACS Appl Mater Inter* **2011**, *3* (9), 3316-3324.

88. Wang, X.; Wang, C.; Cheng, L.; Lee, S.-T.; Liu, Z. Noble metal coated single-walled carbon nanotubes for applications in surface enhanced Raman scattering imaging and photothermal therapy. *J Am Chem Soc* **2012**, *134* (17), 7414-7422.

89. Liu, Q. H.; Wei, L.; Wang, J. Y.; Peng, F.; Luo, D.; Cui, R. L.; Niu, Y.; Qin, X. J.; Liu, Y.; Sun, H.; Yang, J.; Li, Y. Cell imaging by graphene oxide based on surface enhanced Raman scattering. *Nanoscale* **2012**, *4* (22), 7084-7089.

90. Liu, Z. M.; Hu, C. F.; Li, S. X.; Zhang, W.; Guo, Z. Y. Rapid intracellular growth of gold nanostructures assisted by functionalized graphene oxide and its application for surface-enhanced Raman spectroscopy. *Anal Chem* **2012**, *84* (23), 10338-10344.

91. Liu, Z.; Guo, Z.; Zhong, H.; Qin, X.; Wan, M.; Yang, B. Graphene oxide based

surface-enhanced Raman scattering probes for cancer cell imaging. *Phys Chem Chem Phys* **2013**, *15* (8), 2961-2966.

92. Koh, A. L.; Shachaf, C. M.; Elchuri, S.; Nolan, G. P.; Sinclair, R. Electron microscopy localization and characterization of functionalized composite organic-inorganic SERS nanoparticles on leukemia cells. *Ultramicroscopy* **2008**, *109* (1), 111-121.
93. Sun, L.; Sung, K.-B.; Dentinger, C.; Lutz, B.; Nguyen, L.; Zhang, J.; Qin, H.; Yamakawa, M.; Cao, M.; Lu, Y.; Chmura, A. J.; Zhu, J.; Su, X.; Berlin, A. A.; Chan, S.; Knudsen, B. Composite organic-inorganic nanoparticles as Raman labels for tissue analysis. *Nano Lett* **2007**, *7* (2), 351-356.
94. Shachaf, C. M.; Elchuri, S. V.; Koh, A. L.; Zhu, J.; Nguyen, L. N.; Mitchell, D. J.; Zhang, J.; Swartz, K. B.; Sun, L.; Chan, S.; Sinclair, R.; Nolan, G. P. A novel method for detection of phosphorylation in single cells by surface enhanced Raman scattering (SERS) using composite organic-inorganic nanoparticles (COINS). *PLoS ONE* **2009**, *4* (4), e5206.
95. Su, X.; Zhang, J.; Sun, L.; Koo, T.-W.; Chan, S.; Sundararajan, N.; Yamakawa, M.; Berlin, A. A. Composite organic-inorganic nanoparticles (COINS) with chemically encoded optical signatures. *Nano Lett* **2004**, *5* (1), 49-54.
96. Stoddart, P. R.; White, D. J. Optical fibre SERS sensors. *Anal Bioanal Chem* **2009**, *394* (7), 1761-1774.
97. Gilberto, B. Optical fibre nanotaper sensors. *Opt Fiber Technol* **2010**, *16* (6), 331-342.
98. Yang, X.; Gu, C.; Qian, F.; Li, Y.; Zhang, J. Z. Highly sensitive detection of proteins and bacteria in aqueous solution using surface-enhanced Raman scattering and optical fibers. *Anal Chem* **2011**, *83* (15), 5888-5894.
99. Andrade, G. F. S.; Fan, M. K.; Brolo, A. G. Multilayer silver nanoparticles-modified optical fiber tip for high performance SERS remote sensing. *Biosens Bioelectron* **2010**, *25* (10), 2270-2275.
100. Lucotti, A.; Pesapane, A.; Zerbi, G. Use of a geometry optimized fiber-optic surface-enhanced Raman scattering sensor in trace detection. *Appl Spectrosc* **2007**, *61* (3), 260-268.
101. Kitahama, Y.; Itoh, T.; Aoyama, J.-i.; Nishikata, K.; Ozaki, Y. SERRS fiber probe: Fabrication of silver nanoparticles at the aperture of an optical fiber used for SNOM. *Chem Commun* **2009**, (43), 6563-6565.
102. Dykman, L.; Khlebtsov, N. Gold nanoparticles in biomedical applications: Recent advances and perspectives. *Chem Soc Rev* **2012**, *41* (6), 2256-2282.
103. Jokerst, J. V.; Gambhir, S. S. Molecular imaging with theranostic nanoparticles. *Accounts Chem Res* **2011**, *44* (10), 1050-1060.
104. Kaur, S.; Venktaraman, G.; Jain, M.; Senapati, S.; Garg, P. K.; Batra, S. K. Recent trends in antibody-based oncologic imaging. *Cancer Lett* **2012**, *315* (2), 97-111.
105. Huh, Y. S.; Chung, A. J.; Erickson, D. Surface enhanced Raman spectroscopy and its application to molecular and cellular analysis. *Microfluid Nanofluid* **2009**, *6* (3), 285-297.
106. Dreaden, E. C.; El-Sayed, M. A. Detecting and destroying cancer cells in more than one way with noble metals and different confinement properties on the nanoscale. *Accounts Chem Res* **2012**, *45* (11), 1854-1865.

107. Wang, X.; Wang, Y.; Chen, Z. G.; Shin, D. M. Advances of cancer therapy by nanotechnology. *Cancer Res Treat* **2009**, *41* (1), 1-11.

108. Grubisha, D. S.; Lipert, R. J.; Park, H. Y.; Driskell, J.; Porter, M. D. Femtomolar detection of prostate-specific antigen: An immunoassay based on surface-enhanced Raman scattering and immunogold labels. *Anal Chem* **2003**, *75* (21), 5936-5943.

109. Kim, J. H.; Kim, J. S.; Choi, H.; Lee, S. M.; Jun, B. H.; Yu, K. N.; Kuk, E.; Kim, Y. K.; Jeong, D. H.; Cho, M. H.; Lee, Y. S. Nanoparticle probes with surface enhanced Raman spectroscopic tags for cellular cancer targeting. *Anal Chem* **2006**, *78* (19), 6967-6973.

110. Rodriguez-Lorenzo, L.; Fabris, L.; Alvarez-Puebla, R. A. Multiplex optical sensing with surface-enhanced Raman scattering: A critical review. *Anal Chim Acta* **2012**, *745*, 10-23.

111. Alvarez-Puebla, R. A.; Aroca, R. F. Synthesis of silver nanoparticles with controllable surface charge and their application to surface-enhanced Raman scattering. *Anal Chem* **2009**, *81* (6), 2280-2285.

112. Zong, S. F.; Wang, Z. Y.; Yang, J.; Wang, C. L.; Xu, S. H.; Cui, Y. P. An SERS and fluorescence dual mode cancer cell targeting probe based on silica coated Au@Ag core-shell nanorods. *Talanta* **2012**, *97*, 368-375.

113. Maiti, K. K.; Dinish, U. S.; Fu, C. Y.; Lee, J. J.; Soh, K. S.; Yun, S. W.; Bhuvaneswari, R.; Olivo, M.; Chang, Y. T. Development of biocompatible SERS nanotag with increased stability by chemisorption of reporter molecule for *in vivo* cancer detection. *Biosens Bioelectron* **2010**, *26* (2), 398-403.

114. Maiti, K. K.; Samanta, A.; Vendrell, M.; Soh, K. S.; Olivo, M.; Chang, Y. T. Multiplex cancer cell detection by SERS nanotags with cyanine and triphenylmethine Raman reporters. *Chem Commun* **2011**, *47* (12), 3514-3516.

115. Qian, X. M.; Peng, X. H.; Ansari, D. O.; Yin-Goen, Q.; Chen, G. Z.; Shin, D. M.; Yang, L.; Young, A. N.; Wang, M. D.; Nie, S. M. *In vivo* tumor targeting and spectroscopic detection with surface-enhanced Raman nanoparticle tags. *Nat Biotechnol* **2008**, *26* (1), 83-90.

116. Yang, J.; Wang, Z. Y.; Tan, X. B.; Li, J.; Song, C. Y.; Zhang, R. H.; Cui, Y. P. A straightforward route to the synthesis of a surface-enhanced Raman scattering probe for targeting transferrin receptor-overexpressed cells. *Nanotechnology* **2010**, *21* (34), 345101.

117. Narayanan, R.; Lipert, R. J.; Porter, M. D. Cetyltrimethylammonium bromide-modified spherical and cube-like gold nanoparticles as extrinsic Raman labels in surface-enhanced Raman spectroscopy based heterogeneous immunoassays. *Anal Chem* **2008**, *80* (6), 2265-2271.

118. Zedan, A. F.; Moussa, S.; Terner, J.; Atkinson, G.; El-Shall, M. S. Ultrasmall gold nanoparticles anchored to graphene and enhanced photothermal effects by laser irradiation of gold nanostructures in graphene oxide solutions. *ACS Nano* **2012**, *7* (1), 627-636.

119. Zavaleta, C. L.; Smith, B. R.; Walton, I.; Doering, W.; Davis, G.; Shojaei, B.; Natan, M. J.; Gambhir, S. S. Multiplexed imaging of surface enhanced Raman scattering nanotags in living mice using noninvasive Raman spectroscopy. *P Natl Acad Sci USA* **2009**, *106* (32), 13511-13516.

120. Hsieh, H. Y.; Huang, T. W.; Xiao, J. L.; Yang, C. S.; Chang, C. C.; Chu, C. C.; Lo, L. W.; Wang, S. H.; Wang, P. C.; Chieng, C. C.; Lee, C. H.; Tseng, F. G. Fabrication and modification of dual-faced nano-mushrooms for tri-functional cell theranostics: SERS/fluorescence signaling, protein targeting, and drug delivery. *J Mater Chem* **2012**, *22* (39), 20918-20928.

121. Lee, S.; Chon, H.; Yoon, S. Y.; Lee, E. K.; Chang, S. I.; Lim, D. W.; Choo, J. Fabrication of SERS-fluorescence dual modal nanoprobes and application to multiplex cancer cell imaging. *Nanoscale* **2012**, *4* (1), 124-129.

122. Chourpa, I.; Morjani, H.; Riou, J. F.; Manfait, M. Intracellular molecular interactions of antitumor drug amsacrine (m-AMSA) as revealed by surface-enhanced Raman spectroscopy. *Febs Lett* **1996**, *397* (1), 61-64.

123. Seigneuric, R.; Markey, L.; Nuyten, D. S. A.; Dubernet, C.; Evelo, C. T. A.; Finot, E.; Garrido, C. From nanotechnology to nanomedicine: Applications to cancer research. *Cur Mol Med* **2010**, *10* (7), 640-652.

124. Cobley, C. M.; Au, L.; Chen, J.; Xia, Y. Targeting gold nanocages to cancer cells for photothermal destruction and drug delivery. *Expert Opin Drug Deliv* **2010**, *7* (5), 577-587.

125. Fales, A. M.; Yuan, H.; Vo-Dinh, T. Silica-coated gold nanostars for combined surface-enhanced Raman scattering (SERS) detection and singlet-oxygen generation: A potential nanoplatform for theranostics. *Langmuir* **2011**, *27* (19), 12186-12190.

126. Mohs, A. M.; Mancini, M. C.; Singhal, S.; Provenzale, J. M.; Leyland-Jones, B.; Wang, M. D.; Nie, S. M. Hand-held spectroscopic device for *in vivo* and intraoperative tumor detection: Contrast enhancement, detection sensitivity, and tissue penetration. *Anal Chem* **2010**, *82* (21), 9058-9065.

127. Wu, P.; Gao, Y.; Zhang, H.; Cai, C. X. Aptamer-guided silver-gold bimetallic nanostructures with highly active surface-enhanced Raman scattering for specific detection and near-infrared photothermal therapy of human breast cancer cells. *Anal Chem* **2012**, *84* (18), 7692-7699.

128. Song, J. B.; Zhou, J. J.; Duan, H. W. Self-assembled plasmonic vesicles of SERS-encoded amphiphilic gold nanoparticles for cancer cell targeting and traceable intracellular drug delivery. *J Am Chem Soc* **2012**, *134* (32), 13458-13469.

129. Nie, S.; Emory, S. R. Probing single molecules and single nanoparticles by surface-enhanced Raman scattering. *Science* **1997**, *275* (5303), 1102-1106.

130. Sharma, B.; Frontiera, R. R.; Henry, A.-I.; Ringe, E.; Van Duyne, R. P. SERS: Materials, applications, and the future. *Materials Today* **2012**, *15* (1-2), 16-25.

131. Stender, A. S.; Marchuk, K.; Liu, C.; Sander, S.; Meyer, M. W.; Smith, E. A.; Neupane, B.; Wang, G.; Li, J.; Cheng, J. X.; Huang, B.; Fang, N. Single cell optical imaging and spectroscopy. *Chem Rev* **2013**, *113* (4), 2469-2527.

132. Bardhan, R.; Lal, S.; Joshi, A.; Halas, N. J. Theranostic nanoshells: From probe design to imaging and treatment of cancer. *Accounts Chem Res* **2011**, *44* (10), 936-946.

133. Keren, S.; Zavaleta, C.; Cheng, Z.; de la Zerda, A.; Gheysens, O.; Gambhir, S. S. Noninvasive molecular imaging of small living subjects using Raman spectroscopy. *P Natl Acad Sci USA* **2008**, *105* (15), 5844-5849.

134. Stone, N.; Kerssens, M.; Lloyd, G. R.; Faulds, K.; Graham, D.; Matousek, P.

Surface enhanced spatially offset Raman spectroscopic (SESORS) imaging: The next dimension. *Chem Sci* **2011**, *2* (4), 776-780.

135. Stone, N.; Faulds, K.; Graham, D.; Matousek, P. Prospects of deep Raman spectroscopy for noninvasive detection of conjugated surface enhanced resonance Raman scattering nanoparticles buried within 25 mm of mammalian tissue. *Anal Chem* **2010**, *82* (10), 3969-3973.
136. Biju, V.; Itoh, T.; Anas, A.; Sujith, A.; Ishikawa, M. Semiconductor quantum dots and metal nanoparticles: Syntheses, optical properties, and biological applications. *Anal Bioanal Chem* **2008**, *391* (7), 2469-2495.
137. Choi, H. S.; Liu, W. H.; Liu, F. B.; Nasr, K.; Misra, P.; Bawendi, M. G.; Frangioni, J. V. Design considerations for tumour-targeted nanoparticles. *Nat Nanotech* **2010**, *5* (1), 42-47.
138. EY, L.-H.; AO, O.; AP, S.; DL, S.; DS, W.; JH, H.; MC, F.-C.; DO. L. Tunable plasmonic nanoprobes for theranostics of prostate cancer. *Theranostics* **2011**, *1*, 3-17.
139. Fan, M.; Wang, P.; Escobedo, C.; Sinton, D.; Brolo, A. G. Surface-enhanced Raman scattering (SERS) optrodes for multiplexed on-chip sensing of nile blue A and oxazine 720. *Lab on a Chip* **2012**, *12* (8), 1554-1560.
140. Vo-Dinh, T.; Scaffidi, J.; Gregas, M.; Zhang, Y.; Seewaldt, V. Applications of fiber-optics-based nanosensors to drug discovery. *Expert Opin Drug Dis* **2009**, *4* (8), 889-900.
141. Dong, J.; Chen, Q.; Rong, C.; Li, D.; Rao, Y. Minimally invasive surface-enhanced Raman scattering detection with depth profiles based on a surface-enhanced Raman scattering-active acupuncture needle. *Anal Chem* **2011**, *83* (16), 6191-6195.

Chapter 7

Polymeric Nanosystems for Targeted Theranostics

Yongyong Li[1*], *Huiyun Wen*[1], *Haiqing Dong*[1], *Aijun Shen*[2], *Tianbin Ren*[3], *and Donglu Shi*[1,4*]

1 The Institute for Biomedical Engineering and Nano Science, Tongji University School of Medicine, Shanghai 200092, China

2 Department of Medical Imaging, Nantong Tumor Hospital, Nantong University, Nantong 226361, China

3 School of Materials and Engineering, Tongji University, Shanghai 200092, China

4 The Materials Science and Engineering Program, Department of Mechanical and Materials Engineering, College of Engineering and Applied Science, University of Cincinnati, Cincinnati, Ohio 45221, USA

Past several years have seen extensive developments of a variety of theranostic nanosystems. Significant advances on polymeric theranostic nanosystem have been made in image-guided drug delivery taking advantage of the versatility, adaptability and biocompatibility of the rationally designed polymers. In this chapter, the critical issues confronted in the design, synthesis, and characterization of polymeric nanosystem are addressed. An overview is further given on the applications of polymeric nanosystem in biomedicine. In particular, the discussions are focused on some of the key elements of theranostics in terms of fluorescent imaging, drug storage, and controlled delivery. Some novel drug release behavior based on intelligent triggering mechanisms is also discussed.

7.1 Introduction

The cure of the cancer critically depends on the early detection and timely medical treatment. The survival rate of patients with breast cancer is approximately 90 %, if it can be diagnosed at an early stage before occurrence of metastasis to distant tissues, organs or lymph nodes. In contrast, the survival rate significantly decreases to 20%, if discovered at a later stage.[1] The success of early detection of cancer relies on the development of the effective imaging methods that can differentiate the lesions from the normal tissue. It is also important to discriminate cancer cells from the normal cells at the molecular

∗ Corresponding authors: yongyong_li@tongji.edu.cn, shid@ucmail.uc.edu level.[1]

Current imaging techniques include clinical computerized tomography (CT), magnetic resonance imaging (MRI), positron emission tomography (PET), and the nonclinical fluorescent imaging methods that are still being developed.[2,3] Among all these methods, significant improvements are needed for high specificity and sensitivity in early stage of cancer diagnosis. Currently, nanotechnology has served as a powerful tool in developing novel approaches at the genome and proteome levels, and addressing the critical issues in medical diagnostics.[2]

For the patients in the late stage of cancer, chemotherapy is generally the dominant method to treat cancer. It is accomplished by administrating one or more cytotoxic antineoplastic drugs (chemotherapeutic agents) orally or intravenously.[4] However, patients usually suffer from the poor efficacy concomitantly with the serious side effects. These include poor solubility of drugs, short circulation time, low bioavailability, and poor regulation on the drug release. Furthermore, drug resistance is recognized as another significant factor causing inefficient treatment of cancer, therefore leading to high mortality.[4] Drug resistance is a natural biological reaction that occurs commonly for many patients. Its mechanism is not well identified, but believed to be associated with the energy-dependent transporters that can detect and eject anticancer drugs from cells.[5]

To address the critical challenges in early cancer diagnosis and treatment, a new field of "nanomedicine" has emerged in the past two decades.[6] Nanomedicines are defined as the nano-scale technologies specifically applied to human health. It utilizes all novel physical, chemical, and biological properties as a result of the new nanostructures and nanomaterials. The impacts of nanomedicine are clinically profound in the prevention, early diagnosis, and treatment of diseases, especially cancer.[6] Currently, nanomedicine is still at the developmental stage, but extremely active in academic research. The key to success is the joint efforts by researchers from both the physical sciences and medicine. The former searches for the fundamentals that govern the unique behaviors of nanostructures and materials while the latter drives science into the clinically viable ways of curing diseases. However, there has been lacking of interdisciplinary collaborations between the two communities as the importance of nanomedicine is yet to be recognized. Nonetheless, the global nanomedicine market has been estimated to be of a 13.4% growth between 2009 and 2014, exceeding the pharmaceutical markets by nearly 5 % a year.[7]

The concept of nanomedicine has extended into many aspects including but not limited to the following biomedical applications: *in vitro* detection, medical imaging, drug delivery, biomaterials, tissue engineering, and theranostics.[6,7] This chapter is focused on the recent developments of theranostics based on polymeric nanosystems. Theranostics represent a multifunctional system that enables the medical diagnosis, delivery of therapeutic drug, and simultaneous monitoring of the therapy response.[8,9] In past several years, there have been extensive developments of a variety of theranostic nanosystems. Significant advances have been made in image-guided drug delivery taking advantage of unique properties of the polymeric nanosystem. In this chapter, the critical issues in all aspects of polymeric nanosystem design, synthesis, and characterization will be addressed. An overview is given on the applications of polymeric nanosystem in biomedicine. In particular, the discussions are focused on some of the key elements of theranostics in terms of fluorescent imaging, drug storage, and controlled delivery. Some novel drug release systems based on intelligent triggering mechanisms will also be discussed.

7.2 Merits of Polymeric Nanosystem

Polymeric nanosystems are uniquely capable of drug storage and controlled delivery, offering many advantages over conventional anticancer drugs. These include high drug solubility, stability, and efficient delivery over common biological barriers (Fig.7.1).[10] Figure 7.1(a) shows that, upon intravenous administration, conventional anticancer drugs are bodily distributed *via* bloodstream. This can lead to detrimental effects on the normal cells. Drug-carrying nanoparticles can actively or passively target tissues of RES. The net effect is a discrete distribution with decreased adverse effect (shown in blue) compared with the legacy drugs. Figure 7.1(b) is the schematic illustration of the representative polymeric nanoparticle systems.

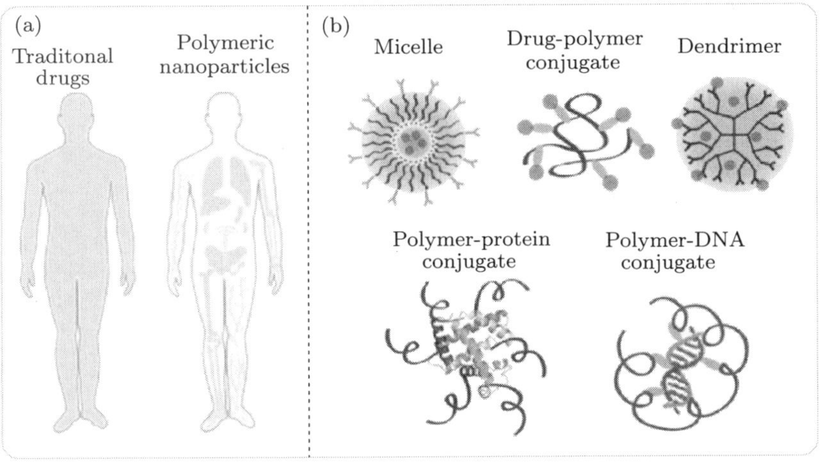

Fig. 7.1 Schematic diagrams showing (a) the difference between conventional medicine and nanotechnology and (b) various types of nanocarrier systems. (color figure at the end of the book)

7.2.1 *Potential Biocompatibility*

Biocompatibility was formally and traditionally defined as "the ability of a material to perform with an appropriate host response in a specific situation".[11] In 2010, Kohane and Langer explained biocompatibility in terms of drug delivery and re-defined biocompatibility as "an expression of the benignity of the relation between a material and its biological environment".[12] In general, it implies moderate toxicity of the nanoparticles in human tissue, blood, and immune systems with only limited adverse immunogenic, thrombogenic, and carcinogenic effects. The nanoparticles are designed to perform the diagnostic and therapeutic functionalities without interfering the normal mechanisms of the healthy organs. Therefore, biocompatibility is an important factor in the design of the nanosystems for drug storage and release.[13]

Polymeric nanoparticle-based therapeutics hold great promise for targeted pharmaceutical carriers which can be readily prepared in many forms such as micelles, polymer-drug conjugates, vesicles, and dendrimers [Fig.7.1(b)].[14] Currently some nanoparticle-bond drug delivery vehicles have already been approved for clinical applications.[15] More than 100 nanoparticle systems have been screened *in vivo* and intended for therapeutic use. The Nanotechnology Characterization Laboratory (NCL) at the National Cancer Institute has done extensive research and determined that the hydrophobicity, size, and surface charge are the dominating factors of biocompatibility.[16] Slight modification of surface chemistry, for instance, can dramatically influence a physiological response. NCL has qualitatively exhibited the relationships between the independent variables such as size, zeta potential, solubility, and the related biocompatibilities.[17] The control over these parameters will determine how long a nanoparticle can circulate before being eliminated by liver, spleen or kidney, and therefore enable its possible clinical applications.

Studies show that the polymeric nanoparticle systems can efficiently cross various physiological barriers and have the ability to mitigate toxicity and side effects in a sustained manner.[18] Albumin-bound paclitaxel (Abraxane®) nanoparticles have been FDA-approved for clinical use in breast cancer therapy, with significantly lower toxic side effects in comparison to free paclitaxel group.[19] "Nanocurcumin", doxorubicin encapsulated polyethylene glycol(PEG)-poly(aspartic acid) micelles and paclitaxel conjugated polylactide have also demonstrated enhanced antitumor activity with reduced toxicity compared with free drug.[20–22]

7.2.2 *Easy Functionality*

The polymers and their derivatives used in the formulation of nanoparticles are highly diverse, including PEG, poly(ε-caprolactone), poly(D, L-lactide), poly(amino acids), and N-(2-hydroxypropyl) methacrylamide copolymers. These also include natural polymers, such as albumin, dextran, gelatin, alginate, collagen, and chitosan.[23–25] These polymers can be functionalized *via* the terminal groups, side groups or the groups in the main chains of different requirements. The mechanism of functionalization can be either chemical (amide reaction, ester formulation, click chemistry) or physical (hydrophobic interaction, coordination, sterocomplex, supramolecular chemistry). The straight forward functionalization of the polymeric nanoparticles makes it unique, compared to other nanoparticles in encapsulation and delivery of various active agents to the desired site.

Several major strategies based on polymers have been developed to provide chemical functional groups for conjugation of targeting ligands on the surfaces of the nanoparticles.[26] These include the combination of controlled radical and ring-opening polymerizations[27], layer-by-layer deposition of polyelectrolyte chains[28], surface-initiated polymerization of high-density polymer brushes[29], block copolymer deposition on colloid surfaces[30], and chemical transformations and supramolecular assembly[31]. These strategies enable ① synthesis of polymers that can respond to environmental triggers, and ② integration of additional functionalities that can stabilize these entities (*e.g.*, crosslinking). Such a versatility in polymer sources and large capability of functionalization makes it possible to develop the polymeric -based nanoparticles for a variety of drug delivery systems.[32]

7.2.3 *Passive and Active Targeting*

Passive targeting refers to the accumulation of drug carrier at tumor tissues as a result of the unique pathophysiological characteristics of the tumor vessels.[33,34] Compared to the normal blood vessels, tumor vessels tend to be very "leaky", which are dilated, disorganized, and poorly aligned with endothelial cells. The "leaky" vascularization, lacking of lymphatic drainage, leads to enlarged gap junctions between the endothelial cells. The polymeric nanoparticles tend to extravasate through these gaps into extravascular spaces and accumulate inside the tumor tissues.[35] In this case, drugs may be released in the extracellular matrix and then diffuse through the tissue. This passive targeting mechanism was originally described by Matsumura and Maeda, known as the EPR effect, for "enhanced permeability and retention".[36] Nanoparticles with the appropriate sizes thus have more opportunities to accumulate at the tumor site *via* the EPR effect.[37] Numerous studies have confirmed the EPR effect of nanoparticles (*e.g.*, polymer-conjugated drugs, polymeric micelles, and dendrimers) with an increase of the therapeutic index and reduced side effects in both academic research and clinical trials on solid tumors.[37,38] In addition to the EPR effect, active targeting is achieved by conjugating or grafting the nanoparticles with the affinity ligands for enhanced specificity of tumor cells. Tumor cells often display increased cell surface expression that may be found at low levels on normal cells (tumor-associated antigens). Certain proteins can also be found exclusively on cancer cell surfaces (tumor-specific antigens).[39] The active targeting is to direct the nanoparticles towards tumor cells through ligand-receptor interactions by their over expressions on the tumor cell surface, thus facilitating drug release inside the cells. Those targeting ligands can be antibodies, peptides, aptamers, sugars, or small molecules. Monoclonal antibodies were first developed to have the ability of binding to specific tumor antigens in 1975.[40] Several monoclonal antibody-based therapeutic agents have been recently approved by FDA.[41] The folic acid is one of the extensively studied small molecules, which can target the membrane-associated folate receptors expressed on the epithelial tumors of various organs.[42] Folate targeting has been used successfully for tumor specific drug delivery in cancers of breast, ovary, brain, and lung.[43] Aptamers are structured by single-stranded DNA or RNA molecules, which are attractive molecules for targeting cancer cells in clinical diagnosis and therapy.[44] Currently, nanoparticle-aptamer conjugates (NP-Apt) have been successfully developed for targeted delivery of therapeutic drug both *in vitro* and *in vivo*. These include DNA aptamer-micelle, aptamer coated paclitaxel-polylactide conjugates, and aptamer-doxorubicin conjugate. Active targeting demonstrates more efficient cellular uptake, high cytotoxicity and tumor regression in cancer therapy.

7.3 Polymer Entities for Nanosystem Design

7.3.1 *PEG Derivatives*

PEG has been reported as the most commonly applied non-ionic hydrophilic polymer in polymer based drug delivery. Surface decoration of polymeric nanoparticles with PEG or "PEGylation", such as micelles and vesicles, is the most preferred method to avoid fast clearance of nanoparticles by the immune system.[45,46] PEGylation refers to the nanoparticle surface modification by covalently linking, entrapping, or adsorbing the PEG chains. PEGylation offers several other advantages:[47,48] ① size increase of the drug molecule, resulting in reduced filtration by kidneys; ② solubility enhancement of the hydrophobic drugs or carriers in aqueous media, reducing aggregation of drugs *in vivo*; ③ protein/opsonization resistance properties by preventing interactions between the nanoparticles surface and blood serum; and ④ improvement of the toxicology profile for reduced side effects. These unique characteristics make PEG a promising method for increasing the therapeutic efficacy of the drug delivery systems in clinical applications. Several PEGylated drugs have been approved by FDA and many other types are being tested in clinical trials.

Many previous studies have indicated PEGylation as an attractive strategy for improved drug delivery.[49–52] Some typical ones include the PEG-protein conjugates, PEG-small molecule drugs conjugates, PEG modified dendrimers, and the polymeric micelles with PEG coronas. In our previous research,[49] a PEGylated nano-graphene oxide has been developed to overcome the severe aggregation of graphene oxide in physiological environments. The PEG shell grafted onto graphene oxide affords the nanocomposite high physiological solubility and stability in circulation. PEG can also selectively detach from graphene oxide and rapidly release an encapsulated payload at tumor-relevant glutathione (GSH) levels upon intracellular GSH stimulation. Camptothecin (CPT) conjugated PEG could self-assemble into nanomicelles, with CPT released at a tumor relevant condition.[51] Tang, Shen and coworkers conjugated curcumin with two short oligo(ethylene glycol) (Curc-OEG) chains *via b*-thioester bonds that are labile in the presence of intracellular glutathione and esterase. Curc-OEG based nanoparticles under aqueous conditions can serve as an anticancer prodrug and a drug carrier.[53] In a similar fashion, Xu, Hagg and coworkers designed a PEG modified pH-responsive dendritic core-shell nanocarriers based on poly(ethylene imine) for polar dyes delivery.[50] The mPEG moieties improved water solubility and the three arms of mPEG gave much greater transport capacities for polar dyes and drugs than their single-arm counterparts. Protein delivery is another important application of PEGylation. Salmaso, Caliceti and coworkers developed non-covalent supramolecular protein/PEG delivery systems using the synthesized PEG-cholane derivatives with PEG of different molecular weights.[51] Their results showed an apparently large hydrophilic size of the protein/PEG associations, which enhanced the protein solubility and stability. The changes in PEG molecular weight and protein/PEG ratio resulted in significant changes in the physicochemical, pharmacokinetic, and pharmacodynamic profiles of the loaded protein. Moreover, the protein/PEG nanocomplex can disassemble and release the protein in the original form without intervention of chemical or enzymatic reactions, and therefore enable long-lasting protein delivery.

7.3.2 *Polypeptide*

Polypeptides are polydisperse structures formed by condensation of amino acid monomers through amide bonds. They are regarded as important building blocks for the design of functional materials for their abilities to form well-defined secondary structures. The self-assembling characters of polypeptide chains are essential to the supramolecular structures for potential biomedical applications.[24] Polypeptides are generally biocompatible and nontoxic. Of particular interest in the polypeptides is the versatile reactive functional groups, such as carboxylic acids, hydroxyl, amino and thiol at side chains that allow for a variety of chemical modifications.[54] All of these features render polypeptides excellent polymers for therapeutic applications. The cationic polylysine (PLL), polyarginine (PArg), the anionic polyasparagine (PAsp), and polyglutamic acid

(PGA) constitute typical part of polypeptides.

Flexible PEG is commonly used to improve the solubility of the α-helical hydrophobic segments in aqueous medium. A number of PEG-polypeptide block copolymers is prepared by the ring-opening polymerization of α-amino acid-N-carboxyanhydrides (NCA) with protective groups from the NH_2-terminated PEG.[55] The particular advantage of this method is the control of the molecular weight ($500 < M_n < 5.0\times10^5$) by the initial monomer/initiator ratio in a narrow molecular weight distribution ($M_w/M_n < 1.2$).[56] In our previous work, we successfully developed a nanomicelle featured with a PEG shell and a poly(ε-benzyloxycarbonyl-lysine) core linked *via* a disulfide linkage (mPEG-SS-PzLL). The disulfide linkage is cleaved under the reduction condition that is equivalent to cell interior, thus capable of intracellular drug delivery.[57] Several polypeptide vesicles were reported by Deming *et al.* composed of lysine-leucine [poly(L-lysine)$_{60}$-blockpoly(L-leucine)$_{20}$, $K_{60}L_{20}$], glutamate-leucine [poly(L-glutamic acid)$_{60}$-block-poly(L-leucine)$_{20}$, $E_{60}L_{20}$] or FITC functionalized poly(L-arginine-*random*-L-lysine)$_{60}$-block-poly (L-leucine)$_{20}$, ($R_{60}L_{20}$) diblock copolypeptide amphiphiles.[58] These vesicles of the size from 50 to 1 000 nm exhibited high stability up to 80 °C, and retained polar contents with retarded leakage. They can also be readily and reproducibly prepared in large quantities. Kataoka *et al.* reported various PEG-polypeptide micelles for drug or gene delivery. These micelles include DOX-conjugated PEG-PAsp for drug delivery[59], thiolated poly(ethylene glycol)-block-poly(L-lysine) (PEG-thioPLL) for DNA delivery[60], PEG-SS-P[Asp(DET)] for gene delivery[61], and c(RGDfK)-PEG-p(Lys-SH) for plasmid DNA delivery[62]. In addition, we have found α-helical conformation of PLL segments playing an important role in controlling the drug release rate, in the micelles formed by amphiphilic poly(L-leucine)-block-poly(ethylene glycol)-block-poly(L-leucine) (PLL-PEG-PLL) hybrid triblock copolymers.[63]

7.3.3 *Biopolymer*

Biopolymer is defined as the biologically-derived polymer with significant advantages over the conventional polymers. The biopolymers include protein, RNA, and DNA. They are mainly known for their biological functions and genetic basis rather than their chemical and physical structures. They have been complicated by easy degradation (such as enzyme degradation) and the ubiquitous hierarchical structures of protein. Only in recent years, researchers have begun to attempt the design of biopolymers for biomedical applications. From a chemical viewpoint, biopolymers share a common repeating structure with the exquisite accuracy on sequence and monomers. In contrast, conventional polymers exhibit low degree of crystallinity and statistical structural distribution, making them unfavorable for systematic research and clinical translation. The structural precision of the biopolymers, however, provides the consistent correlations between chemical properties and hierarchical structures, that is highly preferred in biological performances.

In biopolymers, protein-based nanosystems are particularly interesting and well developed. Some of them have been used in preclinical and clinical applications. Albumin, with a molecular weight of 66.5 kDa and size of 10 nm, is probably the most widely investigated protein. It constitutes the major protein in plasma (35–50 g/L human serum). The human serum albumin (HSA) has a super-long half life (19 days), making it an excellent candidate for nanomedicine design. For targeted imaging, albumin is generally needed to be radiolabeled or conjugated with fluorescent dyes. Tumor uptake in animal models is easily visualized by injecting albumin-labeled contrast agents. The labeled albumin binds on tumors and is distinguishable subcutaneously after post-injection. Recently, Kim *et al.*[64] synthesized gadolinium-labeled-albumin (Gd-albumin), tagged with Evans Blue dye. They observed the time dependence of infusate distributions in the rat dorsal and ventral hippocampus *in vivo*, by dynamic contrast enhanced magnetic resonance imaging (MRI). Dynamic MRI images showed preferentially distributed Gd-albumin within the hippocampus along neuroanatomical structures with less fluid resistance.[64] Due to considerable circulation time, Gd-albumin complex is able to detect malignant lesions aided by EPR effect. Vasovist®, a whole-body MRI image agent has been developed using Gd-albumin complex for detecting malignant lesions in humans.[65,66] Fluorescent imaging is another promising tool for cancer diagnosis. A nanocolloidal albumin-IRDye 800CW, with the size between 10–20 nm, has been synthesized by chemically conjugating fluorescent dye IRDye 800CW onto albumin. The imaging study on the nanocolloid showed a prolonged imaging behavior, whereas a strong decrease of fluorescence was observed from the commercially available counterparts (indocyanine green).

In addition to medical imaging, albumin is an ideal candidate for drug delivery for its high tumor uptake and versatility in storage of anticancer drugs. Abraxane®,[67] is the first clinical nano-formulation approved worldwide over 40 countries for treatment of metastatic breast cancer. It is composed of an albumin-paclitaxel complex with a size on the order of 130 nm. In comparison to its conventional counterparts in a randomized phase Ⅲ study with 460 patients, it has shown a statistically significant higher response rate (33% versus 19%, P =0.001), longer time to tumor progression (23.0 weeks versus 16.9 weeks, P =0.006), and increased survival in the subset of patients who received second-line or greater treatment (65.0 weeks versus 55.7 weeks, P =0.024).[68] The operating mechanism of Abraxane® has also been well documented. The in-depth investigation reveals that the integrated nanoparticles of Abraxane® disassembled rapidly into soluble albumin-paclitaxel complex with an optimal size for the EPR effect. Upon accumulation in tumor, a further albumin transport pathway was mediated by a 60 KDa glycoprotein gp60 which was presumably responsible for the tumor penetration and an even distribution.

7.4 Polymeric Nanosystems for Theranostics

So far, a variety of polymeric nanosystems has been engineered to implement the concept of theranostics, for their biodegradability, biocompatibility, structural flexibility, functional versatility in both synthetic and natural biopolymers.[39,69] In the past two decades, efforts have been made for clinical translation, especially for improved drug delivery. The polymer based systems have been shown to enhance the drug accumulation within the site of the interest. They have also overcome the disadvantages of the conventional pharmaceuticals including dosage limits, side effect, and poor efficacy. When incorporated with the contrast agents, a theranostic system is developed with both functionalities of therapy (drug delivery) and diagnosis (imaging). It allows the real-time monitoring of drug distribution and assessment of therapeutic efficacy for much better controlled treatment. In this fashion, "theranostics" is a newly developed medical concept and a reformation of the conventional therapy.[8,70] In the following section, the "theranostic" polymeric nanosystem is introduced based on the different imaging models, which include magnetic resonance imaging, ultrasound imaging and fluorescent imaging, and the other major techniques currently being developed.

7.4.1 *Polymeric Theranostics for MRI and Drug Delivery*

Magnetic resonance imaging (MRI) is recognized as one of the powerful diagnostic tools for its high resolution, non-exposure to radiation, and noninvasive and functional imaging.[70,71] MRI is able to conveniently distinguish normal tissue from abnormality by altering the relaxation times of water protons in solution. MRI contrast agent can be classified into positive agent (gadolinium complexes for T_1-weighted imaging) and negative contrast [superparamagnetic iron oxide (SPIO) particles for T_2-weighted imaging].

Magnetic nanoparticles (MNPs) based on SPIO exhibit excellent magnetic properties and low cytotoxicity. SPIO can also be used as a carrier for targeted drug delivery. The magnetic hyperthermia has been proposed as one of the effective cancer therapies with the magnetocaloric effect in the region of interest (ROI).[70–73] For *in vivo* study, surface modification of SPIO is required for prolonged blood circulation, prevention of SPIO leakage, drug loading, adherence to the tumor tissue and cells, and conjugation of targeted ligand. Polymers such as dextran, polyethylene glycol (PEG) based micelles are the most widely studied materials for these purposes.[74]

Recently, we have developed a green-synthetic, multifunctional hybrid micelles with shell-embedded magnetic nanoparticles for theranostic applications.[75] The hybrid system is engineered based on complex micelles, self-assembled from amphiphilic block copolymers Pluronic F127 and peptide-amphiphile (Fig.7.2). The magnetic iron oxide nanoparticles were *in situ* grown in the shell of the complex micelles under an ambient condition without any organic solvents. The anticancer drug DOX was also efficiently loaded into the hybrid micelles, thus completing the theranostic nanosystem. The magnetic nanoparticles in the shell were found to significantly retard the DOX release rate, showing the capability of drug release regulation. The cross-linking effect by magnetic nanoparticles was also found to enhance the stability and control the drug diffusion of the hybrid micelles.

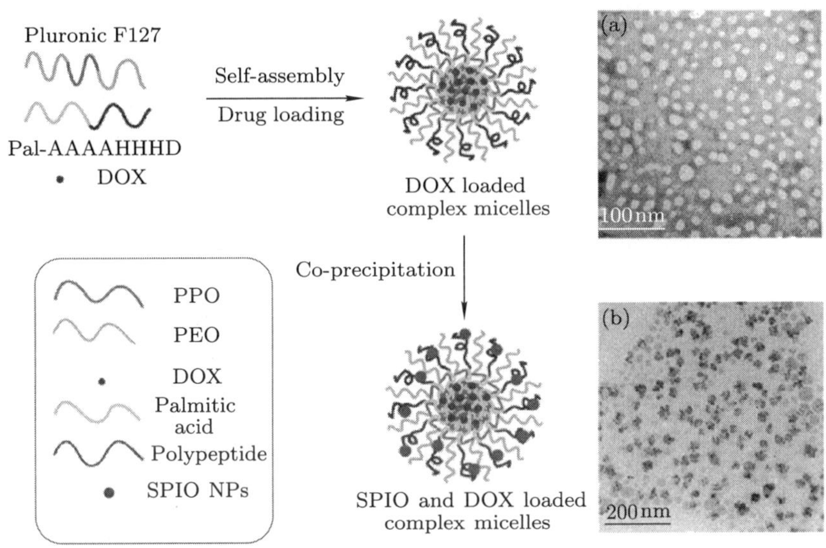

Fig. 7.2 Schematic illustration of the synthesis of the hybrid complex micelles with shell embedded magnetic nanoparticles and TEM images of (a) complex micelles and (b) hybrid micelles.[75] (color figure at the end of the book)

Experimental MRI results *in vitro* and *in vivo* showed a T_2-weighted MR imaging contrast enhancer based on these hybrid micelles. Chen and coworkers[76] synthesized a multifunctional nanocomposite (HAMAFA-b-DMAM -coated SPI-ONPs) which is capable of selective magnetic resonance imaging and pH-sensitive drug release. The release behavior was correlated to the imaging effect of the system, showing the trace of the drug release by MR. Li and coworkers[77] prepared a magnetomicelles by self-assembly of fluorine-containing amphiphilic poly (HFMA-g-PEGMA) copolymers and oleic acid modified Fe_3O_4 nanoparticles. They also investigated these magnetomicelles in potential drug delivery and magnetic resonance imaging. They were able to render the magnetomicells highly stable by incorporation of the extremely hydrophobic fluorine components. The magnetomicelles showed a high transverse relaxivity rates (r_2) of 134.27 $mM^{-1}\cdot s^{-1}$. *In vivo* MRI exhibited high contrast between liver and spleen by these magnetomicelles. Hydrophobic drug-5-fluorouracil was efficiently loaded onto the magnetomicelles with an efficiency of 20.94 wt%.

Gd (Ⅲ)-based chelating complexes are the most common contrast agents in clinical MR imaging, that generate a positive image contrast by increasing longitudinal relaxation rates of surrounding water protons (T_1-weighted imaging).[78] However, most of the clinical contrast agents such as Gd-DTPA, Gd-DOTA, *etc.*, are small molecules with the typical drawbacks of nonspecificity and rapid excretion *in vivo.*[78] The integration of a small molecule Gd complex-based contrast agent with polymeric assemblies can significantly prolong the half-life of the agent and improve the imaging effect. It can also be extended to be an excellent candidate for real-time monitoring of the drug delivery processes. Mixed diblock copolymer micelles as an integrated multifunctional platform was reported for the cancer cell-targeted delivery of chemotherapeutic drugs and MR imaging contrast enhancement under *in vitro* and *in vivo* conditions.[79] In such a system, Gd-DOTA complexes and FA molecules are attached to the hydrophilic outer coronas, and the chemotherapeutic drugs-paclitaxel is encapsulated in the hydrophobic cores. *In vitro* cell viability assays showed that the paclitaxel-loaded micelles can effectively kill HeLa cells in contrast to the drug-free ones. *In vitro* MR imaging experiments indicated dramatic increase of T_1 relaxivity (26.29 $mM^{-1}\cdot s^{-1}$) for mixed micelles compared to their small molecule counterpart, alkynyl-DOTA-Gd (3.12 $mM^{-1}\cdot s^{-1}$).*In vivo* MR imaging experiments in rabbits also showed considerable specificity of the FA-loaded mixed micellar nanoparticles in the interesting tissue. Zhang and coworkers[80] reported similar work on the polymeric micelles and MR T_1-weighted imaging based theranostic platform. Graphene oxide, with a planar structure, has recently received considerable attention for its unique physicochemical and biological properties. For the same purpose of theranostics, we harnessed it to carry the Gd (Ⅲ) chelating complexes and anticancer drug. The loading efficiency of Gd (Ⅲ) chelating complexes was found to be extremely high up to 0.314 mg/mg resulting from the super high surface area of graphene oxide. The DOX release from the nanocomposite showed a pH dependent manner.[81]

7.4.2 *Polymeric Theranostics for US Imaging and Drug Delivery*

Ultrasound (US) imaging represents another clinically important technique as a routine method for effective disease diagnosis and monitoring. Similar to MRI, US imaging avoids the exposure of ionization radiation, but has other advantages of real-time monitoring, dynamic analysis, high availability, and ease of operation.[82] US imaging applies the nano- and micro-bubbles as the contrast imaging agents that enhance the biological differences of the tissue. US imaging relies on the echogenicity difference among different mediums.[82,83] The sound waves are reflected at interfaces such as the one between a small air bubble and tissue. The interfacial behavior can be used to differentiate the disease from the normal tissue. Another intriguing property is the explosive effect of the bubbles triggered by ultrasound wave. The ultrasonic wave can facilitate drug delivery by enhancing permeability of the cell membrane and blood vessel.[82] The prerequisite of US imaging by using bubbles as the contrast agent is to keep them intact and avoid the occurrence of pulmonary embolism which can induce serious life threat. For a bubble, a pocket of air is enclosed in it, protected the membrane. The bubble may burst due to the instability of the membrane resulting in pulmonary embolism. Bubbles are generally formed of lipid, protein, and biodegradable polymer. Lipid and protein based bubbles have been approved for clinical use; while biodegradable polymer based bubbles are at phase Ⅲ stage of the clinical trial.[84] In contrast to lipid and protein, biodegradable polymer based bubbles display an outstanding stability during circulation which is critical for bio-safety. It is also easy to attach targeting ligands to, or load the drugs into these bubbles for their versatile functional groups.

Rapoport *et al.*[85] combined ultrasound imaging with ultrasound-mediated drug delivery for targeted chemotherapy, as a novel strategy in cancer treatment. Bubbles were formed by biodegradable block copolymers PEG-block-poly(L-lactide) (PEG-PLLA) or PEG-block-poly (caprolactone) (PEG-PCL). To minimize the solid core, perfluoropentane (PFP) with a boiling point of 29 °C was used to vaporize the copolymer walls inside, therefore transforming the nanoparticles into nano- and/or micro-bubbles. Lipophilic doxorubicin was selectively encapsulated in the bubbles. In their study, the size and size distribution of the bubbles were well controlled by varying the dosage ratio of PFP to copolymer. The optimal size was achieved by accumulation at tumor site, characterized by the irregular and discontinued vessels. On the other hand, lipophilic doxorubicin was not extravasated into the normal tissues by tight junctions. Tumor uptake of doxorubicin by ultrasound was investigated *in vivo*. 4 h after intravenous injection, the tumor of the mouse was sonicated by 3-MHz ultrasound. The results showed a statistically significant increase in tumor inhibition versus those with or without ultrasound treatment.

A recent system developed by Escoffre *et al.*[86] employed a distinctly different strategy on the polymeric-based theranostics with combined drug delivery and US imaging. The DOX loaded liposome is chemically anchored onto the microbubbles' shell, resulting in a highly integrated system. The different components are respectively responsible for drug encapsulation and US imaging. A covalent thiol-maleimide linkage is applied for integration of different components. At lower mechanical index, MI $\leqslant 0.26$, the best contrast to tissue (CTR=25 dB) was observed. As MI value increases, the microbubbles were irrupted due to high pressure. Bubble destruction led to accelerated drug release. Experimental results on cell apoptosis exposure to the integrated system under sonication showed a 4-fold efficiency versus that of the group treated with only DOX or without sonication.

Gene therapy represents a promising technique to deliver the genetic payload to the site of disease. In this way, the mutated gene is either replaced or altered for bio-function restoration. Borden *et al.*[87] demonstrated a novel layer-by-layer (LBL) assembly that adsorbs multiple layers (repeated DNA and PLL) onto lipid-coated microbubbles, for increased DNA loading capacity. LBL assembly relies on the complementary electrostatic property of the anionic DNA and cationic PLL. Microbubbles were made cationic by coating a layer of cationic lipid before LBL assembly. The DNA loading capacity onto the microbubbles was enhanced by over 10-fold when using five paired layers. Furthermore, the polyelectrolyte multilayers were found to stablize the microbubbles during circulation. Their results showed that the compartmentalized microbubbles can be made as the multifunctional, high-payload ultrasound contrast agents for targeted gene therapy.

7.4.3 *Polymeric Theranostics for Optical Imaging and Drug Delivery*

Optical imaging is one of the effective methods in biomedical studies at the cellular or even molecular level.[88] It provides the real time, noninvasive analysis of the biochemical process of the tissue, live cells and even subcellular organelle. It therefore serves as a significant clinical tool for early stage disease diagnosis, particularly useful on the time dependence of alteration of a specific organ. An optical probe, mostly made of a fluorescent nanoparticle is required for imaging by the confocal scanning microscopy or *in vivo* system. Fluorescent nanoparticle can be made either by attaching the fluorescent dye or nanocrystal [*i.e.*, quantum dots (QDs)] to the nanosystem, or *in situ* processed *via* the chemical or physical route. In the design of the fluorescent system, polymeric nanoparticles are favorable for their biocompatibility and multifunctionality. The polymeric nanosystem can be developed with multiple functional groups for imaging, drug loading, and cell targeting with low cytotoxicity.[89–92] In addition, some polymers are inherently fluorescent, making them ideal candidates for optical imaging.

QDs are semiconductor materials, known for their superb optical properties and applications in optical imaging.[88] QDs exhibit high extinction coefficient, narrow emission, excellent photostability, and high quantum yield, therefore particularly attractive for simultaneous multicolor imaging. Wu *et al.*[90] designed and developed a multifunctional chitosan-poly(methacrylic acid)-CdSe hybrid nanogel, where CdSe nanocrystals were *in situ* grown in a network of cross-linked PMAA. The PMAA networks are semi-interpenetrated by the chitosan chains. Two multifunctional systems characterized by either physically or chemically cross-linked route, showed distinctly different pH sensitive properties. Such a multifunctional system can be used for imaging the B16F10 mouse melanoma cells, while selectively releasing the anticancer drug in the typical abnormal pH range of 5–7.4. A low cytotoxicity was also observed for the chemically cross-linked formulations. However, the physically cross-linked formulations exhibited a non-reversible pH-sensitive photoluminescent property and a prominent cytotoxicity after 24 h treatment. With a similar strategy,[91] this group of researchers further developed a class of core-shell structured hybrid nanogels and demonstrated the highly integrated multifunctional system for simultaneous optical temperature-sensing, cancer cell targeting, fluorescence imaging, and combined chemo-photothermal treatment. The hybrid nanogels consisted of a Ag-Au bimetallic core and a shell of thermo-responsive nonlinear polyethylene glycol (PEG)-based hydrogel.

Despite the photobleaching effect commonly observed, the concern on clinical applications of florescent materials has been on cytotoxicity, especially the long-term biosafety. Manchanda *et al.*[89] simultaneously encapsulated fluorescent dye (ICG) and anticancer drug DOX in the microsphere of poly (DL-lactide-co-glycolicacid) for a potential theranostic system. In a recent work, Chen *et al.*[93] proposed a label free method for theranostics fabricated by abietane-based nanogels which exhibited strong self-fluorescence. The natural abietic acid showed negligible fluorescence in aqueous solution, but a strong fluorescence was generated after a convenient anaerobic thermal treatment. The thermal treatment induced isomerization and fluorescence emission under excitation. Conjugation of folic acid to abietane-based nanogels enabled active targeting and imaging of breast cancer cells. These abietane-based nanogels provide a promising platform for direct cell imaging and disease therapeutics.

Luminescent lanthanide chelates with high color purity and stability represent another type of fluorescent unit and gain great potential to substitute the conventional dyes in bioimaging and biolabeling applications. The Eu(Ⅲ)-based coordination complexes have attracted particular attention for their red emission with little interference. Coordinated with a Eu (Ⅲ)-based chelate, we fabricated a fluorescent, thermosensitive, amphiphilic copolymer micelles for drug delivery and optical imaging.[94] The controlled drug release from the micelles showed a thermosensitive dependent manner. The micelles were capable of tumor cell (A549) internalization and obvious inhibition of the proliferation of A549. *In vivo* study showed uptaking of the self-assembled micelles by the larvae of zebrafish, which is easily tracked and eliminated from the body within several days (Fig.7.3).

7.4.4 *Polymeric Theranostics for Dual-imaging and Drug Delivery*

The information acquired in disease diagnosis depends on the specific technology of imaging due to different imaging mechanisms. Effort has been made

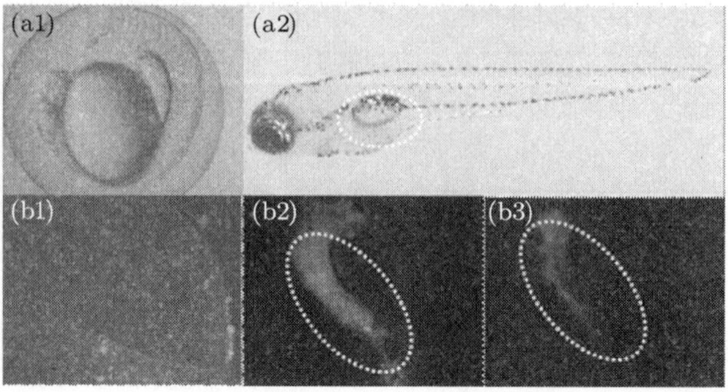

Fig. 7.3 Images of zebrafish embryo (a1), the full body shot of a fish (a2), zebrafish embryo exposure to micelles (b1), larva after treated with micelles for 6 days (b2) and then with fresh water for 3 days (b3). The images of (b1), (b2) and (b3) were obtained by excitation under UV light with a blue background. (color figure at the end of the book)

to develop dual-imaging by combing MRI or US with optical means.[95–97] Dual-imaging may provide even more accurate diagnosis that is particularly advanced when comparing contrast differences between two different modes. Dual-imaging is highly possible by using polymers for their multiple functional groups. It can also be used to monitor the drug delivery process in a real-time fashion.

MRI and optical dual-imaging has received more attention due to their synergetic effect on the imaging performance. MRI is capable of imaging the in-depth tissue while optical imaging is advantageous on micro-structures. Santra *et al.*[95] coated iron oxide nanoparticles with poly(acrylic acid) for developing the biocompatible, multimodal, and theranostic system. For targeted optical/magnetic resonance imaging, the system was functionalized with folate as a biomarker. The lipophilic fluorescent dyes and anticancer drug were encapsulated within the hydrophobic pockets of poly(acrylic acid) matrix for multifunctionalities. *In vitro* drug delivery and imaging experiments confirmed targeted drug delivery and the dual-imaging effect. For combined US and optical imaging, Fokong *et al.*[96] engineered microbubbles composed of poly(butyl cyanoacrylate) based walls, loaded rhodamine-B or anticancer drug coumarin-6, and anti-VEGFR2 antibodies anchored on the surface of the walls. The dual-imaging capabilities of the system were confirmed by the image-guided targeting to the tumor vasculature, an effective evaluation of the therapy process. They also found high levels of model drug release by the VEGFR2-targeted MB with high-mechanical index.

7.5 Perspectives and Concluding Remarks

As evidenced in major research findings, polymers have found promising applications in cancer imaging and therapies. In addition to their native forms such as HPMA polymeric bioconjugates, nanostructures of polymers can be engineered into different forms of micelles, nanospheres, nanogels, vesicles, and dendrimers, *etc.* Their sizes can be tailored and optimized for specific biomedical diagnosis and treatment. Polymer-based nanosystems are rich in functional groups, therefore particularly versatile in structure architecture. Various imaging agents (fluorescent, magnetic, *etc.*), anticancer drugs (DOX, CPT, *etc.*), and targeting ligand (RGD, antibody, *etc.*) can all be chemically or physically incorporated into the polymeric nanosystem making them ideally multifunctional. However, adverse effects resulting from the interactions among all components must be considered in the design and development of the polymeric nanosystems. For example, magnetic nanoparticles can reduce the fluorescent intensity leading to ineffective optical imaging. Drug loading and cell targeting may also interfere with each other in the theranostic process.

In contrast to the conventional polymers, biopolymers will show more advantages in cell targeting and biosafety. They are generally made of more biocompatible unit structures such as amino acid, glucose unit, and nucleotides. Most of them are biodegradable that are friendly to human body. The higher half-life in circulation (19 days for HSA) allows maximized accumulation *via* the EPR effect within the tumor site, facilitating enhanced therapeutical and imaging effects in biopolymer-based nanosystems. Biopolymers share a common repeating structure with the exquisite accuracy on sequence and monomers, whereas conventional polymers generally display a low control and statistical distribution on the structure. These statistical distributions have usually created inconsistencies for systematic studies and further clinical translation. The structural precision of the biopolymers is therefore particularly advantageous for developing structure-property relationships in biological processes.

References

1. Etzioni, R.; Urban, N.; Ramsey, S.; McIntosh, M.; Schwartz, S.; Reid, B.; Radich, J.; Anderson, G.; Hartwell, L. The case for early detection. *Nat Rev Cancer* **2003**, *3* (4), 243-252.
2. Chen, H. M.; Zhen, Z. P.; Todd, T.; Chu, P. K.; Xie, J. Nanoparticles for improving cancer diagnosis. *Mat Sci Eng R* **2013**, *74* (3), 35-69.
3. Brigger, I.; Dubernet, C.; Couvreur, P. Nanoparticles in cancer therapy and diagnosis. *Adv Drug Deliver Rev* **2012**, *64*, 24-36.
4. Blanco, E.; Hsiao, A.; Mann, A. P.; Landry, M. G.; Meric-Bernstam, F.; Ferrari, M. Nanomedicine in cancer therapy: Innovative trends and prospects. *Cancer Sci* **2011**, *102* (7), 1247-1252.
5. Gao, Z. B.; Zhang, L. N.; Sun, Y. J. Nanotechnology applied to overcome tumor drug resistance. *J Control Release* **2012**, *162* (1), 45-55.
6. Yeo, Y. Emerging technology in evaluation of nanomedicine. *Mol Pharmaceut* **2013**, *10* (6), 2091-2092.

7. Chapman, S.; Panaro, N. J.; Hinkal, G. W.; Hook, S. S.; Prabhakar, U.; Ptak, K.; Farrell, D.; Grodzinski, P. Company profile: Kindling translational cancer nanotechnology research. *Nanomedicine* **2012**, *7* (3), 321-325.
8. Svenson, S. Theranostics: Are we there yet? *Mol Pharmaceut* **2013**, *10* (3), 848-856.
9. Palekar-Shanbhag, P.; Jog, S. V.; Chogale, M. M.; Gaikwad, S. S. Theranostics for cancer therapy. *Curr Drug Deliv* **2013**, *10* (3), 357-362.
10. Moros, M.; Mitchell, S. G.; Grazu, V.; de la Fuente, J. M. The fate of nanocarriers as nanomedicines *in vivo*: Important considerations and biological barriers to overcome. *Curr Med Chem* **2013**, *20* (22), 2759-2778.
11. Williams, D. F. On the mechanisms of biocompatibility. *Biomaterials* **2008**, *29* (20), 2941-2953.
12. Kohane, D. S.; Langer, R. Biocompatibility and drug delivery systems. *Chem Sci* **2010**, *1* (4), 441-446.
13. Naahidi, S.; Jafari, M.; Edalat, F.; Raymond, K.; Khademhosseini, A.; Chen, P. Biocompatibility of engineered nanoparticles for drug delivery. *J Control Release* **2013**, *166* (2), 182-194.
14. Haag, R.; Kratz, F. Polymer therapeutics: Concepts and applications. *Angew Chem Int Edit* **2006**, *45* (8), 1198-1215.
15. Duncan, R. Polymer therapeutics as nanomedicines: New perspectives. *Curr Opin Biotech* **2011**, *22* (4), 492-501.
16. McNeil, S. E. Nanoparticle therapeutics: A personal perspective. *Wires Nanomed Nanobi* **2009**, *1* (3), 264-271.
17. McNeil, S. E. Nanotechnology for the biologist. *J Leukocyte Biol* **2005**, *78* (3), 585-594.
18. Pietroiusti, A.; Campagnolo, L.; Fadeel, B. Interactions of engineered nanoparticles with organs protected by internal biological barriers. *Small* **2013**, *9* (9-10), 1557-1572.
19. Miele, E.; Spinelli, G. P.; Tomao, F.; Tomao, S. Albumin-bound formulation of paclitaxel (Abraxane ABI-007) in the treatment of breast cancer. *Int J Nanomedicine* **2009**, *4*, 99-105.
20. Nakanishi, T.; Fukushima, S.; Okamoto, K.; Suzuki, M.; Matsumura, Y.; Yokoyama, M.; Okano, T.; Sakurai, Y.; Kataoka, K. Development of the polymer micelle carrier system for doxorubicin. *J Control Release* **2001**, *74* (1-3), 295-302.
21. Tong, R.; Yala, L. D.; Fan, T. M.; Cheng, J. J. The formulation of aptamer-coated paclitaxel-polylactide nanoconjugates and their targeting to cancer cells. *Biomaterials* **2010**, *31* (11), 3043-3053.
22. Bisht, S.; Feldmann, G.; Soni, S.; Ravi, R.; Karikar, C.; Maitra, A. Polymeric nanoparticle-encapsulated curcumin ("nanocurcumin"): A novel strategy for human cancer therapy. *J Nanobiotechnology* **2007**, *5*, 3.
23. Tong, R.; Cheng, J. J. Anticancer polymeric nanomedicines. *Polym Rev* **2007**, *47* (3), 345-381.
24. Duncan, R. The dawning era of polymer therapeutics. *Nat Rev Drug Discov* **2003**, *2* (5), 347-360.

25. Pasut, G.; Veronese, F. M. Polymer-drug conjugation, recent achievements and general strategies. *Prog Polym Sci* **2007**, *32* (8-9), 933-961.
26. Holgado, M. A.; Martin-Banderas, L.; Alvarez-Fuentes, J.; Fernandez-Arevalo, M.; Arias, J. L. Drug targeting to cancer by nanoparticles surface functionalized with special biomolecules. *Curr Med Chem* **2012**, *19* (19), 3188-3195.
27. Rao, J. P.; Geckeler, K. E. Polymer nanoparticles: Preparation techniques and size-control parameters. *Prog Polym Sci* **2011**, *36* (7), 887-913.
28. Wong, J. E.; Diez-Pascual, A. M.; Richtering, W. Layer-by-layer assembly of polyelectrolyte multilayers on thermoresponsive P(NiPAM-co-MAA) microgel: Effect of ionic strength and molecular weight. *Macromolecules* **2009**, *42* (4), 1229-1238.
29. Edmondson, S.; Osborne, V. L.; Huck, W. T. S. Polymer brushes *via* surface-initiated polymerizations. *Chem Soc Rev* **2004**, *33* (1), 14-22.
30. Spatz, J. P.; Mossmer, S.; Hartmann, C.; Moller, M.; Herzog, T.; Krieger, M.; Boyen, H. G.; Ziemann, P.; Kabius, B. Ordered deposition of inorganic clusters from micellar block copolymer films. *Langmuir* **2000**, *16* (2), 407-415.
31. Zeng, F.; Zimmerman, S. C. Dendrimers in supramolecular chemistry: From molecular recognition to self-assembly. *Chem Rev* **1997**, *97* (5), 1681-1712.
32. Elsabahy, M.; Wooley, K. L. Design of polymeric nanoparticles for biomedical delivery applications. *Chem Soc Rev* **2012**, *41* (7), 2545-2561.
33. Bamrungsap, S.; Zhao, Z. L.; Chen, T.; Wang, L.; Li, C. M.; Fu, T.; Tan, W. H. Nanotechnology in therapeutics: A focus on nanoparticles as a drug delivery system. *Nanomedicine* **2012**, *7* (8), 1253-1271.
34. Misra, R.; Acharya, S.; Sahoo, S. K. Cancer nanotechnology: Application of nanotechnology in cancer therapy. *Drug Discov Today* **2010**, *15* (19-20), 842-850.
35. Danhier, F.; Feron, O.; Preat, V. To exploit the tumor microenvironment: Passive and active tumor targeting of nanocarriers for anti-cancer drug delivery. *J Control Release* **2010**, *148* (2), 135-146.
36. Matsumura, Y.; Maeda, H. A new concept for macromolecular therapeutics in cancer chemotherapy: Mechanism of tumoritropic accumulation of proteins and the antitumor agent smancs. *Cancer Res* **1986**, *46*, 6387-6392.
37. Maeda, H.; Nakamura, H.; Fang, J. The EPR effect for macromolecular drug delivery to solid tumors: Improvement of tumor uptake, lowering of systemic toxicity, and distinct tumor imaging *in vivo*. *Adv Drug Deliver Rev* **2013**, *65* (1), 71-79.
38. Maeda, H.; Bharate, G. Y.; Daruwalla, J. Polymeric drugs for efficient tumor-targeted drug delivery based on EPR-effect. *Eur J Pharmaceut Biopharmaceut* **2009**, *71* (3), 409-419.
39. Park, J. H.; Lee, S.; Kim, J. H.; Park, K.; Kim, K.; Kwon, I. C. Polymeric nanomedicine for cancer therapy. *Prog Polym Sci* **2008**, *33* (1), 113-137.
40. Kohler, G.; Milstein, C. Continuous cultures of fused cells secreting antibody of predefined specificity. *Nature* **1975**, *256* (5517), 495-497.
41. Brannon-Peppas, L.; Blanchette, J. O. Nanoparticle and targeted systems for cancer therapy. *Adv Drug Deliver Rev* **2012**, *64*, 206-212.
42. Sudimack, J.; Lee, R. J. Targeted drug delivery *via* the folate receptor. *Adv Drug Deliver Rev* **2000**, *41* (2), 147-162.

43. Kwon, O. J.; Kang, E.; Choi, J. W.; Kim, S. W.; Yun, C. O. Therapeutic targeting of chitosan-PEG-folate-complexed oncolytic adenovirus for active and systemic cancer gene therapy. *J Control Release* **2013**, *169* (3), 257-265.
44. Levy-Nissenbaum, E.; Radovic-Moreno, A. F.; Wang, A. Z.; Langer, R.; Farokhzad, O. C. Nanotechnology and aptamers: Applications in drug delivery. *Trends Biotechnol* **2008**, *26* (8), 442-449.
45. Ikeda, Y.; Nagasaki, Y. PEGylation technology in nanomedicine. *Adv Polym Sci* **2012**, *247*, 115-140.
46. Aggarwal, P.; Hall, J. B.; McLeland, C. B.; Dobrovolskaia, M. A.; McNeil, S. E. Nanoparticle interaction with plasma proteins as it relates to particle biodistribution, biocompatibility and therapeutic efficacy. *Adv Drug Deliver Rev* **2009**, *61* (6), 428-437.
47. Harris, J. M.; Chess, R. B. Effect of PEGylation on pharmaceuticals. *Nat Rev Drug Discov* **2003**, *2* (3), 214-221.
48. Pasut, G.; Veronese, F. M. State of the art in pegylation: The great versatility achieved after forty years of research. *J Control Release* **2012**, *161* (2), 461-472.
49. Wen, H. Y.; Dong, C. Y.; Dong, H. Q.; Shen, A. J.; Xia, W. J.; Cai, X. J.; Song, Y. Y.; Li, X. Q.; Li, Y. Y.; Shi, D. L. Engineered redox-responsive PEG detachment mechanism in PEGylated nano-graphene oxide for intracellular drug delivery. *Small* **2012**, *8* (5), 760-769.
50. Xu, S.; Luo, Y.; Haag, R. Water-soluble pH-responsive dendritic core-shell nanocarriers for polar dyes based on poly(ethylene imine). *Macromol Biosci* **2007**, *7* (8), 968-974.
51. Salmaso, S.; Bersani, S.; Mastrotto, F.; Tonon, G.; Schrepfer, R.; Genovese, S.; Caliceti, P. Self-assembling nanocomposites for protein delivery: Supramolecular interactions between PEG-cholane and rh-G-CSF. *J Control Release* **2012**, *162* (1), 176-184.
52. Li, X. Q.; Wen, H. Y.; Dong, H. Q.; Xue, W. M.; Pauletti, G. M.; Cai, X. J.; Xia, W. J.; Shi, D. L.; Li, Y. Y. Self-assembling nanomicelles of a novel camptothecin prodrug engineered with a redox-responsive release mechanism. *Chem Commun* **2011**, *47* (30), 8647-8649.
53. Tang, H. D.; Murphy, C. J.; Zhang, B.; Shen, Y. Q.; Sui, M. H.; Van Kirk, E. A.; Feng, X. W.; Murdoch, W. J. Amphiphilic curcumin conjugate-forming nanoparticles as anticancer prodrug and drug carriers: *In vitro* and *in vivo* effects. *Nanomedicine* **2010**, *5* (6), 855-865.
54. Gonzalez-Aramundiz, J. V.; Lozano, M. V.; Sousa-Herves, A.; Fernandez-Megia, E.; Csaba, N. Polypeptides and polyaminoacids in drug delivery. *Expert Opin Drug Deliv* **2012**, *9* (2), 183-201.
55. Ren, T. B.; Xia, W. J.; Dong, H. Q.; Li, Y. Y. Sheddable micelles based on disulfide-linked hybrid PEG-polypeptide copolymer for intracellular drug delivery. *Polymer* **2011**, *52* (16), 3580-3586.
56. Deming, T. J. Polypeptide and polypeptide hybrid copolymer synthesis *via* NCA polymerization. *Adv Polym Sci* **2006**, *202*, 1-18.
57. Wen, H. Y.; Dong, H. Q.; Xie, W. J.; Li, Y. Y.; Wang, K.; Pauletti, G. M.; Shi, D. L. Rapidly disassembling nanomicelles with disulfide-linked PEG shells for glutathione-mediated intracellular drug delivery. *Chem Commun* **2011**, *47* (12), 3550-3552.

58. Holowka, E. P.; Pochan, D. J.; Deming, T. J. Charged polypeptide vesicles with controllable diameter. *J Am Chem Soc* **2005**, *127* (35), 12423-12428.
59. Yokoyama, M.; Okano, T.; Sakurai, Y.; Ekimoto, H.; Shibazaki, C.; Kataoka, K. Toxicity and antitumor activity against solid tumors of micelle-forming polymeric anticancer drug and its extremely long circulation in blood. *Cancer Res* **1991**, *51* (12), 3229-3236.
60. Kakizawa, Y.; Harada, A.; Kataoka, K. Glutathione-sensitive stabilization of block copolymer micelles composed of antisense DNA and thiolated poly(ethylene glycol)-block-poly(L-lysine): A potential carrier for systemic delivery of antisense DNA. *Biomacromolecules* **2001**, *2* (2), 491-497.
61. Takae, S.; Miyata, K.; Oba, M.; Ishii, T.; Nishiyama, N.; Itaka, K.; Yamasaki, Y.; Koyama, H.; Kataoka, K. PEG-detachable polyplex micelles based on disulfide-linked block catiomers as bioresponsive nonviral gene vectors. *J Am Chem Soc* **2008**, *130* (18), 6001-6009.
62. Mickler, F. M.; Vachutinsky, Y.; Oba, M.; Miyata, K.; Nishiyama, N.; Kataoka, K.; Brauchle, C.; Ruthardt, N. Effect of integrin targeting and PEG shielding on polyplex micelle internalization studied by live-cell imaging. *J Control Release* **2011**, *156* (3), 364-373.
63. Hua, S. H.; Li, Y. Y.; Liu, Y.; Xiao, W.; Li, C.; Huang, F. W.; Zhang, X. Z.; Zhuo, R. X. Self-assembled micelles based on PEG-polypeptide hybrid copolymers for drug delivery. *Macromol Rapid Comm* **2010**, *31* (1), 81-86.
64. Kim, J. H.; Astary, G. W.; Nobrega, T. L.; Kantorovich, S.; Carney, P. R.; Mareci, T. H.; Sarntinoranont, M. Dynamic contrast-enhanced MRI of Gd-albumin delivery to the rat hippocampus *in vivo* by convection-enhanced delivery. *J Neurosci Meth* **2012**, *209* (1), 62-73.
65. Essig, M.; Rohrer, M.; Giesel, F.; Tuttenberg, J.; Weber, M. A.; Michaely, H.; Gerigk, L.; Voth, M. Human brain tumor imaging with a protein-binding MR contrast agent: Initial experience. *Eur Radiol* **2010**, *20* (1), 218-226.
66. Yamashita, T.; Takahara, T.; Kwee, T. C.; Kawada, S.; Horie, T.; Inomoto, C.; Hashida, K.; Yamamuro, H.; Myojin, K.; Luijten, P. R.; Imai, Y. Diffusion magnetic resonance imaging with gadofosveset trisodium as a negative contrast agent for lymph node metastases assessment. *Jpn J Radiol* **2011**, *29* (1), 25-32.
67. Elsadek, B.; Kratz, F. Impact of albumin on drug delivery: New applications on the horizon. *J Control Release* **2012**, *157* (1), 4-28.
68. Gradishar, W. J.; Tjulandin, S.; Davidson, N.; Shaw, H.; Desai, N.; Bhar, P.; Hawkins, M.; O'Shaughnessy, J. Phase Ⅲ trial of nanoparticle albumin-bound paclitaxel compared with polyethylated castor oil-based paclitaxel in women with breast cancer. *J Clin Oncol* **2005**, *23* (31), 7794-7803.
69. Khemtong, C.; Kessinger, C. W.; Gao, J. M. Polymeric nanomedicine for cancer MR imaging and drug delivery. *Chem Commun* **2009**, (24), 3497-3510.
70. McCarthy, J. R.; Weissleder, R. Multifunctional magnetic nanoparticles for targeted imaging and therapy. *Adv Drug Deliver Rev* **2008**, *60* (11), 1241-1251.
71. Oghabian, M. A.; Farahbakhsh, N. M. Potential use of nanoparticle based contrast agents in MRI: A molecular imaging perspective. *J Biomed Nanotechnol* **2010**, *6* (3), 203-213.
72. Wang, H. J.; Zhang, S. N.; Liao, Z. Y.; Wang, C. Y.; Liu, Y.; Feng, S. Q.; Jiang, X. G.; Chang, J. PEGlated magnetic polymeric liposome anchored with TAT

for delivery of drugs across the blood-spinal cord barrier. *Biomaterials* **2010**, *31* (25), 6589-6596.

73. Zhang, J.; Misra, R. D. K. Magnetic drug-targeting carrier encapsulated with thermosensitive smart polymer: Core-shell nanoparticle carrier and drug release response. *Acta Biomater* **2007**, *3* (6), 838-850.
74. Zhang, L.; Xue, H.; Gao, C. L.; Carr, L.; Wang, J. N.; Chu, B. C.; Jiang, S. Y. Imaging and cell targeting characteristics of magnetic nanoparticles modified by a functionalizable zwitterionic polymer with adhesive 3,4-dihydroxyphenyl-l-alanine linkages. *Biomaterials* **2010**, *31* (25), 6582-6588.
75. Li, Y. Y.; Ma, J. P.; Zhu, H. Y.; Gao, X. L.; Dong, H. Q.; Shi, D. L. Green synthetic, multifunctional hybrid micelles with shell embedded magnetic nanoparticles for theranostic applications. *ACS Appl Mater Inter* **2013**, *5* (15), 7227-7235.
76. Chen, D. Y.; Xia, X. W.; Gu, H. W.; Xu, Q. F.; Ge, J. F.; Li, Y. G.; Li, N. J.; Lu, J. M. PH-responsive polymeric carrier encapsulated magnetic nanoparticles for cancer targeted imaging and delivery. *J Mater Chem* **2011**, *21* (34), 12682-12690.
77. Li, X. L.; Li, H.; Liu, G. Q.; Deng, Z. W.; Wu, S. L.; Li, P. H.; Xu, Z. S.; Xu, H. B.; Chu, P. K. Magnetite-loaded fluorine-containing polymeric micelles for magnetic resonance imaging and drug delivery. *Biomaterials* **2012**, *33* (10), 3013-3024.
78. Mi, P.; Cabral, H.; Kokuryo, D.; Rafi, M.; Terada, Y.; Aoki, I.; Saga, T.; Takehiko, I.; Nishiyama, N.; Kataoka, K. Gd-DTPA-loaded polymer-metal complex micelles with high relaxivity for MR cancer imaging. *Biomaterials* **2013**, *34* (2), 492-500.
79. Liu, T.; Qian, Y. F.; Hu, X. L.; Ge, Z. S.; Liu, S. Y. Mixed polymeric micelles as multifunctional scaffold for combined magnetic resonance imaging contrast enhancement and targeted chemotherapeutic drug delivery. *J Mater Chem* **2012**, *22* (11), 5020-5030.
80. Zhang, G. D.; Zhang, R.; Wen, X. X.; Li, L.; Li, C. Micelles based on biodegradable poly(L-glutamic acid)-b-polylactide with paramagnetic Gd ions chelated to the shell layer as a potential nanoscale MRI-visible delivery system. *Biomacromolecules* **2008**, *9* (1), 36-42.
81. Shen, A. J.; Li, D. L.; Cai, X. J.; Dong, C. Y.; Dong, H. Q.; Wen, H. Y.; Dai, G. H.; Wang, P. J.; Li, Y. Y. Multifunctional nanocomposite based on graphene oxide for *in vitro* hepatocarcinoma diagnosis and treatment. *J Biomed Mater Res A* **2012**, *100A* (9), 2499-2506.
82. Wang, Y.; Li, X.; Zhou, Y.; Huang, P.; Xu, Y. Preparation of nanobubbles for ultrasound imaging and intracelluar drug delivery. *Int J Pharmaceut* **2010**, *384* (1), 148-153.
83. Smith, M. Long-lived bubbles for ultrasound imaging. *J Mater Chem* **2006**, *16* (15), C26.
84. Bohmer, M. R.; Klibanov, A. L.; Tiemann, K.; Hall, C. S.; Gruell, H.; Steinbach, O. C. Ultrasound triggered image-guided drug delivery. *Eur J Radiol* **2009**, *70* (2), 242-253.
85. Rapoport, N.; Gao, Z. G.; Kennedy, A. Multifunctional nanoparticles for combining ultrasonic tumor imaging and targeted chemotherapy. *J Natl Cancer I* **2007**, *99* (14), 1095-1106.

86. Escoffre, J. M.; Mannaris, C.; Geers, B.; Novell, A.; Lentacker, I.; Averkiou, M.; Bouakaz, A. Doxorubicin liposome-loaded microbubbles for contrast imaging and ultrasound-triggered drug delivery. *IEEE T Ultrason Ferr* **2013**, *60* (1), 78-87.
87. Borden, M. A.; Caskey, C. F.; Little, E.; Gillies, R. J.; Ferrara, K. W. DNA and polylysine adsorption and multilayer construction onto cationic lipid-coated microbubbles. *Langmuir* **2007**, *23* (18), 9401-9408.
88. Deng, D. W.; Chen, Y. Q.; Cao, J.; Tian, J. M.; Qian, Z. Y.; Achilefu, S.; Gu, Y. Q. High-quality $CuInS_2/ZnS$ quantum dots for *in vitro* and *in vivo* bioimaging. *Chem Mater* **2012**, *24* (15), 3029-3037.
89. Manchanda, R.; Fernandez-Fernandez, A.; Nagesetti, A.; McGoron, A. J. Preparation and characterization of a polymeric (PLGA) nanoparticulate drug delivery system with simultaneous incorporation of chemotherapeutic and thermo-optical agents. *Colloid Surface B* **2010**, *75* (1), 260-267.
90. Wu, W.; Shen, J.; Banerjee, P.; Zhou, S. Q. Chitosan-based responsive hybrid nanogels for integration of optical pH-sensing, tumor cell imaging and controlled drug delivery. *Biomaterials* **2010**, *31* (32), 8371-8381.
91. Wu, W. T.; Shen, J.; Banerjee, P.; Zhou, S. Q. Core-shell hybrid nanogels for integration of optical temperature-sensing, targeted tumor cell imaging, and combined chemo-photothermal treatment. *Biomaterials* **2010**, *31* (29), 7555-7566.
92. Chen, T.; Zhao, T.; Wei, D. F.; Wei, Y. X.; Li, Y. Y.; Zhang, H. X. Core-shell nanocarriers with ZnO quantum dots-conjugated Au nanoparticle for tumor-targeted drug delivery. *Carbohyd Polym* **2013**, *92* (2), 1124-1132.
93. Chen, Y.; Wilbon, P. A.; Zhou, J. H.; Nagarkatti, M.; Wang, C. P.; Chu, F. X.; Tang, C. B. Multifunctional self-fluorescent polymer nanogels for label-free imaging and drug delivery. *Chem Commun* **2013**, *49* (3), 297-299.
94. Li, Y. Y.; Cheng, H.; Zhang, Z. G.; Wang, C.; Zhu, J. L.; Liang, Y.; Zhang, K. L.; Cheng, S. X.; Zhang, X. Z.; Zhuo, R. X. Cellular internalization and *in vivo* tracking of thermosensitive luminescent micelles based on luminescent lanthanide chelate. *ACS Nano* **2008**, *2* (1), 125-133.
95. Santra, S.; Kaittanis, C.; Grimm, J.; Perez, J. M. Drug/dye-loaded, multifunctional iron oxide nanoparticles for combined targeted cancer therapy and dual optical/magnetic resonance imaging. *Small* **2009**, *5* (16), 1862-1868.
96. Fokong, S.; Theek, B.; Wu, Z. J.; Koczera, P.; Appold, L.; Jorge, S.; Resch-Genger, U.; van Zandvoort, M.; Storm, G.; Kiessling, F.; Lammers, T. Image-guided, targeted and triggered drug delivery to tumors using polymer-based microbubbles. *J Control Release* **2012**, *163* (1), 75-81.
97. Kim, J.; Kim, H. S.; Lee, N.; Kim, T.; Kim, H.; Yu, T.; Song, I. C.; Moon, W. K.; Hyeon, T. Multifunctional uniform nanoparticles composed of a magnetite nanocrystal core and a mesoporous silica shell for magnetic resonance and fluorescence imaging and for drug delivery. *Angew Chem Int Edit* **2008**, *47* (44), 8438-8441.

Chapter 8

Carbon Nanomaterials for Tumor Targeting Theranostics

João M. M. Leitão[1], *Eliana F. C. Simões*[2], *and*
Joaquim C. G. Esteves da Silva[2*]
1 Centro de Investigação em Química (CIQ-UP), Faculty of Pharmacy, University of Coimbra Polo das Ciências da Saúde, Coimbra 3000-548, Portugal
2 Centro de Investigação em Química (CIQ-UP), Department of Chemistry and Biochemistry, Faculty of Sciences, University of Porto,Rua do Campo Alegre 687, Porto 4169-007, Portugal

Carbon nanomaterials are the focus of an intense research in several areas of science. Due to their apparent low toxicity, when compared to other nanomaterials, carbon based nanoparticles found particularly important applications in biomedicine. Indeed, carbon shows many nanoallotropic forms like carbon nanotubes, graphene, carbon dots, fullerene and nanodiamonds. Some of them show fluorescence in the visible wavelength range, like graphene and carbon dots, but all have in common being nanosized with the corresponding properties. The main theranostics applications of carbon based nanomaterials (the fluorescent nanoallotropes) are for improving tumor imaging, allowing earlier detection, and specific drug delivery allowing high treatment efficient and reduced side effects. This chapter will describe the carbon nanomaterials mostly used for tumor targeting theranostics, their properties and the engineering strategies used to modify them. Also, the most successful biomedical applications of the carbon nanomaterials will be described.

8.1 Introduction

Theranostic is the term used to refer the combination of therapy and diagnostic of disease. The cancer therapy and diagnostic has been the main field of research of theranostic applications namely the utilization of nanoparticles in theranostic.[1–3] The development of earlier diagnostic methodologies and more efficient and safer therapeutic agents is of crucial importance in cancer therapy. The selective cancer cell targeting is essential in order to avoid any unwanted damage to the normal cells and corresponding secondary health effects. Concurrent *in vivo* bioimaging allows the detection of tumors at a much earlier stage, which may improve the success of treatment, and provide information

* Corresponding author: jcsilva@fc.up.pt

about its efficacy. The number and type of the applications in theranostics have been multiplied and diversified. Figure 8.1 illustrates some of theranostic applications in therapy and diagnostic.

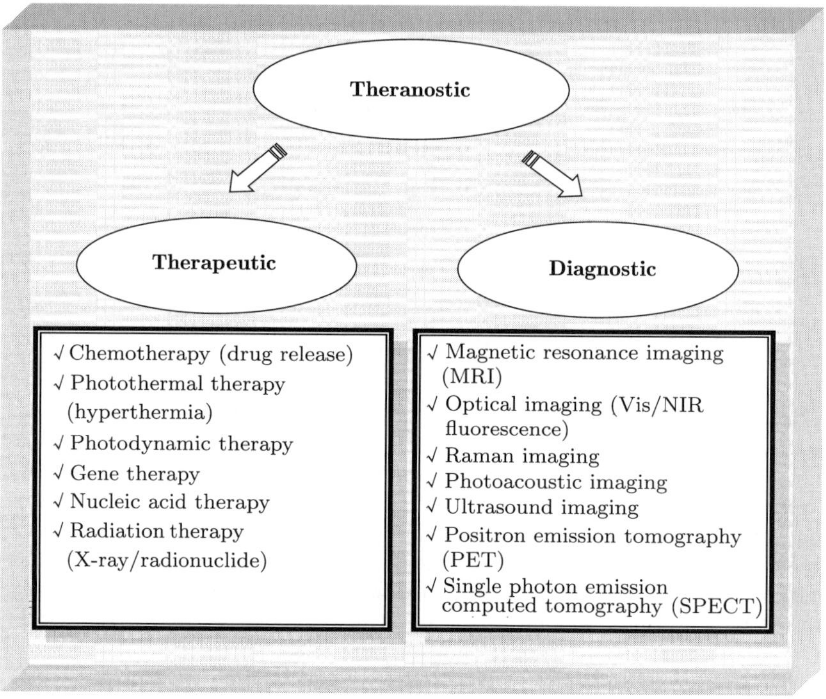

Fig. 8.1 Theranostic applications.

In the last few years, there was a growing interest in the study of the nanomaterials as both therapeutic and diagnostic options for the cancer treatment and many other diseases due to the advantages associated with the nanomaterials including spatially and temporally controllable drug release, localizing therapy and avoiding invasive treatment.[4–8] The theranostic nanoparticles may include a wide range of materials such as polymeric nanoparticles, emulsions, nanocapsules, nanospheres, micelles, liposomes, dendrimers, quantum dots, carbon nanotubes and fullerenes.[2,3] It is possible to use nanocarriers that are capable of active targeting using cell-specific ligands and transport multiple therapeutic and imaging agents for accurate diagnosis and controlled drug delivery. These materials are being used to encapsulate or solubilize chemotherapeutic agents for improved drug delivery *in vivo* or to provide unique optical, magnetic and electrical properties for imaging.[6]

Carbon is capable of forming many allotropes due to its unique valency. Besides the bulk allotropic forms of carbon (diamond, graphite and amorphous carbon), different nanoallotropic forms of carbon as the fullerene forms [fullerene (FL), carbon nanotubes (CNT), graphene (GP)], nanodiamonds (ND) and carbon dots (CD) have also been discovered.[9,10] Having versatile properties, these different nano forms of carbon play important roles in different fields of application. The nano forms of carbon are gaining relevance in the tumor target theranostics.[10] Most of the tumor theranostics applications found in the literature uses CNT.[3–8] Other applications are also found using graphene, carbon dots, fullerene and nanodiamonds. Figure 8.2 depicts the variety of carbon nanomaterials used in tumor targeting theranostic.

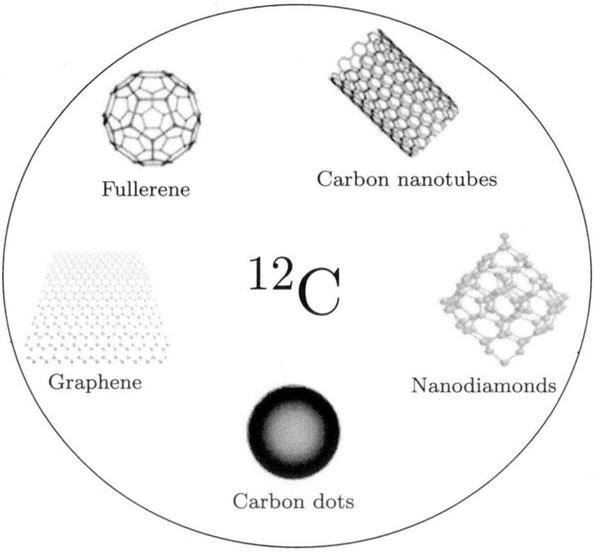

Fig. 8.2 Carbon nanomaterials used in tumor targeting theranostics.

The potential of carbon based nanoparticles in biomedical applications is enormous because the major component, carbon, is a nontoxic element. The carbon based nanomaterials consequently have lower toxicity compared to other nanomaterials. Furthermore, the availability of carbon nanoparticles for conjugation with other materials makes it possible to exploit their attributes in cancer theranostics. The photothermal properties of some carbon nanomaterials, such as the CNT, also allows them to be used in conjunction with near infrared radiation and lasers to thermally ablate cancer cells. This chapter presents mainly the carbon nanomaterials used as theranostic agents and their bioanalytical and biomedical applications.

8.2 Tumor Theranostic Applications of Carbon Nanomaterials

8.2.1 *Tumor Theranostics Applications of Carbon Nanotubes*

Carbon nanotubes (CNTs) are a member of the structural family of fullerenes with a cylindrical nanostructure constituted with carbon atoms arranged in condensed aromatic rings.[11] Raw CNTs have highly hydrophobic surfaces and are not soluble in aqueous solutions. Their name is derived from their long, hollow structure with their walls formed by a one-atom-thick concentric graphene sheet. CNTs, depending on their formation by one or more graphene sheets, are classified respectively as single-walled carbon nanotubes (SWCNTs), with 0.4–2.0 nm of diameter and a few μm of length, or as multi-walled carbon nanotubes (MWCNTs), with diameter up to 100 nm and a length from 1 to several μm. The graphene sheets are rolled at specific and discrete ("chiral") angles, and the combination of the rolling angle and radius decide the properties of CNT. Individual CNT naturally tend to align themselves into "ropes" held together by van der Waals forces, more specifically, pi-stacking.[12,13]

Due to their high length-to-diameter ratio, high mechanical strength, large surface areas, chemical and thermal stability, electronic and optical properties, CNTs become valuable for nanotechnology, electronics, optics and other fields of materials science and technology.[11,14] Their unique mechanical strength are due to the strength of the chemical bonds of CNT entirely composed of sp^2 bonds, similar to those of graphite, stronger than the sp^3 bonds found in alkanes and diamond.[14] These cylindrical CNTs have been constructed with a length-to-diameter ratio significantly larger than any other material with a very large surface area, which makes CNT amenable for high sensitivity molecular detection and recognition in biological systems. Moreover, SWCNTs have very high optical absorbance in the NIR range where the biological systems are transparent.[12] Nevertheless, some CNT toxicity and biodegradable issues are still under discussion.[8,11,15]

The CNT functionalization decreases the cytotoxicity, improves the biocompatibility, increases the solubility and gives the opportunity to append drugs molecules, proteins or genes, for the construction of delivery systems. The CNT functionalization can be done using several different processes. Functionalization can be done at the covalent ends or at the defect sites with an initial oxidation procedure. On the other hand, covalent sidewall functionalization can be achieved where reagents with high chemical reactivity are involved, due to carbon bonds with lower reactivity. On the contrary, a noncovalent functionalization through an amphiphilic substance or ionic interaction has the advantage of no destruction of the conjugation in the CNT.[11–13,16] Besides the functionalization with one substance, CNT can also be multifunctionalized allowing the immobilization of different types of bioactive molecules such as peptide, proteins, nucleic acids and small drugs, to make them able to act in different ways simultaneously.[17]

Intense research was performed with CNT in the theranostics field that has been reflected in the high number of reviews featuring this carbon nanomaterial.[2–8,12,13,16–19] In cancer theranostic, CNT have been used in photothermal[4,6,8,16,17,19], photodynamic[19], radio frequency ablation and photoacustic therapy[17]; in nucleic acid or gene[2,4,7,8,12,13,19] and drug delivery[2–5,7,8,12,13,16–18] and in diagnosis as imaging agents[2,4–8,12]. Although, most of the theranostics applications with CNT involved drug delivery coupled with different types of imaging techniques, but in some cases other kinds of therapy were performed. Due to their inherent characteristics, the multifuncionalization of CNT has been extensively used to generate complex nanocomposites with different properties. Among the several theranostics applications developed with CNT, the most representative applications will be discussed in this chapter. Figure 8.3 presents the main cancer theranostics application with CNT.

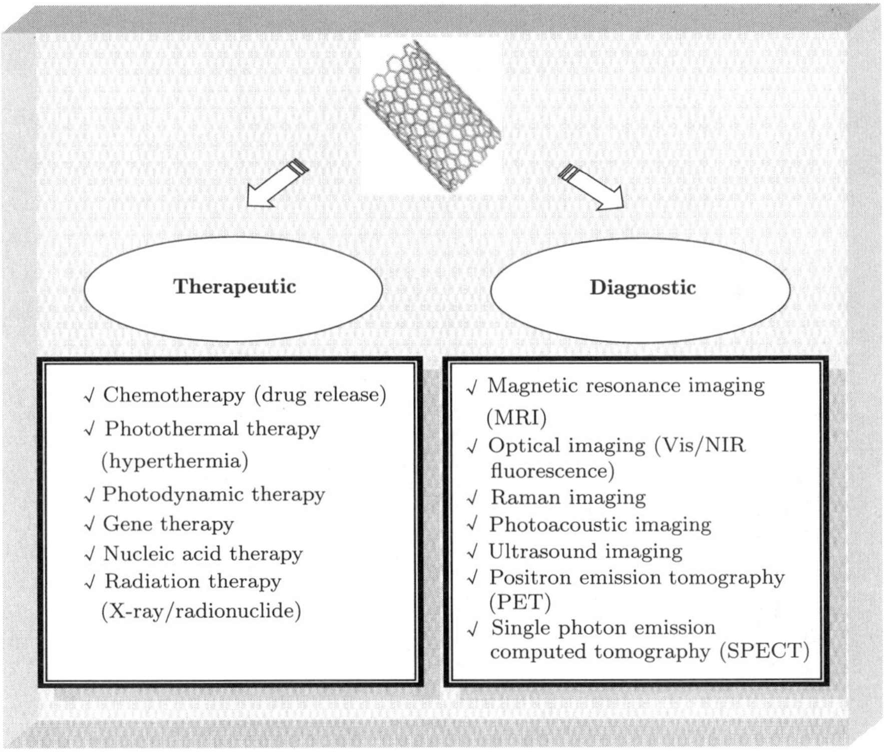

Fig. 8.3 Cancer theranostics application of carbon nanotubes.

Some complex multifunctionalized CNT for tumour theranostics applications have been established.[20–24] While SWCNT were used in four of these multifunctionalized CNT,[20–22,24] MWCNT has also been used in one case.[23] Water soluble pristine SWCNT that were noncovalently functionalized by single stranded DNA (ssDNA) have been modified with silver (Ag) or gold (Au) and then coated with folic acid (FA) conjugated lipoic acid modified polyethylenoglicol (PEG-FA). In the SWCNT-ssDNA-Au-PEG-FA, the over expressed FA cancer cell gives the selectivity and Au provokes the enhancement of Raman signal of SWCNT. Strong surface plasmon resonance absorption allows the use of these nanocomposites in Raman imaging and photothermal ablation of KB and HeLa cancer cells.[20]

Water soluble SWCNT were conjugated with Cy3 labeled with ssDNA and covalently functionalized with a mixture of phospholipids PEG-FA and PEG fluorescein. The multifunctionalized nanocomposites was used to target cancer HeLa cells by the FA and for optical bioimaging by the fluorescein probe and NIR photothermal therapy based in the intense SWCNT NIR absorption.[24] Magnetic SWCNT conjugating doxorubicin (DOX) and transferrin (TF) with silica coated CdTe quantum dots (SiO_2-CdTe) linked to SWCNT filled with Fe_3O_4 were synthesized. This SWCNT nanocomposite provides a Fe_3O_4 and TF dual-targeted DOX delivery system to HeLa cancer cells with the possibility of visible fluorescence optical imaging by the SiO_2-CdTe fluorescent probe.[21] Oxidized SWCNTs were also functionalized with DOX anticancer drug, 5(6)-carboxyfluorescein fluorescent probe and a carcinoembryonic antigen (CEA) monoclonal antibody for targeting cancer cells. DOX was noncovalently linked to SWCNT and monoclonal antibody and 5(6)-carboxyfluorescein were covalently linked using bovine serum albumin (BSA) as multifunctional linker. The synthesized nanocomposite was used for DOX delivery and visible optical imaging for human cancer colon cells.[22]

In the theranostics application with multifunctionalized MWCNT, Das *et al.* conjugated oxidized MWCNT with the fluorescent label Alexa Fluor (AF), a Technitium radionuclide (^{99m}Tc), tumor targeting molecule (FA) and an anticancer drug methotrexate (MTX). The ^{99m}Tc was coordinated and the other three compounds were covalently functionalized at MWCNT surface. For the MTX functionalization, a glycidylated MWCNT-FA-AF was previously synthesized. The MWCNT-FA-AF-MTX was used for selectively targeting lung (A549) and breast cancer (MCF7) cells *in vitro* and *in vivo* by the FA and for multimodal visible optical and radio imaging by respectively the AF and ^{99m}Tc.[23]

Simpler SWCNT nanocomposites[25] or nonfunctionalized SWCNT[26] were used for theranostics tumor targeting. SWCNT, functionalized only with a mixture of two PEGylated phospholipids 1,2-distearoylphosphatidylethanolamine-methyl-polyethyleneglycol and poly(maleic anhydride-alt-1-octadecene)-poly(ethylene glycol) methyl ether, were used for *in vivo* tumor NIR optical imaging and for photothermal ablation of tumor with a referred high tumor uptake.[25] Nonfunctionalized SWCNT were used in tissue-mimicking materials and the response to the enhancement of the dielectric and temperature of that materials can be attributed to the utilization of these SWCNT for simultaneous microwave detection and hyperthermic treatment of breast cancer.[26]

Similar CNT nanocomposites were also developed for therapeutic or diagnostic applications with potential for cancer theranostics applications. Several kind of SWCNT or MWCNT nanocomposites of more or less complexity were developed for DOX[27,28], cyclosporine (CY)[29], etopoxide (ETO)[30] and gemcitabine (GEM)[31] drug delivery; DNA[32,33] and small interfering RNA (siRNA)[34,35] delivery and PET and Raman imaging[36]. Similar CNT nanomaterial, with a different conical shape designed as nanohorns, were used as multifunctionalized nanocomposites for photodynamic and hyperthermic therapy.[37]

8.2.2 *Tumor Theranostics Applications of Graphene*

Graphene (GP) is a two dimensional material, and similar to CNT, is composed of sp^2-bonded carbon atoms arranged in a regular hexagonal pattern. GP has a tunable planar structure (sheet, ribbon, triangle and hexagon), with a thickness from 0.34 to 100 nm and lateral size from 0.3 to 10 μm and a mono or multilayer structure.[38] The main derivatives of GP are the graphene oxide (GO) and reduced graphene oxide (RGO). GO can be considered as consisting of individual sheets of GP decorated with oxygen functional groups on both the basal planes and edges.[39] The physicochemical features of GP and their derivatives are characterized by a low aspect ratio, larger surface area, and overall better dispersibility in most solvents compared to CNT. Colloidal dispersions of individual GP or other derivatives sheets can easily be engineered without metallic impurities, with high stability and less aggregation.[38]

A quite important property of GP nanostructures is that they are not fiber-shaped. Theoretically this characteristic offers significant advantages in terms of toxicity and biocompatibility over inhomogeneous dispersions of fiber-shaped CNT. However, studies that directly compare GP and CNT are uncommon, making comparative considerations of their overall safety and risk assessment challenging.[38] Their unique mechanical, optical and electronic properties, being different from the CNT, allow their application in different areas including quantum physics, nanoelectronic devices and for nanocomposites materials. GO is soluble in water and has been used for the detection of biomolecules.[40–42]

An intense research is being done in the last years with the GP in the theranostics field and some reviews dealing with theranostics applications of GP were published.[6,40–45] In cancer theranostic, GP has been used in photothermic[6,40,42–44] and photodynamic therapy[6,43]; in nucleic acid or gene therapy[43–45]; drug delivery[40–45]; and, in diagnosis as imaging agents[41–45]. Figure 8.4 presents the main cancer theranostics applications of GP.

Most of the GP theranostics applications for cancer involved multifunction-

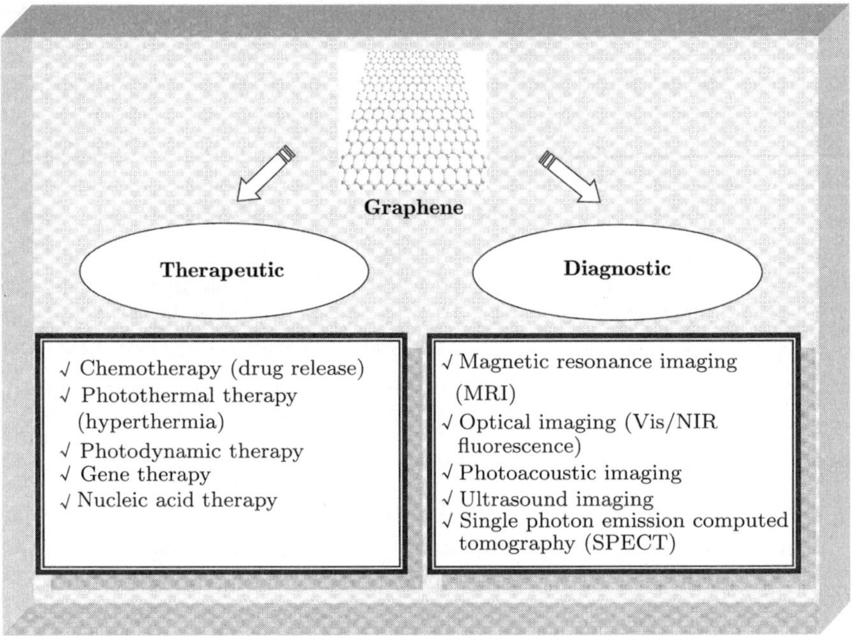

Fig. 8.4 Cancer theranostic applications of graphene.

alized single layer GO.[46–50] GO was linked to hyaluronic acid conjugated with chlorin e6 (Ce6) by non-covalent functionalization and it was used for NIR fluorescence and photothermal and photodynamic induced therapy of A549 cancer cells.[46] In other application, aminated PEG, of molecular weight of 6 000 g (PEG6000N), and FA as the targeting ligands were initially covalently linked to GO and then the GO-PEG-FA were noncovalently and covalently functionalized, respectively with a gadolinium (Gd) salt, as a MRI probe, and DOX, for drug delivery where the GO-PEG6000N-FA-Gd-DOX nanocomposites were tested in HepG2 cancer cells.[47] Moreover, GO was covalently functionalized with branched PEG and with an antibody anti-CD20, for selective targeting, for DOX drug delivery and it was also used for NIR optical imaging of B and T cancer cells.[48]

Multifunctionalized GO for theranostics applications using GO magnetic nanocomposites were also synthesized with iron oxide nanoparticles.[49,50] One of the GO linked to iron oxide was covalently linked with branched PEG and DOX and used for targeted drug delivery and photothermal therapy in $4T_1$ cancer cells and *in vivo* MRI using the nanocomposites as the T_2 contrast agent.[49] The other GO linked with iron oxide was noncovalently functionalized with poly (maleic anhydride-alt-1-octadecene) (PMA-PEG) and used for photothermal therapy and MRI, visible optical and photoacoustic imaging.[50]

A different kind of multifunctional GO nanocomposites was synthesized by coating poly(acrylic acid) (PAA) microcapsules, containing Au nanoparticles, with GO by an electrostatic layer by layer technique. The Au-PAA-GO nanocomposites were used *in vivo* photothermal therapy, due to the strong NIR absorption of GO, and *in vivo* bimodal ultrasound and computed tomography imaging, due to the enhancement of ultrasound imaging by the PAA microcapsules and the use of Au nanoparticles as contrast agent, respectively.[51]

Simpler nanocomposites were developed for cancer theranostics applications with single layer RGO.[52,53] RGO were only functionalized with bovine serum albumin (BSA) and the synthesized RGO-BSA nanocomposites were used for photothermal therapy of MCF-7 cancer cells and for *in vivo* photoacoustic and ultrasounds imaging.[52] In another application, RGO was linked with CdSe/ZnS QD after surface adsorption of poly(L-lysine) amphiphlic polypeptide. The synthesized RGO-CdSe/ZnS nanocomposites were also used for photothermal therapy of MCF-7 and HeLa cancer cells, and were used for *in vivo* visible optical imaging. Graphene sheets of different size, linked with FA, were also evaluated for their internalization in two lines of cancer cells.[53]

Nonetheless, some GO nanocomposites, constituted by single layer GO, were developed only for therapeutic or diagnostic applications with potential for cancer theranostics applications.[54–66] Some of these nanocomposites were developed for DOX[54,55], adryamicincamptothecin analog[56] and iressa drug delivery[57]. Others were developed for DNA[58], siRNA[59] and single stranded RNA (ssRNA)[60] delivery. Some nanocomposites were developed for visible[61] and NIR[62] optical imaging. Several nanocomposites were developed for photothermal[54,64] and photodynamic[65,66] therapy. With RGO, a photothermal therapeutic[67] and PET imaging diagnostic applications[68] were also developed.

8.2.3 *Tumor Theranostics Applications of Carbon Dots*

Carbon dots (CDs) are small photoluminescent carbon nanoparticles usually of dimensions less than 10 nm, suspending in water or in other solvents. Their properties are similar to the classical nanocrystalline semiconductors quantum dots (QD) but the CDs had been developed as a biocompatible and nontoxic alternative.

All the theranostics applications involving CDs are of functionalized CDs.[69–71] In one application, CD were passivated with aminated polyethileno glycol of molecular weight 2 000 g (PEG2000N) and was linked to the CD-PEG2000N chlorin e6 (Ce6).[69] In another case, CDs of silica nanocomposites (CD-SiO_2) was coated with PEG900N.[70] In a third application, CDs were passivated with branched polyethylenimine (BPEI).[71] For the CD-PEG2000N-Ce6, the possibility of simultaneously enhanced photosensitizer fluorescence detection, by a fluorescence resonance energy transfer (FRET) mechanism between the CDs and Ce6 in MGC803 cancer cells, and of *in vivo* visible optical imaging and photodynamic therapy of gastric cancer was demonstrated.[69] For the other two CD applications, their diagnostic ability by visible bioimaging and the capability of drug[70] and gene[71] delivery were exhibited after functionalization with the anticancer drug DOX in HeLa and vector/pDNA complexes (pDNA) in COS7 and HepG2 cancer cells. The capability of DOX drug release was done by evaluating ratiometric changes of the fluorescence of CDs (blue emission) and DOX (red emission) at the cytoplasm and nucleus, and the capability of gene transfection of the CD-BPEI-pDNA nanocomposites was verified. Figure 8.5 schematically shows the cancer theranostics applications of CD.

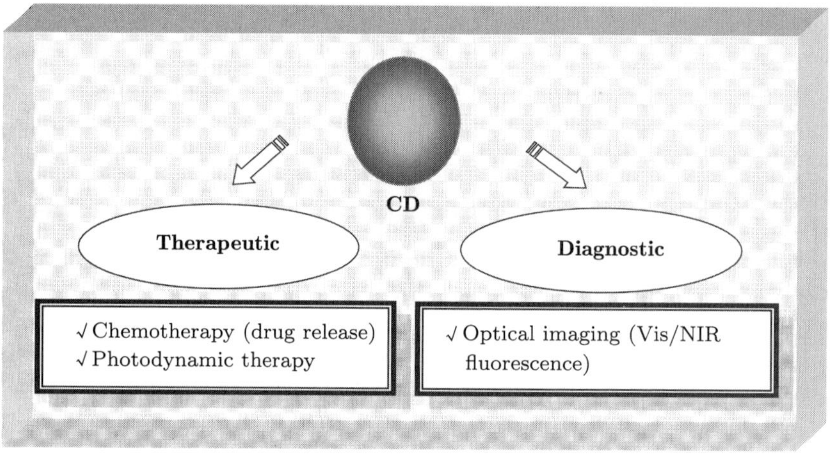

Fig. 8.5 Cancer theranostic applications of carbon dots.

Cytotoxicity assays suggested that mesoporous carbon dots (MCD) could be used for drug delivery. The MCDs served as effective carriers for camptothecin which efficiently inhibited the growth of MCF-7 cancer cells after its sustained release therein.[72] Iron based carbon magnetic nanoparticle DOX drug delivery systems were also reported.[73]

8.2.4 *Tumor Theranostics Applications of Fullerene*

Fullerences (FLs) are three dimensional polyhedron forms of carbon with a molecular formula of C_{2n} ($n = 30 - 41$) and a diameter around 1 nm. They are composed of carbon in the form of a hollow sphere, ellipsoid or tube. Spherical FLs are called buckyballs. Their structure is similar to graphite, which is composed of stacked graphene sheets of linked hexagonal rings, but they may also contain pentagonal (or sometimes heptagonal) rings.[74] Their poor solubility and the difficulty in purifying and processing have impaired the application of FLs, and functionalization is required in order to modify their physical and chemical properties.[75] Studies on functionalization of FLs for various applications in the field of biomedical sciences have been significantly increased. The ultimate goal towards employing these functionalized fullerenes involves their use in the diagnosis and in the therapy of cancer and others human diseases.[75–77]

The preparation of specific multifunctionalized derivatives, mostly of C_{60}, with organic addends for biomedical applications is currently undergoing in several research groups.[76,77] Two fullerenes derivatives have recently gained relevance in the therapy and diagnostic of cancer,[77] the polyhydroxy fullerenes (fullerenols, PHFL)[78] and the endohedral metallofullerenes (EMFL)[75,79]. The water soluble FLOH, with a molecular formula $C_{60}(OH)_n$ (n = 12–26), are considered to be excellent antioxidants due to their free radical scavenging ability and it is expected that they have a large biomedical use.[78] The EMFL, a new class of hybrid molecules, are formed by encapsulation of metallic species inside fullerene cages. They exhibit unique properties that differ from those of empty fullerenes due to the presence of metals and their hybridization effects *via* electron transfer. The introduction of metallic species inside fullerene cages provides a novel platform for developing future nanomedicines with multiple functions. The most important advantage that these EMFL offer is the cage protection from the toxic metal ions inside the large surface area of the fullerene, associated with the possibility of multiple hydrophilic functionalization with the attachment of some tissue-targeting ligands. For example, these EMFL have the possibility to accommodate one or more gadollinium atoms to obtain gadofullerenes and, due to their own characteristics, these gadofullerenes can be used for MRI.[75–78,79] In cancer therapy and diagnostic the functionalized FL have been used for: drug[76] and DNA[75,76] delivery; in photothermal[77,78,80], photodynamic[76], photoacoustic[77,78] and radio therapy[77,79] in MRI[75–78,79] and, as chemotherapeutic pharmaceutical drugs[77,79]. The FL usually used is the buckyball C_{60} fullerene. In Fig.8.6, some cancer theranostics applications of FL are presented.

Few theranostic applications have been developed until now with FL or its derivatives.[80,81,82] PHFL $C_{60}(OH)_n$ ($n = 18 - 22$) functionalized with chitosan (CS) were used for *in vivo* photothermal therapy and photoacoustic imaging. Besides these PHFL-CS nanocomposites, some other nanocomposites with fluorescein isotiocianate doped silica nanoparticles were synthesized in different ways.[80] C_{60} FL was functionalized with HA in the presence of lithium hydroxide. The synthesized C_{60}-HA nanocomposites was used for *in vivo* photothermal therapy, for optical NIR fluorescence imaging and in HCT 116 and KB cancer cells.[81] EMFL with trigadolinium nitride (Gd_3N) inside a C_{80} FL was functionalized with ^{177}Lu isotope after previously linked with bifunctional 2-(4-aminobenzyl)-1,4,7,10-tetraazacyclododecane-1,4,7,10-tetraacetic acid (p-NH2-Bn-DOTA, DOTA). The synthesized radiolabeled metallofullerene was used for radiation therapy and for MRI and radioactive labeling in U87MG and GBM_{12} cancer cell lines.[82]

Most of the FL, PHFL or EMFL nanocomposites for cancer applications

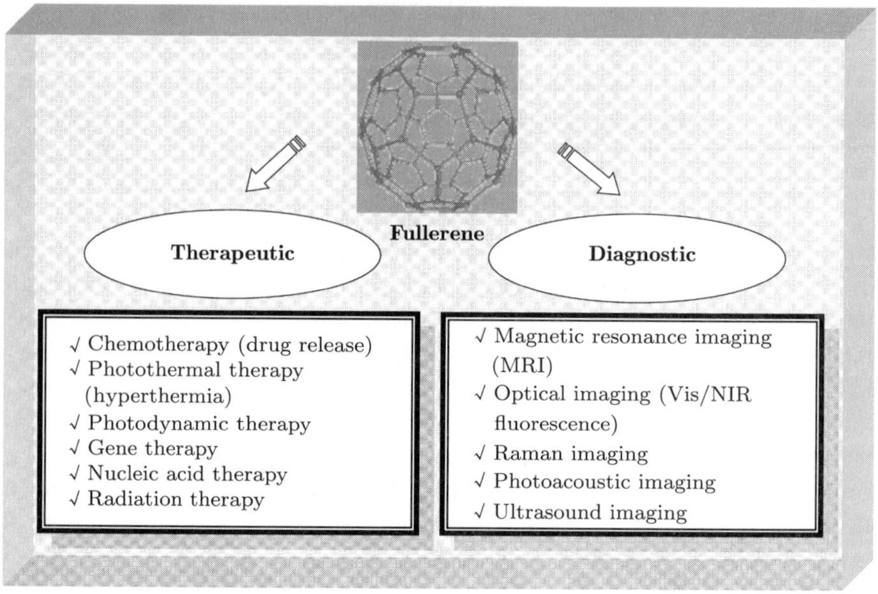

Fig. 8.6 Cancer theranostic applications of fullerene.

used C_{60} only for therapeutic or diagnostic application with potential for cancer theranostics.[83–87] They were used for DOX[83,85], paclitaxel (PTX)[87] and docetaxel[84] drug delivery and photodynamic therapy[83,85,87]. Besides these applications, some FLs and their derivatives with potential in chemotherapy as active chemotherapeutic pharmaceutical drug was also demonstrated.[88–90] The therapeutic activity was demonstrated with C_{60}[88] and with gadolinium metallofullerenol $[Gd@C_{82}(OH)_{22}]_n$[89–91] for *in vivo* and *in vitro* cancer cells. Furthermore, $[Gd@C_{82}(OH)_{22}]_n$ nanoparticles overcome tumor resistance to cisplatin by increasing its intracellular accumulation through the mechanism of restoring defective endocytosis.[92]

8.2.5 *Tumor Theranostics Applications of Nanodiamonds*

Nanodiamonds (NDs) are monodisperse nanometre-sized diamond particles with an irregular tetrahedral structure containing 46 carbon atoms. NDs can be produced by detonation, high-pressure, high temperature, ultrananocrystalline diamonds and single/polycrystalline diamond films. The NDs have a size around 4 to 5 nm and are soluble in water and they are a chemically inert material with low cytotoxicity, good biocompatibility and a large surface area.[93,94]

Based on ND's chemical, biological, and physical (optical, mechanical, electrical, thermal) properties, they can be used in several biomedical applications such as drug delivery, sensing and imaging.[93–98] They can be functionalized with different types of pharmaceutical drugs and can be suspended easily in water, which makes them important drug delivery systems.[99] Nanocrystalline diamond films have also been used in the coating of biomedical implants.[100]

In theranostics applications, NDs nanocomposites act as a chemotherapeutic pharmaceutical drug[101] or as a drug delivery system[102–105]. They were used for the diagnostic application in visible wavelength range. When ND acts as chemotherapeutic pharmaceutical drug, the fluorescent nanomaterial was covalently functionalized with chlorotoxin peptide BmK CT, known as active to glioma cancer cells. The activity of the synthetized ND-BmK CT nanocomposites and the visible optical imaging capability were demonstrated in C6 glioma cancer cells.[101]

For the nanocomposites used for drug delivery, an ND multifunctional PTX drug delivery agent was obtained by covalently linking a fluorescently labeled oligonucleotide PTX-DNA conjugate strands and a monoclonal antibody anti-EGFR (mAbs) as targeting ligand. The synthesized nanocomposites were used in MCF7 and MDA-MB-231 cancer cells.[103] Simpler nanocomposites functionalized with fluorescein and DOX were synthetized for the DOX drug delivery and visible optical imaging and used in human breast adenocarcinoma. It was demonstrated for this nanocomposites that the liberation of PTX drug depends on the pH.[103] For the other ND nanocomposites that act as a drug delivery system, fluorescent ND were functionalized with a porous silica coating using tetraethyl orthosilicate (TEOS) and posterior coupled with PEG-polyethylenimine (PEI) copolymer. The synthetized nanocomposites were used for visible optical bioimaging of HeLa cancer cells and their potential for drug delivery was investigated after linking a red fluorescent dye.[105] Figure 8.7 presents the principal cancer theranostic applications of NDs.

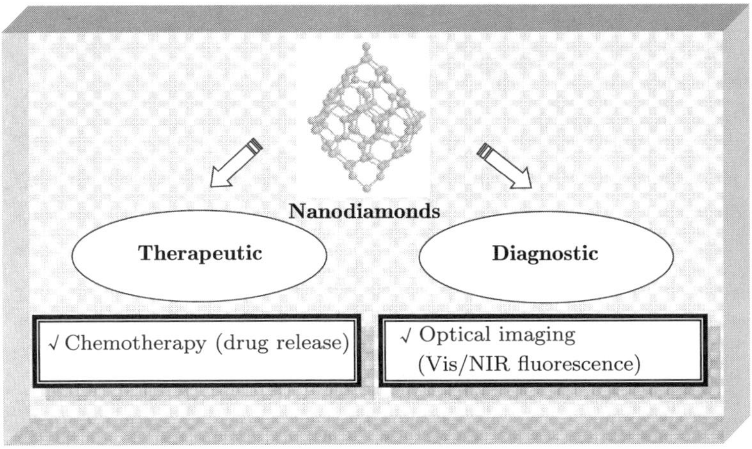

Fig. 8.7 Cancer theranostic applications of nanodiamonds.

Besides the described applications of NDs, some therapeutic or diagnostic applications with different nanocomposites were also developed with potential use in cancer theranostics applications[106–112] including DOX[106,107,109], PTX[110], 10-hydroxycamptothecin (HCPT)[111] drug delivery, radio-frequency hyperthermic[108] therapy and visible optical imaging[109,112].

8.3 Future Perspectives

Different carbon nanoallotropes forms are being developed for human theranostic applications. The large majority of theranostic applications that have been developed are focused on cancer diseases. Indeed, nanomaterials in general are particularly suited for cancer theranostics because the success of the cancer treatment is highly dependent on the yearly diagnostics and on the selective cytotoxic drug delivery. Carbon nanomaterials have the nanosized properties coupled with a potential low toxicity and biocompatibility and, consequently, are highly promising nanomaterials for cancer theranostics. Nevertheless, some of the carbon nanoallotropes raised some toxicity questions, *e.g.*, CNT, which must be investigated and clarified further.

The intense research observed in the last decade about the use of carbon nanomaterials for *in vitro* cancer theranostics set up the basic strategies. In diagnostics, the development of intrinsically high fluorescent nanomaterials, coupled with the paramagnetic materials and/or specific functionalization, improved earlier tumor detection methodologies. In drug delivery, they have an important role as agents for the selective drug delivery of classic cytotoxic drugs or gene/protein oriented therapies. Indeed, besides the major advances that are occurring in cancer therapies that resulted from the identification of the oncogenes pathways associated with selected cancers, nanosized carbon materials will have an important role in the development of these new therapies.

Acknowledgments

Financial support from Fundação para a Ciência e Tecnologia (FCT, Lisbon) project PEST and a PhD grant to Eliana Simões SFRH/BD/81074/2011 is acknowledged.

References

1. Kelkar, S. S.; Reineke, T. M. Theranostics: Combining imaging and therapy. *Bioconjug Chem* **2011**, *22* (10), 1879-1903.
2. Duncan, R.; Gaspar, R. Nanomedicine(s) under the microscope. *Mol Pharm* **2011**, *8* (6), 2101-2141.
3. Svenson, S. Theranostics: Are we there yet? *Mol Pharm* **2013**, *10* (3), 848-856.

4. Choi, K. Y.; Liu, G.; Lee, S.; Chen, X. Theranostic nanoplatforms for simultaneous cancer imaging and therapy: Current approaches and future perspectives. *Nanoscale* **2012**, *4* (2), 330-342.
5. Menon, J. U.; Jadeja, P.; Tambe, P.; Vu, K.; Yuan, B.; Nguyen, K. T. Nanomaterials for photo-based diagnostic and therapeutic applications. *Theranostics* **2013**, *3* (3), 152-166.
6. Shibu, E. S.; Hamada, M.; Murase, N.; Biju, V. Nanomaterials formulations for photothermal and photodynamic therapy of cancer. *J Photoch Photobio C* **2013**, *15*, 53-72.
7. Ahmed, N.; Fessi, H.; Elaissari, A. Theranostic applications of nanoparticles in cancer. *Drug Discov Today* **2012**, *17* (17-18), 928-934.
8. Huang, H. C.; Barua, S.; Sharma, G.; Dey, S. K.; Rege, K. Inorganic nanoparticles for cancer imaging and therapy. *J Control Release* **2011**, *155* (3), 344-357.
9. Terrones, M.; Terrones, H. The carbon nanocosmos: Novel materials for the twenty-first century. *Philos Trans A Math Phys Eng Sci* **2003**, *361* (1813), 2789-2806.
10. Liu, Z.; Liang, X. J. Nano-carbons as theranostics. *Theranostics* **2012**, *2* (3), 235-237.
11. Moradi, O.; Yari, M.; Zare, K.; Mirza, B.; Najafi, F. Carbon nanotubes: A review of chemistry principles and reactions. *Fuller Nanotub Car N* **2012**, *20* (2), 138-151.
12. Liu, Z.; Tabakman, S.; Welsher, K.; Dai, H. J. Carbon nanotubes in biology and medicine: *In vitro* and *in vivo* detection, imaging and drug delivery. *Nano Res* **2009**, *2* (2), 85-120.
13. Zhang, W. X.; Zhang, Z. Z.; Zhang, Y. G. The application of carbon nanotubes in target drug delivery systems for cancer therapies. *Nanoscale Res Lett* **2011**, *6*.
14. Yang, W.; Ratinac, K. R.; Ringer, S. P.; Thordarson, P.; Gooding, J. J.; Braet, F. Carbon nanomaterials in biosensors: Should you use nanotubes or graphene? *Angew Chem Int Ed Engl* **2010**, *49* (12), 2114-2138.
15. Bianco, A.; Kostarelos, K.; Prato, M. Making carbon nanotubes biocompatible and biodegradable. *Chem Commun (Camb)* **2011**, *47* (37), 10182-10188.
16. Zhang, Y.; Bai, Y. H.; Yan, B. Functionalized carbon nanotubes for potential medicinal applications. *Drug Discov Today* **2010**, *15* (11-12), 428-435.
17. Lamanna, G.; Battigelli, A.; M é nard-Moyon, C.; Bianco, A. Multifunctionalized carbon nanotubes as advanced multimodal nanomaterials for biomedical applications. *Nanotechnol Rev* **2012**, *1*, 17-29.
18. Boncel, S.; Muller, K. H.; Skepper, J. N.; Walczak, K. Z.; Koziol, K. K. Tunable chemistry and morphology of multi-wall carbon nanotubes as a route to non-toxic, theranostic systems. *Biomaterials* **2011**, *32* (30), 7677-7686.
19. Ji, S. R.; Liu, C.; Zhang, B.; Yang, F.; Xu, J.; Long, J. A.; Jin, C.; Fu, D. L.; Ni, Q. X.; Yu, X. J. Carbon nanotubes in cancer diagnosis and therapy. *Bba-Rev Cancer* **2010**, *1806* (1), 29-35.
20. Wang, X.; Wang, C.; Cheng, L.; Lee, S. T.; Liu, Z. Noble metal coated single-walled carbon nanotubes for applications in surface enhanced raman scattering imaging and photothermal therapy. *J Am Chem Soc* **2012**, *134* (17), 7414-7422.

21. Chen, M. L.; He, Y. J.; Chen, X. W.; Wang, J. H. Quantum dots conjugated with Fe_3O_4-filled carbon nanotubes for cancer-targeted imaging and magnetically guided drug delivery. *Langmuir* **2012**, *28* (47), 16469-16476.

22. Heister, E.; Neves, V.; Tilmaciu, C.; Lipert, K.; Beltran, V. S.; Coley, H. M.; Silva, S. R. P.; McFadden, J. Triple functionalization of single-walled carbon nanotubes with doxorubicin, a monoclonal antibody, and a fluorescent marker for targeted cancer therapy. *Carbon* **2009**, *47* (9), 2152-2160.

23. Das, M.; Datir, S. R.; Singh, R. P.; Jain, S. Augmented anticancer activity of a targeted, intracellularly activatable, theranostic nanomedicine based on fluorescent and radiolabeled, methotrexate-folic acid-multiwalled carbon nanotube conjugate. *Mol Pharm* **2013**, *10* (7), 2543-2557.

24. Kam, N. W. S.; O'Connell, M.; Wisdom, J. A.; Dai, H. J. Carbon nanotubes as multifunctional biological transporters and near-infrared agents for selective cancer cell destruction. *Proc Natl Acad Sci USA* **2005**, *102* (33), 11600-11605.

25. Robinson, J. T.; Welsher, K.; Tabakman, S. M.; Sherlock, S. P.; Wang, H.; Luong, R.; Dai, H. High performance *in vivo* near-IR (>1 μm) imaging and photothermal cancer therapy with carbon nanotubes. *Nano Res* **2010**, *3* (11), 779-793.

26. Mashal, A.; Sitharaman, B.; Li, X.; Avti, P. K.; Sahakian, A. V.; Booske, J. H.; Hagness, S. C. Toward carbon-nanotube-based theranostic agents for microwave detection and treatment of breast cancer: Enhanced dielectric and heating response of tissue-mimicking materials. *IEEE T Bio-Med Eng* **2010**, *57* (8), 1831-1834.

27. Liu, Z.; Fan, A. C.; Rakhra, K.; Sherlock, S.; Goodwin, A.; Chen, X.; Yang, Q.; Felsher, D. W.; Dai, H. Supramolecular stacking of doxorubicin on carbon nanotubes for *in vivo* cancer therapy. *Angew Chem Int Ed Engl* **2009**, *48* (41), 7668-7672.

28. Ali-Boucetta, H.; Al-Jamal, K. T.; McCarthy, D.; Prato, M.; Bianco, A.; Kostarelos, K. Multiwalled carbon nanotube-doxorubicin supramolecular complexes for cancer therapeutics. *Chem Commun (Camb)* **2008**, (4), 459-461.

29. Hadidi, N.; Kobarfard, F.; Nafissi-Varcheh, N.; Aboofazeli, R. PEGylated single-walled carbon nanotubes as nanocarriers for cyclosporin a delivery. *AAPS PharmSciTech* **2013**, *14* (2), 593-600.

30. Chen, C.; Xie, X. X.; Zhou, Q.; Zhang, F. Y.; Wang, Q. L.; Liu, Y. Q.; Zou, Y.; Tao, Q.; Ji, X. M.; Yu, S. Q. EGF-functionalized single-walled carbon nanotubes for targeting delivery of etoposide. *Nanotechnology* **2012**, *23* (4), 045104.

31. Yang, F.; Jin, C.; Yang, D.; Jiang, Y.; Li, J.; Di, Y.; Hu, J.; Wang, C.; Ni, Q.; Fu, D. Magnetic functionalised carbon nanotubes as drug vehicles for cancer lymph node metastasis treatment. *Eur J Cancer* **2011**, *47* (12), 1873-1882.

32. Singh, R.; Pantarotto, D.; McCarthy, D.; Chaloin, O.; Hoebeke, J.; Partidos, C. D.; Briand, J.-P.; Prato, M.; Bianco, A.; Kostarelos, K. Binding and condensation of plasmid DNA onto functionalized carbon nanotubes: Toward the construction of nanotube-based gene delivery vectors. *J Am Chem Soc* **2005**, *127* (12), 4388-4396.

33. Liu, Y.; Wu, D. C.; Zhang, W. D.; Jiang, X.; He, C. B.; Chung, T. S.; Goh, S. H.; Leong, K. W. Polyethylenimine-grafted multiwalled carbon nanotubes for

secure noncovalent immobilization and efficient delivery of DNA. *Angew Chem Int Ed Engl* **2005**, *44* (30), 4782-4785.

34. Bartholomeusz, G.; Cherukuri, P.; Kingston, J.; Cognet, L.; Lemos, R. Jr.; Leeuw, T.; Gumbiner-Russo, L.; Weisman, R. B.; Powis, G. *In vivo* therapeutic silencing of hypoxia-inducible factor 1 Alpha (HIF-1α) using single-walled carbon nanotubes noncovalently coated with siRNA. *Nano Res* **2009**, *2* (4), 279-291.
35. Podesta, J. E.; Al-Jamal, K. T.; Herrero, M. A.; Tian, B.; Ali-Boucetta, H.; Hegde, V.; Bianco, A.; Prato, M.; Kostarelos, K. Antitumor activity and prolonged survival by carbon-nanotube-mediated therapeutic siRNA silencing in a human lung xenograft model. *Small* **2009**, *5* (10), 1176-1185.
36. Liu, Z.; Cai, W.; He, L.; Nakayama, N.; Chen, K.; Sun, X.; Chen, X.; Dai, H. *In vivo* biodistribution and highly efficient tumour targeting of carbon nanotubes in mice. *Nat Nanotechnol* **2007**, *2* (1), 47-52.
37. Zhang, M.; Murakami, T.; Ajima, K.; Tsuchida, K.; Sandanayaka, A. S.; Ito, O.; Iijima, S.; Yudasaka, M. Fabrication of ZnPc/protein nanohorns for double photodynamic and hyperthermic cancer phototherapy. *Proc Natl Acad Sci USA* **2008**, *105* (39), 14773-14778.
38. Bussy, C.; Ali-Boucetta, H.; Kostarelos, K. Safety considerations for graphene: Lessons learnt from carbon nanotubes. *Accounts Chem Res* **2012**, *46* (3), 692-701.
39. Chen, D.; Feng, H.; Li, J. Graphene oxide: Preparation, functionalization, and electrochemical applications. *Chem Rev* **2012**, *112* (11), 6027-6053.
40. Bao, H.; Pan, Y.; Li, L. Recent advances in graphene-based nanomaterials for biomedical applications. *Nano Life* **2012**, *02* (01), 1230001-1230015.
41. Shen, H.; Zhang, L. M.; Liu, M.; Zhang, Z. J. Biomedical applications of graphene. *Theranostics* **2012**, *2* (3), 283-294.
42. Mao, H. Y.; Laurent, S.; Chen, W.; Akhavan, O.; Imani, M.; Ashkarran, A. A.; Mahmoudi, M. Graphene: Promises, facts, opportunities, and challenges in nanomedicine. *Chem Rev* **2013**, *113* (5), 3407-3424.
43. Yang, K.; Feng, L. Z.; Shi, X. Z.; Liu, Z. Nano-graphene in biomedicine: Theranostic applications. *Chem Soc Rev* **2013**, *42* (2), 530-547.
44. Feng, L.; Wu, L.; Qu, X. New horizons for diagnostics and therapeutic applications of graphene and graphene oxide. *Adv Mater* **2013**, *25* (2), 168-186.
45. Zhang, Y.; Nayak, T. R.; Hong, H.; Cai, W. B. Graphene: A versatile nanoplatform for biomedical applications. *Nanoscale* **2012**, *4* (13), 3833-3842.
46. Cho, Y.; Kim, H.; Choi, Y. A graphene oxide-photosensitizer complex as an enzyme-activatable theranostic agent. *Chem Commun (Camb)* **2013**, *49* (12), 1202-1204.
47. Shen, A. J.; Li, D. L.; Cai, X. J.; Dong, C. Y.; Dong, H. Q.; Wen, H. Y.; Dai, G. H.; Wang, P. J.; Li, Y. Y. Multifunctional nanocomposite based on graphene oxide for *in vitro* hepatocarcinoma diagnosis and treatment. *J Biomed Mater Res A* **2012**, *100* (9), 2499-2506.
48. Sun, X. M.; Liu, Z.; Welsher, K.; Robinson, J. T.; Goodwin, A.; Zaric, S.; Dai, H. J. Nano-graphene oxide for cellular imaging and drug delivery. *Nano Res* **2008**, *1* (3), 203-212.

49. Ma, X. X.; Tao, H. Q.; Yang, K.; Feng, L. Z.; Cheng, L.; Shi, X. Z.; Li, Y. G.; Guo, L.; Liu, Z. A functionalized graphene oxide-iron oxide nanocomposite for magnetically targeted drug delivery, photothermal therapy, and magnetic resonance imaging. *Nano Res* **2012**, *5* (3), 199-212.
50. Yang, K.; Hu, L.; Ma, X.; Ye, S.; Cheng, L.; Shi, X.; Li, C.; Li, Y.; Liu, Z. Multimodal imaging guided photothermal therapy using functionalized graphene nanosheets anchored with magnetic nanoparticles. *Adv Mater* **2012**, *24* (14), 1868-1872.
51. Jin, Y.; Wang, J.; Ke, H.; Wang, S.; Dai, Z. Graphene oxide modified PLA microcapsules containing gold nanoparticles for ultrasonic/CT bimodal imaging guided photothermal tumor therapy. *Biomaterials* **2013**, *34* (20), 4794-4802.
52. Sheng, Z.; Song, L.; Zheng, J.; Hu, D.; He, M.; Zheng, M.; Gao, G.; Gong, P.; Zhang, P.; Ma, Y.; Cai, L. Protein-assisted fabrication of nano-reduced graphene oxide for combined *in vivo* photoacoustic imaging and photothermal therapy. *Biomaterials* **2013**, *34* (21), 5236-5243.
53. Hu, S. H.; Chen, Y. W.; Hung, W. T.; Chen, I. W.; Chen, S. Y. Quantum-dot-tagged reduced graphene oxide nanocomposites for bright fluorescence bioimaging and photothermal therapy monitored *in situ*. *Adv Mater* **2012**, *24* (13), 1748-1754.
54. Qin, X. C.; Guo, Z. Y.; Liu, Z. M.; Zhang, W.; Wan, M. M.; Yang, B. W. Folic acid-conjugated graphene oxide for cancer targeted chemo-photothermal therapy. *J Photoch Photobio B* **2013**, *120*, 156-162.
55. Kurapati, R.; Raichur, A. M. Graphene oxide based multilayer capsules with unique permeability properties: Facile encapsulation of multiple drugs. *Chem Commun (Camb)* **2012**.
56. Wu, J.; Wang, Y. S.; Yang, X. Y.; Liu, Y. Y.; Yang, J. R.; Yang, R.; Zhang, N. Graphene oxide used as a carrier for adriamycin can reverse drug resistance in breast cancer cells. *Nanotechnology* **2012**, *23* (35), 355101.
57. Liu, Z.; Robinson, J. T.; Sun, X. M.; Dai, H. J. PEGylated nanographene oxide for delivery of water-insoluble cancer drugs. *J Am Chem Soc* **2008**, *130* (33), 10876-10877.
58. Feng, L.; Yang, X.; Shi, X.; Tan, X.; Peng, R.; Wang, J.; Liu, Z. Polyethylene glycol and polyethylenimine dual-functionalized nano-graphene oxide for photothermally enhanced gene delivery. *Small* **2013**, *9* (11), 1989-1997.
59. Yang, X. Y.; Niu, G. L.; Cao, X. F.; Wen, Y. K.; Xiang, R.; Duan, H. Q.; Chen, Y. S. The preparation of functionalized graphene oxide for targeted intracellular delivery of siRNA. *J Mater Chem* **2012**, *22* (14), 6649-6654.
60. Zhang, L. M.; Wang, Z. L.; Lu, Z. X.; Shen, H.; Huang, J.; Zhao, Q. H.; Liu, M.; He, N. Y.; Zhang, Z. J. PEGylated reduced graphene oxide as a superior ssRNA delivery system. *J Mater Chem B* **2013**, *1* (6), 749-755.
61. Qian, J.; Wang, D.; Cai, F. H.; Xi, W.; Peng, L.; Zhu, Z. F.; He, H.; Hu, M. L.; He, S. Observation of multiphoton-induced fluorescence from graphene oxide nanoparticles and applications in *in vivo* functional bioimaging. *Angew Chem Int Ed Engl* **2012**, *51* (42), 10570-10575.
62. Yue, Z.; Lv, P.; Yue, H.; Gao, Y.; Ma, D.; Wei, W.; Ma, G. Inducible graphene oxide probe for high-specific tumor diagnosis. *Chem Commun (Camb)* **2013**, *49* (37), 3902-3904.

63. Hong, H.; Zhang, Y.; Engle, J. W.; Nayak, T. R.; Theuer, C. P.; Nickles, R. J.; Barnhart, T. E.; Cai, W. *In vivo* targeting and positron emission tomography imaging of tumor vasculature with (66)Ga-labeled nano-graphene. *Biomaterials* **2012**, *33* (16), 4147-4156.

64. Yang, K.; Wan, J.; Zhang, S.; Tian, B.; Zhang, Y.; Liu, Z. The influence of surface chemistry and size of nanoscale graphene oxide on photothermal therapy of cancer using ultra-low laser power. *Biomaterials* **2012**, *33* (7), 2206-2214.

65. Li, F.; Park, S.; Ling, D.; Park, W.; Han, J. Y.; Na, K.; Char, K. Hyaluronic acid-conjugated graphene oxide/photosensitizer nanohybrids for cancer targeted photodynamic therapy. *J Mater Chem B* **2013**, *1* (12), 1678-1686.

66. Cho, Y.; Choi, Y. Graphene oxide-photosensitizer conjugate as a redox-responsive theranostic agent. *Chem Commun (Camb)* **2012**, *48* (79), 9912-9914.

67. Robinson, J. T.; Tabakman, S. M.; Liang, Y.; Wang, H.; Casalongue, H. S.; Vinh, D.; Dai, H. Ultrasmall reduced graphene oxide with high near-infrared absorbance for photothermal therapy. *J Am Chem Soc* **2011**, *133* (17), 6825-6831.

68. Shi, S.; Yang, K.; Hong, H.; Valdovinos, H. F.; Nayak, T. R.; Zhang, Y.; Theuer, C. P.; Barnhart, T. E.; Liu, Z.; Cai, W. Tumor vasculature targeting and imaging in living mice with reduced graphene oxide. *Biomaterials* **2013**, *34* (12), 3002-3009.

69. Huang, P.; Lin, J.; Wang, X.; Wang, Z.; Zhang, C.; He, M.; Wang, K.; Chen, F.; Li, Z.; Shen, G.; Cui, D.; Chen, X. Light-triggered theranostics based on photosensitizer-conjugated carbon dots for simultaneous enhanced-fluorescence imaging and photodynamic therapy. *Adv Mater* **2012**, *24* (37), 5104-5110.

70. Lai, C. W.; Hsiao, Y. H.; Peng, Y. K.; Chou, P. T. Facile synthesis of highly emissive carbon dots from pyrolysis of glycerol; gram scale production of carbon dots/mSiO_2 for cell imaging and drug release. *J Mater Chem* **2012**, *22* (29), 14403-14409.

71. Liu, C.; Zhang, P.; Zhai, X.; Tian, F.; Li, W.; Yang, J.; Liu, Y.; Wang, H.; Wang, W.; Liu, W. Nano-carrier for gene delivery and bioimaging based on carbon dots with PEI-passivation enhanced fluorescence. *Biomaterials* **2012**, *33* (13), 3604-3613.

72. Gu, J.; Su, S.; Li, Y.; He, Q.; Shi, J. Hydrophilic mesoporous carbon nanoparticles as carriers for sustained release of hydrophobic anti-cancer drugs. *Chem Commun (Camb)* **2011**, *47* (7), 2101-2103.

73. Ma, Y. H.; Manolache, S.; Denes, F. S.; Thamm, D. H.; Kurzman, I. D.; Vail, D. M. Plasma synthesis of carbon magnetic nanoparticles and immobilization of doxorubicin for targeted drug delivery. *J Biomat Sci-Polym E* **2004**, *15* (8), 1033-1049.

74. Kroto, H. W.; Heath, J. R.; Obrien, S. C.; Curl, R. F.; Smalley, R. E. C-60-buckminsterfullerene. *Nature* **1985**, *318* (6042), 162-163.

75. Nakamura, E.; Isobe, H. Functionalized fullerenes in water: The first 10 years of their chemistry, biology, and nanoscience. *Accounts Chem Res* **2003**, *36* (11), 807-815.

76. Partha, R.; Conyers, J. L. Biomedical applications of functionalized fullerene-based nanomaterials. *Int J Nanomed* **2009**, *4*, 261-275.

77. Chen, Z.; Ma, L.; Liu, Y.; Chen, C. Applications of functionalized fullerenes in tumor theranostics. *Theranostics* **2012**, *2* (3), 238-250.
78. Grobmyer, S. R.; Krishna, V. Minimally invasive cancer therapy using polyhydroxy fullerenes. *Eur J Radiol* **2012**, *81* (S1), S51-S53.
79. Lu, X.; Feng, L.; Akasaka, T.; Nagase, S. Current status and future developments of endohedral metallofullerenes. *Chem Soc Rev* **2012**, *41* (23), 7723-7760.
80. Krishna, V.; Singh, A.; Sharma, P.; Iwakuma, N.; Wang, Q.; Zhang, Q.; Knapik, J.; Jiang, H.; Grobmyer, S. R.; Koopman, B.; Moudgil, B. Polyhydroxy fullerenes for non-invasive cancer imaging and therapy. *Small* **2010**, *6* (20), 2236-2241.
81. Kwag, D. S.; Park, K.; Oh, K. T.; Lee, E. S. Hyaluronated fullerenes with photoluminescent and antitumoral activity. *Chem Commun (Camb)* **2013**, *49* (3), 282-284.
82. Wilson, J. D.; Broaddus, W. C.; Dorn, H. C.; Fatouros, P. P.; Chalfant, C. E.; Shultz, M. D. Metallofullerene-nanoplatform-delivered interstitial brachytherapy improved survival in a murine model of glioblastoma multiforme. *Bioconjug Chem* **2012**, *23* (9), 1873-1880.
83. Fan, J.; Fang, G.; Zeng, F.; Wang, X.; Wu, S. Water-dispersible fullerene aggregates as a targeted anticancer prodrug with both chemo- and photodynamic therapeutic actions. *Small* **2013**, *9* (4), 613-621.
84. Kotelnikov, A. I.; Rybkin, A. Y.; Khakina, E. A.; Kornev, A. B.; Barinov, A. V.; Goryachev, N. S.; Ivanchikhina, A. V.; Peregudov, A. S.; Martynenko, V. M.; Troshin, P. A. Hybrid photoactive fullerene derivative-ruboxyl nanostructures for photodynamic therapy. *Org Biomol Chem* **2013**, *11* (26), 4397-4404.
85. Lucafo, M.; Pacor, S.; Fabbro, C.; Da Ros, T.; Zorzet, S.; Prato, M.; Sava, G. Study of a potential drug delivery system based on carbon nanoparticles: Effects of fullerene derivatives in MCF7 mammary carcinoma cells. *J Nanopart Res* **2012**, *14* (4).
86. Kwag, D. S.; Oh, N. M.; Oh, Y. T.; Oh, K. T.; Youn, Y. S.; Lee, E. S. Photodynamic therapy using glycol chitosan grafted fullerenes. *Int J Pharm* **2012**, *431* (1-2), 204-209.
87. Zakharian, T. Y.; Seryshev, A.; Sitharaman, B.; Gilbert, B. E.; Knight, V.; Wilson, L. J. A fullerene-paclitaxel chemotherapeutic: Synthesis, characterization, and study of biological activity in tissue culture. *J Am Chem Soc* **2005**, *127* (36), 12508-12509.
88. Didenko, G.; Prylutska, S.; Kichmarenko, Y.; Potebnya, G.; Prylutskyy, Y.; Slobodyanik, N.; Ritter, U.; Scharff, P. Evaluation of the antitumor immune response to C-60 fullerene. *Materialwiss Werkst* **2013**, *44* (2-3), 124-128.
89. Meng, H.; Xing, G.; Blanco, E.; Song, Y.; Zhao, L.; Sun, B.; Li, X.; Wang, P. C.; Korotcov, A.; Li, W.; Liang, X. J.; Chen, C.; Yuan, H.; Zhao, F.; Chen, Z.; Sun, T.; Chai, Z.; Ferrari, M.; Zhao, Y. Gadolinium metallofullerenol nanoparticles inhibit cancer metastasis through matrix metalloproteinase inhibition: Imprisoning instead of poisoning cancer cells. *Nanomedicine* **2012**, *8* (2), 136-146.
90. Chen, C. Y.; Xing, G. M.; Wang, J. X.; Zhao, Y. L.; Li, B.; Tang, J.; Jia, G.; Wang, T. C.; Sun, J.; Xing, L.; Yuan, H.; Gao, Y. X.; Meng, H.; Chen, Z.; Zhao, F.; Chai, Z. F.; Fang, X. H. Multi hydroxylated [Gd@C-82(OH)(22)](n)

nanoparticles: Antineoplastic activity of high efficiency and low toxicity. *Nano Lett* **2005**, *5* (10), 2050-2057.

91. Meng, H.; Xing, G.; Sun, B.; Zhao, F.; Lei, H.; Li, W.; Song, Y.; Chen, Z.; Yuan, H.; Wang, X.; Long, J.; Chen, C.; Liang, X.; Zhang, N.; Chai, Z.; Zhao, Y. Potent angiogenesis inhibition by the particulate form of fullerene derivatives. *ACS Nano* **2010**, *4* (5), 2773-2783.
92. Liang, X. J.; Meng, H.; Wang, Y.; He, H.; Meng, J.; Lu, J.; Wang, P. C.; Zhao, Y.; Gao. X.; Sun, B.; Chen, C.; Xing, G.; Shen, D.; Gottesman, M. M.; Wu, Y.; Yin, J. J.; Jia, L. Metallofullerene nanoparticles circumvent tumor resistance to cisplatin by reactivating endocytosis. *Proc Natl Acad Sci USA* **2010**, *107*, 7449-7454.
93. Schrand, A. M.; Hens, S. A. C.; Shenderova, O. A. Nanodiamond particles: Properties and perspectives for bioapplications. *Crit Rev Solid State* **2009**, *34* (1-2), 18-74.
94. Mochalin, V. N.; Shenderova, O.; Ho, D.; Gogotsi, Y. The properties and applications of nanodiamonds. *Nat Nanotechnol* **2012**, *7* (1), 11-23.
95. Man, H. B.; Ho, D. Diamond as a nanomedical agent for versatile applications in drug delivery, imaging, and sensing. *Phys Status Solidi A* **2012**, *209* (9), 1609-1618.
96. Kaur, R.; Badea, I. Nanodiamonds as novel nanomaterials for biomedical applications: Drug delivery and imaging systems. *Int J Nanomedicine* **2013**, *8*, 203-220.
97. Say, J. M.; van Vreden, C.; Reilly, D. J.; Brown, L. J.; Rabeau, J. R.; King, N. J. C. Luminescent nanodiamonds for biomedical applications. *Biophys Rev* **2011**, *3* (4), 171-184.
98. Hui, Y. Y.; Cheng, C. L.; Chang, H. C. Nanodiamonds for optical bioimaging. *J Phys D Appl Phys* **2010**, *43* (37).
99. Zhu, Y.; Li, J.; Li, W.; Zhang, Y.; Yang, X.; Chen, N.; Sun, Y.; Zhao, Y.; Fan, C.; Huang, Q. The biocompatibility of nanodiamonds and their application in drug delivery systems. *Theranostics* **2012**, *2* (3), 302-312.
100. Bajaj, P.; Akin, D.; Gupta, A.; Sherman, D.; Shi, B.; Auciello, O.; Bashir, R. Ultrananocrystalline diamond film as an optimal cell interface for biomedical applications. *Biomed Microdevices* **2007**, *9* (6), 787-794.
101. Fu, Y. J.; An, N.; Zheng, S. H.; Liang, A. H.; Li, Y. Q. Bmk CT-conjugated fluorescence nanodiamond as potential glioma-targeted imaging and drug. *Diam Relat Mater* **2012**, *21*, 73-76.
102. Zhang, X. Q.; Lam, R.; Xu, X. Y.; Chow, E. K.; Kim, H. J.; Ho, D. Multimodal nanodiamond drug delivery carriers for selective targeting, imaging, and enhanced chemotherapeutic efficacy. *Adv Mater* **2011**, *23* (41), 4770-4775.
103. Yan, J. J.; Guo, Y.; Altawashi, A.; Moosa, B.; Lecommandoux, S.; Khashab, N. M. Experimental and theoretical evaluation of nanodiamonds as pH triggered drug carriers. *New J Chem* **2012**, *36* (7), 1479-1484.
104. von Haartman, E.; Jiang, H.; Khomich, A. A.; Zhang, J. X.; Burikov, S. A.; Dolenko, T. A.; Ruokolainen, J.; Gu, H. C.; Shenderova, O. A.; Vlasov, I. I.; Rosenholm, J. M. Core-shell designs of photoluminescent nanodiamonds with porous silica coatings for bioimaging and drug delivery. I: Fabrication. *J Mater Chem B* **2013**, *1* (18), 2358-2366.

105. Prabhakar, N.; Nareoja, T.; von Haartman, E.; Karaman, D. S.; Jiang, H.; Koho, S.; Dolenko, T. A.; Hanninen, P. E.; Vlasov, D. I.; Ralchenko, V. G.; Hosomi, S.; Vlasov, I. I.; Sahlgren, C.; Rosenholm, J. M. Core-shell designs of photoluminescent nanodiamonds with porous silica coatings for bioimaging and drug delivery. II: Application. *Nanoscale* **2013**, *5* (9), 3713-3722.

106. Wang, D. X.; Tong, Y. L.; Li, Y. Q.; Tian, Z. M.; Cao, R. X.; Yang, B. S. PEGylated nanodiamond for chemotherapeutic drug delivery. *Diam Relat Mater* **2013**, *36*, 26-34.

107. Orel, V. E.; Shevchenko, A. D.; Bogatyreva, G. P.; Leshchenko, O. V.; Romanov, A. V.; Rykhal's'kii, O. Y.; Dzyatkovskaya, I. I.; Nikolov, N. A.; Dzyatkovskaya, N. N.; Shchepotin, I. B. Magnetic characteristics and anticancer activity of a nanocomplex consisting of detonation nanodiamond and doxorubicin. *J Superhard Mater* **2012**, *34* (3), 179-185.

108. Vaijayanthimala, V.; Cheng, P. Y.; Yeh, S. H.; Liu, K. K.; Hsiao, C. H.; Chao, J. I.; Chang, H. C. The long-term stability and biocompatibility of fluorescent nanodiamond as an *in vivo* contrast agent. *Biomaterials* **2012**, *33* (31), 7794-7802.

109. Huang, S. S.; Shao, J. Q.; Gao, L. F.; Qi, Y. Z.; Ye, L. Adsorption of cathepsin B-sensitive peptide conjugated DOX on nanodiamonds. *Appl Surf Sci* **2011**, *257* (20), 8617-8622.

110. Liu, K. K.; Zheng, W. W.; Wang, C. C.; Chiu, Y. C.; Cheng, C. L.; Lo, Y. S.; Chen, C.; Chao, J. I. Covalent linkage of nanodiamond-paclitaxel for drug delivery and cancer therapy. *Nanotechnology* **2010**, *21* (31), 315106.

111. Li, J.; Zhu, Y.; Li, W. X.; Zhang, X. Y.; Peng, Y.; Huang, Q. Nanodiamonds as intracellular transporters of chemotherapeutic drug. *Biomaterials* **2010**, *31* (32), 8410-8418.

112. Chang, I. P.; Hwang, K. C.; Chiang, C. S. Preparation of fluorescent magnetic nanodiamonds and cellular imaging. *J Am Chem Soc* **2008**, *130* (46), 15476-15481.

Chapter 9

Lipopolyplex Nanoparticles for Tumor Targeting Theranostics

*Fengying Dai and Xin Zhang**
Institute of Process Engineering, Chinese Academy of Sciences, Beijing 100190, China

At the present time, the difficulties in the clinical application of gene therapy issue from the lack of effective gene vectors that are safe and tumor-selective. Cationic lipids (lipoplexes) and cationic polymers (polyplexes) emerge as a great potential alternative to viral vectors in the last decade. However, most nonviral gene carriers still suffer from inefficiency dilemma in clinical malignant cancer therapeutic. Understanding the enhancement of specific target gene delivery and a combination of diagnosis and treatment of lipopolyplex mediated gene delivery would be beneficial. The cationic nonviral gene delivery systems which include target therapeutic and combined imaging application will be the major focus of this chapter. The first section describes properties of some typical cationic liposome and cationic polymer gene carriers. In the next section, representative research of various targeting modification including targeting gene transfer, targeting gene delivery and targeting gene release of cationic lipoplexes and polyplexes will be discussed. The last section of this chapter will then show recent applications on a combination of cationic lipopolyplexes for gene therapy and theranostic applications.

9.1 Introduction

Gene therapy is a promising approach for the treatment of malignant tumors. The main obstacle for the clinical application of gene therapy for malignant tumors is the lack of safe and tumor-selective gene vectors. The accurate design of appropriate carriers is a prerequisite for the successful delivery of the gene materials into target cells. Although it has good transaction efficiency, virus vectors mediated gene transport still faces some tough challenges. Multiple significant problems, such as short-term transgene expression, inability to persist in host cells and toxicity of early clinical trials using recombinant viral vectors have been reported.[1] In 1999, direct injection of an adenovirus vector into the hepatic artery caused a patient to die of systemic inflammatory response syndrome trigger by viral vectors. In view of the observed problems with viral vectors, research focus on the design of nonviral gene delivery systems, in-

* Corresponding author: xzhang@home.ipe.ac.cn

cluding systems composed of cationic lipids (lipoplexes) and cationic polymers (polyplexes) has been emerged in large numbers in the last decade.[2] Nonviral gene carriers may also cause a slight inflammatory response, but specific recognition risks are essentially nonexistent, making the carriers less hazardous in terms of antigen-specific immune response.

However, most of the nonviral gene carriers show therapeutic effect inferior to viral vectors in *in vivo* test. A better understanding of approaches currently being taken to improve specific target gene treatment and combination of diagnosis and treatment of nonviral gene delivery systems would certainly be beneficial. In this chapter, a broad focus from catioinc nonviral gene delivery systems to include target therapeutic and combined imaging application will be highlighted. Concerns arising from the current state of target cationic nonviral gene delivery research will be summarized along with a discussion on the avenues for future various theranonstic applications evolving around the use of cationic lipoplexes and polyplexes.

9.2 Nonviral Gene Vectors

9.2.1 *Cationic Liposomes*

In the last few years, cationic gene carriers have gained an increasing interest. One of the most popular applications has been the cationic liposome (CL) among the various nonviral vectors. A considerable amount of clinical trials involving CL-based vectors have been approved and carried out in the treatment of various gene disorder diseases and multiple malignant tumor.[3–7] Lipids based vectors are less biologically hazardous compared to the viral vectors and worthy of being widely used in the clinical. Figure 9.1 shows typical structures of lipoplexes composed of amphiphilic lipid self-assembling and gene material in water that are frequently used for gene delivery.

The tranfection characters of the lipoplexes are influenced by many elements. These factors, including the lipid structure and composition of liposomes, the preparatory condition of liposomes, the component ratio of lipid/nucleic and the overall particle size of lipoplexes, have been found to influence the transfection efficiency. The chemical structure of cationic lipid moiety, among the other associated factors, shows a significant impact on both the transfection activity and the biological hazard of lipoplexes. The type of cationic lipid molecule is considered to have a close relationship with the phase properties of the lipoplexes as well as their interactions with the membrane lipids.

Previous research also proposed that the phase structure of liposome affect the transfection activity of the lipoplexes. It has been proven that some helper lipid such as dioleoyl phosphatidyl ethanolamine (DOPE) and dioleoyl

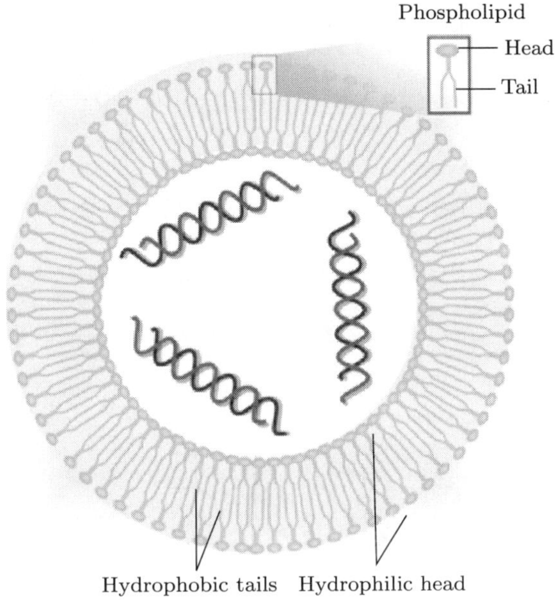

Fig. 9.1 Typical structure of lipoplex for gene delivery.

phosphatidyl choline (DOPC) could help to maintain suspension stability of lipoplexes and facilitate further cell membrane penetration and endosomal disrupting capacity.[8,9] Structures of some of the representative neutral helper lipids used in liposomal nonviral gene formations were shown in Fig.9.2.

Strategies to manage lipoplexes-based gene delivery include enhancing the stability and circulation time in the blood stream, targeting to specific tissues and cells, and facilitating intracellular trafficking. The most commonly used CL gene delivery system working in the long circulation modified-release manner include PEG-conjugated lipoplexes.[10] However, PEGylation should be designed and handled carefully as PEG decoration may impede the intracellular uptake efficiency and intracellular trafficking of lipoplexes to some extent, and therefore the gene transfection efficiency of lipoplexes might be compromised.[11]

9.2.2 *Cationic Polymers*

Polymers, bearing positive charges on their backbone and/or as side chains, are considered as cationic polymers. Negatively charged molecules may combine with cationic polymers by self-assembly method based on electrostatic binding interactions. These unique physicochemical properties render cationic polymers appealing for gene transfer applications. Cationic polymers form polyelectrolyte complexes with nucleic acids (DNA, RNA and oligonucleotides), provide protection from enzymatic degradation and facilitate cellular uptake and endo/lysosomal escape, thus, becoming excellent candidates for gene de-

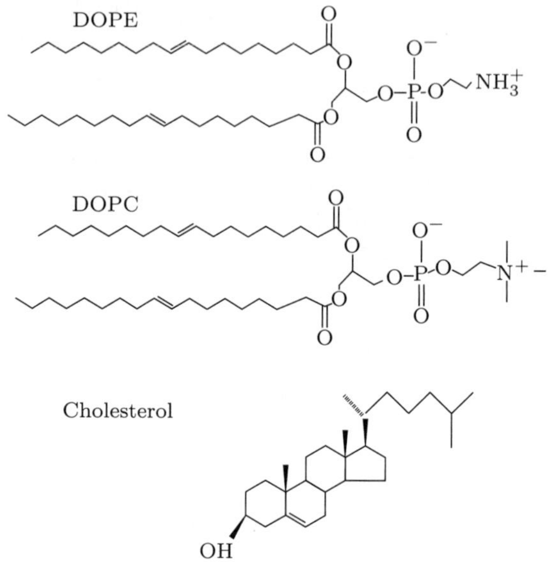

Fig. 9.2 Neutral helper lipids DOPE, DOPC and cholesterol used in liposomal nonviral gene formations.

livery. Figure 9.3 represents a typical strcuture of polyplexes formed by electrostatic interactions between polycations and DNA.

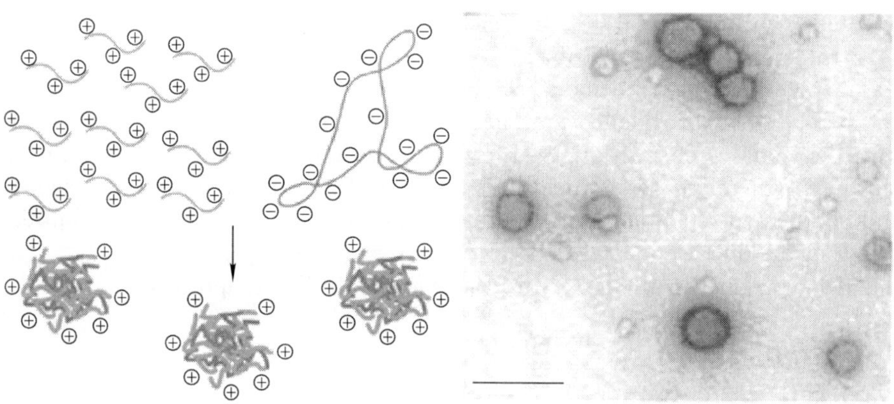

(a) (b)

Fig. 9.3 (a) Polyplexes are formed by electrostatic interactions between polycations and DNA. (b) Transmission electron micrograph of polyplexes comprising plasmid DNA and polyethylenimine (PEI). Scale bar = 200 nm. Image is adapted with permission from Ref. 12. Copyright (2005), rights managed by Nature Publishing Group.

In comparison to viruses and cationic liposomes gene vectors, the cationic polymeric vectors show several distinct advantages, such as high safety, low cost of synthesis and easy scale-up. Therefore, the design and synthesis of cationic polymeric gene vectors have attracted significantly growing attention. The gene transport properties of cationic polymers are highly dependent on the polymeric chain structure. Most of the cationic polymers bear amine functionality that can be protonated. The relative numbers of the protonable amines influenced gene condense and transfer efficiency. In addition, the molecular weight, flexibility, H-bond formation, hydrophobic interactions, amine group and its neighboring functionalities also could affect the gene delivery performance. A wide range of cationic polymers have been investigated for gene transfer and gene therapy. Among the natural and synthetic cationic amino polymers, polyethylenimine (PEI), poly-L-(lysine) (PLL), poly[2-(N,N-dimethylamino)ethyl methacrylate](PDMAEMA) and chitosan (CS) had been extensively studied (Fig.9.4).

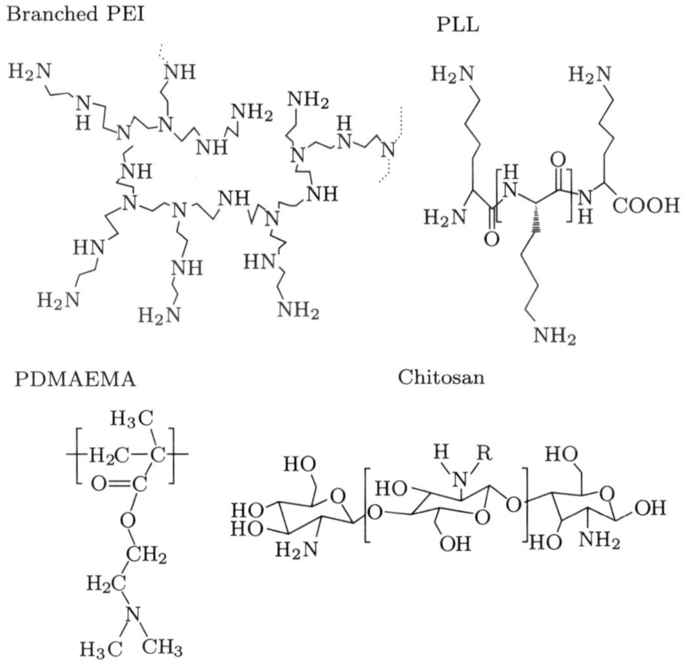

Fig. 9.4 Structure of some representative cationic polymers.

9.2.2.1 *Polyethylenimine (PEI)*

Among different cationic polymers, polyethylenimine (PEI) is one of the most extensively investigated one for the administration of a wide array of nucleotide-based therapies, including DNA, siRNA, and oligonucleotides. One of the most prominent features of PEI as cationic gene vector is the high amino group density. Branched PEI contains primary, secondary and tertiary amines in a ratio of 1:2:1. At physiological pH, the polymer has a high amount of amine functionalities and is protonated to only approximately 25%. Nearly a third of the atoms was nitrogen in branched PEI, and contains primary, secondary, and tertiary amine groups, that could be protonated in a wide range pH values. These structural characters greatly facilitate the compact condensation and transport of genetic materials to pass through various cellular barriers, and target the nucleus of cells.[13–15]

Excluding the molecular architecture factor, the molecular weight of PEI can greatly impact the gene delivery performance of polyplex. Generally, the higher molecular weight (M_w) PEI facilitate to form complexes with a more compact structure and provide more effective protection of gene materials, while a lower molecular weight PEI may lose capacity for condensation of gene materials. Many investigations had proved that the PEI with branched structures showed significantly higher performance in transfection experiment than normal PEI. Since the branched PEI has the ability to capture a large amount of protons in an acidic environment, it can be considered as an ideal buffer for variations in pH.[12,16] This buffer capacity is a useful property in the endosomal escape route (Fig.9.5). As a consequence, branched PEI have frequently served as the controls to check the efficiency of new carrier materials in the field of fundamental research.

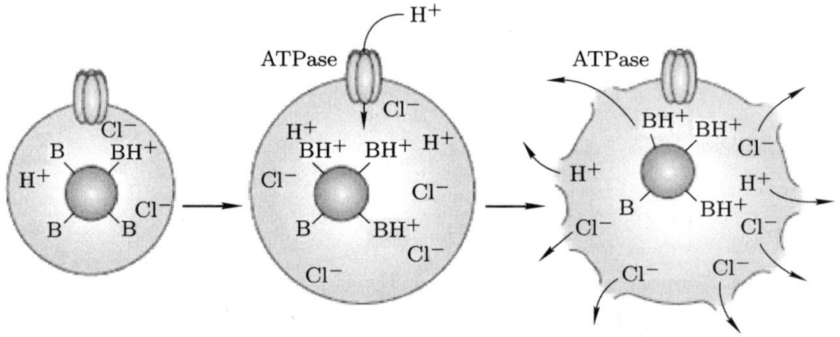

Fig. 9.5 Schematic of the proton-sponge mechanism. Protonation of the proton-sponge polymer (green) causes increased influx of protons (and counter-ions) into endocytic vesicles. Increasing osmotic pressure causes the vesicle to swell and rupture. Image is adapted with permission from Ref. 12. Copyright (2005), rights managed by Nature Publishing Group. (color figure at the end of the book)

9.2.2.2 *Poly-L-lysine (PLL)*

PLL is a cationic homopolymer of the amino acid L-lysine and is composed of a large number of protonable primary amines in physiological environment, so they can condense negatively charged nucleic acid molecules through electrostatic interaction. Synthesis of PLL was obtained by conversion of a primary amine protected L-lysine monomer into the cyclic N-carboxy-(N-benzyloxy carbonyl)-L-lysine anhydride.[17] The molecular weight of PLL can be controlled by the use of specific monomer to initiator feed ratio. As a typical cationic gene carrier, PLL is the earliest agent of polyamide acid used for cationic non-viral gene delivery. Owing to the amine groups in the polymer chain, PLL can electrostatically interact with gene materials to form compact complexes. However, another disincentive to transfection lies on the fact that only a part of the primary amine groups of PLL is protonated at physiological pH putting a negative effect on buffering capacity of lipoplexes and consequently leading to a low transfection efficiency. Moreover, high molecular weight PLL is cytotoxic and shows a tendency to aggregate and precipitate at higher physiological salt concentration.

9.2.2.3 *Poly[2-(N,N-dimethylamino)ethyl Methacrylate) (PDMAEMA)*

As an amphiphilic macromolecular cationic gene carrier, PDMAEMA was commonly prepared by radical polymerization of 2-(dimethylamino)ethyl methacrylate monomer. PDMAEMA was first used as cationic DNA carrier in COS-7 and OVCAR-3 cells transfection investigate by Hennink *et al.* It was confirmed that PDMAEMA polymer revealed molecular weight and polymer/DNA weight ratio related transfection efficiency, and the optimum PDMAEMA complexes showed higher transfer activity than PLL carrier.[18] Up to the date, multiple attempts were performed to improve the transfection activity of PDMAEMA lipoplexes. Some typical structural modifications, including the introduction of tertiary amino groups, to enhance the buffer capacity effect of PDMAEMA perplexes and introduction of guanidinium groups to facilitate lipid membrane penetration effect of PDMAEMA polyplexe were performed.[19,20] By making these operations, transition efficiency of PDMAEMA polyplexes could be improved.

9.2.2.4 *Chitosan (CS)*

Chitosan is deactivated products of natural polysaccharide source which obtained by the alkaline deacetylation of chitin. As one of the natural gene vectors, the application of chitosan as cationic gene carrier made significant progress due to its non-toxicity and low immunogenicity. According to the reported investigation, some important factors, such as the molecular weight, degree of deacetylation, charge ratio of chitosan to gene materials, tend to have influence on transfection efficiency of CS polyplexes. Some researchers have found that effective structural modifications of chitosan, for instance quaternization and hydrophobic modification, could improve transfection efficiency on CS polyplexes effectively.[21,22]

9.2.3 *In Vivo Barriers to Cationic Nonviral Gene Delivery*

The systemic administration of nucleotide delivery complexes through blood stream remains the most common method in gene therapeutics. The nonviral gene vector/ nucleotide complexes for systemic injection must navigate a series of *in vivo* barriers before engaging into the target cellular machinery. Following intravenous administration of complexes, nucleic acid molecules firstly encounter the risk of enzymatic cleavage. Endogenous nucleases rapidly degrade nucleic acids in the serum. The next challenge starts as danger of rapid excretion of complexes by reticuloendothelial system (RES). As complexes meet blood components as well as extracellular negatively charged proteins, they can form larger size aggregates *via* nonspecific adsorption which are removed by phagocytes. Phagocytic cells like macrophages and monocytes efficiently remove any larger foreign material and are considered to be an important immunological barrier. In addition, high physiological salt concentration can also support aggregation of complexes which eventually leads to phagocytosis. Another major obstacle in the delivery of the complexes to tumor is extravasation from the blood stream. Although particles larger than 5 nm do not typically cross endothelial membranes, certain tissues of some tumors, permit the entry of delivery particles of up to 200 nm in diameter. After overcoming the blood endothelial wall barrier, the complexes arrive in the vicinity of the tumor cell. Once complex diffuses through the tumor extracellular matrix, it has to pass through several physical barriers including cellular membrane, endosome/lysosome and nucleus barrier. Due to the relatively large size and negative charges, naked nucleic acid cannot cross these biological membranes. Nucleic acid and cationic lipopolyplexes can promote endocytosis, especially in the presence of specific ligands. In cytoplasm, complexes should escape from the endosomes before ending up into the lysosome where the pH significantly drops in order to avoid the degradation of nucleic acid. Finally, siRNA must be released to the cellular machinery if the siRNA was formulated with delivery agents. For DNA delivery system, nucleic acid must be transported to the nucleus where endomitosis occurs.

9.3 Tumor Targeting Nonviral Gene Therapeutic

Nonviral gene delivery contributes to clinical reality needs to be stable, nontoxic, efficient and specific. In order to achieve efficient gene delivery, it will be useful to modify complexes with targeting moiety for binding to the surface of target cells. Such inactive and active target complexes have greater opportunities to specifically recognize and bind to tumor site in systemic circulation. This binding will lead to an increase in the efficiency of transfection and a reduction in the toxicity of the vector. So far, a variety of different targeting strategies have been characterized in order to improve the binding of positively charged complexes to cells with more inactive and active targeting delivery.[10,11,15,17,20,23–25]

9.3.1 *Inactive Target*

Successful delivery of exogenous gene into tumor systems that have developed natural barriers and adopted a defense system can be extremely difficult. Systemic gene delivery strategies encounter multiple hurdles including capture by the mononuclear phagocyte system, glomerular filtration and excretion, and cellular and subcellular barriers. The therapeutic gene penetration effect was associated with vessel permeability of solid tumor site. Most solid tumors contain a rich highly permeable vascular system. Such fenestrated vascular structures ensure a sufficient supply of nutrients and oxygen for rapid tumor growth. In normal tissues, lymphatic system can clear efficiently the exogenous molecules or particles that leak into the blood vessel. However, the insufficient clearance of the leakage in tumor blood vessel of the lymphatic system leads to accumulation of macromolecules or particles in tumor tissues. This unique phenomenon is known as enhanced permeability and retention (EPR) effect and its fundamental mechanism is exploited for passive targeting of nanoparticles into solid tumor. Nevertheless, the architecture of large tumors is highly heterogeneous. That highly vascularize tissue located at the outer part of a tumor shows an EPR effect, while the central area posses typical vascular tissues, resulting in a diminished EPR effect. Although many studies on nanoparticles have been performed on the basis of the EPR effects, the *in vivo* data have revealed that the tumor specificity of the nanoparticles was only slightly improved compared to the controls in many cases.

A basic and most important strategy to boost effectiveness is to prolong the blood circulation time of gene medicines, providing more time and opportunity for their cellular interaction. Various factors, including electric charge, size and surface composition of complexes may impose an influence on the circulation efficiency of gene therapeutic particles. After intravenously injected into a blood vessel, particles smaller than 5 nm are removed from blood by rapid renal clearance through kidney, whereas large micro sized particles are filtrated mechanically by sinusoids and cleared by the reticuloendothelial system (RES) of liver and spleen. Therefore, complexes with 10–500 nm sizes can remain in the circulation for an extended period of time and can be employed in rational strategies for passive target gene delivery.

Binding to the negatively charged plasma proteins is the primary mechanism for RES to recognize the circulating complexes. Macrophages, such as the Kupffer cells in the liver, recognize the opsonized antiparticle *via* the scavenger receptor. Hydrophilic polymers with highly flexible chains, such as polyethylene glycol(PEG), polyvinyl alcohol (PVA), poly(vinyl pyrrolidone) (PVP) represent efficient agents for complexes surface shielding against unintended interactions with the plasma protein.[23] PEGylation is one of the most predominant method used to reduce the opsonization and aggregation of nonviral vectors and minimize the clearance by RES, leading to a prolonged circulation lifetime *via* so called "steric stabilization". PEG coating outside the surfaces of the cationic nonviral vectors can prevent complexes from non-specific binding with blood components, reduce RES uptake, decrease biological toxicity, and increase the binding affinity with the target cells. PEGylated complexes are therefore often referred to as "stealth" complexes.[11]

The prolonged blood circulation time is a critical factor for any drug delivery strategy aiming at solid tumors. In this respect, the use of a hydrophilic component, like PEG, is unavoidable. However, the PEGylation was not always a contributing factor in gene therapeutic. Although the development of PEG-derivatized gene vector has brought significant advances to the design of complexes for intravenous administration, the experience accumulated over the past decade has revealed some important limitations of PEGylation.[25] After the PEGylation, nucleic acid loaded particles accumulate inside a tumor area, where they may encounter ineffectiveness of the released gene cargo into tumor cells. Most of the cationic complexes have to be taken up by a cell *via* an endocytic pathway. The presence of PEG shielding on surface significantly hinders the cellular uptake and the endosomal escape of the complexes, because it limits the immediate contact between cationic gene vector and negative membranes. Therefore, the design of cationic gene vector aimed at the successful systemic delivery of nucleic acid to solid tumor should rely on a balance between the system circulation efficiency and favorable cellular and subcellular bioavailability.

For a PEGylation lipoplex system, the incorporation of fusogenic lipids turns out to be a useful strategy to enhance the intracellular bioavailability. Amphiphilic lipid molecules such as 1,2-dioleoyl-snglycero-3- phosphoethanolamine (DOPE) containing a protonatable acidic group among phosphoethanolamine molecules were introduced into PEGylation lipoplexes.[26,27] In the neutral physiological pH, stable lipoplexes are formed and the carboxylic groups of the amphiphiles are protonated upon progressive acidification of the endoscopic, followed by the occurrence of instability as the phosphoethanolamine molecules revert into their inverted hexagonal phase. This process promotes the lipoplexesl destabilization, which further facilitates the escape of gene cargo from the lysosome.

In addition, an interesting approach relies on the development of a new cleavable PEG derivatized nanoparticles sensitive to the pH and redox potential. Stimuli-sensitive PEG cleavage from the complexes surfaces is beneficial to overcome the PEG limitations, and improves the transfection ability of the complexes toward cancer cells.[28–30] The aforementioned methods illustrate a deep understanding of the PEGylation of complexes on gene cargo delivery activity for the rational development of nucleic acid transfer vector. Indeed, a balance between prolonged blood circulation times and high cell transfection efficiency is mandatory to achieve significant therapeutic results.

9.3.2 *Specific Target Gene Delivery*

Galactose and mannose were mostly investigated for cell-specific receptor-mediated endocytosis in the carbohydrate related target gene delivery. Cationic gene delivery system modified with galactose and mannose moieties leads to hepatic cell target delivery.[12,31] As a matter of fact, the first ever report about specific agalactose modified PLL carrier targeted to the asialoglyco protein receptors on the hepatocyte cell membrane.[32] After that, target research based on the galactose and mannose upsurge in cationic gene transfer field. Galactose and mannose have been mostly exploited as glycoliands owing to their hepatic and immune system cell targeting capability. Glycotargeted versions of the typical cationic polymers have also been investigated.[33]

As cancer causes abnormal upregulation of many cell surface molecules, this anomalous cell provides many useful receptors for peptide targeting. The highly proliferative nature of tumor tissue allows for easier access to growth factors. Specialty peptide target strategy is based on excessive proliferation of tumor cell. A large number of studies has been performed to design target nonviral gene delivery vehicles coupled with peptide targeting ligands.[34] Peptide targeted gene delivery has therefore seen much success *in vitro* and *in vivo* models of cancer states.

Elevated levels of epidermal growth factor receptor (EGFR) are found on several types of solid tumors and consequently strategies for targeting EGFR have been designed. Growth factor target has been designed based on the feature of abnormal increasing of growth factor receptor expression in tumors and tumor vasculature. Fibroblast growth factor receptors (FGFR) and fibroblast growth factor (FGF) abnormal increasing are typical examples that are caused by overgrowth of tumor. FGFR have been shown to be over expressed in various cancer cells, including colon, ovarian, renal, and lung carcinomas. While FGF is a known mitogen which stimulates cell division when bound to FGFR, EGFR was proven to be a promising therapeutic target agent to cancerous cells and tumors where angiogenic vessels grow rapidly to feed the surrounding tumor tissues.[35–37] Kim *et al.*[37] have selected coupled luteinizing hormone-releasing hormone (LHRH) peptide analogue as a cancer targeting ligand to the distal end of the poly(ethylene glycol) (PEG)-siRNA conjugate. The siRNA-PEG-LHRH conjugate self-assembled to form nanosized micelles upon mixing with poly(ethylenimine) (PEI) *via* ionic interactions. Complexes with LHRH exhibited superior cellular uptake capacity than those without LHR in LHRH receptor over expressing ovarian cancer cells (A2780), resulting in an enhancement of gene silencing efficiency *via* receptor-mediated endocytosis (Fig.9.6).

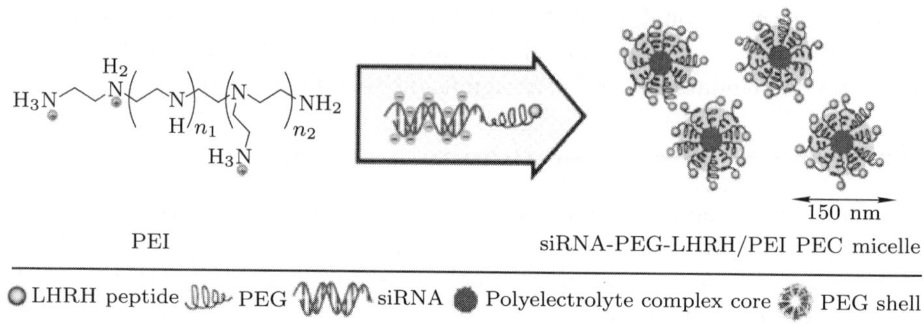

Fig. 9.6 A schematic diagram of the preparation for siRNA-PEG-LHRH/PEI PEC micelles in aqueous phase. Image is adapted with permission from Ref. 37. Copyright (2008) American Chemical Society.

The second promising group of target focuses on integrins target for specific delivery of gene therapeutics. In many tumor cells and tumor endothelium, the $\alpha_v\beta_3$, $\alpha_v\beta_5$, and $\alpha_5\beta_1$ expression is upregulated abnormally. A common targeting approach involves the use of arginine-glycine-aspartic acid sequence (RGD) peptide motif binding to integrin receptors, which are over expressed on tumor cells. Almost half of the known integrins bind RGD peptide. RGD peptides mediated high affinity integrin binding has allowed the development of synthetic gene delivery systems that can penetrate into the cell by integrin dependent pathway.[38–40]

Mickler *et al.*[41] revealed that RGD ligands enhance the internalization efficiency of complexes into integrin expressing HeLa cells without altering important uptake pathway of the complexes. Both targeted as well as uncharged micelles are predominantly internalized *via* clathrin mediated endocytosis (Fig.9.7). The number of PEG shielding of polyplexes imposes an important influence on their targeting specificity. Selective endocytosis of integrin targeted micelles occurs at high PEG shielding, whereas low PEG shielding disfavor the targeted internalization indicated by the comparative study of target and nontarget complexes.

9.3.3 *Stimuli-sensitive Gene Delivery*

Effective gene delivery requires dual modal interactions between nucleic acid and the cationic vectors. In fact, some conflicting and challenging demands are encountered by complexes after systemic injection. A neutral and hydrophilic shielding layer on the outside of complexes is preferred during systemic circulation. However, this shielding layer is generally considered to impede efficient cellular uptake, while endosomal escape can disassemble at the intracellular targets. The alternative involves polycations with large molecular weight and high cationic density condensed nucleic acids to form stable polyplexes under phys-

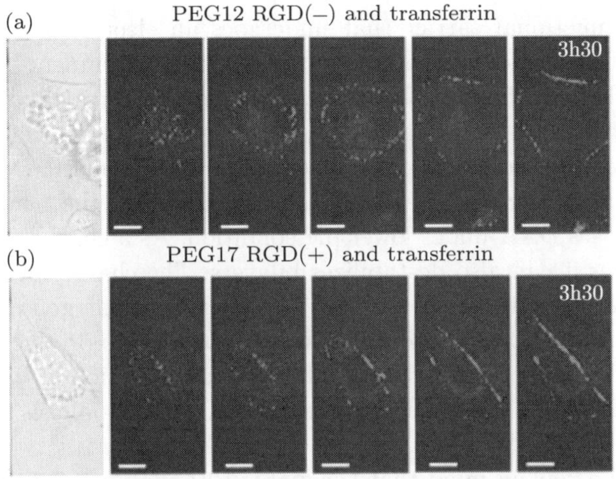

Fig. 9.7 Chlorpromazine inhibition of clathrin mediated endocytosis. HeLa cells were preincubated with 10 μg/mL chlorpromazine for 30 min before addition of Cy5 labeled PEG12 RGD(−) (a) or PEG17 RGD(+) (b) micelles to the cell medium. Cells were evaluated 3.5 h after micelle addition by spinning disk confocal microscopy. Scale bar: 10 μm. Reprinted from Ref. 41. Copyright (2011), with permission from Elsevier.

iological conditions, and result in high transfection efficacies. However, these high molecular weight carriers raise the problem of potential toxicity than their low molecular weight analogues. One of the practical strategies to solve these dilemmas is to design intelligent nonviral vectors to undergo molecular transformation triggered by a specific exogenous stimulus such as changes in pH, redox potential and so on. Stimuli-responsive polymer carriers have been extensively studied through dynamic covalent linkages or non-covalent interaction.[33] In this section, we focus on pH-sensitive reversible shielding and redox potential sensitive strategies have been developed in non-viral intelligent nucleic acid delivery systems.

9.3.3.1 *pH-sensitive Gene Delivery*

Environment sensitive liposomes have been intensively studied over the past decade for delivery of nucleic acid. One of the well known examples of acidic lysosome environment sensitive lipid was the dioleoyl phosphatidyl ethanolamine (DOPE). DOPE molecules could facilitate destabilization of the endosomal membrane and efficiently release the cargo into the cytoplasm. DOPE molecules form a stable lipid bilayer at a physiological pH of 7, while at acidic pHs of 5–6, there has been a structure of hexagonal-II structure formed, inducing the fusion of liposome with the endosomal membrane and facilitating endosomal escape of nucleic acid.[42]

Charge-reversal or charge-switching carriers for gene delivery are a type of pH-sensitive functional carrier that undergoes an electrostatic transition intracellularly from cationic to anionic in an acidic environment to improve the release of nucleic acid from the complex. The charge-reversal lipid has a cationic ammonium condensed nucleic acid, lipophilic acyl chains to form a bilayer, and benzyl estersat the terminus of the acyl chains for enzymatic hydrolysis. This charge-reversal carrier has the ability to bind DNA in the neutral condition and then releases DNA under lower pH condition. As a charged switching amphiphile carrier it can also destabilizes bilayers. The charge reversal lipid and plasmid DNA forms the lipoplex initial status of in system gene delivery. After entering the cell, the terminal ester linkages hydrolyze to produce an ionic amphiphiles and the anionic amphiphiles facilitates disruption of the lipid bilayer of the lipoplexes and the efficient release of DNA from lipoplexes for subsequent transcription.[43]

It should be kept in mind that the stable PEGylation of gene carriers may not always enhance nucleic acid delivery. At the cellular level, PEGylation has negative influence on internalization efficiency of weakening of the positive charge of complexes. Besides interference with cellular internalization, PEGylation has been found to interfere endosomal escape and cytoplasm release of nucleic acid, which impairs the expected clinical outcomes of gene therapeutic inevitable. Therefore, protective PEG chains should be removable after they arrive in target cells. Stealth PEG layers of nonviral vectors can be designed to be disarmed in response to a mildly acidic endosomal pH, when linked *via* acid-cleavable hydrazone and acetal linkages, respectively. The hydrazone and acetal linkages are fairly stable at a pH of about 7.3–7.5 in systemic circulation, but easily hydrolyzed at a pH of 6 or below of a tumor mass or endocytic vacuole. As a consequence, the occurrence of dePEGylation of complexes at subacid tumor microenvironment and a significantly higher gene transfection efficiency obtained from this changeable PEGylated.[44]

A pH-triggered small interfering RNA (siRNA) delivery system to liver for treatment of hepatitis B virus infections *in vivo* was developed by Kolli *et al.*[45] The pH response property of lipoplexes was derived from the oxime formation by the combination of an aminoxy functional group and an aldehyde. Oxime decomposition under mildly acidic conditions might contribute to the pH-triggered release of the entrapped siRNA. The pH-triggered PEGylated siRNA lipoplexes, maintaining stable structure in 80% serum, can release siRNA payload at pH 5.5 within 30 min. This capability allows functional delivery to cultured murine hepatocyte cells *in vitro*. The study also confirmed that pH-triggered siRNA delivery nanoparticles typically enter cells by clathrin-coated pit endocytosis, while functional delivery requires membrane fusion events. According to the body distribution studies, more than 70% of the pH-triggered lipoplexes are found in liver hepatocytes after being injected. The pH-triggered lipoplexes show siRNA delivery to murine liver effecting maximum knockdown 48 h post administration from a single dose proven by Pharmaco dynamic experiments. Comparison group of controllers lipoplexes requires 96 h and two doses to demonstrate the same effect. This pH-triggered siRNA delivery method is a developing approach towards *in vivo* siRNA-therapeutic for the treatment of liver diseases.[45]

9.3.3.2 *Redox Potential Sensitive Gene Delivery*

The next approach entails the introduction of reducible bonds into cationic carrier to facilitate destabilization of complexes and release of nucleic acid from complexes into cytosome. The rationale of reducible response gene delivery lie in the fact that the relatively high concentrations of reductive species including glutathione, the enzymes thioredoxin and glutaredoxin cytoplasm. Once internalized in the cell, complexes possessing a redox potential sensitive disulfide bond within its structure will undergo disulfide bridge cleavage.

As an effective stimuli-sensitive gene delivery strategy, reducing modification is not only limited to gene carrier level, but also located in modification

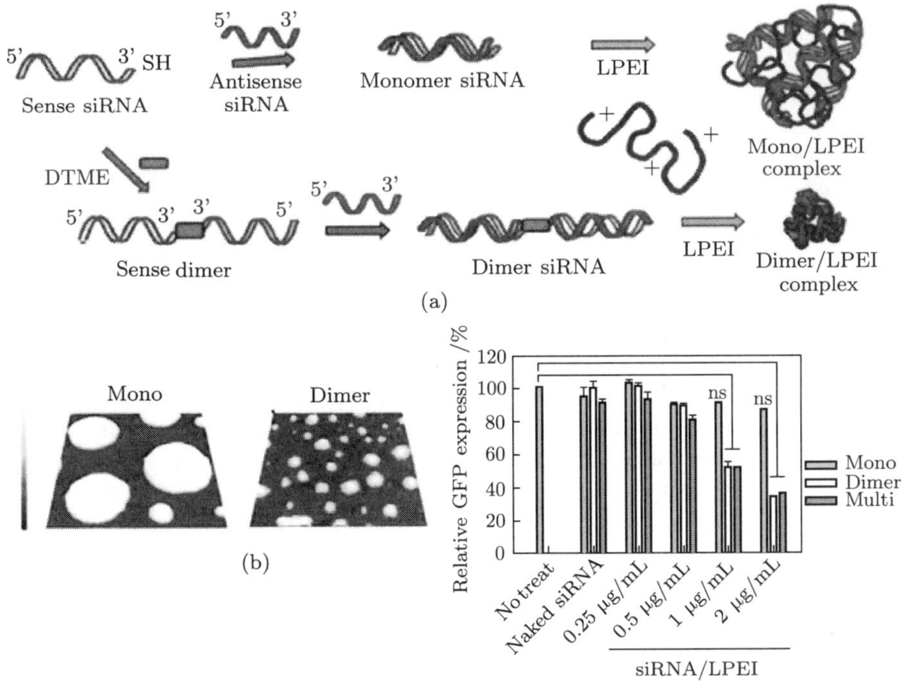

Fig. 9.8 (a) Synthesis of reducible dimeric siRNA for gene silencing. (b) AFM images of the siRNA complexes at an N/P ratio of 20 (size bar 100 nm). (c) Gene silencing using PEGylated dimeric siRNA.[46] Image is adapted with permission from Ref. 46. Copyright (2011) American Chemical Society.

of nucleic acid. Synthesis of Dimeric siRNA targeting GFP and VEGF genes linked by a cleavable disulfide bond in the reducible dimerized siRNAs has a superior performance in intracellular delivery and gene silencing as compared to the monomeric siRNA at the same N/P ratio. Dimerized siRNAs targeting GFP and VEGF formed complexes with linear polyethylenimine (LPEI). Dimeric siRNA/LPEI complexes showed greatly enhanced cellular uptake and gene silencing effects in various cell lines compared to the monomer siRNA/LPEI complexes (Fig.9.8). The author interprets the superior gene silencing efficiencies of Dimeric siRNA/LPEI complexes were creation of the two combining powers: the more compact and stable siRNA complexes and the conjugation strategy of reducible siRNA dimmers. The dimerized siRNA complex system holds great potential for *in vivo* systemic gene therapy.[46]

9.4 Agents Combined Imaging and Gene Delivery

Combination of gene delivery system and molecular imaging agent has power potential for efficient treatment and monitoring the prognosis of incurable diseases. Traditional analytical methods performed in gene transfer efficiency evaluation can only give very limited information and failed in providing tissue distribution information. Obviously, these studies would simply be unacceptable in the clinical practice that would put the patient at risk. Both cellular and subcellular levels and *in vivo* gene expressions should be studied in detail in the clinical area with the appropriate approach. Introduction of real-time imaging of gene expression in living subjects has been studied extensively in biomedical research in the last ten years.[17,47,48] Combing imaging/diagnostics of gene technologies can offer a deep understanding about the biological activity and pharmacokinetics of gene drugs. Transgenic methods based on associated imaging are undoubtedly changing the way of research in animal models, and a substantial growth of the use of imaging technologies have become readily accessible. In this scenario, theranostic with the integration diagnosis and treatment could play an essential role both in preclinical and clinical gene therapy research. It could collect gene therapeutic information *in vivo* by defining the location, magnitude and time-variation of gene expression over time, but would also lead to a better understanding of the tissue distribution of nucleic acid *in vivo*.

Three different technologies have been adapted to measure and monitor the process of gene transport in recent years: optical imaging, magnetic resonance imaging and radionuclide imaging. Reporter genes with bioluminescence and fluorescence signatures are low-cost alternatives for real-time analysis of gene expression in animal experiments. However, the diagnosis effect of optical imaging methods is limited for some disadvantages. The optical signal from gene delivery system may be hindered by absorption and scattering action *in vivo*, limiting the sensitivity of depth imaging. Magnetic resonance imaging (MRI) and positron emission tomography (PET) imaging has also been used to characterize biodistrtibution of various radiolabele gene substrates *in vivo* imaging. The advantages of MR for the imaging of gene expression include the excellent three-dimensional spatial resolution and its high spatial resolution at imaging. PET could give more sensitive analytical information than MRI in which compounds labeled with positron emitting radioisotopes to measure biological processes. In this context, the continuous developments of optical imaging, MR imaging and PET imaging of gene therapy were focused to determine the location, magnitude and time-variation of transfection efficiency *in vivo.*

9.4.1 *Optical Imaging*

Optical imaging is one of the more common modalities used in gene delivery research. Optical imaging utilizes photons emitted from bioluminescent or fluorescent probes. Noninvasive optical imaging of reporter gene expression has substantial advantages, including ease of use, lack of any radioactive probe, relatively low background signal, relative simplicity of use, and low cost relative to radionuclide and MR approaches. The most well known example would be to use a handful of well characterized imaging reporter genes for indirect imaging of therapeutic gene expression.

Bioluminescence imaging relies on photons emitted *via* the luciferase-luciferin reaction as the source of contrast. Biologically, luciferase uses energy from adenosine triphosphate to convert its substrate, luciferin to oxyluciferin with the emission of the photons that can then be detected by bioluminescence imaging. Bioluminescence imaging requires administration of secondary substrates, such as luciferin, while fluorescent optical imaging does not require any substrate and does not appear to interfere with cell growth and function.[49]

Fluorescence imaging uses a fluorescent protein that is excited with external illumination, and the emission is subsequently detected. Fluorescence does not require a reporter substrate, because the reporter fluorescent protein is first excited by external illumination and then emits light at a different wavelength that can be externally detected. Fluorescence emission is much more intense than bioluminescence, and there are reporters available in the whole spectral range of light that can also be used for fluorescence microscopy imaging. The major limitations of fluorescence imaging are the contribution of autofluorescence of tissue and poor penetration of light through tissue. Therefore, fluorescence imaging is mainly suitable for application in small animals.

According to research report by Mizuno *et al.*[50] dry powder containing both ICG and pCMV-Luc was useful as a dual imaging system to visualize pulmonary delivery and gene expression in mice. Complexes compose of the plasmid DNA pCMV-Luc encode luciferase as a reporter gene, chitosan as a gene carrier, indocyanine green (ICG) as label for fluorescent imaging were adapted in this study. Both the pulmonary delivery of the gene agent and the gene expression properties in the lung was visualized and evaluated in mice (Fig.9.9). In the mice that administered pDNA-ICGdp, pulmonary delivery was visualized with fluorescent imaging and the level of gene expression was evaluated with luminescent imaging in the same animal. The relationship between luciferase activity and fluorescent strength shown in Fig.9.9(c) suggests that more complexes delivered in the lung resulted in a higher gene expression.[50]

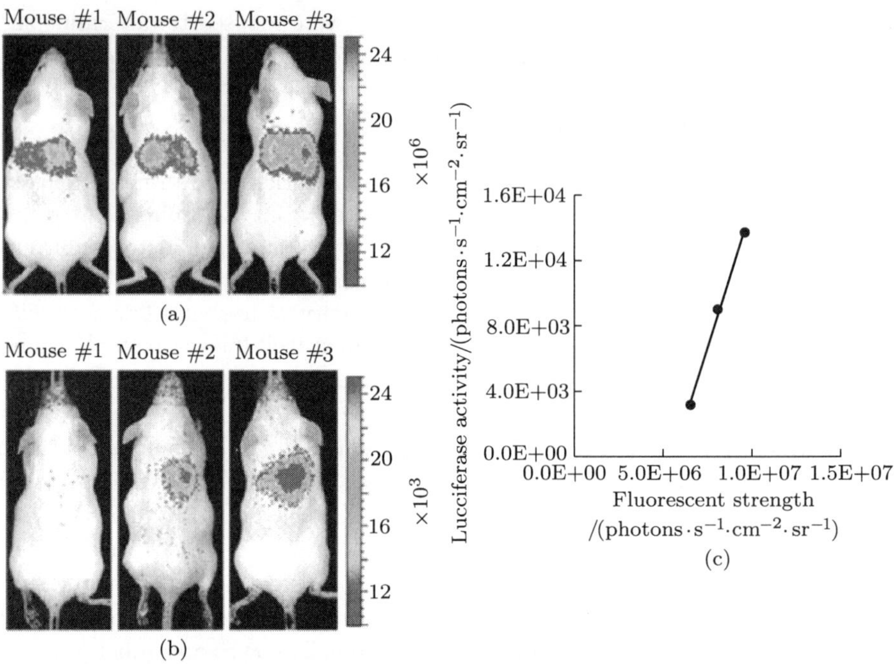

Fig. 9.9 Dual imaging of the pulmonary delivery and gene expression. (a) The fluorescence was detected in the whole body 20 min after administration of the dry powder. (b) The luciferase activity was measured 9 h post-administration. (c) The fluorescence measured 20 min after administration of the dry powder remarkably correlated to the bioluminescence after 9 h. Reprinted from Ref. 50. Copyright (2009), with permission from Elsevier.

For *in vivo* optical imaging, probes emitting in the near infrared fluorescence (NIRF) spectrum are favored because NIRF has emission wavelengths of approximately 600–1 000 nm, which are not well absorbed in biological tissues. The NIR (700–900 nm) window has the advantages of reduced autofluorescence, reduced tissue scattering, and greater depth of penetration, which is the most suitable for *in vivo* imaging.[30] Polymethines such as Cy5.5 are still the most common NIR fluorophores used for fluorescence imaging *in vivo*.[51]

9.4.2 *Magnetic Resonance Imaging (MRI)*

MRI has been one of the most powerful noninvasive diagnostic imaging methods and provides detailed anatomical information about tumor diagnosis. The precession of the response of water hydrogen nuclei to an applied magnetic field is the basis for MRI signal. After application of radio frequency pulses, the relaxation process through which the nuclei return to the original aligned state can be utilized to form an image. MRI techniques for the detection of gene expression are essential for future clinical application of gene therapy. A number of MR imaging genes transfection has been developed in the last few years.[52–54] However, research on MR imaging gene therapy is still in its infancy in clinical application research.

Imaging of gene expression using MRI may be achieved mainly through two ways, namely imaging of the translated protein introduced by the reporter gene or imaging of enhancement spectroscopic signatures on post contrast images. One of the first instances was transferrin receptor-dependent accumulation used to image a xenografted tumor expressing an ectopically expressed transferrin receptor reporter gene. The second method for detecting gene expression *in vivo* relies on the incorporation of MRS-visible contrast compounds into a gene delivery system. Both of these MR-based approaches are being developed to monitor *in vivo* gene expression.

MRI has a huge advantage in providing high spatial resolution reporter gene expression image compared to the optional imaging. However, to obtain such high resolution image in diagnosis time, very high concentrations of substances with appropriate paramagnetic properties must be introduced into the target cells. The magnetic properties of tissues affect MR images, and differences in magnetic contrast can be used to enhance the differentiation between tissues. Contrast agents are used to shorten the relaxation parameter (T_1 and T_2) of water. Paramagnetic molecules such as gadolinium (Gd) could accumulate in the interstitial spaces and lead to an increase in a fast clearance. Derivatives of gadolinium contrast agents are unfavorable for imaging performances at the cellular level.[55] Conversely, magnetic iron-oxide imaging contrast agents allow easy entry into target cells and can be accumulated permanently in target tissue; it is generally more useful to image diagnosis. Superparamagnetic iron oxide (SPIO) particles consist of a crystalline iron oxide core and a shell of hydrophilic polymer has been developed for nucleic acid transfection, including the use of the gene delivery agent. Earlier research showed that it is possible to noninvasively characterize gene expression in tumors. Transferring receptor reporter gene was adopted to monitor endostatin gene expression using transferrin conjugate of ultra small supramagnetic iron oxide nanoparticle (Tf-USPIO) as magnetic resonance (MR) reporter probe. After intravenous injection of the Tf-USPIO conjugate, there was a more pronounced decrease in T_2 relaxation time in tumors treated with Tf-USPIO than in tumors treated with retrovirus vector group.[56]

A nanocarrier of polyethyleneglycol-grafted polyethylenimine and superparamagnetic iron oxide (PEG-g-PEI-SPION) was used for siRNA delivery in gastric cancer. At the appropriate N/P ratio of PEG-g-PEI-SPION/siRNAs, effective down-regulating of the CD44v6 expression of gastric carcinoma cell line SGC-7901 occurred *in vitro*. PEG-g-PEI-SPION reveals enhancement effect as a highly efficient contrast agent for *in vivo* MRI scanning.[57]

A recent report exhibited the progress of dual-purpose chitosan and polyethylenimine (PEI) coated magnetic micelles (CP-mag-micelles) composed of monodisperse hydrophobic superparamagnetic iron oxide nanoparticles (SPIONs) loaded into the cores of micelles that are self-assembled from a block copolymer of poly (D, L-lactide) (PLA) and monomethoxy polyethylene glycol (mPEG) (Fig.9.10). The T_2 relaxivity of mag-micelles was similar to that of CP-mag-micelles confirming that coating with cationic polymers did not impose negative influence on magnetism of the theranostic nanoparticles. The

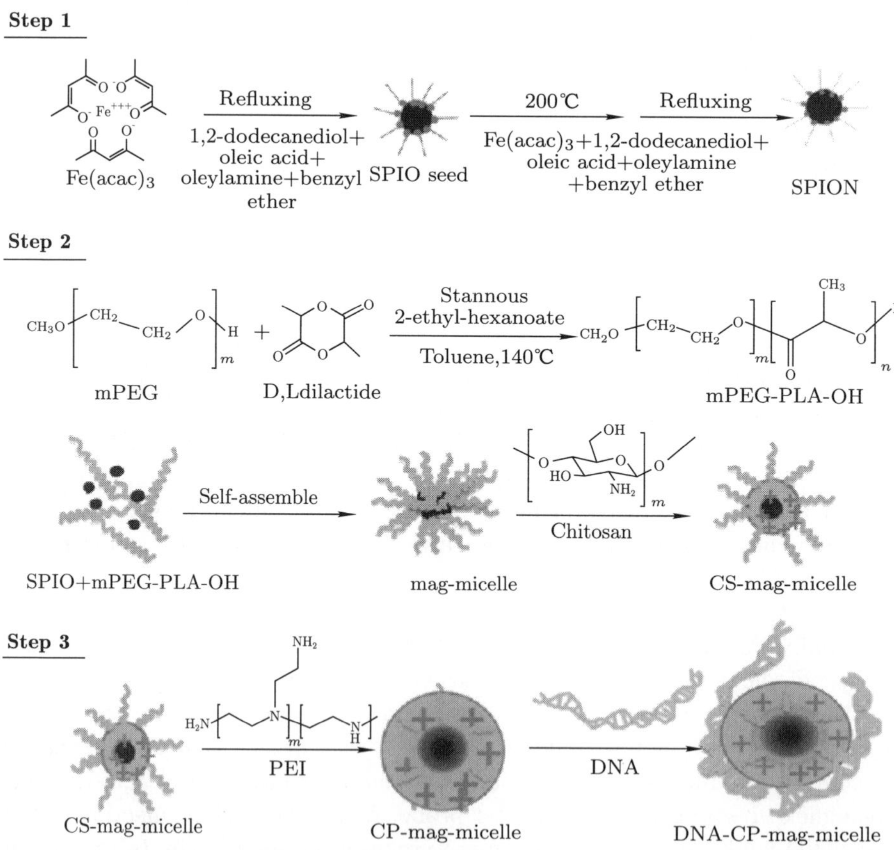

Fig. 9.10 Schematic illustration of the dual-purpose CP-mag-micelle synthesis. Reprinted from Ref. 52. Copyright (2012), with permission from Elsevier.

structure of SPIONs loaded in the core of an mPEG-PLA micelle coated with cationic polymers provides reinforcement on DNA delivery efficient and enhanced MRI potential. These results indicate that CP-mag-micelles are a promising carrier for gene delivery and for MRI diagnostic both *ex vivo* and *in vivo*.[52]

9.4.3 *Positron Emission Tomography (PET)*

PET is a tomographical imaging technique in which compounds labeled with positron emitting radioisotopes are used to detect biologically active molecules *in vivo*. The sensitivity of PET imaging is higher and the speed of imaging is relatively rapid (min) compared to MR imaging. PET imaging facilitated the development and testing of gene therapy trials with the capability by monitoring the biodistribution in body and quantitatively monitoring the magnitude and duration of gene expression.[48] PET imaging possessed the potential of the translation from small-animal to human imaging of human gene therapy.

Pharmacokinetics study of lipoplexes was studied using a novel and favorable positron emitter-labeled siRNA prepared by amino group-modification of N-succinimidyl 4-[fluorine-18] fluorobenzoate ([18F]SFB). Real-time analysis of siRNA trafficking was performed by using positron emission tomography (PET). Cationic liposome/[18F]-labeled siRNA lipoplexes and [^{18}F]-labeled siRNA control was injected to mice. The body distribution of the label was imaged by PET. The lipoplexes mediated a long circulation and tended to accumulate in the lungs while the naked siRNA suffered from cleared quite rapidly from the bloodstream and excreted by kidneys. A similar biodistribution of the cationic liposome/siRNA complexes was confirmed by the use of a near-infrared fluorescence imaging system. The results indicate that PET imaging of siRNA provided important information about the development of clinical siRNA therapeutic.[58]

9.5 Summary

This chapter does not exhaustively review the area of cationic nonviral gene delivery, but rather illustrates the target gene delivery modalities and combination of gene therapeutic and imaging. Novel synthetic methods are being developed that broaden the scope of gene delivery and enable the exploration of target gene therapeutic in malignant tumors. Probably the best recent example of a synthetic approach that had a major impact on a wide variety of scientific areas is the therapeutic method. The power of this technique is in the fact that it can obtain information about of gene delivery and preclinical and clinical image diagnose simultaneously, including optional imaging, MRI and PET. Additionally, therapeutic is useful for an extensive screening of gene carriers in preclinical processes. This target gene theranostic method has found broad application in gene therapy science. However, substantial progress in this field is yet to be achieved. In order to deepen the comprehension of gene delivery process mediated by cationic carrier, design and synthesis of the most influential target cationic gene delivery system that has been reported in recent years were summarized in this chapter. It is hoped that this could facilitate a better understanding of potential preclinical and clinical applications of human gene therapy.

References

1. Thomas, C. E.; Ehrhardt, A.; Kay, M. Progress and problems with the use of viral vectors for gene therapy. *Nat Rev Genet* **2003**, *4* (5), 346-358.
2. Gao, X.; Kim, K. S.; Liu, D. X. Nonviral gene delivery: What we know and what is next. *Aaps J* **2007**, *9* (1), E92-E104.
3. Davidson, B. L.; Paulson, H. L. Molecular medicine for the brain: Silencing of disease genes with RNA interference. *Lancet Neurol* **2004**, *3* (3), 145-149.
4. Ding, H. L.; Schwarz, D. S.; Keene, A.; Affar, E. B.; Fenton, L.; Xia, X. A.; Shi, Y.; Zamore, P. D.; Xu, Z. S. Selective silencing by RNAi of a dominant allele that causes amyotrophic lateral sclerosis. *Aging Cell* **2003**, *2* (4), 209-217.
5. Takeshita, F.; Ochiya, T. Therapeutic potential of RNA interference against cancer. *Cancer Sci* **2006**, *97* (8), 689-696.
6. Hadj-Slimane, R.; Lepelletier, Y.; Lopez, N.; Garbay, C.; Raynaud, F. Short interfering RNA (siRNA), a novel therapeutic tool acting on angiogenesis. *Biochimie* **2007**, *89* (10), 1234-1244.
7. Li, S. D.; Chen, Y. C.; Hackett, M. J.; Huang, L. Tumor-targeted delivery of siRNA by self-assembled nanoparticles. *Mol Ther* **2008**, *16* (1), 163-169.
8. Hafez, I. M.; Maurer, N.; Cullis, P. R. On the mechanism whereby cationic lipids promote intracellular delivery of polynucleic acids. *Gene Ther* **2001**, *8* (15), 1188-1196.
9. Felgner, P. L.; Gadek, T. R.; Holm, M.; Roman, R.; Chan, H. W.; Wenz, M.; Northrop, J. P.; Ringold, G. M.; Danielsen, M. Lipofection: A highly efficient, lipid-mediated DNA-transfection procedure. *P Natl Acad Sci USA* **1987**, *84* (21), 7413-7417.
10. de Lima, M. C. P.; Neves, S.; Filipe, A.; Duzgunes, N.; Simoes, S. Cationic liposomes for gene delivery: From biophysics to biological applications. *Curr Med Chem* **2003**, *10* (14), 1221-1231.
11. Song, L. Y.; Ahkong, Q. F.; Rong, Q.; Wang, Z.; Ansell, S.; Hope, M. J.; Mui, B. Characterization of the inhibitory effect of PEG-lipid conjugates on the intracellular delivery of plasmid and antisense DNA mediated by cationic lipid liposomes. *Bba-Biomembranes* **2002**, *1558* (1), 1-13.
12. Pack, D. W.; Hoffman, A. S.; Pun, S.; Stayton, P. S. Design and development of polymers for gene delivery. *Nat Rev Drug Discov* **2005**, *4* (7), 581-593.
13. Akinc, A.; Thomas, M.; Klibanov, A. M.; Langer, R. Exploring polyethylenimine-mediated DNA transfection and the proton sponge hypothesis. *J Gene Med* **2005**, *7* (5), 657-663.

14. Zou, S. M.; Erbacher, P.; Remy, J. S.; Behr, J. P. Systemic linear polyethylenimine (L-PEI)-mediated gene delivery in the mouse. *J Gene Med* **2000**, *2* (2), 128-134.

15. Russ, V.; Elfberg, H.; Thoma, C.; Kloeckner, J.; Ogris, M.; Wagner, E. Novel degradable oligoethylenimine acrylate ester-based pseudodendrimers for *in vitro* and *in vivo* gene transfer. *Gene Ther* **2008**, *15* (1), 18-29.

16. Sonawane, N. D.; Szoka, F. C.; Verkman, A. S. Chloride accumulation and swelling in endosomes enhances DNA transfer by polyamine-DNA polyplexes. *J Biol Chem* **2003**, *278* (45), 44826-44831.

17. Ahn, C. H.; Chae, S. Y.; Bae, Y. H.; Kim, S. W. Synthesis of biodegradable multi-block copolymers of poly(L-lysine) and poly(ethylene glycol) as a non-viral gene carrier. *J Control Release* **2004**, *97* (3), 567-574.

18. van de Wetering, P.; Cherng, J. Y.; Talsma, H.; Crommelin, D. J. A.; Hennink, W. E. 2-(dimethylamino)ethyl methacrylate based (co)polymers as gene transfer agents. *J Control Release* **1998**, *53* (1-3), 145-153.

19. Dubruel, P.; Christiaens, B.; Vanloo, B.; Bracke, K.; Rosseneu, M.; Vandekerckhove, J.; Schacht, E. Physicochemical and biological evaluation of cationic polymethacrylates as vectors for gene delivery. *Eur J Pharm Sci* **2003**, *18* (3-4), 211-220.

20. Funhoff, A. M.; van Nostrum, C. F.; Lok, M. C.; Fretz, M. M.; Crommelin, D. J.; Hennink, W. E. Poly(3-guanidinopropyl methacrylate): A novel cationic polymer for gene delivery. *Bioconjug Chem* **2004**, *15* (6), 1212-1220.

21. Klotzbach, T.; Watt, M.; Ansari, Y.; Minteer, S. D. Effects of hydrophobic modification of chitosan and Nafion on transport properties, ion-exchange capacities, and enzyme immobilization. *J Membrane Sci* **2006**, *282* (1-2), 276-283.

22. Fernandez-Megia, E.; Novoa-Carballal, R.; Quinoa, E.; Riguera, R. Conjugation of bioactive ligands to PEG-grafted chitosan at the distal end of PEG. *Biomacromolecules* **2007**, *8* (3), 833-842.

23. Yang, X. Z.; Du, J. Z.; Dou, S.; Mao, C. Q.; Long, H. Y.; Wang, J. Sheddable ternary nanoparticles for tumor acidity-targeted siRNA delivery. *ACS Nano* **2012**, *6* (1), 771-781.

24. Knorr, V.; Ogris, M.; Wagner, E. An acid sensitive ketal-based polyethylene glycol-oligoethylenimine copolymer mediates improved transfection efficiency at reduced toxicity. *Pharm Res-Dord* **2008**, *25* (12), 2937-2945.

25. Shim, M. S.; Kwon, Y. J. Acid-responsive linear polyethylenimine for efficient, specific, and biocompatible siRNA delivery. *Bioconjugate Chem* **2009**, *20* (3), 488-499.

26. Cullis, P. R.; de Kruijff, B. Lipid polymorphism and the functional roles of lipids in biological membranes. *Biochim Biophys Acta* **1979**, *559* (4), 399-420.

27. Zelphati, O.; Szoka, F. C. Jr. Mechanism of oligonucleotide release from cationic liposomes. *Proc Natl Acad Sci USA* **1996**, *93* (21), 11493-11498.

28. Murthy, N.; Campbell, J.; Fausto, N.; Hoffman, A. S.; Stayton, P. S. Design and synthesis of pH-responsive polymeric carriers that target uptake and enhance the intracellular delivery of oligonucleotides. *J Control Release* **2003**, *89* (3), 365-374.

29. Kwon, Y. J. Before and after endosomal escape: Roles of stimuli-converting siRNA/polymer interactions in determining gene silencing efficiency. *Accounts Chem Res* **2012**, *45* (7), 1077-1088.

30. Yhee, J. Y.; Koo, H.; Lee, D. E.; Choi, K.; Kwon, I. C.; K, K. Multifunctional chitosan nanoparticles for tumor imaging and therapy. *Adv Polym Sci* **2011**, *243*, 139-162.

31. Kim, T. H.; Jiang, H. L.; Nah, J. W.; Cho, M. H.; Akaike, T.; Cho, C. S. Receptor-mediated gene delivery using chemically modified chitosan. *Biomed Mater* **2007**, *2* (3), S95-S100.

32. Cristiano, R. J.; Roth, J. A. Molecular conjugates: A targeted gene delivery vector for molecular medicine. *J Mol Med-Jmm* **1995**, *73* (10), 479-486.

33. Du, F. S.; Wang, Y.; Zhang, R.; Li, Z. C. Intelligent nucleic acid delivery systems based on stimuli-responsive polymers. *Soft Matter* **2010**, *6* (5), 835-848.

34. McKenzie, D. L.; Kwok, K. Y.; Rice, K. G. A potent new class of reductively activated peptide gene delivery agents. *J Biol Chem* **2000**, *275* (14), 9970-9977.

35. Kim, S. H.; Jeong, J. H.; Lee, S. H.; Kim, S. W.; Park, T. G. Local and systemic delivery of VEGF siRNA using polyelectrolyte complex micelles for effective treatment of cancer. *J Control Release* **2008**, *129* (2), 107-116.

36. Kim, S. H.; Jeong, J. H.; Kim, T. I.; Kim, S. W.; Bull, D. A. VEGF siRNA delivery system using arginine-grafted bioreducible poly(disulfide amine). *Mol Pharmaceut* **2009**, *6* (3), 718-726.

37. Kim, S. H.; Jeong, J. H.; Lee, S. H.; Kim, S. W.; Park, T. G. LHRH receptor-mediated delivery of siRNA using polyelectrolyte complex micelles self-assembled from siRNA-PEG-LHRH conjugate and PEI. *Bioconjugate Chem* **2008**, *19* (11), 2156-2162.

38. Hoyer, J.; Neundorf, I. Peptide vectors for the nonviral delivery of nucleic acids. *Accounts Chem Res* **2012**, *45* (7), 1048-1056.

39. Levine, R. M.; Scott, C. M.; Kokkoli, E. Peptide functionalized nanoparticles for nonviral gene delivery. *Soft Matter* **2013**, *9* (4), 985-1004.

40. Wagner, E. Polymers for siRNA delivery: Inspired by viruses to be targeted, dynamic, and precise. *Accounts Chem Res* **2012**, *45* (7), 1005-1013.

41. Mickler, F. M.; Vachutinsky, Y.; Oba, M.; Miyata, K.; Nishiyama, N.; Kataoka, K.; Brauchle, C.; Ruthardt, N. Effect of integrin targeting and PEG shielding on polyplex micelle internalization studied by live-cell imaging. *J Control Release* **2011**, *156* (3), 364-373.

42. Zhang, X. X.; McIntosh, T. J.; Grinstaff, M. W. Functional lipids and lipoplexes for improved gene delivery. *Biochimie* **2012**, *94* (1), 42-58.

43. Prata, C. A.; Zhao, Y.; Barthelemy, P.; Li, Y.; Luo, D.; McIntosh, T. J.; Lee, S. J.; Grinstaff, M. W. Charge-reversal amphiphiles for gene delivery. *J Am Chem Soc* **2004**, *126* (39), 12196-12197.

44. Meyer, M.; Wagner, E. pH-responsive shielding of non-viral gene vectors. *Expert Opin Drug Deliv* **2006**, *3* (5), 563-571.

45. Kolli, S.; Wong, S. P.; Harbottle, R.; Johnston, B.; Thanou, M.; Miller, A. D. pH-triggered nanoparticle mediated delivery of siRNA to liver cells *in vitro* and *in vivo*. *Bioconjug Chem* **2013**, *24* (3), 314-332.

46. Chung, H. J.; Hong, C. A.; Lee, S. H.; Jo, S. D.; Park, T. G. Reducible siRNA dimeric conjugates for efficient cellular uptake and gene silencing. *Bioconjugate Chem* **2011**, *22*, 299-306.
47. Bell, J. D.; Taylor-Robinson, S. D. Assessing gene expression *in vivo*: Magnetic resonance imaging and spectroscopy. *Gene Ther* **2000**, *7*, 1259-1264.
48. Kim, E. M.; Jeong, H. J.; Park, I. K.; Cho, C. S.; Bom, H. S.; Kim, C. G. Monitoring the effect of PEGylation on polyethylenimine *in vivo* using nuclear imaging technique. *Nuclear Medicine and Biology* **2004**, *31* (6), 781-784.
49. Hauck, E. S.; Zou, S.; Scarfo, K.; Nantz, M. H.; Hecker, J. G. Whole animal *in vivo* imaging after transient, nonviral gene delivery to the rat central nervous system. *Molecular Therapy* **2008**, *11*, 1857-1864.
50. Mizuno, T.; Mohri, K.; Nasu, S.; Danjo, K.; Okamoto, H. Dual imaging of pulmonary delivery and gene expression of dry powder inhalant by fluorescence and bioluminescence. *J Control Release* **2009**, *134* (2), 149-154.
51. Koo, H.; Huh, M. S.; Sun, I. C.; Yuk, S. H.; Choi, K.; Kim, K.; Kwon, I. C. *In vivo* targeted delivery of nanoparticles for theranosis. *Accounts Chem Res* **2011**, *44*, 1018-1028.
52. Wang, C.; Ravi, S.; Martinez, G. V.; Chinnasamy, V.; Raulji, P.; Howell, M.; Davis, Y.; Mallela, J.; Seehra, M. S.; Mohapatra, S. Dual-purpose magnetic micelles for MRI and gene delivery. *J Control Release* **2012**, *163*, 82-92.
53. Wan, Q.; Xie, L.; Gao, L.; Wang, Z.; Nan, X.; Lei, H.; Long, X.; Chen, Z. Y.; He, C. Y.; Liu, G.; Liu, X.; Qiu, B. Self-assembled magnetic theranostic nanoparticles for highly sensitive MRI of minicircle DNA delivery. *Nanoscale* **2013**, *5*, 744-752.
54. Robinson, J. T.; Tabakman, S. M.; Liang, Y. Y.; Wang, H. L.; Casalongue, H. S.; Vinh, D.; Dai, H. J. Ultrasmall reduced graphene oxide with high near-infrared absorbance for photothermal therapy. *J Am Chem Soc* **2011**, *133* (17), 6825-6831.
55. Shi, S.; Yang, K.; Hong, H.; Valdovinos, H. F.; Nayak, T. R.; Zhang, Y.; Theuer, C. P.; Barnhart, T. E.; Liu, Z.; Cai, W. Tumor vasculature targeting and imaging in living mice with reduced graphene oxide. *Biomaterials* **2013**, *34* (12), 3002-3009.
56. Wang, K.; Wang, K.; Shen, B.; Huang, T.; Sun, X.; Li, W.; Jin, G.; Li, L.; Bu, L.; Li, R.; Wang, D.; Chen, X. MR reporter gene imaging of endostatin expression and therapy. *Mol Imaging Biol* **2010**, *12*, 520-529.
57. Chen, Y.; Lian, G.; Liao, C.; Wang, W.; Zeng, L.; Qian, C.; Huang, K.; Shuai, X. Characterization of polyethylene glycol-grafted polyethylenimine and superparamagnetic iron oxide nanoparticles (PEG-g-PEI-SPION) as an MRI-visible vector for siRNA delivery in gastric cancer *in vitro* and *in vivo*. *J Gastroenterol* **2013**, *48* (7), 809-821.
58. Hatanaka, K.; Asai, T.; Koide, H.; Kenjo, E.; Tsuzuku, T.; Harada, N.; Tsukada, H.; Oku, N. Development of double-stranded siRNA labeling method using positron emitter and its *in vivo* trafficking analyzed by positron emission tomography. *Bioconjugate Chem* **2010**, *21*, 756-763.

Chapter 10

Microbubbles for Tumor Targeting Theranostics

*Daming Yong, Xuejing Wang, Lei Wang, and Xiaojun Han**
State Key Laboratory of Urban Water Resource and Environment, School of Chemical Engineering and Technology, Harbin Institute of Technology, Harbin 150001, China

Ultrasonography has been used for over 50 years as a diagnostic tool for the evaluation of tumor growth and dissemination in the body. Compared with other diagnostic methods, ultrasound is a safe, simple, noninvasive and low cost imaging technique, although its resolution is not comparable to that of magnetic resonance imaging. Contrast-enhanced ultrasound employing microbubbles is now widely used to diagnose a number of diseases in the clinic due to its better resolution, especially for tumor targeting theranostics. Ultrasound contrast agents are not only effective in ultrasonic imaging but also as important tools for drug or gene delivery by disrupting microbubbles locally using ultrasound beams. In the last decade, microbubbles have been widely used as ultrasound contrast agents, drug delivery systems and gene transfection tools for tumor targeting theranostics. In this chapter, we summarize the fabrication methods for microbubbles, their applications in tumor diagnostic, drug delivery and therapeutic fields and predict the future directions of tumor theranostics using microbubbles.

10.1 Introduction

The structure of typical microbubbles (MBs) is shown in Fig.10.1. Microbubbles are gas-filled particles with a size ranging from 0.5–10 μm, which consist of a gas core surrounded by a stabilizing shell. The gas core occupies most of the total volume of particle with a shell against the surrounding aqueous medium. The gases, such as air[1], nitrogen[2], rare gas[3], perfluorocarbon[4], are used as the gas core. Among these commonly used gases, perfluorocarbon is usually known as the best choice because of its high backscatter intensity resulting from strong hydrophobicity. Many different materials have been used as the shell, including lipids[5], proteins[6] and polymers[7], *etc.* Microbubbles are generally fabricated by sonication, mechanical agitation, or microfluidic devices.

In vivo noninvasive tumor imaging has significantly contributed to cancer preclinical research and clinical applications. There are some significant advances in noninvasive medical imaging during last decades. Ultrasound is one

* Corresponding author: hanxiaojun@hit.edu.cn

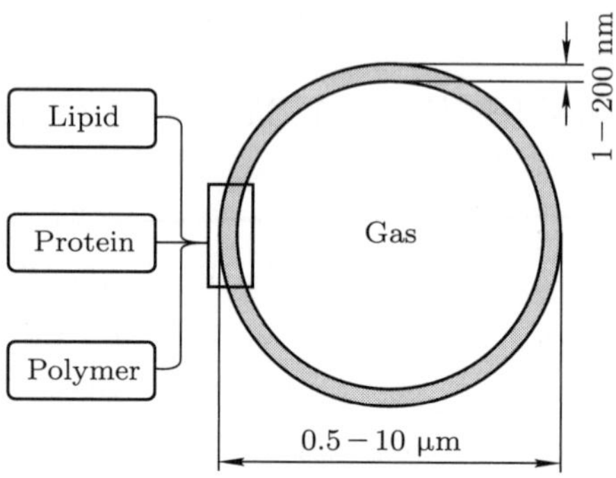

Fig. 10.1 The structure of typical MBs with different shell compositions.

of the most frequently used tomographic clinical imaging modalities, due to its advantages of low cost, real-time and no radiation. Similar to most other methods for medical imaging, ultrasound has undergone revolutionary changes in the improvement in spatial resolution. It provides three-dimensional information, and permits evaluation of dynamic physiologic processes such as blood flow or tissue motion. In terms of cancer imaging, ultrasound is usually used for many routine screening examinations, which includes characterizing liver lesions, investigating the urogenital tract, and detecting non symptomatic cancer in prostate, uterus, ovary, bladder and kidney. It is also an excellent method to characterize the shape and inner architecture of lymph nodes. With the development of Doppler imaging, tissue mechanical properties with ultrasound elastography[8–11] and blood flow velocity can be measured with Doppler[12–14] to assess further tissue property and therapy response. However, the Doppler is not sensitive enough at clinically used frequencies to characterize very small blood vessels with minimal flow velocity.[14,15] Therefore, contrast agents are required to enhance the sensitivity.

In 1968, Gramiak and Shah first demonstrated that agitated saline containing air bubbles could improve aortic root and cardiac chamber delineation after intracardiac injection.[16] The first generation microbubbles (Echovist®) were simple air microspheres, lacking a protecting shell, which survived only for a few seconds following intravenous injection. In 1985, Detrano *et al.*[17] and Vitarelli *et al.* [18] developed smaller microbubbles which allowed a transpulmonary passage from right to left heart and systemic circulation to be possible. The microbubbles (Albunex®, Levovist®) still suffered from instable air gas cores because the air would disappear less than 5 minutes.[19] Second generation microbubbles (Optison®, SonoVue®) were stabilized with shells and filled with low diffusive gases to prolong MB circulation half-life to more than 15 min. Clinical ultrasound systems use those stable MBs to opacify blood filled cavities and to visualize tissue perfusion. Apart from the development of contrast agents, the ultrasound technology and signal processing also improved dramatically during the last decades. Harmonic imaging is enabled by the acoustic echo waves including higher and lower frequencies than the main pulse frequency due to the nonlinearity in wave propagation and scattering.[20–22] Although the "harmonic ultrasound imaging" technique allowed better visualization of the soft and deeper tissues, the quality of the images was still highly user and objective dependent. This drawback promoted the development of 3D-scan modes.[23] The first device with 3D-scan mode was used for clinical breast imaging by controlling the ultrasound transducer *via* a precise servomotor.[24] The hardware development significantly contributed to the clinical acceptance of ultrasound imaging.

On the other hand, ultrasound contrast agents are also important tools for drug or gene delivery. More recently, microbubbles have been widely used as a promising carrier for targeted drug and gene delivery for tumor targeting theranostics. The major issue of systemic chemotherapeutic drug application is off-target side-effects. Thus, it is desirable to have better organ and tissue specific chemotherapeutic strategies. More recently, cancer gene therapy is a promising way to treat a variety of malignancies by inducing tumor cells to generate gene products for killing cancer cells without damaging healthy tissues. However, cancer gene therapy has met with bottleneck of the difficulty of safe and effective delivery of DNA into the cytoplasm of tumor cells. To solve these problems, ultrasound targeted microbubble destruction (UTMD) was introduced to drug delivery field. UTMD is a promising method with possibly higher gene and drug transfer efficiency as well as high organ and tissue-specificity.[25] Microbubbles oscillate at their resonance frequency. High ultrasound pressure and high amplitude oscillation result in microbubble destruction, which increases the cell membrane permeability around ruptured microbubbles. The underlying mechanisms involve several phenomena including the formation of micropores in cell membranes due to cavitation and formation of micro-jets by high amplitude oscillations.[26–28] The major advantage of ultrasound-assisted drug and gene delivery is its possibility to release drug in a controlled manner by the energy field of the ultrasound beam.

In the contents below, we will first briefly review the preparation of microbubbles, and then discuss the pertinent progress in microbubble application in tumor targeting theranostics in three aspects, *i.e.*, molecular ultrasound imaging, drug delivery and gene delivery.

10.2 The Preparation of Microbubbles

10.2.1 *Shell Materials*

The shell outside the bubbles protects them from fusing together, and also improves their stability during storage and *in vivo*. The commonly used shell materials are lipids, biocompatible polymers and proteins.

Currently, the most popular microbubble for a blood pool contrast agent is surrounded by a lipid monolayer. Lipid-stabilized microbubbles are easy to manufacture, biocompatible, and echogenic. Lipid monolayer protected microbubbles is usually fabricated as described below. Lipids are mixed with the water-soluble dry reagents, such as salt, carbohydrate or ethyleneglycol and sealed in a vial containing fluorinated gas atmosphere. Water is added to the vial immediately prior to use. After shaking lipid shell coated microbubbles are generated in the form of aqueous dispersion.

Protein-protected microbubbles are mostly made of albumin by sonication. During sonication, microbubbles of air are formed with a shell of albumin. Suslick *et al.*[6] produced the air-filled proteinaceous microbubbles by inter-protein crosslinking of cysteine residues. The principal cross-linking agent is superoxide created by the extremely high temperatures during acoustic cavitation. Some albumin stabilized microbubble formulations include: Albunex® (Molecular Biosystems, San Diego, CA, USA; and Mallinckrodt, St. Louis, MO, USA), QuantisonTM (Quadrant Ltd., Nottingham, UK) and Optison® (GE Healthcare). Porter *et al.*[4,29,30] further developed formulations of albumin coated microbubbles containing perfluorobutane-exposed sonicated dextrose albumin (PESDA). A mixture of avidin and albumin has also been used as the shell of microbubbles.[31] The incorporation of avidin in the microbubble shell allowed for biotin-mediated coupling of antibodies for the purpose of site selective imaging. Other proteins have been used to stabilize microbubbles due to their amphipathic nature. An example is to use lysozyme for the production of microbubbles. Cavalieri *et al.*[32] found those microbubbles were very stable, and the lysozyme maintained its enzymatic activity over several months. The composite structure of the microbubble shell has been observed by confocal microscopy (Fig.10.2), where small protein particles on the surface were visualized.

Polymer-shelled microbubbles are usually fabricated by a variety of emulsion methods, *e.g.* double-emulsion technique.[33] An aqueous phase is dispersed in organic solvent containing polymers to form water-in-oil emulsions, which were then dispersed in the excess of water phase to generate double emulsion. Organic solvent is then evaporated to form the polymer shell coating the aqueous core of the initial dispersion. After water is removed by lyophilization or spray drying, fluorinated gas or air can be introduced in the internal cavity inside the polymer shell. This method applies to the water-insoluble polymers. Single emulsion was also used to prepare microbubbles. Oil-in-water emulsions containing a polymer in a very volatile organic solvent and a moderately volatile organic nonsolvent[34,35] were fabricated first. When the volatile component evaporates, the polymer is forced to form the shell around the nonsolvent core. Further evaporation process, such as lyophilization, is subse-

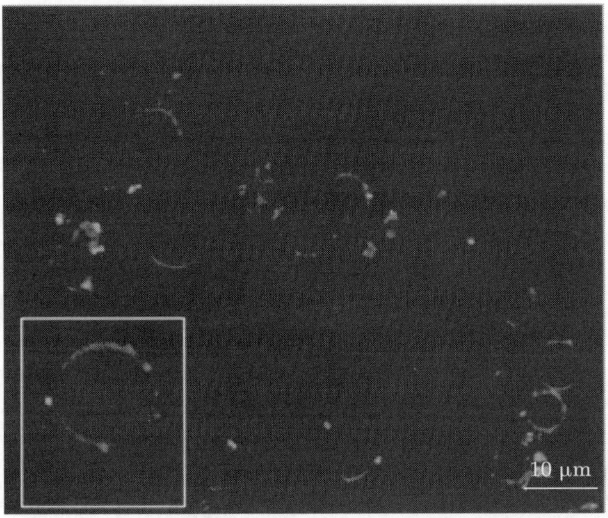

Fig. 10.2 Confocal laser scanning microscopy (CLSM) image of lysozyme microbubbles. The inset shows clusters of proteins adsorbed on the surface of microbubbles. Reprinted from Ref. 32. Copyright (2008) with permission from American Chemical Society.

quently applied to remove the nonsolvent to leave the empty internal core for refilling gas. An alternative technique to prepare polymer-based shells is to use layer-by-layer (LBL) deposition technique. An existing electrostatically charged bubble shell is used as a template for sequential deposition of positively and negatively charged water-soluble polymers on the surface.[36–38] Olbrich *et al.*[39] developed two methods using polyelectrolyte complexes to produce the air-filled PBCA microparticles (AF-PBCA-MP), and the microscopic characterizations of AF-PBCA-MP were shown in Fig.10.3.

10.2.2 *Microbubble Fabrication*

Gas-filled and polymer/lipid/protein coated microbubbles have been well established in the last 20–30 years. Over the years a number of microbubble reagents have been commercialized as diagnostic aids. They can be fabricated by various methods, such as sonication, high shear emulsification, ink jet printing, electrohydrodynamic atomization and microfluidic processing.

The most commonly used method for microbubble preparation is sonication, which involves dispersing gas or liquid in a suspension of a desirable coating material using high intensity ultrasound.[40–42] The gas or liquid is emulsified to form a suspension of micro-droplets/bubbles coated with proteins or surfactants which improves their stability. Although this method is very simple and cheap, it produces too large microbubbles which are not suitable for intravenous injection.

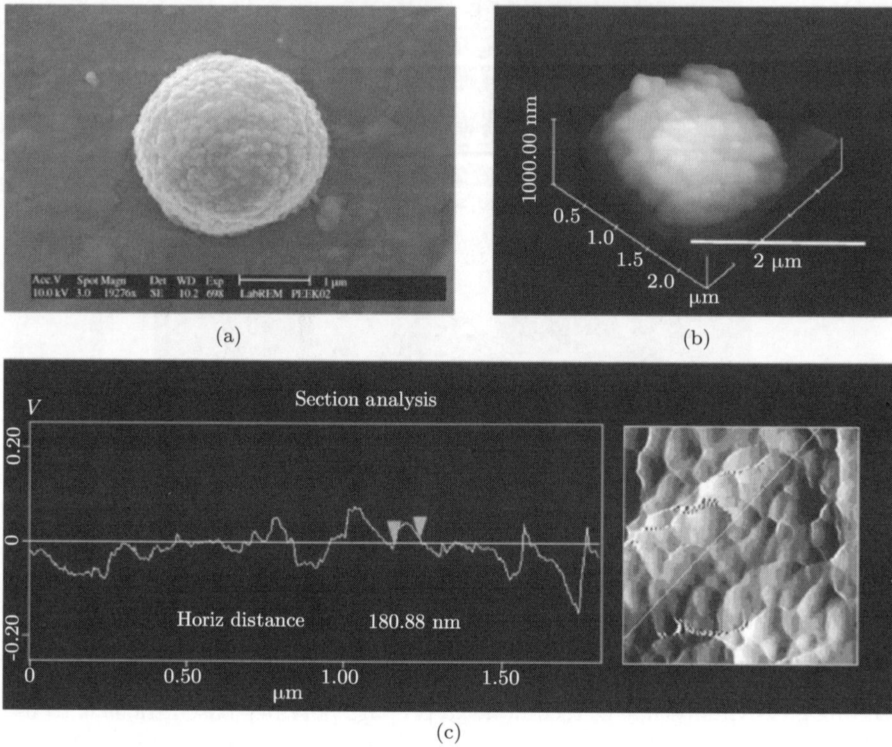

Fig. 10.3 (a) ESEM image of a superficially hydrolyzed one-step AF-PBCA-MP. (b) AFM image of a superficially hydrolyzed one-step AF-PBCA-MP. (c) AFM image of the surface of a superficially hydrolyzed one-step AF-PBCA-MP with the determination of the size of the nanoparticulate structure of the polymer shell, size of the nanoparticle with two arrows: 180.88 nm. Reprinted from Ref. 39. Copyright (2006) with permission from Elsevier.

A second commonly used method, particularly for preparing polymer coated microbubbles, is to emulsify the polymer in an aqueous suspension by high shear stirring.[43,44] If the polymer solvent is sufficiently volatile, it will begin to evaporate, leading to the polymer to precipitate onto the surface of the droplets to form coated liquid filled microspheres. The microspheres are washed to remove excess solvent and then freeze dried to produce gas filled shells.

The third technique employed to improve microbubble uniformity is ink jet printing.[45] The bubbles are prepared from a piezoelectric crystal embedded chamber containing a stainless steel nozzle. By varying the voltage across the crystal, pressure pulses are generated in the liquid. With each pulse a droplet is forced out of the nozzle to form a microbubble.

Another technique which has recently been applied to microbubble preparation is coaxial electrohydrodynamic atomization (CEHDA).[46,47] In CEHDA, a coaxial jet of two fluids is formed and then atomised to form uniform droplets. When the fluids are immiscible, it is possible to encapsulate one fluid inside another. Farook *et al.*[46] described the scheme of microbubble preparation by CEHDA. As shown in Fig.10.4, two needles were connected to the same applied voltage relative to an earthed ring electrode placed 12 mm below the outer needle. The inner needle is placed ~2 mm above the tip of the outer needle. The outer and inner diameters of the ring electrodes were 20 and 15 mm respectively. The air and the glycerol flow rates were controlled by high precision Harvard syringe pumps using plastic syringes. Liquid and air flowing out the needles were monitored on a screen. The bubbles were collected in glycerol filled glass container below the ring electrode, and a cloud of microbubbles could be seen floating in glycerol after bubble formation.

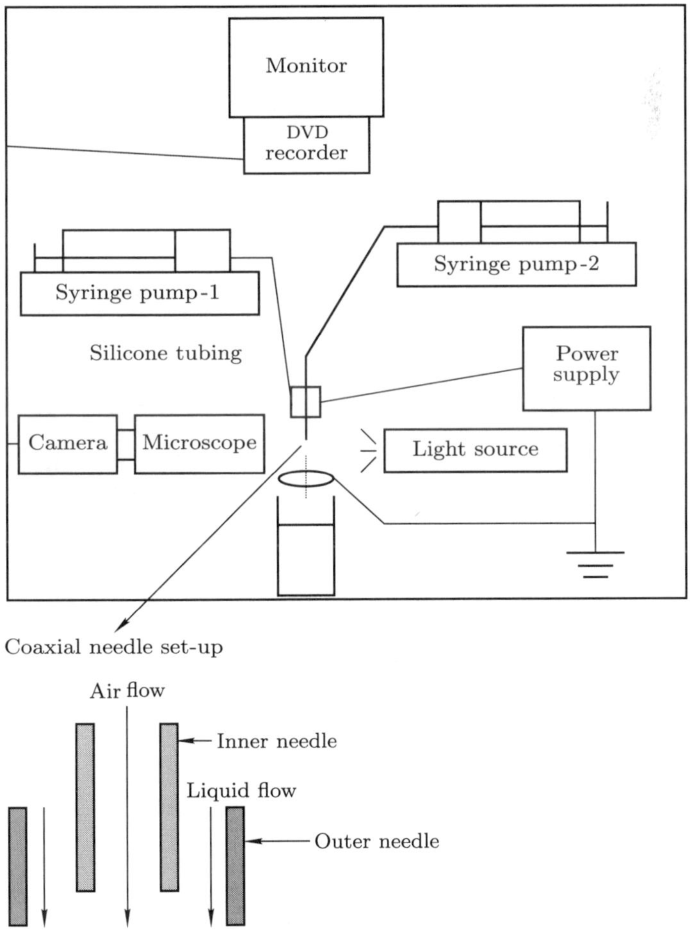

Fig. 10.4 Equipment rig and needle arrangement used for co-axialelectro hydrodynamic atomization. Reprinted from Ref. 46. Copyright (2007) with permission from Elsevier.

Microbubbles can also be fabricated with a high precision over their size and dispersity using microfluidic devices.[48] However, unlike CEHDA apparatus, microfluidic devices only offer operational control under very limited pressure and flow rate conditions. These methods are well established for the preparation of monodisperse liquid droplets as well as microbubble suspensions.[49,50] To date there are two main types of microfluidic device which have been used for bubble preparation: flow focusing device produced using soft lithography techniques[49] and mechanically assembled setup consisting of capillaries in a polymeric block arranged to form a T-junction.[50] In both cases, the essential feature of the device is an orifice where a column of gas impinges upon a liquid flow (top left image in Fig.10.5). Subsequently, at a certain distance from the orifice, the gas-liquid interface becomes unstable and bubbles are formed by a "pinch-off" process (bottom left image in Fig.10.5).[51] The microbubbles are monodispersed, as seen from bottom right image in Fig.10.5.

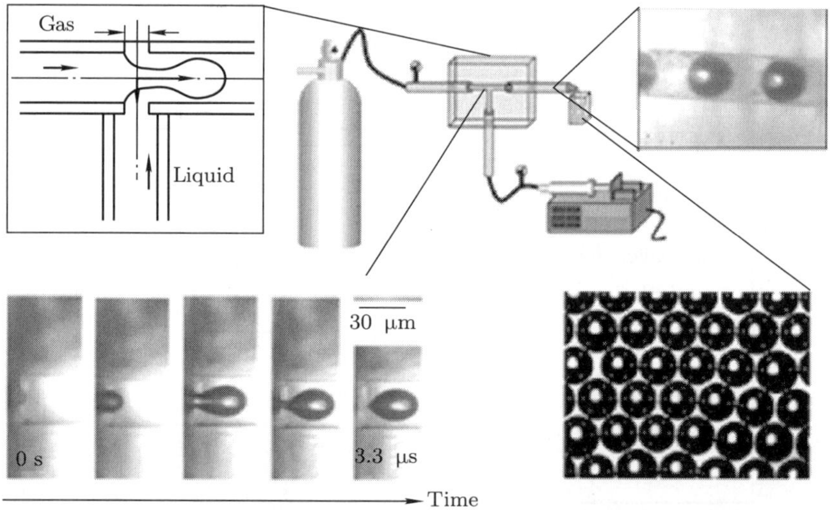

Fig. 10.5 Microbubble preparation by a microfluidic T-junction device. High speed camera frames of microbubble formation (bottom left image) and the monodisperse suspension formed (bottom right image). Reprinted from Ref. 51. Copyright (2008) with permission from American Chemical Society.

10.2.3 *Methods of Loading the Therapeutic Agent*

The therapeutic entity, such as drug and gene, can be loaded by several methods including encapsulating them into the microbubble, attaching them to the microbubble shell through a covalent bond, or binding them to the inner layer of the oil phase.[52] It is relatively simple to bind the therapeutic agents directly to the surface of the shell, but the drawback is that the drug can easily be removed in the serum. Conversely, therapeutic agents encapsulated into the microbubble are stable, but the fabricating process is too complex to be controlled.[52] Figure

10.6 shows various physicochemical concepts to load a bioactive substance on different microbubbles.[53] The more common loading methods currently used are to bind the drugs to the inner layers of microbubbles.[54] The microbubble has great potential to carry the drugs and release them at target area.

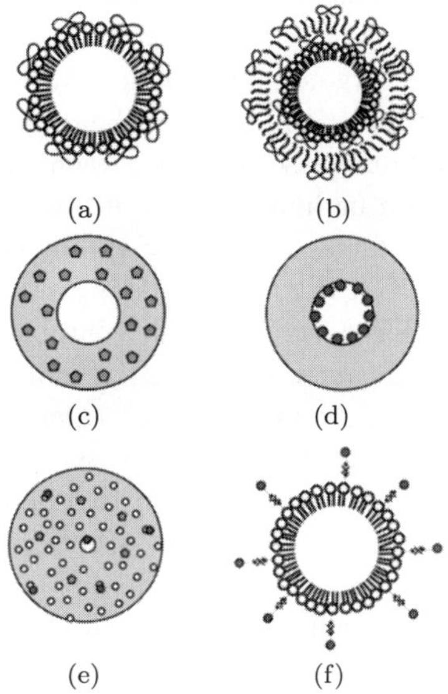

Fig. 10.6 Schematic representation of loading strategies of drugs and genes on microbubbles. (a) Noncovalently binding of DNA to the surface of cationic lipid microbubbles. (b) Multilayered structure based on a lipid microbubble sequentially coated with DNA and poly-L-lysine layers. Polymeric microbubbles with central hollow core surrounded by a thick polymeric shell. (c) Polymeric microbubble loaded with hydrophobic drug loaded in the shell phase. (d) Polymeric microbubble with hydrophilic drug loaded in the internal void. (e) Internal structure of polymeric microbubbles: water-phase is dispersed through the polymer matrix, forming upon lyophilization a plurality of cavities distributed over the particle volume. (f) Attachment of liposomes or nanoparticles to the surface of microbubbles through biotin-avidin-biotin bridging system. The drawing is not to scale. Reprinted from Ref. 53. Copyright (2008) with permission from Elsevier.

10.3 Application of Microbubbles for Tumor Targeting Theranostics

10.3.1 *Microbubble-based Molecular Ultrasound Imaging of Tumor*

Angiogenesis is one of the hallmarks of cancer which is caused by the sprouting of new vessels from pre-existing ones.[55] Some of the well known pro-angiogenic factors secreted by the tumor cells include the vascular endothelial growth factor (VEGF), basic fibroblast growth factor (bFGF), platelet derived growth factor (PDGF) and the transforming growth factor-β_1 (TGF-β_1). These pro-angiogenic factors are the indicators for an advanced tumor stage and a poor prognosis in various human cancers.[56,57] Thus these factors are considered to be targets for ligand modified MBs, consequently for detecting and characterizing tumor tissues.

Molecular imaging with targeted ultrasound contrast agents is illustrated in Fig.10.7. Bubbles are modified with particular ligand for specific receptor. During the circulation, these bubbles selectively attach to the receptor. Therefore the receptor/disease site can be imaged using ultrasound technique. The most popular molecular markers for assessment of tumor angiogenesis and therapy effects by ultrasound are VEGFR-2[58–60] and $\alpha_v\beta_3$ integrin. Rychak *et al.*[58] and Willmann *et al.*[59] have both reported that microbubbles modified with antibodies against VEGFR-2 significantly accumulated more than unspecific control microbubbles within subcutaneously implanted tumors. Willmann *et al.*[59] evaluated contrast agent-enhanced ultrasonography with microbubbles targeted to VEGFR-2 for imaging tumor angiogenesis in two murine tumor models. It is noted that VEGF-carrying microbubbles gives better contrast. Delineation of tumor vasculature can be achieved within 10 minutes after intravenous (IV) administration of the VEGF-carrying bubbles, while the accumulation of control (VEGF-free) bubbles to tumor are low. Further, Willmann *et al.*[59] and Palmowski *et al.*[61] found good correlationship between the molecular ultrasound signal from VEGFR-2 targeted microbubbles and the VEGFR-2 expression assessed by immunochemistry.

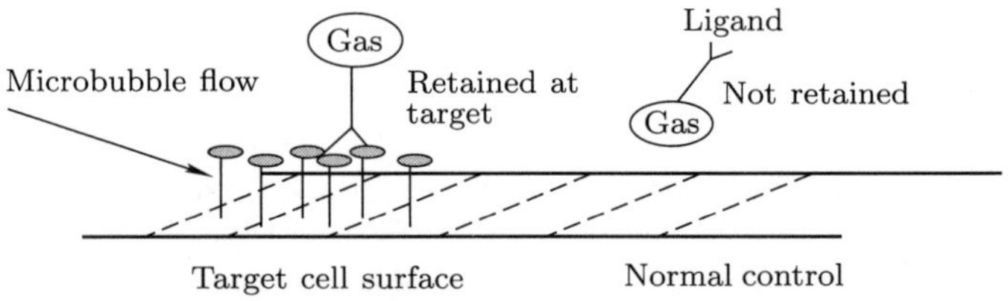

Fig. 10.7 Schematic representation of molecular imaging with targeted ultrasound contrast agents.

The $\alpha_v\beta_3$ integrin is another target for molecular imaging, which is overexpressed on activated tumor endothelium.[62] Therefore it has been targeted by microbubbles modified with cyclic RGD peptides or specific antibodies in many studies with different tumor models.[63,64] The idea of using RGD peptides to target $\alpha_v\beta_3$ is inspired by the fact that integrin $\alpha_v\beta_3$ interacts with RGD-containing proteins, such as vitronectin, fibronectin, and lamin. Dumy *et al.*[65] reported the modular synthesis and biological characterization of template assembled cyclopeptides as a multimeric system for targeting and endocytosis of cells expressing $\alpha_v\beta_3$ integrin (Fig.10.8). The alternative ligand for $\alpha_v\beta_3$ integrin is echistatin, a viper venom disintegrin containing an RGD sequence. Echistatin-modified microbubbles can be used to image activated endotheliumin subcutaneously implanted matrigel plugs enriched with fibroblast growth factor-2 (FGF-2).[66] Ellegala *et al.* also used Echistatin to image $\alpha_v\beta_3$ expression in human glioma cells implanted intracerebrally in rats.[67] Microbubble retention within tumor microvessels was as long as 28 days for $\alpha_v\beta_3$-targeted but not for control microbubbles. Contrast ultrasound imaging performed 10 minutes after injection of $\alpha_v\beta_3$-targeted microbubbles indicated intense signal enhancement within the tumors and lower intensity enhancement in surrounding nonneoplastic brain parenchyma. Willmann *et al.*[68] engineered a small disulfide-constrained cystine knottin peptide for targeting $\alpha_v\beta_3$ integrin. They showed higher accumulation of specific knotting peptide modified microbubbles around $\alpha_v\beta_3$ integrin positive mouse angiosarcoma endothelial cells than that of control microbubbles. Furthermore, higher signal intensity was measured using dual-targeted microbubbles against both VEGFR-2 and $\alpha_v\beta_3$ integrin than using single-targeted ones.[69] Palmowski *et al.*[61] also measured this dual-targeted microbubble accumulation in squamous cell carcinoma xenografts in mice using the quantitative volumetric ultrasound scanning technique. They observed a dramatic increase of VEGFR-2 and $\alpha_v\beta_3$ integrin expression during tumor growth and a considerable decrease in both marker densities after matrix metalloproteinase inhibitor treatment. Histologic data suggested that the increasing VEGFR-2 and $\alpha_v\beta_3$ integrin concentrations are related to an up-regulation of the expression by the endothelial cells, whereas the decrease under therapy is more related to the decreasing relative vessel density.[61]

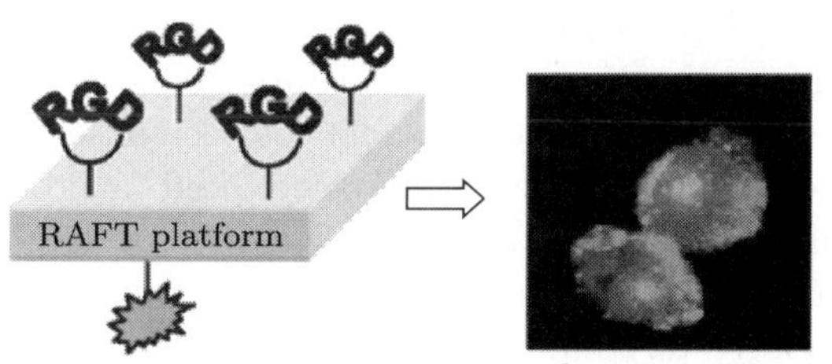

Fig. 10.8 Scheme of template assembled cyclopeptides as a multimeric system for targeting and endocytosis of cells expressing $\alpha_v\beta_3$ integrin. Reprinted from Ref. 65. Copyright (2004) with permission from American Chemical Society.

Apart from the markers mentioned above, other markers have been also used to design functionalized microbubbles against tumor vasculature. CD105 (endoglin) is a cell membrane glycoprotein involved in vascular development and remodeling.[70,71] Another marker is P/E-selectin. Lindner and colleagues used sulfo-Lewis-x-conjugated microbubbles to target P/E-selectin in subcutaneously implanted tumors.[72]

All of aforementioned studies show that molecular ultrasound is well suitable to depict different expression patterns of angiogenic marker molecules in different tumor types during tumor growth. Thus, it helps to non-invasively predict the tumor stage or tumor aggressiveness.

10.3.2 *Microbubbles for Drug Delivery*

Nowadays, chemotherapy is still a principal approach for clinical cancer treatment. However, most of anticancer drugs have poor solubility in body fluid or blood, significant toxicity and side-effects. Appropriate drug carriers may concomitantly improve the therapeutic effect and reduce the toxicity of these drugs through a combination of tumor targeted delivery and site-specific controlled release. In this sense, UTMD offers broad potential as a site-specific and non-invasive delivery tool for controlled release of drug in the target area. Most importantly, it facilitates the access of the drug to intracellular compartments of cells at the target area by increasing capillary and cell membrane permeability, which results in lower undesired systemic side-effects by reducing drug doses.

UTMD-assisted chemotherapeutic drug delivery has been used to deliver various chemotherapeutic agents in different tumor entities. In 1998 Unger *et al.*[40] demonstrated reduced toxicity of paclitaxel delivered by acoustically active lipospheres in mice. Using a similar MB type modified with a targeting ligand for $\alpha_v\beta_3$ integrin, paclitaxel was successfully delivered to the endothelium and interstitium of chicken chorioallantoic membrane vasculature. Tartis *et al.*[73] studied the therapeutic effects of paclitaxel-containing ultrasound contrast agents. Furthermore, significantly reduced tumor volume in a murine xenografted ovarian carcinoma model has been observed by ultrasound-mediated drug release from paclitaxel-loaded lipid microbubbles.[74] Using bleomycin, a significant tumor growth inhibition of human gingival squamous carcinoma cell tumors in mice was realized by sonoporation.[75] Fan *et al.*[76] introduced novel VEGF-targeting MBs that significantly further enhanced targeted drug release and reduce tumor progression in a rat model. The concept of targeting drug-loaded MBs is illustrated in Fig.10.9. After the purification of biotinylated BCNU-MBs, they were then incubated with fluorescein labeled avidin for 10 min at room temperature. The MBs were subsequently mixed with biotin-labeled rat anti-mouse CD309 (VEGF-R2, Flk-1) monoclonal antibody for 10 min. The VEGF-BCNU-MBs were obtained after removing excess biotinylated antibody. They were found to be accumulated in tumor cell site for drug releasing by trigger after injecting them *in vivo*. The treatment effect of targeted drug release *via* microbubbles was also confirmed by the observation of considerable regression of ovarian, breast, and orthotopic pancreatic tumors in mice.[77–82]

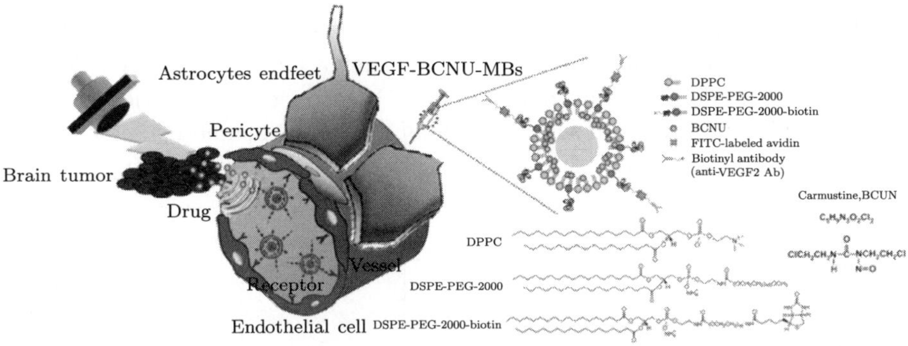

Fig. 10.9 The concept of antibody-based and tumor vascular targeting-guided drug delivery. Reprinted from Ref. 76. Copyright (2013) with permission from Elsevier.

10.3.3 *Microbubbles for Gene Delivery*

Gene therapy is an alternative approach for cancer treatment. Similar to chemotherapeutic drug delivery system, the main challenges are first the site selective release of therapeutic nucleic acid, and second the effective transportation of them across the cell membranes. Virus based vector system is a solution for effective delivery of therapeutic nucleic acid to intracellular spaces. But it may cause significant problems to the patient in the clinic, since it induces adverse immune responses or moderate immune responses precluding repeated administration of the vector. Given the potential advantages associated with the use of ultrasound-induced, transient cell membrane permeability, MBs would be a suitable way to deliver therapeutic nucleic acids.

A significant reduction of tumor size in mice was found using herpes simplex virus thymidine kinase (HSV-TK) plasmid DNA. This suicide gene method based on the observation that HSV-TK-mediated phosphorylation of the prodrug ganciclovir causes cell death.[83] Using this approach combined with UTMD technique, Zhou *et al.* observed reduction of tumor growth and improvement of survival time in a mouse hepatoma model.[84] Similarly the reduction of tumor cells in a mouse ovarian cancer model was realized using liposomal bubbles and ultrasound exposure for targeted gene delivery.[85] Fig.10.10(a) showed the effect of a single delivery of interleukin-12 (IL-12) gene. Gene delivery using bubble liposomes [solid triangle in Fig.10.10(a)], ultrasound [solid diamond in Fig.10.10(a)] or Lipofectamine 2000 [open diamand in Fig.10.10(a)] showed no apparent anti-tumor effect. The similar negative result was observed for pCMV-Luc intratumoral delivery using bubble liposomes and ultrasound [open square in Fig.10.10(a)]. However, the tumor size was significantly reduced using bubble liposomes and ultrasound to deliver pCMV-IL-12 [open circle in Fig.10.10(a)]. The drawback of single gene delivery is that the complete regression was not observed. Thus the repetitive IL-12 gene therapy was done and proved to be more effective in obtaining complete regression [open circle in Fig.10.10(b)].

Apart from ultrasound, other physical stimulation has also been demonstrated in gene deliver using microbubbles. Vlaskou *et al.*[86] fabricated magnetic microbubbles through self-assembly of the lipids, magnetic nanoparticles (MNPs) and nucleic acids (Fig.10.11). They used three different types of nu-

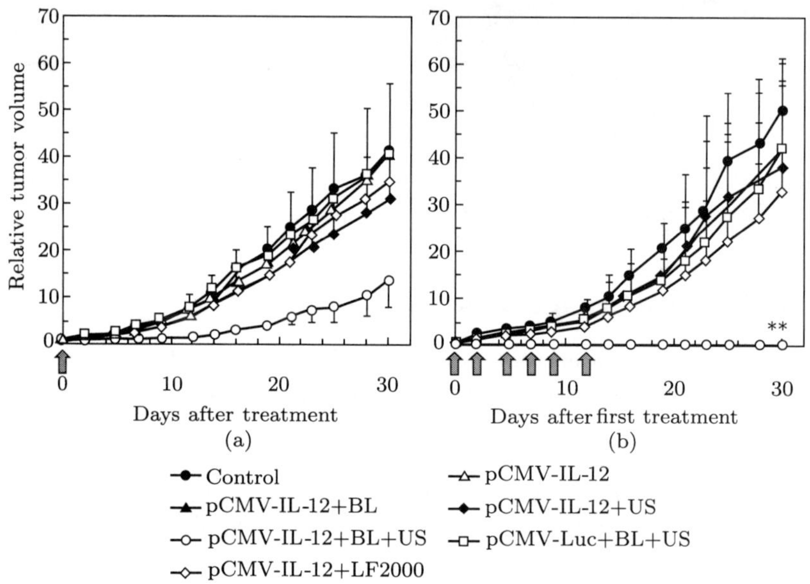

Fig. 10.10 Anti-tumor effect of IL-12 gene delivery. (a) Single gene therapy. (b) Repetitive gene therapy. Reprinted from Ref. 85. Copyright (2010) with permission from Elsevier.

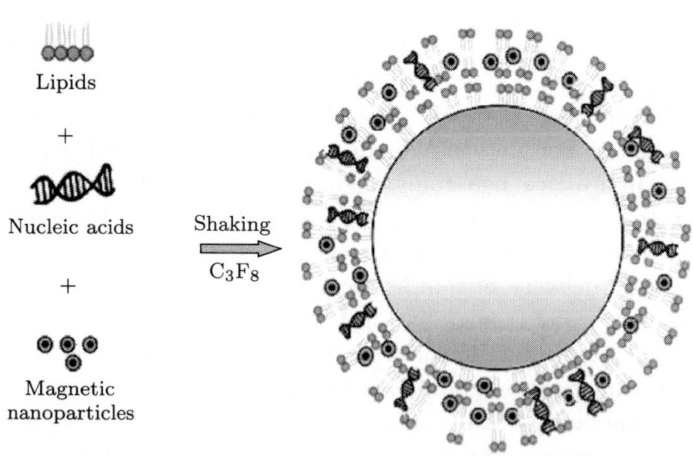

Fig. 10.11 Schematic representation of the self-assembly of the lipids, MNPs, and nucleic acids into magnetic, acoustically active lipospheres upon shaking of the components in an aqueous medium in the presence of perfluoropropane. Reprinted from Ref. 86. Copyright (2010) with permission from Wiley.

cleic acids to study the delivery efficiency in cultured adherent cells. The results showed that when a magnetic field was applied, efficient delivery of these nucleic acids delivery was achieved.

10.4 Future Aspects

Microbubbles have been demonstrated to be effective contrast agents for molecular imaging, and a carrier for drug/gene delivery in cancer preclinical diagnostic and clinical applications. However, this conclusion is drawn based on small animal models. Thus, in the next step, their validity in clinics, large animal studies are required. Another obstacle on the way to future clinical applications of microbubbles is its limited loading capacity. Although, by using multilayer or additional oillayer, progress towards increased loading capacities of microbubbles has been made, further improvement by novel microbubble formulations is still required. Finally, microbubbles are usually prepared from lipids, albumin, sugar, and a few biodegradable polymers. Limited library of candidate materials cannot meet the increasing demands on better microbubble performance, including higher stability, higher echo-enhancement, drug controlled release, organ or tissue targeting and so on. Novel functionalized biomedical macromolecules may provide more possibilities and choices for improving the performance of microbubbles.

References

1. Correas, J. M.; Bridal, L.; Lesavre, A.; Mejean, A.; Claudon, M.; Helenon, O. Ultrasound contrast agents: Properties, principles of action, tolerance, and artifacts. *Eur Radiol* **2001**, *11* (8), 1316-1328.
2. Grocott, H. P.; Sato, Y.; Homi, H. M.; Smith, B. E. The influence of xenon, nitrous oxide and nitrogen on gas bubble expansion during cardiopulmonary bypass. *Eur J Anaesthesiol* **2005**, *22* (5), 353-358.
3. Okitsu, K.; Suzuki, T.; Takenaka, N.; Bandow, H.; Nishimura, R.; Maeda, Y. Acoustic multibubble cavitation in water: A new aspect of the effect of a rare gas atmosphere on bubble temperature and its relevance to sonochemistry. *J Phys Chem B* **2006**, *110* (41), 20081-20084.
4. Porter, T. R.; Iversen, P. L.; Li, S. P.; Xie, F. Interaction of diagnostic ultrasound with synthetic oligonucleotide-labeled perfluorocarbon-exposed sonicated dextrose albumin microbubbles. *J Ultras Med* **1996**, *15* (8), 577-584.
5. Tinkov, S.; Coester, C.; Serba, S.; Geis, N. A.; Katus, H. A.; Winter, G.; Bekeredjian, R. New doxorubicin-loaded phospholipid microbubbles for targeted tumor therapy: *In-vivo* characterization. *J Control Release* **2010**, *148* (3), 368-372.
6. Grinstaff, M. W.; Suslick, K. S. Air-filled proteinaceous microbubbles: Synthesis of an echo-contrast agent. *P Natl Acad Sci USA* **1991**, *88* (17), 7708-7710.
7. Kwan, J. J.; Borden, M. A. Lipid monolayer dilatational mechanics during microbubble gas exchange. *Soft Matter* **2012**, *8* (17), 4756-4766.

8. Trabulsi, E. J.; Sackett, D.; Gomella, L. G.; Halpern, E. J. Enhanced transrectal ultrasound modalities in the diagnosis of prostate cancer. *Urology* **2010**, *76* (5), 1025-1033.
9. Thittai, A. K.; Yamal, J. M.; Mobbs, L. M.; Kraemer-Chant, C. M.; Chekuri, S.; Garra, B. S.; Ophir, J. Axial-shear strain elastogrphy for breast lesion classification: Further results from *in vivo* data. *Ultrasound Med Biol* **2011**, *37* (2), 189-197.
10. Athanasiou, A.; Tardivon, A.; Tanter, M.; Sigal-Zafrani, B.; Bercoff, J.; Eux, T. D.; Gennisson, J. L.; Fink, M.; Neuenschwander, S. Breast lesions: Quantitative elastography with supersonic shear imaging-preliminary results. *Radiology* **2010**, *256* (1), 297-303.
11. Tanter, M.; Bercoff, J.; Athanasiou, A.; Deffieux, T.; Gennisson, J. L.; Montaldo, G.; Muller, M.; Tardivon, A.; Fink, M. Quantitative assessment of breast lesion viscoelasticity: Initial clinical results using supersonic shear imaging. *Ultrosound Med Biol* **2008**, *34* (9), 1373-1386.
12. Martinoli, C.; Derchi, L. E.; Rizzatto, G.; Solbiati, L. Power Doppler sonography: General principles, clinical applications, and future prospects. *Eur Radiol* **1998**, *8* (7), 1224-1235.
13. Goertz, D. E.; Yu, J. L.; Kerbel, R. S.; Burns, P. N.; Foster, F. S. High-frequency Doppler ultrasound monitors the effects of antivascular therapy on tumor blood flow. *Cancer Res* **2002**, *62* (22), 6371-6375.
14. Cosgrove, D.; Lassau, N. Imaging of perfusion using ultrasound. *Eur J Nucl Med Mol Imaging* **2010**, *37*, S65-S85.
15. Graubner, T.; Fritzsch, T.; Hauff, P.; Kaiser, W. A. Contrast medium supported imaging of tumorous vessels *via* colour Doppler and power Doppler mode in animal experiment. *Rofo-Fortschr Rontg* **1996**, *165* (6), 563-567.
16. Gramiak, R.; Shah, P. M. Echocardiography of the aortic root. *Invest Radiol* **1968**, *3* (5), 356-366.
17. Detrano, R.; Salcedo, E. E.; Yiannikas, J.; Moodie, D. S. Contrast two-dimensional echocardiography in the diagnosis of adult congenital heart-disease. *Clev Clin J Med* **1985**, *52* (2), 229-238.
18. Vitarelli, A.; Scapato, A.; Arata, L. Contrast two-dimensional echocardiography in mitral and aortic-insufficiency. *Clin Res* **1985**, *33* (2), A235.
19. Kabalnov, A.; Klein, D.; Pelura, T.; Schutt, E.; Weers, J. Dissolution of multicomponent microbubbles in the bloodstream: 1. Theory. *Ultrasound Med Bio* **1998**, *24* (5), 739-749.
20. Forsberg, F.; Liu, J. B.; Shi, W. T.; Furuse, J.; Shimizu, M.; Goldberg, B. B. *In vivo* pressure estimation using subharmonic contrast microbubble signals: Proof of concept. *IEEE T Ultroson FERR* **2005**, *52* (4), 581-583.
21. Kollmann, C. New sonographic techniques for harmonic imaging: Underlying physical principles. *Eur J Radiol* **2007**, *64* (2), 164-172.
22. Thomas, J. D.; Rubin, D. N. Tissue harmonic imaging: Why does it work? *J Am Soc Echocardiog* **1998**, *11* (8), 803-808.
23. Chen, H.; Varghese, T. Three-dimensional canine heart model for cardiac elastography. *Med Phys* **2010**, *37* (11), 5876-5886.
24. Sankar, D.; Thomas, T. A new fast fractal modeling approach for the detection of microcalcifications in mammograms. *J Digit Imaging* **2010**, *23* (5), 538-546.

25. Newman, C. M. H.; Bettinger, T. Gene therapy progress and prospects: Ultrasound for gene transfer. *Gene Ther* **2007**, *14* (6), 465-475.
26. Taniyama, Y.; Tachibana, K.; Hiraoka, K.; Namba, T.; Yamasaki, K.; Hashiya, N.; Aoki, M.; Ogihara, T.; Yasufumi, K.; Morishita, R. Local delivery of plasmid DNA into rat carotid artery using ultrasound. *Circulation* **2002**, *105* (10), 1233-1239.
27. Van Wamel, A.; Bouakaz, A.; Versluis, M.; De Jong, N. Micromanipulation of endothelial cells: Ultrasound-microbubble-cell interaction. *Ultrasound Med Bio* **2004**, *30* (9), 1255-1258.
28. Bouakaz, A.; Versluis, M.; de Jong, N. High-speed optical observations of contrast agent destruction. *Ultrasound Med Bio* **2005**, *31* (3), 391-399.
29. Porter, T. R.; LeVeen, R. F.; Fox, R.; Kricsfeld, A.; Xie, F. Thrombolytic enhancement with perfluorocarbon-exposed sonicated dextrose albumin microbubbles. *Am Heart J* **1996**, *132* (5), 964-968.
30. Ronderos, R. E.; Boskis, M.; Chung, N.; Cornell, D. B.; Escudero, E. M.; Ha, J. W.; Charlante, C.; Rim, S. J.; Portis, M.; Fabris, N.; Camilletti, J.; Mele, A. A.; Otero, F.; Porter, T. R. Correlation between myocardial perfusion abnormalities detected with intermittent imaging using intravenous perfluorocarbon microbubbles and radioisotope imaging during high-dose dipyridamole stress echo. *Clin Cardiol* **2002**, *25* (3), 103-111.
31. Korpanty, G.; Grayburn, P. A.; Shohet, R. V.; Brekken, R. A. Targeting vascular endothelium with avidin microbubbles. *Ultrasound Med Bio* **2005**, *31* (9), 1279-1283.
32. Cavalieri, F.; Ashokkumar, M.; Grieser, F.; Caruso, F. Ultrasonic synthesis of stable, functional lysozyme microbubbles. *Langmuir* **2008**, *24* (18), 10078-10083.
33. El-Sherif, D. M.; Wheatley, M. A. Development of a novel method for synthesis of a polymeric ultrasound contrast agent. *J Biomed Mater Res A* **2003**, *66A* (2), 347-355.
34. Bohmer, M. R.; Schroeders, R.; Steenbakkers, J. A. M.; de Winter, S.; Duineveld, P. A.; Lub, J.; Nijssen, W. P. M.; Pikkemaat, J. A.; Stapert, H. R. Preparation of monodisperse polymer particles and capsules by ink-jet printing. *Colloid Surface A* **2006**, *289* (1-3), 96-104.
35. Ottoboni, S.; Short, R. E.; Kerby, M. B.; Tickner, E. G.; Steadman, E.; Ottoboni, T. B. Characterization of the *in vitro* adherence behavior of ultrasound responsive double-shelled microspheres targeted to cellular adhesion molecules. *Contrast Media Mol I* **2006**, *1* (6), 279-290.
36. Borden, M. A.; Caskey, C. F.; Little, E.; Gillies, R. J.; Ferrara, K. W. DNA and polylysine adsorption and multilayer construction onto cationic lipid-coated microbubbles. *Langmuir* **2007**, *23* (18), 9401-9408.
37. De Geest, B. G.; Skirtach, A. G.; Mamedov, A. A.; Antipov, A. A.; Kotov, N. A.; De Smedt, S. C.; Sukhorukov, G. B. Ultrasound-triggered release from multilayered capsules. *Small* **2007**, *3* (5), 804-808.
38. Lentacker, I.; De Geest, B. G.; Vandenbroucke, R. E.; Peeters, L.; Demeester, J.; De Smedt, S. C.; Sanders, N. N. Ultrasound-responsive polymer-coated microbubbles that bind and protect DNA. *Langmuir* **2006**, *22* (17), 7273-7278.

39. Olbrich, C.; Hauff, P.; Scholle, F.; Schmidt, W.; Bakowsky, U.; Briel, A.; Schirner, M. The *in vitro* stability of air-filled polybutylcyanoacrylate microparticles. *Biomaterials* **2006**, *27* (19), 3549-3559.
40. Unger, E. C.; McCreery, T. P.; Sweitzer, R. H.; Caldwell, V. E.; Wu, Y. Q. Acoustically active lipospheres containing paclitaxel: A new therapeutic ultrasound contrast agent. *Invest Radiol* **1998**, *33* (12), 886-892.
41. Zhao, Y. Z.; Liang, H. D.; Mei, X. G.; Halliwell, M. Preparation, characterization and *in vivo* observation of phospholipid-based gas-filled microbubbles containing hirudin. *Ultrasound Med Biol* **2005**, *31* (9), 1237-1243.
42. Christiansen, C.; Kryvi, H.; Sontum, P. C.; Skotland, T. Physical and biochemical-characterization of albunex (TM), a new ultrasound contrast agent consisting of air-filled albumin microspheres suspended in a solution of human albumin. *Biotechnol Appl Bioc* **1994**, *19*, 307-320.
43. Bjerknes, K.; Dyrstad, K.; Smistad, G.; Agerkvist, I. Preparation of polymeric microcapsules: Formulation studies. *Drug Dev Ind Pharm* **2000**, *26* (8), 847-856.
44. Jiang, B. B.; Gao, C. Y.; Shen, J. C. Polylactide hollow spheres fabricated by interfacial polymerization in an oil-in-water emulsion system. *Colloid and Polym Sci* **2006**, *284* (5), 513-519.
45. Kukizaki, M.; Goto, M. Spontaneous formation behavior of uniform-sized microbubbles from Shirasu porous glass (SPG) membranes in the absence of water-phase flow. *Colloid Surface A* **2007**, *296* (1-3), 174-181.
46. Farook, U.; Zhang, H. B.; Edirisinghe, M. J.; Stride, E.; Saffari, N. Preparation of microbubble suspensions by co-axial electrohydrodynamic atomization. *Med Eng Phys* **2007**, *29* (7), 749-754.
47. Farook, U.; Stride, E.; Edirisinghe, M. J.; Moaleji, R. Microbubbling by co-axial electrohydrodynamic atomization. *Med Biol Eng Comput* **2007**, *45* (8), 781-789.
48. Whitesides, G. M. The origins and the future of microfluidics. *Nature* **2006**, *442* (7101), 368-373.
49. Hettiarachchi, K.; Talu, E.; Longo, M. L.; Dayton, P. A.; Lee, A. P. On-chip generation of microbubbles as a practical technology for manufacturing contrast agents for ultrasonic imaging. *Lab Chip* **2007**, *7* (4), 463-468.
50. Pancholi, K. P.; Farook, U.; Moaleji, R.; Stride, E.; Edirisinghe, M. J. Novel methods for preparing phospholipid coated microbubbles. *Eur Biophys J Biophy* **2008**, *37* (4), 515-520.
51. Pancholi, K.; Stride, E.; Edirisinghe, M. Dynamics of bubble formation in highly viscous liquids. *Langmuir* **2008**, *24* (8), 4388-4393.
52. Escoffre, J. M.; Zeghimi, A.; Novell, A.; Bouakaz, A. *In-vivo* gene delivery by sonoporation: Recent progress and prospects. *Curr Gene Ther* **2013**, *13* (1), 2-14.
53. Hernot, S.; Klibanov, A. L. Microbubbles in ultrasound-triggered drug and gene delivery. *Adv Drug Deliver Rev* **2008**, *60* (10), 1153-1166.
54. Ferrara, K.; Pollard, R.; Borden, M. Ultrasound microbubble contrast agents: Fundamentals and application to gene and drug delivery. *Annu Rev Biomed Eng* **2007**, 415-447.
55. Kareva, I.; Hahnfeldt, P. The emerging "hallmarks" of metabolic reprogramming and immune evasion: Distinct or linked? *Cancer Res* **2013**, *73* (9), 2737-2742.

56. Reinmuth, N.; Parikh, A. A.; Ahmad, S. A.; Liu, W. B.; Stoeltzing, O.; Fan, F.; Takeda, A.; Akagi, M.; Ellis, L. M. Biology of angiogenesis in tumors of the gastrointestinal tract. *Microsc Res Techniq* **2003**, *60* (2), 199-207.

57. Srivastava, A.; Laidler, P.; Davies, R. P.; Horgan, K.; Hughes, L. E. The prognostic-significance of tumor vascularity in intermediate-thickness (0.76–4.0 mm thick) skin melanoma: A quantitative histologic study. *Am J Pathol* **1988**, *133* (2), 419-423.

58. Rychak, J. J.; Graba, J.; Cheung, A. M. Y.; Mystry, B. S.; Lindner, J. R.; Kerbel, R. S.; Foster, F. S. Microultrasound molecular imaging of vascular endothelial growth factor receptor 2 in a mouse model of tumor angiogenesis. *Mol Imag* **2007**, *6* (5), 289-296.

59. Willmann, J. K.; Paulmurugan, R.; Chen, K.; Gheysens, O.; Rodriguez-Porcel, M.; Lutz, A. M.; Chen, I. Y.; Chen, X.; Gambhir, S. S. US imaging of tumor angiogenesis with microbubbles targeted to vascular endothelial growth factor receptor type 2 in mice. *Radiology* **2008**, *246* (2), 508-518.

60. Liu, D.; Liu, F.; Liu, Z.; Wang, L.; Zhang, N. Tumor specific delivery and therapy by double-targeted nanostructured lipid carriers with anti-VEGFR-2 antibody. *Mol Pharm* **2011**, *8* (6), 2291-2301.

61. Palmowski, M.; Huppert, J.; Ladewig, G.; Hauff, P.; Reinhardt, M.; Mueller, M. M.; Woenne, E. C.; Jenne, J. W.; Maurer, M.; Kauffmann, G. W.; Semmier, W.; Kiesslin, F. Molecular profiling of angiogenesis with targeted ultrasound imaging: Early assessment of antiangiogenic therapy effects. *Mol Cancer Ther* **2008**, *7* (1), 101-109.

62. Avraamides, C. J.; Garmy-Susini, B.; Varner, J. A. Integrins in angiogenesis and lymphangiogenesis. *Nat Rev Cancer* **2008**, *8* (8), 604-617.

63. Jun, H. Y.; Park, S. H.; Kim, H. S.; Yoon, K. H. Long residence time of ultrasound microbubbles targeted to integrin in murine tumor model. *Acad Radiol* **2010**, *17* (1), 54-60.

64. Weller, G. E. R.; Wong, M. K. K.; Modzelewski, R. A.; Lu, E. X.; Klibanov, A. L.; Wagner, W. R.; Villanueva, F. S. Ultrasonic imaging of tumor angiogenesis using contrast microbubbles targeted *via* the tumor-binding peptide arginine-arginine-leucine. *Cancer Res* **2005**, *65* (2), 533-539.

65. Boturyn, D.; Coll, J. L.; Garanger, E.; Favrot, M. C.; Dumy, P. Template assembled cyclopeptides as multimeric system for integrin targeting and endocytosis. *J Am Chem Soc* **2004**, *126* (18), 5730-5739.

66. Leong-Poi, H.; Christiansen, J.; Klibanov, A. L.; Kaul, S.; Lindner, J. R. Noninvasive assessment of angiogenesis by ultrasound and microbubbles targeted to alpha(v)-integrins. *Circulation* **2003**, *107* (3), 455-460.

67. Ellegala, D. B.; Poi, H. L.; Carpenter, J. E.; Klibanov, A. L.; Kaul, S.; Shaffrey, M. E.; Sklenar, J.; Lindner, J. R. Imaging tumor angiogenesis with contrast ultrasound and microbubbles targeted to alpha(v)beta(3). *Circulation* **2003**, *108* (3), 336-341.

68. Willmann, J. K.; Kimura, R. H.; Deshpande, N.; Lutz, A. M.; Cochran, J. R.; Gambhir, S. S. Targeted contrast-enhanced ultrasound imaging of tumor angiogenesis with contrast microbubbles conjugated to integrin-binding knottin peptides. *J Nucl Med* **2010**, *51* (3), 433-440.

69. Willmann, J. K.; Lutz, A. M.; Paulmurugan, R.; Patel, M. R.; Chu, P.; Rosenberg, J.; Gambhir, S. S. Dual-targeted contrast agent for US assessment of tumor angiogenesis *in vivo*. *Radiology* **2008**, *248* (3), 936-944.

70. Korpanty, G.; Carbon, J. G.; Grayburn, P. A.; Fleming, J. B.; Brekken, R. A. Monitoring response to anticancer therapy by targeting microbubbles to tumor vasculature. *Clin Cancer Res* **2007**, *13* (1), 323-330.

71. Fonsatti, E.; Altomonte, M.; Nicotra, M. R.; Natali, P. G.; Maio, M. Endoglin (CD105): A powerful therapeutic target on tumor-associated angiogenetic blood vessels. *Oncogene* **2003**, *22* (42), 6557-6563.

72. Lindner, J. R.; song, J.; Christiansen, J.; Klibanov, A. L.; Xu, F.; Ley, K. Ultrasound assessment of inflammation and renal tissue injury with microbubbles targeted to P-selectin. *Circulation* **2001**, *104* (17), 2107-2112.

73. Tartis, M. S.; McCallan, J.; Lum, A. F. H.; LaBell, R.; Stieger, S. M.; Matsunaga, T. O.; Ferrara, K. W. Therapeutic effects of paclitaxel-containing ultrasound contrast agents. *Ultrasound Med Biol* **2006**, *32* (11), 1771-1780.

74. Xing, W.; Gang, W. Z.; Yong, Z.; Yi, Z. Y.; Shan, X. C.; Tao, R. H. Treatment of xenografted ovarian carcinoma using paclitaxel-loaded ultrasound microbubbles. *Acad Radiol* **2008**, *15* (12), 1574-1579.

75. Iwanaga, K.; Tominaga, K.; Yamamoto, K.; Habu, M.; Maeda, H.; Akifusa, S.; Tsujisawa, T.; Okinaga, T.; Fukuda, J.; Nishihara, T. Local delivery system of cytotoxic agents to tumors by focused sonoporation. *Cancer Gene Ther* **2007**, *14* (4), 354-363.

76. Fan, C. H.; Ting, C. Y.; Liu, H. L.; Huang, C. Y.; Hsieh, H. Y.; Yen, T. C.; Wei, K. C.; Yeh, C. K. Antiangiogenic-targeting drug-loaded microbubbles combined with focused ultrasound for glioma treatment. *Biomaterials* **2013**, *34* (8), 2142-2155.

77. Gao, Z.; Kennedy, A. M.; Christensen, D. A.; Rapoport, N. Y. Drug-loaded nano/microbubbles for combining ultrasonography and targeted chemotherapy. *Ultrasonics* **2008**, *48* (4), 260-270.

78. Rapoport, N.; Gao, Z. G.; Kennedy, A. Multifunctional nanoparticles for combining ultrasonic tumor imaging and targeted chemotherapy. *J National Cancer Institute* **2007**, *99* (14), 1095-1106.

79. Rapoport, N.; Kennedy, A. M.; Shea, J. E.; Scaife, C. L.; Nam, K. H. Ultrasonic nanotherapy of pancreatic cancer: Lessons from ultrasound imaging. *Mol Pharm* **2010**, *7* (1), 22-31.

80. Rapoport, N.; Nam, K. H.; Gupta, R.; Gao, Z. G.; Mohan, P.; Payne, A.; Todd, N.; Liu, X.; Kim, T.; Shea, J.; Scaife, C.; Parker, D. L.; Jeong, E. K.; Kennedy, A. M. Ultrasound-mediated tumor imaging and nanotherapy using drug loaded, block copolymer stabilized perfluorocarbon nanoemulsions. *J Control Release* **2011**, *153* (1), 4-15.

81. Rapoport, N. Y.; Kennedy, A. M.; Shea, J. E.; Scaife, C. L.; Nam, K. H. Controlled and targeted tumor chemotherapy by ultrasound-activated nanoemulsions/microbubbles. *J Control Release* **2009**, *138* (3), 268-276.

82. Rapoport, N. Y.; Nam, K. H.; Gao, Z.; Kennedy, A. Application of ultrasound for targeted nanotherapy of malignant tumors. *Acoust Phys* **2009**, *55* (4-5), 594-601.

83. Aoi, A.; Watanabe, Y.; Mori, S.; Takahashi, M.; Vassaux, G.; Kodama, T. Herpes simplex virus thymidine kinase-mediated suicide gene therapy using nano/microbubbles and ultrasound. *Ultrasound Med Biol* **2008**, *34* (3), 425-434.

84. Zhou, S. J.; Li, S. W.; Liu, Z. J.; Tang, Y.; Wang, Z. G.; Gong, J. P.; Liu, C. G. Ultrasound-targeted microbubble destruction mediated herpes simplex virus-thymidine kinase gene treats hepatoma in mice. *J Exp Clin Canc Res* **2010**, *29*.

85. Suzuki, R.; Namai, E.; Oda, Y.; Nishiie, N.; Otake, S.; Koshima, R.; Hirata, K.; Taira, Y.; Utoguchi, N.; Negishi, Y.; Nakagawa, S.; Maruyama, K. Cancer gene therapy by IL-12 gene delivery using liposomal bubbles and tumoral ultrasound exposure. *J Control Release* **2010**, *142* (2), 245-250.

86. Vlaskou, D.; Mykhaylyk, O.; Krotz, F.; Hellwig, N.; Renner, R.; Schillinger, U.; Gleich, B.; Heidsieck, A.; Schmitz, G.; Hensel, K.; Plank, C. Magnetic and acoustically active lipospheres for magnetically targeted nucleic acid delivery. *Adv Funct Mater* **2010**, *20* (22), 3881-3894.

Chapter 11

Upconversion Nanomaterials for Tumor Targeting Theranostics

*Wei Wang**

Lab of Functional and Biomedical Nanomaterials, College of Materials Science and Engineering, Qingdao University of Science and Technology, Qingdao 266042, China

Upconversion refers to nonlinear optical processes in which the sequential absorption of two or more photons leads to the emission of light at shorter wavelength than the excitation wavelength. For bio-applications, upconverting materials are superior to the traditional fluorescent agents since they provide remarkably high optical contrast because of the absence of autofluorescence and decreased light scattering. The near infrared-to-visible (NIR-to-VIS) or NIR-to-NIR upconversion fluorescence is also a good solution to the problems, such as photo damage and possible phototoxicity to living organisms, caused by ultraviolet (UV) radiation. In this chapter, mechanism of upconversion fluorescence and bio-applications of upconversion nanomaterials in bioimaging, tumor photodynamic therapy and drug/gene delivery are outlined.

11.1 Introduction

Bionanotechnology is an emerging field promoted by applications of nanomaterials in biomedical fields. Benefit from their small scale comparable to cellular structures, nanostructured materials have become particularly useful tools in drug delivery, bioimaging, biosensors, gene transfection and biomimetics.

Upconversion (UC) is a phenomenon first observed by F. Auzel in 1966.[1,2] It refers to nonlinear optical processes in which the sequential absorption of two or more photons leads to the emission of light at shorter wavelength than the excitation wavelength (anti-Stokes type emission). Three basic mechanisms are energy transfer upconversion (ETU), excited-state absorption (ESA) and photon avalanche (PA).[3–5] Upconversion should be distinguished from two-photon absorption and second-harmonic generation.

ETU [Fig.11.1(b)] is the most commonly observed mechanism in UC nanomaterials. It involves both sensitizer and activator as two types of luminescent centers embedded in the crystal matrix. Being excited by pump photons, the

* Corresponding author: wangwei@qust.edu.cn

sensitizers transfer energy to the adjacent activator *via* a non-radiative way and return to the ground state. The activators are then excited to higher energy level and radiatively emit UC fluorescence after they return to the ground state. ESA [Fig.11.1(a)] involves sequential absorption of photons within the levels of a given ion. Firstly, a given ion is excited to intermediate state after it absorbs one lower-energy photon. Then the already excited ion is excited to higher state after it absorbs the second photon and finally emits UC fluorescence when it returns to the ground state. PA [Fig.11.1(c)] requires a large number of ions at intermediate state level and hence requires high pumping power. Intermediate state ions are formed *via* cross relaxation energy transfer between the excited ions and ground state ions, and are excited to higher levels after absorption of pumping photons. UC fluorescence occurs when the excited ions return to the ground level through radiation.

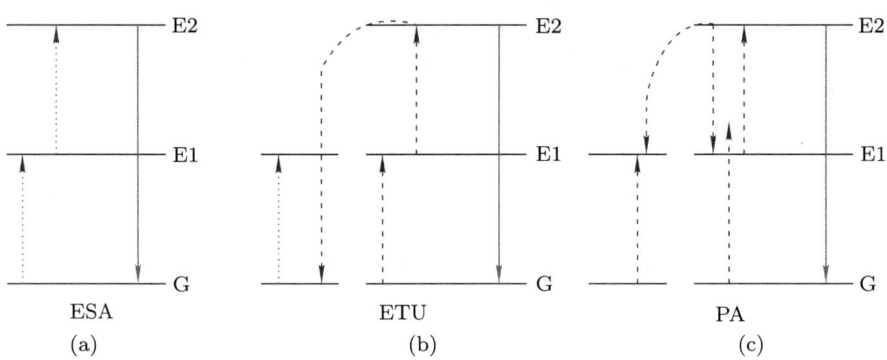

Fig. 11.1 Mechanism of upconversion process.

Upconversion fluorescence used for biological purpose needs to be highly efficient, thus UC nanomaterials that exhibit efficient ETU process are dominantly developed.

Photoexcitation at a certain wavelength in the NIR followed by luminescence at a shorter wavelength in the VIS is called NIR-to-VIS photon upconversion. This is a rather unusual process since lower energy photons are "converted" to higher energy photons. At least two NIR photons are required to generate one VIS photon. Upconversion can only occur in materials in which multiphonon relaxation processes are not predominant, thus allowing more than one metastable excited state. In rare-earth compounds, the 4f or 5f electrons are efficiently shielded and thus not strongly involved in the metal-to-ligand bonding. As a consequence, electron-phonon coupling to f-f transitions is reduced, and multiphonon relaxation processes are less competitive. The phenomenon of upconversion is therefore most common and best studied in materials containing lanthanide ions. But there are also transition-metal systems and rare-earth/transition-metal combinations which show this phenomenon.

11.2 Upconversion Nanomaterials for Tumor Imaging

11.2.1 *Upconversion Fluorescent Imaging*

Cell imaging is becoming a requisite technique for cell biology, developmental biology, cancer biology, drug discovery and many other related biomedical research laboratories. Powerful cellular imaging systems facilitate the integration of complex biology into the screening process, and allow the quantitative analysis of cellular events and visualization of relevant cellular phenotypes.

In recent years, the emergence of newly-developed cellular imaging systems established on nanomaterials greatly promotes the advancement of bioimaging technologies, including optical/fluorescent imaging (FI), magnetic resonance imaging (MRI), positron emission tomography (PET), X-ray computerized tomography (CT) and so on.

Fluorescent imaging is one of most significant applications of luminescent nanoparticles in bioanalysis for detection and sensing. Till now organic dyes are the most widely used labeling agents in fluorescent imaging technique. However, dye molecules usually suffer from photobleaching and quenching due to interactions with solvent molecules and reactive species such as oxygen or ions dissolved in solution when they are exposed to a variety of harsh environments. Consequently, inorganic luminescent nanoparticles which may circumvent some of these limitations and act as alternatives to the organic dyes in biomedical applications are attracting much attention. Compared with organic dyes, inorganic fluorescent nanoparticles usually exhibit sharper spectral peaks with large Stokes shift and better stability.

Most nanoparticle-based fluorescent imaging agents can be subdivided into two categories: down conversion and upconversion fluorescent nanoparticles.

For bioimaging, down conversion fluorescence has many disadvantages including possible phototoxicity to living organisms, low light-penetration depth and interference from autofluorescence of many biological samples under UV radiation. The NIR-to-VIS or NIR-to-NIR upconversion fluorescence is a good solution to the problems because photo damage from NIR wavelengths to living organisms is minimal and extremely low autofluorescence is generated in biological samples under NIR excitation. Consequently, the upconversion nanoparticles (UCNPs) emitting NIR-to-VIS or NIR-to-NIR fluorescence are believed to be superior to traditional bioimaging agents with down conversion.

11.2.1.1 *RE Fluorides*

RE fluorides, in particular, REF_3 and $AREF_4$ (RE = rare earth; A = Li, Na, K) material, generally possess higher upconversion efficiency than the other upconversion host materials (oxide, oxyfluoride, vanadates, oxysulfide, *etc*), due to their low phonon energies ($<400\ cm^{-1}$) which minimized multiphononassissted nonradiative losses in the intermediate state and the emitting state of RE ions.[6] Hence, a range of chemical approaches have been being explored to fabricate high-quality small-sized RE-doped fluorides.

$NaYF_4$:Yb, Er(Tm) UCNP is one of the most frequently studied UC nanomaterial due to its efficient upconversion fluorescence that can be pumped with an inexpensive and common 980 nm diode laser source. A large number of papers have reported the application of $NaYF_4$:Yb, Er(Tm) UCNP as a fluorescent imaging agent which offers excellent prospects for the development of new non-invasive strategies for the bioimaging *in vivo* and cancer therapy.

The first demonstration of use of upconversion fluorophores for cellular and tissue imaging was done by Zhang *et al.*[7] in 2008. They demonstrated the fabrication of polyethylenimine (PEI) coated $NaYF_4$:Yb, Er nanoparticles which produce very strong upconversion fluorescence when excited at 980 nm by a NIR laser. The nanoparticles were shown to be stable in PBS, non-toxic to bone marrow stem cells, and resistant to photo-bleaching. The nanoparticles delivered into some cell lines or injected intradermally and intramuscularly into some tissues either near the body surface or deep in the body of rats showed visible fluorescence, when exposed to a 980 nm NIR laser.

Tm^{3+}-doped UCNPs with NIR-to-NIR upconversion have been proved to have a long penetration depth. In 2008, Nyk *et al.*[8] reported high contrast, photoluminescence imaging that utilized NIR-to-NIR upconversion transition in Tm^{3+} and Yb^{3+} doped $NaYF_4$ nanocrystals (20–30 nm). In this study, both the excitation wavelength of 975 nm and the upconverted emission at around 800 nm are in the NIR range, which is an ideal case for deeper tissue penetration. Consequently, upconversion fluorescent imaging provided remarkably high optical contrast because of absence of autofluorescence and decreased light scattering. The intense NIR luminescence peaking at $\sim$ 800 nm obtained from the mouse injected with the $NaYF_4$:Tm^{3+},Yb^{3+} UCNPs was detectable even through the skin without hair removal (Fig.11.2).

Drawbacks of $NaYF_4$:Yb,Er(Tm) UCNPs lie within 2 aspects. On one hand, most of $NaYF_4$:Yb,Er(Tm) UCNPs are hydrophobic because they are prepared by using oleic acid for controlling their shapes and sizes at high temperature, thus not suitable for biomedical applications. Many efforts have been devoted to modify the surface of $NaYF_4$:Yb,Er(Tm) UCNPs by various functional groups, such as polyethylene glycol (PEG) diacid[9], hexanedioic acid (HDA)[10], citrate[11], carboxyl[12], poly (acrylic acid) (PAA)[13], amphiphilic polymers[14], and phospholipid[15]. Surface ligand oxidation is an alternative method to convert hydrophobic UCNP to hydrophilic ones, in which the capped oleic acid ligands are oxidized to azelaic acid groups by using the Lemieux-von Rudloff reagent. The oxidized UCNPs still possessed an unchanging shape, phase, composition, and luminescent properties.[16–18]

Another drawback of $NaYF_4$:Yb,Er(Tm) UCNPs is that it is hard to ob-

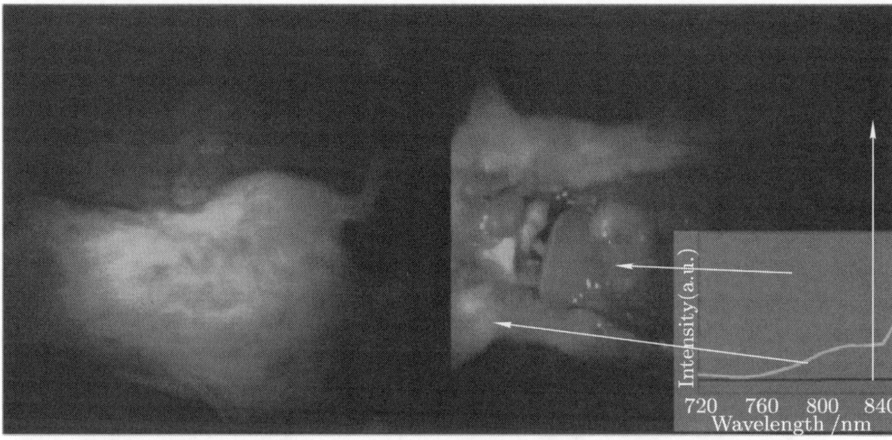

Fig. 11.2 Whole body images of mouse injected i.v. with UCNPs; intact mouse (left), same mouse after dissection (right). The red color indicates emission from UCNPs, green and black show background as indicated by the arrows. The inset presents the PL spectra corresponding to the spectrally unmixed components of the multispectral image obtained with the Maestro system. Reprinted with permission from Ref. 8. Copyright (2008) American Chemical Society. (color figure at the end of the book)

tain ideal nanocrystals with smaller size (less than 10 nm) and bright upconversion luminescence.[19] Other fluorides such as Na_xScF_{3+x}[20], $NaYbF_4$[21], $KGdF_4$[22], CaF_2[23,24], $NaLuF_4$[25–27], and $BaLuF_5$[28] which can be made smaller with brighter UCL are fabricated as alternative host materials to $NaYF_4$. (α-$NaYbF_4$:Tm^{3+})/CaF_2 core/shell nanoparticles that exhibit highly efficient NIRin (980 nm)-NIRout (800 nm) upconversion for high contrast and deep bioimaging were fabricated by Han *et al.*[29] By suppressing surface quenching effects *via* heteroepitaxial growth of a biocompatible CaF_2 shell, a very high quantum yield of 0.6 $\pm$ 0.1%, which was a 35-fold increase in the intensity of UC photoluminescence (PL) from the core, was obtained under low power density excitation ($\sim$0.3 W/cm^2). Due to the bright fluorescence, exceptional contrast (signal-to-background ratio of 310) small-animal whole-body UC PL imaging was realized. High-contrast UC PL imaging of deep tissues (Fig.11.3) was also demonstrated, using a nanoparticle-loaded synthetic fibrous mesh wrapped around rat femoral bone and a cuvette with aqueous dispersion of nanoparticles covered with a 3.2 cm thick animal tissue (pork).

11.2.1.2 *Lanthanide-doped Semiconductors*

Compared to dye molecules, semiconductors are more photochemically stable and have narrower, tunable emission spectra, and are more metabolically stable. However, issues of toxicity (generally related to heavy metal ions such as

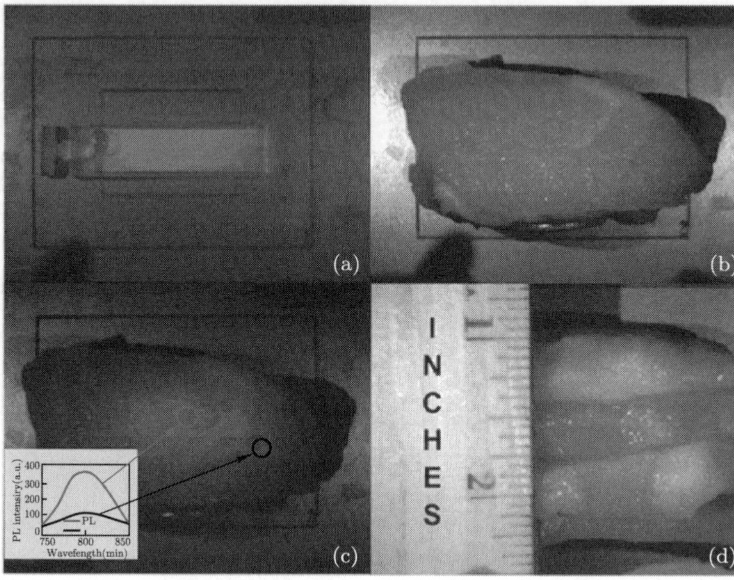

Fig. 11.3 (a) UC PL bright-field image of a cuvette filled with a suspension of the core/shell nanoparticles. (b) Bright-field image of a cuvette covered with pork tissue with a quarter coin stood aside showing its thickness. (c) Merged UC PL/bright-field image of the cuvette covered with pork tissue. (d) Bright-field image of the pork tissue (side view). The inset in (c) shows the spectra obtained from the circled areas. Reprinted with permission from Ref. 29. Copyright (2012) American Chemical Society. (color figure at the end of the book)

Pb^{2+} or Cd^{2+}), photo-oxidation, and difficult surface conjugation chemistry associated with QDs limit their applications. A protective shell of insulating material or semiconductor is introduced to overcome these problems, which indeed complicates the fabricating process.

Among all semiconductor materials, Zn-based II-VI semiconductors have been the subject of intense interest because of their potential wide-ranging applications in pigments, photonic crystal devices and short-wavelength optoelectronic devices. Among these applications, there are several important features of Zn-based II-VI semiconductors for bioimaging: ① superior chemical stability and photostability *in vitro* or *in vivo*; ② resistance to photobleaching; ③ high absorbency and low-toxicity; ④ adequate dispersibility in the biological environment and so on. These properties make Zn-based II-VI semiconductor nanomaterials excellent candidates as biocompatible nanoplatforms for bioimaging applications. As mentioned above, applications of many semiconductor materials, such as Cd compounds, are highly limited in biological systems due to toxicity consideration. Fortunately, the excellent biocompatibilities of ZnS and ZnO with outstanding optical properties have given them some new applications in biomedication and bioanalysis.

Rare-earth ions with some excellent physical or chemical properties have attracted widespread attention along with the application of Zn-based II-VI semiconductors in bioanalysis. For example, some rare-earth ions (Er^{3+}, Tb^{3+}, Eu^{3+}, *etc*) exhibit intense narrow-band intra-4f luminescence in a wide range of hosts, and many previous study have shown that these rare-earth ions can enhance and modify the optical and other performances of semiconductors. For example, in comparison to undoped semiconductor nanoparticles, the Tb^{3+} or Eu^{3+} doped nanoparticles can acquire higher fluorescence quantum yields and penetration depth. Expressly, the introduction of some special rare-earth ions (Tm^{3+}, Er^{3+}) to host materials can obtain NIR-to-VIS upconversion fluorescence, which can avoid the overlap of emission from nanomaterials and autofluorescence from tissues.

Although there are plenty of reports of unconverted emission from quantum dots,[30,31] there is few reports on semiconductors emitting upconversion luminescence used for biological applications.

Most recently, our group investigated the preparation of ZnS:Er,Yb,Gd and ZnO:Er,Yb,Gd nanoparticles with upconversion fluorescence *via* a facile fabrication method.[32] The results show that these nanoparticles can successfully label the human hepatocellular carcinoma (HepG2) cells (Fig.11.4) and present low toxicity even at the concentration up to 5 mg/mL.

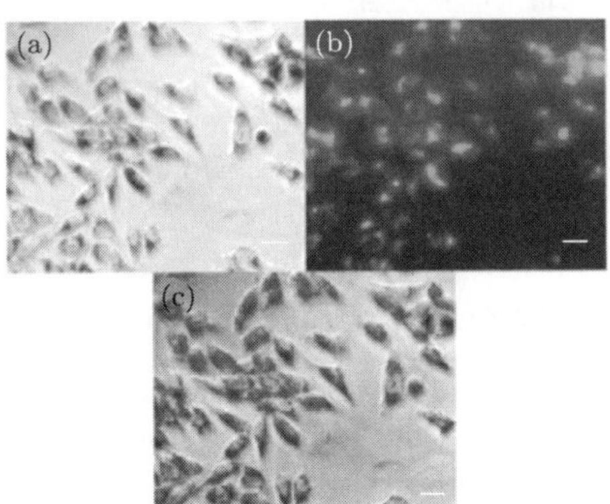

Fig. 11.4 Optical microscopy images of HepG2 cells incubated with ZnS:Tb,Gd nanoparticles: (a) phase contrast image, (b) fluorescence image and (c) overlay of fluorescence and phase contrast images. Scale bar = 20 μm. Reprinted with permission from Ref. 32. Copyright (2013) The Royal Society of Chemistry. (color figure at the end of the book)

Additionally, for the nanoparticles calcinated at high temperatures, NIR-to-VIS upconversion fluorescence was obtained. Figure 11.5 shows the PL spectra evolution of ZnS/ZnO:Er,Yb,Gd nanoparticles for 980 nm excitation as a function of sintering temperature, which contain two peaks characteristic of Er^{3+} sensitized by Yb^{3+} and center at 550 nm in green region and 662 nm in red region, respectively. According to previous report, these two emission bands are root respectively in the $^4F_{9/2} \rightarrow {}^4I_{15/2}$ and $^4S_{3/2} \rightarrow {}^4I_{15/2}$ transition of the Er^{3+} ions, and this process can proceed as a result of efficient Yb^{3+} sensitization and energy transfer upconversion. Moreover sintering performed at higher temperatures greatly enhanced the fluorescence of the nanoparticles. The ZnS/ZnO:Er,Yb,Gd nanoparticles emitting NIR-to-VIS fluorescence are believed to be superior to traditional ZnS/ZnO-based bioimaging agents with down conversion because photo damage from NIR wavelengths to living organisms is minimal and extremely low autofluorescence is generated in biological samples under NIR excitation.

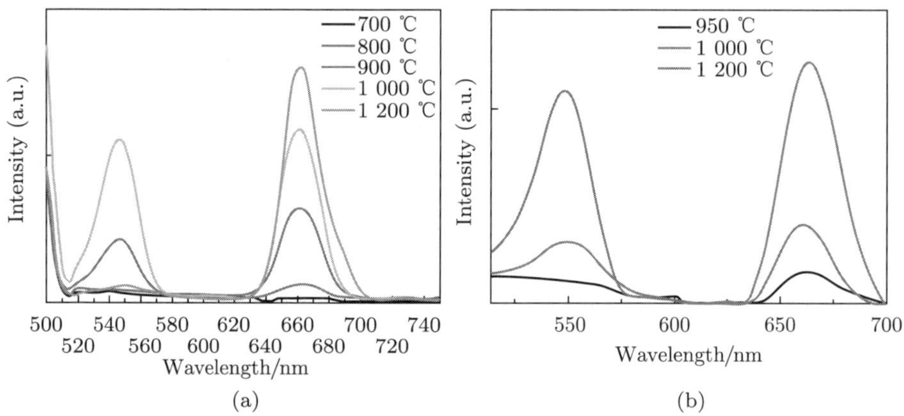

Fig. 11.5 PL spectra of (a) ZnO:Er,Yb,Gd and (b) ZnS:Er,Yb,Gd nanoparticles with varied sintering temperatures (λ_{EX} = 980 nm). Adapted with permission from Ref. 32. Copyright (2013) The Royal Society of Chemistry. (color figure at the end of the book)

11.2.2 *Fluorescent-MRI/CT/PET Multi-model Imaging*

Nanotechnology holds tremendous potential for early detection, accurate diagnosis, and personalized treatment of cancer. In this regard, a number of molecular imaging techniques, such as FI, MRI, ultrasound imaging (USI), PET, CT and others have been reported for imaging of *in vitro* and *in vivo* biological specimens. Different imaging techniques are, in general, complementary rather than competitive and the choice of imaging modality depends primarily on the specific question that is needed to be addressed. Therefore, various multimodal imaging probes combining different imaging modalities have been developed for more accurate imaging and diagnosis. For example, nanoparticles combining FI modality and MRI modality can offer the advantages of fluorescence imaging, *i.e.*, high sensitivity, along with the high spatial resolution anatomic imaging capability of MRI.

Research on magnetic nanoparticles is another typical example of bionanotechnology. They have been extensively investigated due to their excellent magnetic properties and biocompatibility and have shown great potential in biomedical applications such as magnetic separation, magnetic resonance imaging, targeted drug delivery, and hyperthermia treatment of cancer. At present, FI-MRI bi-functional nanoparticles are the most developed multi-modal imaging agent since excellent spatial resolution and high depth for *in vivo* imaging provided by MRI technique finely compensate the disadvantage of FI imaging without disturbing the high sensitivity and spatial resolution for *in vitro* imaging of FI.[19]

UCNPs could also be further engineered to acquire additional MRI functionality by introducing magnetic entity such as Gd, Mn, Fe_3O_4, and so on.

Owing to the large magnetic moment, the materials doped with Gd^{3+} may present excellent paramagnetic behavior. The magnetic property of Gd ions arises from the seven unpaired inner 4f electrons which are closely bound with the nucleus and shielded by the outer closed-shell electrons $5s^25p^6$ from the crystal field. By introducing Gd^{3+} in the synthesis of UCNPs, the particles could be used as MRI contrast agents as the Gd^{3+} ions in the host materials can accelerate longitudinal (T_1) relaxation of water protons and exert bright contrast in regions where the nanoprobes localize.

F. Y. Li's group devoted their efforts to fabrication of RE fluorides upconverting fluorescent nanocrystals. Based on their phase transfer and separation mechanism occurring at the interfaces of the liquid, solid and solution phases (LSS), they have developed a series of RE fluorides nanocrystals with different hosts including $NaYF_4$[33–41], $NaLaF_4$[42], LaF_3[43,44], $BaYF_5$[45–49], BaY_2F_8[50], Ba_2YF_7[51], Ba_2YbF_7[52], YF_3[33,34], GdF_3[35], CaF_2[36,37], SrF_2[38,39], KY_3F_{10}[40] and KMF_3 (M = Mn, Zn, Cd or Mg)[53]. Multifunctional imaging agents are further explored based on these RE fluorides nanocrystals. For example, $NaLuF_4$:Yb^{3+},Tm^{3+}@SiO_2-GdDTPA core-shell structured nanoparticles[54] were designed and synthesized for near-infrared to near-infrared (NIR-to-NIR) UCL, CT and T_1-enhanced magnetic resonance (MR) trimodality *in vivo* imaging. The property of NIR-to-NIR UCL was found to enhance the high signal-to-noise ratio (SNR) of *in vivo* imaging of small animal. Biological toxicity evaluated by methyl thiazolyl tetrazolium (MTT) assay and histological analysis of viscera sections indicated that SiO_2 layer as shell makes the lanthanide-based nanoparticles to be relatively less toxic for small animals. Moreover, changes in the MR signal intensity (SI) in liver, one of the main probe-uptake organs, monitored at timed intervals, showed clearly enhanced (+37.9%) and reduced SI (–38.1%) values of the T_1- and T_2-models, respectively, after 30 min, demonstrating the potential of the probe as a dual-weighted MR contrast agent *in vivo*. In contrast, Gd-DTPA and SPION could only work better in either a T_1- or T_2-model.

Wang *et al.*[55] fabricated ~7 nm highly uniform and monodisperse Tm^{3+},Yb^{3+}:BaF_2/Gd^{3+},Nd^{3+}:SrF_2 active-core/active-shell nanocubes. In this architecture, the Gd^{3+},Nd^{3+}:SrF_2 shell not only benefits the enhancement of the NIR to NIR upconversion luminescence of the Tm^{3+},Yb^{3+}:BaF_2 core, but also acts as the host to realize the NIR to NIR down conversion luminescence of Nd^{3+} dopants and the paramagnetism of Gd^{3+} ones. These hydrophobic core/shell nanocrystals are further converted into hydrophilic ones using thioglycolic acid as the surface ligand.

Zhao *et al.*[56] fabricated small-sized Mn-doped $NaYF_4$:Yb/Er nanoparticles with strong red UC emission. Paramagnetic Mn^{2+} ions are able to provide a second imaging capacity as MRI probes. Furthermore, Mn^{2+} ions disturb the transition possibilities between green and red emissions of Er^{3+} and facilitate the occurrence of red emission, which is believed to be more favorable than green emission for *in vivo* imaging since the dark red emission falls within the "optical window" that could afford the deep tissue penetration. An emission color output from green to red can be achieved by rational controlling the Mn^{2+}-doping level.

ZnS: Tb, Gd nanoparticles prepared by our group exhibit longitudinal (r_1) and transverse (r_2) relaxation rates of 39.46 and 51.67 $mM^{-1}{\cdot}s^{-1}$, respectively.[32] The ratio of r_2/r_1 is 1.3, demonstrating that the ZnS:Tb,Gd nanoparticles could serve as effective T_1-MRI contrast agents. Notably, the resulting r_1 value was much larger than the commercially available Gd-DTPA (r_1=4.5 $mM^{-1}{\cdot}s^{-1}$),[57] and currently reported Gd-base nanoparticles (r_1=17.7–35.8 $mM^{-1}{\cdot}s^{-1}$),[58–60] Gd_2O_3 spheres (r_1=22.2 $mM^{-1}{\cdot}s^{-1}$) and Gd-doped ZnO quantum dots (r_1=16 $mM^{-1}{\cdot}s^{-1}$)[61] measured using a 3 T NMR system.

Bifunctional nanoparticles based on $NaYF_4$ with combined optical and magnetic properties were synthesized *via* a simple layer-by-layer (LBL) self-assembly strategy by Liu *et al.*[62] Ultrasmall superparamagnetic iron oxide (Fe_3O_4) nanoparticles (IONPs) were adsorbed on the surface of a $NaYF_4$-based UCNP by electrostatic attraction to form a UCNP-IONP complex. A thin gold shell was then formed on top of the UCNP-IONP complex by seed-induced reduction growth. The authors found that the Fe_3O_4 IONPs layer not only afforded magnetic properties but also significantly reduced the luminescence quenching effects of the gold nanostructure to UCNPs. *In vivo* UCL imaging and MR imaging were successfully performed on a mouse after the nanoparticles were intracutaneously injected into one rear paw of the mouse.

Core-shell Fe_3O_4@$NaLuF_4$: Yb, Er/Tm nanostructure that combines upconversion luminescent, magnetic and X-ray attenuation properties was fabricated by Li group,[63] and successfully used for *in vivo* T_2-MR, CT and *in vitro* UCL images of tumor-bearing mice. Due to the presence of large Fe_3O_4 nanoparticle core, the core-shell nanostructure has relatively large size (330 nm in diameter).

Very recently, ^{153}Sm element that can emit medium-energy beta particles is introduced into UCNPs to develop additional CT imaging functionality.

$NaLuF_4$:^{153}Sm, Yb, Tm UCL nanoparticles[64] synthesized by a facile one-step hydrothermal method were found to be suitable for long-term single photon emission computed tomography (SPECT) due to the long half-life time (46.3 h, 1.93 days) of ^{153}Sm. The introduction of small amount of radioactive $^{153}Sm^{3+}$ can hardly vary the upconversion luminescence properties of the nanoparticles. *In vivo* UCL imaging and *in vivo* three-dimensional SPECT whole body imaging of a Kunming mouse was successfully performed after intravenous injection. By means of a combination of UCL and SPECT imaging *in vivo*, the distribution of the nanoparticles in the mouse body was studied, and the results indicated that these particles were mainly accumulated in the liver and spleen.

Degradable ^{153}Sm-doped $Gd(OH)_3$ nanorod as SPECT imaging and MR imaging agent was synthesized and found to be degraded under acidic condition *in vivo*, which is believed to be more easily excreted and more safer than the undegradable nanomaterials for *in vivo* application.[65] By using SPECT imaging, the long-term *in vivo* biodistribution of the $Gd(OH)_3$ nanorods was investigated. The results showed that most of the nanorods were excreted from the bodies of mice very rapidly. The uptake and retention of the nanorods took place primarily in the liver, spleen and lung. The mice survived for 150 days without any apparent adverse effects to their health even with intravenously injected amount of 100 mg/kg of the nanorods.

In order to combine PET imaging modality, upconversion nanomaterials can be labeled by ^{18}F, which is the most widely used radionuclide, to fabricate multifunctional nanoprobes. Based on the strong binding between Y^{3+} and F^-, ^{18}F-labeled $Gd^{3+}/Yb^{3+}/Er^{3+}$ co-doped $NaYF_4$ nanoparticles[66] were fabricated for multimodality PET, MRI, and laser scanning upconversion luminescence (UCL) imaging. Successful *in vivo* PET/MR whole-body imaging of small animals and *ex vivo* UCL imaging experiments concerning the biodistribution of ^{18}F-labeled lanthanide nanoparticles in small animals indicated the effectiveness of the nanoparticle as a multimodality nanoprobe.

Recently, Gd-, Yb-, and Lu-doped nanostructures are used as excellent CT contrast agents due to their stronger X-ray attenuation and higher atomic number than those of routinely used small iodinated molecules.

The Yb^{3+} in the lattice of $NaYbF_4$ nanoparticles functions not only as an excellent sensitizer contributing to the strong NIR-fluorescent emissions for its large NIR absorption cross-section, but also a promising CT contrast medium due to its high X-ray absorption coefficiency. For example, PEGylated $NaYbF_4$:Tm^{3+} nanopaticle[67] of less than 20 nm in diameter was explored as an efficient contrast media for dualmodal CT and NIR-fluorescent imaging. Remarkably, experimental results showed that the $NaYbF_4$:Tm^{3+} nanopaticles were about 75% higher than clinical iobitridol in CT contrast efficacy. In addition, these nanoparticles could be easily excreted mainly *via* feces without detectable remnant in the animal bodies.

Very recently, a multifunctional nanoprobe based on PEGylated Gd_2O_3: Yb^{3+}, Er^{3+} nanorods (denoted as PEG-UCNPs)[68] combining *in vivo* UCL, MR and CT imaging modality into one nano-entity was constructed *via* a large-scale method. Due to the higher amount of Gd in a single nanorod, a lower dose of PEG-UCNPs is necessary for MR imaging. For CT imaging, well-defined PEG-UCNPs offer a much higher contrast efficacy than small iodinated agents in clinic (iobitridol) due to the strong X-ray attenuation characteristics of Gd and Yb elements. Distributions of iobitridol were tracked by CT imaging before injection (control) as well as 2 min, 10 min, 0.5 h, 1 h, 2 h, 3 h and 24 h post-injection. PEG-UCNPs were believed to have unique imaging advantages since the characteristic hepatic imaging acquired with PEG-UCNPs could not be obtained with clinical iodinated molecules even under larger dosages. Compared with previous lanthanide-doped fluoride, the PEG-UCNPs nanoprobes exhibited more advantages, such as easy decomposition inside macrophage cells after RES uptake and nearly total excretion from the mouse body within a month.

11.3 Tumor Photodynamic Therapy

11.3.1 *Mechanism of Photodynamic Therapy*

Photodynamic therapy (PDT) is a treatment modality based on the photochemical reactions mediated by the interaction of photosensitizers (PS, also called PDT drugs) with specific light and oxygen.

During PDT process, a PS molecule is excited to a high-energy state after radiation of light. Energy transfer from the excited PS molecule to neighboring oxygen or other substrate molecules generates singlet oxygen ($^1O_2$) or other reactive oxygen species (ROS), which can destroy tumors by multifactorial mechanisms, including directly inducing cancer cell death by necrosis and/or apoptosis, destruction of tumor vasculatures as an anti-angiogenesis effect, and also the stimulation of the host immune system to recognize, track down and destroy any remaining tumor cells, in contrast to radiotherapy and chemotherapy which are mostly immunosuppressive.

11.3.2 *Applications of Upconverting Nanoparticles in Photodynamic Therapy*

Although PDT is a minimally invasive therapeutic modality approved by many countries for numerous clinical cancer treatments, some challenges still persist, including limited penetration depth of the excitation light, poor water solubility of photosensitizers and so on.[69,70] Therefore, many efforts have focused on the NIR light triggered PDT with excitation wavelength shifted to the NIR region for deeper tissue treatment. NIR light triggered PDT could effectively reduce the photosensitivity to ambient light and improve the safety of PDT treatment.[71–74] Unfortunately, most of the available PS only absorbs visible light rather than NIR light. With the help of UCNPs, NIR light can be converted to visible light that can activate the PDT effect of PS. In other words, UCNPs serve as light nanotransducers in the NIR light triggered PDT. In order to further enhance the efficiency of PDT treatment, PS drugs may be combined with the UCNPs. The particles of nanometer size also facilitate the delivery and excretion of the proposed photosensitizer as a PDT drug. With all the above-mentioned fascinating properties, UCNPs therefore hold great potential in biological applications, especially as a nanotransducer for solving the problems that traditional PDT has faced, and for inspiring the fabrication of UCNP/PS-based PDT drugs.

In the past few years, UCNPs have been widely explored as novel drugs.[75–77] The first report on UCNPs triggered PDT was performed by Zhang and colleagues in the year 2007 based on the mechanism shown in Fig.11.6.[78] The UCNPs used in this study are $NaYF_4$:Yb^{3+},Er^{3+} photo upconverting nanoparticles (denoted as PUNPs), which were coated by a porous silica shell. Merocyanine 540 (MC540) with the maximum absorption at 540 nm was used as PS molecule, and doped into the silica layer during the coating process. *In vitro* PDT tests were performed to confirm the photodynamic cytotoxicity of the MC540-coated UCNPs toward MCF-7/AZ breast cancer cells. A mouse monoclonal antibody, anti-MUC1/episialin, was covalently attached to the silica-coated nanoparticles in order to bring them close to the target cancer cells through antigen-antibody interaction with good specificity and versatility. When excited by an infrared (974 nm) source, UCNPs emitted strong visible bands around 537 and 635 nm. The upconverting light was then absorbed by MC540 and generated ROS to kill the MCF-7/AZ breast cancer cells. Images of MCF-7/AZ cells before IR irradiation and after 43 min of IR irradiation are shown in Fig.11.7(d) demon-

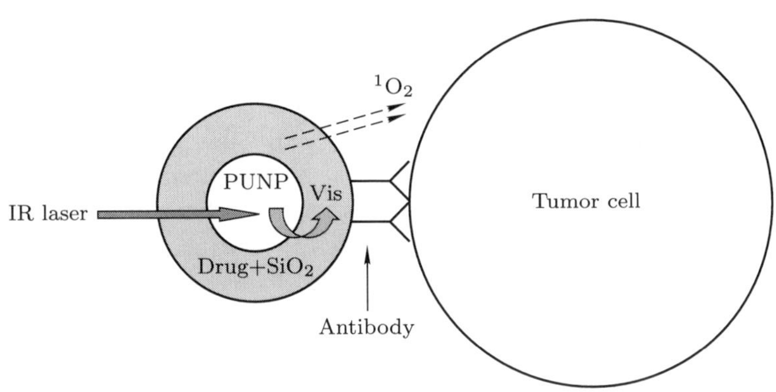

Fig. 11.6 Schematic of the design of the versatile photosensitizer based on PUNPs. Dimensions not drawn to scale. Reprinted with permission from Ref. 78. Copyright (2007) American Chemical Society.

strated increased fluorescence over time under IR irradiation, indicating the dying process of the cell.

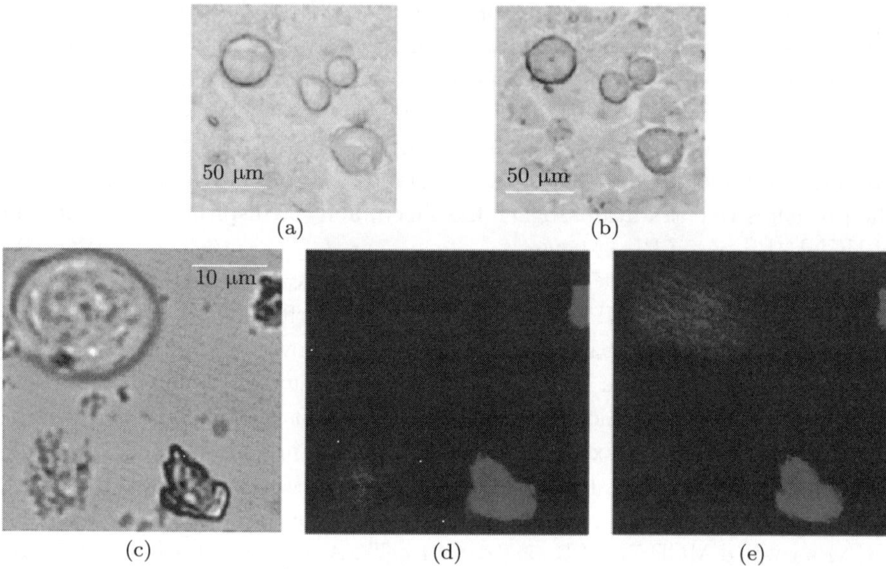

Fig. 11.7 (a) MCF-7/AZ cells incubated with Trypan Blue and antibody-functionalized MC540-coated PUNPs modified with antibody on surface. (b) MCF-7/AZ cells in (a) after 974 nm laser irradiation for 36 min. (c) Optical image of MCF-7/AZ cells incubated with PI and MC 540-coated PUNPs modified with antibody on surface. (d) Fluorescence image of MCF-7/AZ cells in (c) before 974 nm laser irradiation. (e) Fluorescence image of MCF-7/AZ cells in (c) after 974 nm laser irradiation for 43 min. Reprinted with permission from Ref. 78. Copyright (2007) American Chemical Society. (color figure at the end of the book)

The first *in vivo* UCNP-based PDT study in animal experiments was demonstrated by Liu's group.[79] They non-covalently loaded chlorin e6 (Ce6), a photosensitizer, on PEGylated amphiphilic polymer-coated UCNPs which could enter cancer cells. After being exposed to the 980-nm light at 0.5 W/cm^2 for 30 min, the UCNP-Ce6 generated $^1O_2$ and induced 4T1 murine breast cancer cell death. For animal experiments, they directly injected UCNP-Ce6 (20 mg/mL UCNP, ~1.5 mg/mL Ce6) into 4T1 murine breast cancer tumors grown on BABL/C mice. After 980-nm NIR laser excitation for 30 min (with 1-min interval for each minute to avoid heating) at 0.5 W/cm^2, excellent tumor regression effect was observed for UCNP-Ce6 injected tumors, while neither UCNP-Ce6 injection by itself nor NIR-laser irradiation alone was able to delay the tumor growth. It was further uncovered that UCNPs after PDT treatment were gradually cleared out from mouse organs, without rendering appreciable toxicity to the treated animals.

The first quantification of the cell-killing capabilities of the UCNP/PS composite nanoparticles formulated for PDT was conducted by Prud'homme's group.[76] They have synthesized 100 nm β-$NaYF_4$: Yb^{3+}, Er^{3+} UCNPs and utilized meso-tetraphenyl porphine (TPP) as the PS. TPP loading of 10 wt% with respect to UCNP crystal was achieved *via* the Flash NanoPrecipitation (FNP) process. In order to avoid nanoparticle aggregation and ensure nanoparticle stability in water, PBS buffer, and culture medium containing serum proteins, biocompatible poly(ethylene glycol-block-(DL)lactic acid) block copolymers (PEG-b-PLA) was coated on the nanoparticle surface. Then the β-$NaYF_4$: Yb^{3+}, Er^{3+}/TPP nanocomposites were conducted to perform PDT on human cervical cancer cell (HeLa cell) line. Experimental results showed that exposure with 134 W/cm^2 of 978 nm light for 45 min resulted in 75% HeLa cell death.

Other examples on the application of UCNP/PS composite nanomaterials for PDT include ZnPc-FA-PEI/$NaYF_4$: Yb^{3+}, Er^{3+} nanoparticles[80,81], UCP-TPP-PEG particles[82], $NaYF_4$: Yb, Er/$NaGdF_4$-Ce6 core-shell UCNPs[83], $NaYF_4$: Er, Yb@ SiO_2-ZnPc nanoparticles[84,85] and so on.

Generally, the area of UCNP/PS nanocomposites based on the $AREF_4$ (RE=rare earth; A = Li, Na, K) material is still very much in its developing stage, some remaining problems of $AREF_4$ such as poor water-solubility, low biocompatibilities and the poor mechanical properties have limited the development of $AREF_4$-based-UCNP/PS nanoparticles to some extent. Some studies have shown that the coating of SiO_2 can improve the biological application performance of $AREF_4$. But unfortunately, not only the "core-shell" method is rather tedious in the sense that they involve synthesizing the functional nanocrystals, and then putting them into one entity, the efficiency of PDT for these particles is also low because the nonporous or mesoporous silica shell blocks egress from the particles of the reactive $^1O_2$ species generated.[84]

For applications, zinc oxide (ZnO) is well-recognized as a biocompatible multifunctional material because there are several important features of ZnO for the applications in bioimaging and biomedicine: ① superior chemical stability and photostability *in vitro* or *in vivo*; ② resistance to photobleaching; ③ high absorbency and low-toxicity; ④ adequate dispersibility in the biological environment and so on. ZnO is also a wide band gap semiconductor (3.37 eV) with high exciton binding energy (60 meV), which leads to efficient excitonic blue and near-UV emission. These properties make ZnO nanomaterials excellent candidates as biocompatible nanoplatforms for bioimaging applications such as DNA delivery, MRI and sensing. So far, there are limited examples of ZnO based UCNPs and UCNP/PS drugs for applications of bioimaging and PDT to the best of our knowledge. Zhang and coworkers[86] reported the ZnO:Er^{3+} nanocrystals, and investigated the effect of annealing on the upconversion luminescence of products, unfortunately, there was no further research on biological applications of the UCNPs.

Our group synthesized a series of RE-doped (RE = Er^{3+}, Yb^{3+}, Gd^{3+}) ZnO particles with well controlled phase, shape, and size by a facile homogeneous precipitation method.[87] The as-synthesized optical and magnetic bi-functional ZnO:Er,Yb,Gd particles have high efficient upconversion luminescence and excellent magnetic behavior. In addition, by using well-selected PDT drug of water-soluble methylene blue (MB), we have succeeded in developing a kind of UCNP/MB-based PDT drug, ZnO:Er,Yb,Gd-MB, which can kill the HepG2 cells effectively triggered by NIR light. In principle, when the PS is close enough to UCNP, the smaller the wavelength mismatch between PS absorption and

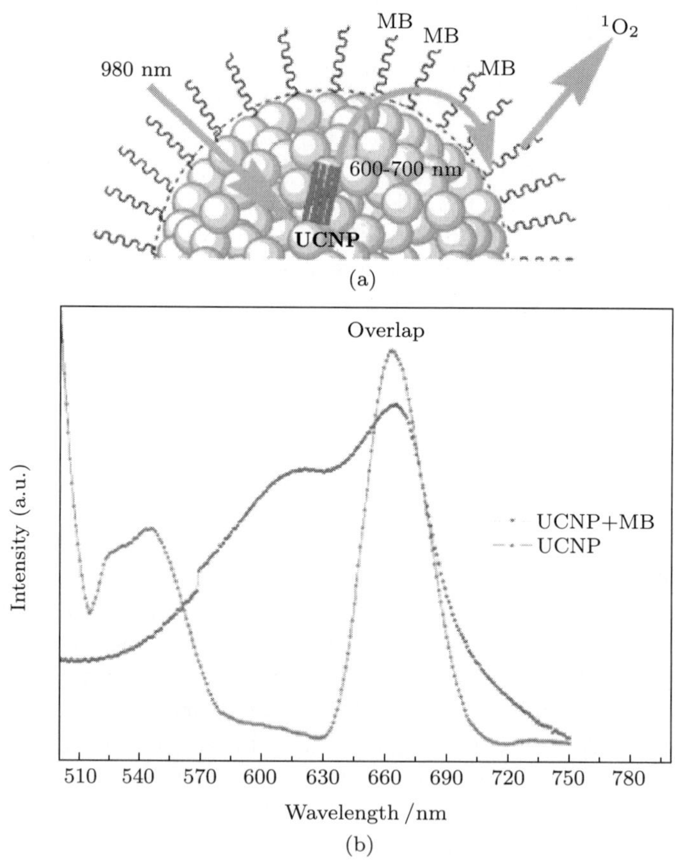

Fig. 11.8 (a) Schematic illustration showing how PDT works using UCNPs-MB. On exposure to NIR light, the nanoparticles convert NIR light to visible light, which will activate the MB to produce reactive oxygen species to kill cancer cells. (b) Overlap between ZnO:Er,Yb,Gd-MB absorption and ZnO:Er,Yb,Gd UCNPs emission: red-band emission of upconversion fluorescence spectrum of ZnO:Er,Yb,Gd under NIR-laser excitation (red line); UV/Vis absorption spectrum of ZnO:Er,Yb,Gd-MB in water located at 665 nm (blue line). Reprinted with permission from Ref. 87. Copyright (2013) American Chemical Society. (color figure at the end of the book)

UCNP emission is, the higher the $^1O_2$ generation efficiencies will be.[88] Due to its well-matched absorption (665 nm) with ZnO:Er,Yb,Gd UCNPs emission (662 nm), MB is selected as PS in our study. MB has been actively used in the microbiological and pharmacological fields before being considered as a drug for NIR-laser-triggered PDT. The quantum yield of $^1O_2$ generation of MB is high and it has almost negligible dark toxicity.[29,89] Figure 11.8 clearly shows the spectra overlap between MB absorption and ZnO: Er, Yb, Gd red emission peak with an absorption/emission mismatch estimated to be less than 5 nm.

The *in vitro* PDT effects of the ZnO:Er,Yb,Gd-MB particles were assessed by cell viability. After incubation with different concentrations of ZnO:Er, Yb, Gd-MB for 12 h, the HepG2 cells were irradiated with NIR light for 3 min and incubated for an additional 12 h. Compared to the control group without NIR light irradiation, the cell viability in the NIR light irradiated groups decreased dramatically with an increase in the concentration of ZnO:Er,Yb,Gd-MB samples [Fig.11.9(a)]. Figure 11.9(c) demonstrates the differences in HepG2 cell morphology with (zone 2) and without (zone 1) NIR light irradiation. As is shown, the incubation with ZnO:Er,Yb,Gd-MB for 12 h did not cause significant changes on the cells (zone 1), however, after a further exposition to 980 nm laser for a period of time the cell morphology was notably changed and obvious cell apoptosis can be observed (zone 2). In order to differentiate the dead and lives cells, the trypan blue (TB), which can detect the integrity of the cell membrane, was used to stain the dead cells. Figure 11.9(d) further demonstrates

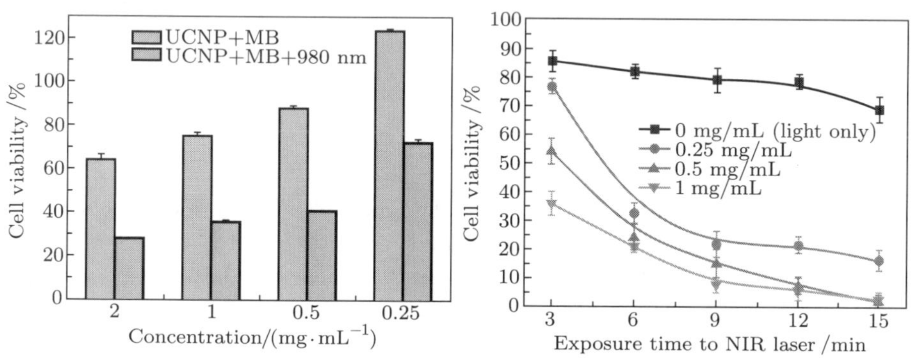

(a) (b)

Fig. 11.9 (a, b) Cell viability data of HepG2 cells after various treatments indicated as evaluated by the standard MTT assay. (c) Optical microscopy image of HepG2 cells incubated with ZnO:Er,Yb,Gd-MB particles for 12 h (zone 1) and further exposed to NIR laser for a period of time (zone 2). (d) The corresponding TB stained cells image of (c). Reprinted with permission from Ref. 87. Copyright (2013) American Chemical Society. (color figure at the end of the book)

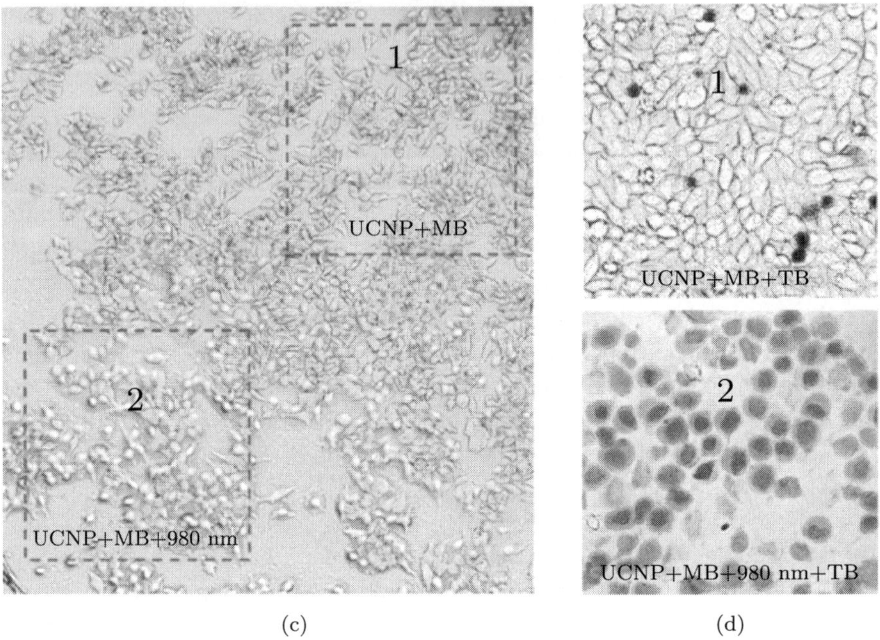

Fig. 11.9 (Continued).

the NIR-induced PDT effects of ZnO:Er,Yb,Gd-MB UCNPs. These results suggest that death of cells is dominated by the PDT effect of MB activated by the ZnO:Er,Yb,Gd UCNPs. Due to the deep tissue penetration of NIR light, ZnO:Er,Yb,Gd-MB particles possess great potential for *in vivo* PDT of cancer.

11.4 Drug/Gene Delivery

Nanoscaled drug/gene carrier is one of the most studied subjects related to nanoparticles due to their small scale and large surface to volume ratio. In the past decades, a great number of "smart" nanocarriers have been explored that can achieve high targeting efficiency at the tumor site and reduce side effects *via* ligand-, pH- or temperature-responsive mechanism to regulate the drug/gene release.

In order to utilize the upconversion fluorescence of UCNPs, drugs that are loaded on the UCNP nanocarriers are those can be photo activated (UV or visible light activated).

In the year 2013, Chen and colleges fabricated a new type of drug carrier based on NIR light-responsive mechanism.[90] They firstly prepared Fe_3O_4@C nanospheres and then deposited Ag nanoparticles onto the surface of Fe_3O_4@C nanospheres to obtain Fe_3O_4@C@Ag hybrid nanoparticles with a diameter of 200 nm. Under radiation of NIR light, the radiative recombination of sp-band electrons and d-band holes generates fluorescence, and could be enhanced by 4–6 orders magnitude due to the surface plasmons in Ag nanoparticles. This NIR triggered fluorescence can accelerate the esterization between carboxyl group of Fe_3O_4@C@Ag nanoparticles and hydroxy group of the anti-tumor drug doxorubicin (DOX). A high DOX loading content of 997 mg/g was achieved due to the chemical bond formed during esterification. Moreover, the formed chemical bonds could be broken with NIR to promote drug release since esterification for drug loading is reversible, and hence NIR-regulated drug release was realized. Measurements showed that the nanocarriers could also be used as MRI contrast agents and fluorescent probes.

Zhang *et al.*[91] reported gene delivery based on NIR-to-UV UCNPs. Nanocrystals made of $NaYF_4$:Yb^{3+}, Tm^{3+} coated with thin layers of mesoporous silica which can emit in the UV-visible region were used for this study. DNA/siRNA caged with 4,5-dimethoxy-2-nitroacetophenone (DMNPE) was loaded into the mesopores. Under excitation by an NIR laser at 980 nm, UV light centered at 350 nm emitted from the $NaYF_4$:Yb^{3+}, Tm^{3+} nanocarriers, which coincided well with the wavelength needed to uncage DMNPE from the nucleic acids and made the nanocarriers ideal transducers for activation. After that, controlled gene expression and specific gene silencing of DNA and siRNA happened, respectively. To prove the versatility of upconversion nanocrystals in actual applications, this system was then further used to activate photocaged nucleic acids in animal models. Cells transfected with upconversion nanocrystals containing photocaged plasmid were loaded into a PDMS device and transplanted into mice. The cells in the device were then activated using NIR light through the skin and PDMS layer, and efficient activation was observed even at that depth. All these results proved that this technique has enormous potential in the fields such as gene therapy for controlled and specific gene delivery/knockdown.

11.5 Dye Sensitized Efficiency of UC

Due to the extremely weak and narrowband NIR absorption, practical applications of UCNPs are highly constrained by low conversion efficiencies.[92,93] Fortunately, upconversion luminescence of nanoparticles may be dramatically enhanced by using organic dye molecules as sensitizers that can enhance the absorption of infrared light by the nanoparticles. In 2012, Zou *et al.*[94] used an organic NIR dye cyanine as an antenna for the β-$NaYF_4$:Yb,Er nanoparticles in which the upconversion occurs. They first prepared oleylamine-coated β-$NaYF_4$:Yb,Er nanoparticles showing green-yellow upconversion fluorescence with an average core diameter of 16 nm *via* a known method from a mixture of trifluoroacetates in oleylamine. Then a commercially available cyanine dye, IR-780, was attached to the surface of the nanoparticles and the central chlorine atom in IR-780 was substituted to prepare a carboxylic acid-functionalized derivative. The resulting carboxylated dye absorbs light mainly between 650 and 850 nm with a maximum at 806 nm in $CHCl_3$. The extinction coefficient of IR-806 at 806 nm is 390 $L^{-1}\cdot g^{-1}\cdot cm^{-1}$, which is $\sim 5 \times 10^6$ times higher than that of β-$NaYF_4$:Yb,Er nanoparticles at 975 nm (7×10^{-5} $L^{-1}\cdot g^{-1}\cdot cm^{-1}$). The overall upconversion by the dye-sensitized nanoparticles is dramatically enhanced (by a factor of ∼3 300) as a result of increased absorptivity and overall broadening of the absorption spectrum of the upconverter.

Another feature of dye-sensitized upconversion nanoparticles lies within that they can be excited by 800 nm light. Conventional upconversion nanoparticles are excited by 980 nm which can result in degraded imaging resolution due to water absorption of 980 nm light and cell/tissue damage caused by 980 nm-induced overheating. In contrast, water absorption and sample overheating caused by 800 nm are much less efficient than by 980 nm. Consequently, dye-sensitized upconversion nanoparticles not only emit stronger luminescence but also can overcome the overheating problem with enhanced imaging resolution.

Very recently, Song *et al.*[95] fabricated PMMA opal photonic crystal/$NaYF_4$: Yb^{3+}, Tm^{3+}/Er^{3+} nanocrystal composites with tremendous improvement in upconversion luminescence under infrared 980 nm excitation. Photonic crystal (PC) is a kind of material with a periodically varying refractive index which can control the propagation of electromagnetic waves. It is frequently used to modulate the emission wavelength, direction, radiative rate and intensity of some luminescent species. In this study, PMMA opal photonic crystals (OPCs) were prepared by growing PMMA beads onto a glass substrate firstly and then dropped into the cyclohexane solution dissolving $NaYF_4$:Yb^{3+},Tm^{3+} upconversion nanocrystals (UCNCs). After the evaporation of cyclohexane, PMMA/$NaYF_4$:Yb^{3+}, Tm^{3+}/Er^{3+} composites were fabricated. By controlling the diameters of PMMA beads during polymerization, the photonic stop band (PSB) of OPCs was finely tuned. The authors found that the UCL enhancement factor (EF) depended strongly on the PSB, and OPC with PSB of 980 nm was the optimum, which yielded an EF of 32, indicating that the coupling of OPC with the excitation field was the strongest. After further investigation of the effect of incident light on the EF, the authors concluded that the universal enhancement of upconversion luminescence in PMMA/$NaYF_4$:Yb^{3+},Tm^{3+} OPCs was mainly due to the random scattering of PMMA spheres to light, and the extra enhancement of the sample with a PSB of 980 nm was due to the PC effect. Luminescence of cells incubated with $NaYF_4$:Yb^{3+},Tm^{3+} UCNCs was successfully enhanced by using the prepared PMMA OPCs as reflection substrates.

11.6 Conclusions

Upconversion nanoparticles are nanoscaled particles that are able to convert two or more low-energy pump photons from the NIR spectral region (700–1 000 nm) to a higher-energy output photon with a shorter wavelength. They have found potential applications in tumor targeting theranostics including bioimaging, thermal dynamic therapy and drug/gene delivery. For fluorescent bioimaging, upconversion nanoparticles are advantageous over the traditional down converting ones because the NIR-light-excited upconversion luminescence imaging exhibits improved tissue penetration depth, uses particles with higher photochemical stability, and is free of auto-fluorescence. Thus, it affords remarkably improved *in vivo* sensitivity. By introducing other functional entities, upconversion nanoparticles can be fabricated into fluorescent-MRI/CT/PET multi-model imaging probes combining different imaging modalities.

For tumor photodynamic therapy, upconverting nanoparticles can act as light transducer to convert NIR light to visible light that can activate the PDT effect of photosentisizers. Upconverting nanoparticles also are "smart" nanocarriers for drugs or genes that can be photo activated based on the light control release mechanism.

Further progress in utilizing upconversion processes has been constrained by low conversion efficiencies, largely due to the extremely weak and narrowband near infrared absorption of the nanoparticles used. Consequently, fabrication of brighter upconverting nanoparticles with high conversion efficiencies remains a challenge in realizing upconverting nanoparticles clinical applications.

References

1. Auzel, F. Compteur quantique par transfert d'energie entre deux ions de terres reres dans un tungtate mixte et dans un verre. *C R Acad Sci Paris B* **1966**, *262*, 1016-1019.
2. Auzel, F. Compteru quantique par transfer d'energie entre de Yb^{3+} a Tm^{3+} dans un tungstate mixte et dans verre germanate. *C R Acad Sci Paris B* **1966**, *263*, 819-821.
3. Haase, M.; Schäfer, H. Upconverting nanoparticles. *Angew Chem Int Edit* **2011**, *50* (26), 5808-5829.
4. Auzel F. Upconversion and anti-Stokes processes with f and d ions in solids. *Chem Rev* **2004**, *104* (1), 139-174.
5. Gamelin, D. R.; Güdel, H. U. Design of luminescent inorganic materials: New photophysical processes studied by optical spectroscopy. *ACC Chem Res* **2000**, *33* (4), 235-242.
6. Hao, S. W.; Chen, G. Y.; Yang, C. H. Sensing using rare-earth-doped upconversion nano-particles, *Theronostics* **2013**, *3* (5), 331-345.
7. Chatterjeea, D. K.; Rufaihaha, A. J.; Zhang, Y. Upconversion fluorescence imaging of cells and small animals using lanthanide doped nanocrystals. *Biomaterials*

2008, *29* (7), 937-943.

8. Nyk, M.; Kumar, R.; Ohulchanskyy, T. Y.; Bergey, E. J.; Prasad, P. N. High contrast *in vitro* and *in vivo* photoluminescence bioimaging using near infrared to near infrared up-conversion in Tm^{3+} and Yb^{3+} doped fluoride nanophosphors. *Nano Lett* **2008**, *8* (11), 3834-3838.
9. Shun, Y. G.; Chow, G. Synthesis of hexagonal-phase $NaYF_4$:Yb,Er and $NaYF_4$:Yb,Tm nanocrystals with efficient up-conversion fluorescence. *Adv Funct Mater* **2006**, *16* (18), 2324-2329.
10. Zhang, Q. B.; Song, K.; Zhao, J. W.; Kong, X. G.; Sun, Y. J.; Liu, X. M.; Zhang, Y. L.; Zeng, Q. H.; Zhang, H. Hexanedioic acid mediated surface-ligand-exchange process for transferring $NaYF_4$:Yb/Er (or Yb/Tm) up-converting nanoparticles from hydrophobic to hydrophilic. *J Colloid Interface Sci* **2009**, *336* (1), 171-175.
11. Cao, T. Y.; Yang, T. S.; Gao, Y.; Yang, Y.; Hu, H.; Li, F. Y. Water-soluble $NaYF_4$:Yb/Er upconversion nanophosphors: Synthesis, characteristics and application in bioimaging. *Inorg Chem Commun* **2010**, *13* (3), 392-394.
12. Yang, J. P.; Shen, D. K.; Li, X. M.; Li, W.; Fang, Y.; Wei, Y.; Yao, C.; Tu, B.; Zhang, F.; Zhao, D. Y. One-step hydrothermal synthesis of carboxyl-functionalized upconversion phosphors for bioapplications. *Chem Eur J* **2012**, *18* (43), 13642-13650.
13. Naccache, R.; Vetrone, F.; Mahalingam, V.; Cuccia L. A.; Capobianco, J. A. Controlled synthesis and water dispersibility of hexagonal phase $NaGdF_4:Ho^{3+}/Yb^{3+}$ nanoparticles. *Chem Mater* **2009**, *21* (4), 717-723.
14. Budijono, S. J.; Shan, J. N.; Yao, N.; Miura, Y.; Hoye, T.; Austin, R. H.; Ju Y. G.; Prud'homme, R. K. Synthesis of stable block-copolymer-protected $NaYF_4:Yb^{3+},Er^{3+}$ up-converting phosphor nanoparticles. *Chem Mater* **2009**, *22* (2), 311-318.
15. Li, L. L.; Zhang, R. B.; Yin, L. L.; Zheng, K. Z.; Qin, W. P.; Selvin P. R.; Lu, Y. Biomimetic surface engineering of lanthanide-doped upconversion nanoparticles as versatile bioprobes. *Angew Chem Int. Ed* **2012**, *51* (25), 6121-6125.
16. Chen, Z. G.; Chen, H. L.; Hu, H.; Yu, M. X.; Li, F. Y.; Zhang, Q.; Zhou, Z. G.; Yi, T.; Huang, C. H. A versatile synthesis strategy for carboxylic acid-functionalized upconverting nanophosphors as biological labels. *J Am Chem Soc* **2008**, *130* (10), 3023-3029.
17. Zhou, H. P.; Xu, C. H.; Sun, W.; Yan, C. H. Clean and flexible modification strategy for carboxyl/aldehyde-functionalized upconversion nanoparticles and their optical applications. *Adv Funct Mater* **2009**, *19* (24), 3892-3900.
18. Zhou, J.; Sun, Y.; Du, X. X.; Xiong, L. Q.; Hu, H.; Li, F. Y. Dual-modality *in vivo* imaging using rare-earth nanocrystals with near-infrared to near-infrared (NIR-to-NIR) upconversion luminescence and magnetic resonance properties. *Biomaterials* **2010**, *31* (12), 3287-3295.
19. Gu, Z. J.; Yan, L.; Tian, G.; Li, S. J.; Chai, Z. F.; Zhao, Y. L. Recent advances in design and fabrication of upconversion nanoparticles and their safe theranostic applications. *Adv Mater* **2013**, *25* (28), 3758-3779.
20. Teng, X.; Zhu, Y. H.; Wei, W.; Wang, S. C.; Huang, J. F.; Naccache, R.; Hu, W. B.; Tok, A. Y.; Han, Y.; Zhang, Q. C.; Fan, Q. L.; Huang, W.; Capobianco,

J. A.; Huang, L. Lanthanide-doped Na_xScF_{3+x} nanocrystals: Crystal structure evolution and multicolor tuning. *J Am Chem Soc* **2011**, *134* (20), 8340-8343.

21. Chen, G. Y.; Ohulchanskyy, T. Y.; Kumar, Rajiv.; Ågren, H.; Prasad, P. N. Ultrasmall monodisperse $NaYF_4:Yb^{3+}/Tm^{3+}$ nanocrystals with enhanced near-infrared to near-infrared upconversion photoluminescence. *ACS Nano* **2010**, *4* (6), 3163-3168.
22. Wong, H. T.; Vetrone, F.; Naccache, R.; Lai, H.; Chan, W.; Hao, J. H.; Capobianco, J. A. Water dispersible ultra-small multifunctional $KGdF_4:Tm^{3+},Yb^{3+}$ nanoparticles with near-infrared to near-infrared upconversion. *J Mater Chem* **2011**, *21* (21), 16589-16596.
23. Wang, G. F.; Peng, Q.; Li, Y. D. Upconversion luminescence of monodisperse $CaF_2:Yb^{3+}/Er^{3+}$ nanocrystals. *J Am Chem Soc* **2009**, *131* (40), 14200-14201.
24. Dong, N. N.; Pedroni, M.; Piccinelli, F.; Conti, G.; Sbarbati, A.; Ramírez-Hernández, J. E.; Maestro, L. M.; Iglesias-de la Cruz, M. C.; Sanz-Rodriguez, F.; Juarranz, A.; Chen, F.; Vetrone, F.; Capobianco, J. A.; Solé, J. G.; Bettinelli, M.; Jaque, D.; Speghini, A. NIR-to-NIR two-photon excited CaF_2: Tm^{3+},Yb^{3+} nanoparticles: Multifunctional nanoprobes for highly penetrating fluorescence bio-imaging. *ACS Nano* **2011**, *5* (11), 8665-8671.
25. Liu, Q.; Sun, Y.; Yang, T. S.; Feng, W.; Li, C. G.; Li, F. Y. Sub-10 nm hexagonal lanthanide-doped $NaLuF_4$ upconversion nanocrystals for sensitive bioimaging *in vivo*. *J Am Chem Soc* **2011**, *133* (43), 17122-17125.
26. Shi, F.; Wang, J. S.; Zhai, X. S.; Zhao, D.; Qin, W. P. Facile synthesis of β-$NaLuF_4$:Yb/Tm hexagonal nanoplates with intense ultraviolet upconversion luminescence. *CrystEngComm* **2011**, *13* (11), 3782-3787.
27. Yang, D. M.; Dai, Y. L.; Ma, P. G.; Kang, X. J.; Cheng, Z. Y.; Li, C. X.; Lin, J. One-step synthesis of small-sized and water-soluble $NaREF_4$ upconversion nanoparticles for *in vitro* cell imaging and drug delivery. *Chem-Eur J* **2013**, *19* (8), 2685-2694.
28. Sarkar, S.; Meesaragandla, B.; Hazra, C.; Mahalingam, V. Sub-5 nm Ln^{3+}-doped $BaLuF_5$ nanocrystals: A platform to realize upconversion *via* interparticle energy transfer (IPET). *Adv Mater* **2013**, *25* (6), 856-860.
29. Chen, G. Y.; Shen, J.; Ohulchanskyy, T. Y.; Patel, N. J.; Kutikov, A.; Li, Z. P.; Song, J.; Pandey, R. K.; Ågren, H.; Prasad, P. N.; Han, G. (α-$NaYbF_4:Tm^{3+}$)/CaF_2 core/shell nanoparticles with efficient near-infrared to near-infrared upconversion for high-contrast deep tissue bioimaging. *ACS Nano* **2012**, *6* (9), 8280-8287.
30. Santos, J. R.; Vasilevskiy, M. I.; Filonovich, S. A. Cascade upconversion of photoluminescence in quantum dot ensembles. *Phys Rev B* **2008**, *78* (24), 245422.
31. Baumgartner, A.; Chaggar, A.; Patane, A.; Eaves, L.; Henini, M. Upconversion electroluminescence in InAs quantum dot lightemitting diodes. *Appl Phys Lett* **2008**, *92* (9), 091121-091121-3.
32. Wei, X. J.; Wang, W.; Chen, K. Z. Preparation and characterization of ZnS:Tb,Gd and ZnS:Er,Yb,Gd nanoparticles for bimodal magneticfluorescent imaging. *Dalton Trans* **2013**, *42* (5), 1752-1759.
33. Zhang, F.; Wan, Y.; Yu, T.; Zhang, F. Q.; Shi, Y. F.; Xie, S. H.; Li, Y. G.; Xu, L.; Tu, B.; Zhao, D. Y. Uniform nanostructured arrays of sodium rare-earth

fluorides for highly efficient multicolor upconversion luminescence. *Angew Chem Int Ed* **2007**, *46* (42), 7976-7979.

34. Zhang, F.; Li, J.; Shan, J.; Xu, L.; Zhao, D. Y. Shape, size, and phase-controlled rare-earth fluoride nanocrystals with optical up-conversion properties. *Chem-Eur J* **2009**, *15* (41), 11010-11019.
35. Zhang, F.; Zhao, D. Y. Synthesis of uniform rare earth fluoride ($NaMF_4$) nanotubes by *in situ* ion exchange from their hydroxide [$M(OH)_3$] parents. *ACS Nano* **2009**, *3* (1), 159-164.
36. Xu, Z. H.; Li, C. X.; Yang, P. P.; Zhang, C. M.; Huang, S. S.; Lin, J. Rare earth fluorides nanowires/nanorods derived from hydroxides: Hydrothermal synthesis and luminescence properties. *Cryst Growth Des* **2009**, *9* (11), 4752-4758.
37. Yang, L. W.; Han, H. L.; Zhang, Y. Y.; Zhong, J. X. White emission by frequency up-conversion in Yb^{3+}-Ho^{3+}-Tm^{3+} triply doped hexagonal $NaYF_4$ nanorods. *J Phys Chem C* **2009**, *113* (44), 18995-18999.
38. Zhang, Y. Y.; Yang, L. W.; Xu, C. F.; Zhong, J. X.; Sun, C. Q. Sensitized deep-ultraviolet up-conversion emissions of Gd^{3+} *via* Tm^{3+} and Yb^{3+} in hexagonal $NaYF_4$ nanorods. *Appl Phys B: Lasers Opt* **2009**, *98* (2), 243-247.
39. Li, J. J.; Yang, L. W.; Zhang, Y. Y.; Zhong, J. X.; Sun, C. Q.; Chu, P. K. Pump-power tunable white upconversion emission in lanthanide-doped hexagonal $NaYF_4$ nanorods. *Opt Mater* **2011**, *33* (6), 882-887.
40. Chen, D. Q.; Huang, P.; Yu, Y. L.; Huang, F.; Yang, A. P.; Wang, Y. S. Dopant-induced phase transition: A new strategy of synthesizing hexagonal upconversion $NaYF_4$ at low temperature. *Chem Commun* **2011**, *47* (20), 5801-5803.
41. Wang, Y. H.; Cai, R. X.; Liu, Z. H. Controlled synthesis of $NaYF_4$:Yb,Er nanocrystals with upconversion fluorescence *via* a facile hydrothermal procedure in aqueous solution. *CrystEngComm* **2011**, *13* (6), 1772-1774.
42. Wang, L. Y.; Li, P.; Li, Y. D. Down- and up-conversion luminescent nanorods. *Adv Mater* **2007**, *19* (20), 3304-3307.
43. Hu, H.; Chen, Z. G.; Cao, T. Y.; Zhang, Q.; Yu, M. X.; Li, F. Y.; Yi, T.; Huang, C. H. Hydrothermal synthesis of hexagonal lanthanide-doped LaF_3 nanoplates with bright upconversion luminescence. *Nanotechnology* **2008**, *19* (37), 375702.
44. De, G. J. H.; Qin, W. P.; Wang, W. H.; Gui, B. Infrared-to-ultraviolet upconversion luminescence of $La_{0.95}Yb_{0.49}Tm_{0.01}F_3$ nanostructures. *Opt Commun* **2009**, *282* (14), 2950-2953.
45. Zhang, C. M.; Ma, P. A.; Li, C. X.; Li, G. G.; Huang, S. S.; Yang, D. M.; Shang, M. M.; Kang, X. J.; Lin, J. Controllable and white upconversion luminescence in $BaYF_5:Ln^{3+}$ (Ln =Yb, Er, Tm) nanocrystals. *J Mater Chem* **2011**, *21* (3), 717-723.
46. Sun, J. Y.; Xian, J. B.; Du, H. Y. Hydrothermal synthesis of $BaYF_5:Yb^{3+}/Er^{3+}$ upconversion luminescence submicrospheres by a surfactant-free aqueous solution route. *J Phys Chem Solids* **2011**, *72* (3), 207-213.
47. Sun, J. Y.; Xian, J. B.; Xia, Z. G.; Du, H. Y. Synthesis of well oil-dispersible $BaYF_5:Yb^{3+}/Er^{3+}$ nanocrystals with green upconversion fluorescence. *J Rare Earths* **2010**, *28* (2), 219-221.

48. Sun, J. Y.; Xian, J. B.; Zhang, X. Y.; Du, H. Y. Size and shape controllable synthesis of oil-dispersible $BaYF_5:Yb^{3+}/Er^{3+}$ upconversion fluorescent nanocrystals. *J Alloys Compd* **2011**, *509* (5), 2348-2354.
49. Mo, M.; Yang, L. W.; Ren, G. Z.; Xu, C. F.; Lin, J. G.; Yang, Q. B. Solvothermal synthesis, cubic structure and multicolor upconversion emission of ultrasmall monodisperse lanthanide-doped $BaYF_5$ nanocrystals. *J Lumin* **2011**, *131* (7), 1482-1486.
50. Wang, G. F.; Peng, Q.; Li, Y. D. Synthesis and up conversion luminescence of $BaY_2F_8:Yb^{3+}/Er^{3+}$ nanobelts. *Chem Commun* **2010**, *46* (40), 7528-7529.
51. Chuai, X. H.; Zhang, D. S.; Zhao, D.; Zheng, K. Z.; He, C. F.; Shi, F.; Wang, L. L.; Chen, H.; Qin, W. P. Synthesis and characterization of $Yb^{3+},Tm^{3+}:Ba_2YF_7$ nanocrystalline with efficient upconversion fluorescence. *Mater Lett* **2011**, *65* (15-16), 2368-2370.
52. Xu, C. F.; Ma, M.; Zeng, S. J.; Ren, G. Z.; Yang, L. W.; Yang, Q. B. Solvothermal synthesis of monodisperse ultrasmall cubic-structure Ba_2YbF_7 nanocrystals with intense upconversion. *J Alloys Compd* **2011**, *509* (30), 7943-7947.
53. Zeng, J. H.; Xie, T.; Li, Z. H.; Li, Y. D. Monodispersed nanocrystalline fluoroperovskite up-conversion phosphors. *Cryst Growth Des* **2007**, *7* (12), 2774-2777.
54. Xia, A.; Chen, M.; Gao, Y.; Wu, D. M.; Feng, W.; Li, F. Y. Gd^{3+} complex-modified $NaLuF_4$-based upconversion nanophosphors for trimodality imaging of NIR-to-NIR upconversion luminescence, X-Ray computed tomography and magnetic resonance. *Biomaterials* **2012**, *33* (21), 5394-5405.
55. Chen, D. Q.; Yu, Y. L.; Huang, F.; Lin, H.; Huang, P.; Yang, A. P.; Wang, Z. X.; Wang, Y. S. Lanthanide dopant-induced formation of uniform sub-10 nm active-core/active-shell nanocrystals with near-infrared to near-infrared dual-modal luminescence. *J Mater Chem* **2012**, *22* (6), 2632-2640.
56. Tian, G.; Gu, Z. J.; Zhou, L. J.; Yin, W. Y.; Liu, X. X.; Yan, L.; Jin, S.; Ren, W. L.; Xing, G. M.; Li, S. J.; Zhao, Y. L. Mn^{2+} dopant-controlled synthesis of $NaYF_4$:Yb/Er upconversion nanoparticles for *in vivo* imaging and drug delivery. *Adv Mater* **2012**, *24* (9), 1226-1231.
57. Sasaki, M.; Shi, E.; Kanbara, Y.; Ehara, S. Enhancement effects and relaxivities of gadolinium-DTPA at 1.5 versus 3 Tesla: A phantom study. *Magn Reson Med Sci* **2005**, *4* (3), 145-149.
58. Rieter, W. J.; Taylor, K. M. L.; An, H. Y.; Lin, W. L.; Lin, W. B. Nanoscale metal-organic frameworks as potential multimodal contrast enhancing agents. *J Am Chem Soc* **2006**, *128* (18), 9024-9025.
59. Huang, C. C.; Liu, T. Y.; Su, C. H.; Lo, Y. W.; Chen, J. H.; Yeh, C. S. Superparamagnetic hollow and paramagnetic porous Gd_2O_3 particles. *Chem Mater* **2008**, *20* (12), 3840-3848.
60. Li, I. F.; Su, C. H.; Sheu, H. S.; Chiu, H. C.; Lo, Y. W.; Lin, W. T.; Chen, J. H.; Yeh, C. S. $Gd_2O(CO_3)_2{\cdot}H_2O$ particles and the corresponding Gd_2O_3: Synthesis and applications of magnetic resonance contrast agents and template particles for hollow spheres and hybrid composites. *Adv Funct Mater* **2008**, *18* (5), 766-776.
61. Liu, Y. L.; Ai, K. L.; Yuan, Q. H.; Lu, L. H. Fluorescence-enhanced gadolinium-doped zinc oxide quantum dots for magnetic resonance and fluorescence imaging.

Biomaterials **2011**, *32* (4), 1185-1192.

62. Cheng, L.; Yang, K.; Li, Y. G.; Chen, J. H.; Wang, C.; Shao, M. W.; Lee, S.-T.; Liu, Z. Facile preparation of multifunctional upconversion nanoprobes for multimodal imaging and dual-targeted photothermal therapy. *Angew Chem Int Ed* **2011**, *50* (32), 7385-7390.
63. Zhu, X. J.; Zhou, J.; Chen, M.; Shi, M.; Feng, W.; Li, F. Y. Core-shell Fe_3O_4@$NaLuF_4$:Yb,Er/Tm nanostructure for MRI, CT and upconversion luminescence tri-modality imaging. *Biomaterials* **2012**, *33* (18), 4618-4627.
64. Yang, Y.; Sun, Y.; Cao, T. Y.; Peng, J. J.; Liu, Y.; Wu, Y. Q.; Feng, W.; Zhang Y. J.; Li, F. Y. Hydrothermal synthesis of $NaLuF_4$:^{153} Sm,Yb,Tm nanoparticles and their application in dual-modality upconversion luminescence and SPECT bioimaging. *Biomaterials* **2013**, *34* (3), 774-783.
65. Yang, Y.; Sun, Y.; Liu, Y.; Peng, J. J.; Wu, Y. Q.; Zhang, Y. J.; Feng, W.; Li, F. Y. Long-term *in vivo* biodistribution and toxicity of $Gd(OH)_3$ nanorods. *Biomaterials* **2013**, *34* (2), 508-515.
66. Zhou, J.; Yu, M. X.; Sun, Y.; Zhang, X. Z.; Zhu, X. J.; Wu, Z. H.; Wu, D. M.; Li, F. Y. Fluorine-18-labeled $Gd^{3+}/Yb^{3+}/Er^{3+}$ co-doped $NaYF_4$ nanophosphors for multimodality PET/MR/UCL imaging. *Biomaterials* **2011**, *32* (4), 1148-1156.
67. Lee, S. J.; Koo, H.; Jeong, H.; Huh, M. S.; Choi, Y.; Jeong, S. Y.; Byun, Y.; Choi, K.; Kim, K.; Kwon, I. C. Comparative study of photosensitizer loaded and conjugated glycol chitosan nanoparticles for cancer therapy. *J Controlled Release* **2011**, *152* (1), 21-29.
68. Liu, Z.; Pu, F.; Huang, S.; Yuan, Q.; Ren, J.; Qu, X. Long-circulating Gd_2O_3:Yb^{3+}, Er^{3+} up-conversion nanoprobes as high-performance contrast agents for multi-modality imaging. *Biomaterials* **2013**, *34* (6), 1712-1721.
69. Chatterjee, D. K.; Fong, L. S.; Zhang, Y. Nanoparticles in photodynamic therapy: An emerging paradigm. *Adv Drug Deliver Rev* **2008**, *60* (15), 1627-1637.
70. Bechet, D.; Couleaud, P.; Frochot, C.; Viriot, M. L.; Guillemin, F.; Barberi-Heyob, M. Exploring and exploiting carotenoid accumulation in dunaliella salina for cell-factory applications. *Trends Biotechnol* **2008**, *26* (11), 631-638.
71. Detty, M. R.; Gibson, S. L.; Wagner, S. J. Current clinical and preclinical photosensitizers for use in photodynamic therapy. *J Med Chem* **2004**, *47* (16), 3897-3915.
72. Lu, T.; Shao, P.; Mathew, I.; Sand, A.; Sun, W. F. Synthesis and photophysics of benzotexaphyrin: A near-infrared emitter and photosensitizer. *J Am Chem Soc* **2008**, *130* (47), 15782-15783.
73. Ji, H. T.; Chien, L. T.; Lin, Y. H.; Chien, H. F.; Chen, C. T. 5-ALA mediated photodynamic therapy induces autophagic cell death *via* AMP-activated protein kinase. *Mol Cancer* **2010**, *9* (1), 91-101.
74. Cui, S. S.; Chen, H. Y.; Zhu, H. Y.; Tian, J. M.; Chi, X. M.; Qian, Z. Y.; Achileuf, S.; Gu, Y. Q. Amphiphilic chitosan modified upconversion nanoparticles for *in vivo* photodynamic therapy induced by near-infrared light. *J Mater Chem* **2012**, *22* (11), 4861-4873.
75. Lee, S. J.; Koo, H.; Jeong, H.; Huh, M. S.; Choi, Y.; Jeong, S. Y.; Byun, Y.; Choi, K.; Kim, K.; Kwon, I. C. Comparative study of photosensitizer loaded and conjugated glycol chitosan nanoparticles for cancer therapy. *J Controlled Release* **2011**, *152* (1), 21-29.

76. Shan, J.; Budijono, S. J.; Hu, G.; Yao, N.; Kang, Y.; Ju, Y.; Prud'homme, R. K. PEGylated composite nanoparticles containing upconverting phosphors and meso tetraphenyl porphine (TPP) for photodynamic therapy. *Adv Funct Mater* **2011**, *21* (13), 2488-2495.

77. Li, C.; Xu, Z.; Yang, D.; Cheng, Z.; Hou, Z.; Lin, J. Well-dispersed KRE_3F_{10} (RE= Sm-Lu, Y) nanocrystals: Solvothermal synthesis and luminescence properties. *CrystEngComm* **2012**, *14* (2), 670-678.

78. Zhang, P.; Steelant, W.; Kumar, M.; Scholfield, M. Versatile photosensitizers for photodynamic therapy at infrared excitation. *J Am Chem Soc* **2007**, *129* (15), 4526-4527.

79. Wang, C.; Tao, H. Q.; Cheng, L.; Liu, Z. Near-infrared light induced *in vivo* photodynamic therapy of cancer based on upconversion nanoparticles. *Biomaterials* **2011**, *32* (26), 6145-6154.

80. Chatterjee, D. K.; Zhang, Y. Upconverting nanoparticles as nanotransducers for photodynamic therapy in cancer cells. *Nanomedicine:NMB* **2008**, *3* (1), 73-82.

81. Wang, C.; Cheng, L.; Liu, Y. M.; Wang, X. J.; Ma, X. X.; Deng, Z. Y.; Li, Y. G.; Liu, Z. Imaging-guided pH-sensitive photodynamic therapy using charge reversible upconversion nanoparticles under near-infrared light. *Adv Funct Mater* **2013**, *23* (24), 3077-3086.

82. Ungun, B.; Prud'homme, R. K.; Budijono, S. J.; Shan, J. N.; Lim, S. F.; Ju, Y. G.; Austin, R. Nanofabricated upconversion nanoparticles for photodynamic therapy. *Opt Express* **2009**, *17* (1), 80-86.

83. Park, Y. I.; Kim, H. M.; Kim, J. H.; Moon, K. C.; Yoo, B.; Lee, K. T.; Lee, N.; Choi, Y.; Park, W.; Ling, D. S.; Na, K.; Moon, W. K.; Choi, S. H.; Park, H. S.; Yoon, S.-Y.; Suh, Y. D.; Lee, S. H.; Hyeon, T. Theranostic probe based on lanthanide-doped nanoparticles for simultaneous *in vivo* dual-modal imaging and photodynamic therapy. *Adv Mater* **2012**, *24* (42), 5755-5761.

84. Guo, H. C.; Qian, H. S.; Idris, N. M.; Zhang, Y. Singlet oxygen-induced apoptosis of cancer cells using upconversion fluorescent nanoparticles as a carrier of photosensitizer. *Nanomedicine: NBM* **2010**, *6* (3), 486-495.

85. Idris, N. M.; Gnanasammandhan, M. K.; Zhang, J.; Ho, P. C.; Mahendran, R.; Zhang, Y. *In vivo* photodynamic therapy using upconversion nanoparticles as remote-controlled nanotransducers. *Nat Med* **2012**, *18* (10), 1580-1585.

86. Wang, X.; Kong, X.; Yu, Y.; Sun, Y. J.; Zhang, H. Effect of annealing on upconversion luminescence of $ZnO:Er^{3+}$ nanocrystals and high thermal sensitivity. *J Phys Chem C* **2007**, *111* (41), 15119-15124.

87. Wei, X. J.; Wang, W.; Chen, K. Z. ZnO:Er,Yb,Gd particles designed for magnetic-fluorescent imaging and near-infrared light triggered photodynamic therapy. *J Phys Chem C* **2013**, *117* (45), 23716-23729.

88. Wang, F.; Banerjee, D.; Liu, Y. S.; Chen, X. Y.; Liu, X. G. Upconversion nanoparticles in biological labeling, imaging, and therapy. *Analyst* **2010**, *135* (8), 1839-1854.

89. Redmond, R. W.; Gamlin, J. N. A compilation of singlet oxygen yields from biologically relevant molecules. *Photochem Photobiol* **1999**, *70* (4), 391-475.

90. Chen, J.; Guo, Z.; Wang, H.-B.; Gong, M.; Kong, X.-K.; Xia, P.; Chen, Q.-W. Multifunctional Fe_3O_4@C@Ag hybrid nanoparticles as dual modal imaging

probes and near-infrared light-responsive drug delivery platform. *Biomaterials* **2013**, *34* (2), 571-581.

91. Jayakumar, M. K. G.; Idris, N. M.; Zhang, Y. Remote activation of biomolecules in deep tissues using near-infrared-to-UV upconversion nanotransducers. *P Natl Acad Sci USA* **2012**, *109* (22), 8483-8488.
92. Xie, X. J.; Liu, X. G. Upconversion goes broadband. *Nat Mater* **2012**, *11* (10), 842-843.
93. Wang, F.; Deng, R. R.; Wang, J.; Wang, Q. X.; Han, Y.; Zhu, H. M.; Chen, X. Y.; Liu, X. G. Tuning upconversion through energy migration in core-shell nanoparticles. *Nat Mater* **2011**, *10* (12), 968-973.
94. Zou, W. Q.; Visser, C.; Maduro, J. A.; Pshenichnikov, M. S.; Hummelen, J. C. Broadband dye-sensitized upconversion of near-infrared light. *Nat Photonics* **2012**, *6* (8), 560-564.
95. Yin, Z.; Zhu, Y. S.; Xu, W.; Wang, J.; Xu, S.; Dong, B.; Xu, L.; Zhang, S.; Song, H. W. Remarkable enhancement of upconversion fluorescence and confocal imaging of PMMA Opal/$NaYF_4$:Yb^{3+},Tm^{3+}/Er^{3+} nanocrystals. *Chem Commun* **2013**, *49* (36), 3781-3783.

Chapter 12

Functional Dendrimers as Nanoscale Theranostic Vehicles for Cancer Treatment

*Kui Luo and Zhongwei Gu**

National Engineering Research Center for Biomaterials, Sichuan University, Chengdu 610064, China

With nanotechnology advancements in the field of biomaterials, functional nanoparticles have been designed as vehicles for cancer diagnosis and therapy. Early clinical results demonstrated that nanoscale imaging probes and nanoparticles therapeutics can result in increased accumulation in tumor and enhanced anticancer efficacy as well as reduced side effects due to the passive targeting features *via* enhanced permeability and retention (EPR) effect. Based on the results, theranostic nanomedicine, which combines therapeutic and diagnostic capabilities in a single entity for personalized medicine, has emerged. There are ongoing efforts on concurrent molecular imaging of biomarkers, delivery of therapeutic agents, and monitoring of the vehicles' behaviors *in vivo* and therapeutic outcomes. Among the emergent formulations, polymer-based theranostic vehicles have showed great potential towards clinical impact for the foreseeable future. Dendrimers are a class of well defined macromolecules with a central core, low-density branching units and high-density surface functional groups. Compared to other polymeric counterparts, dendrimers are nanosized and symmetrical, with precise structures, low polydispersity, controllable molecular size as well as highly adaptable and flexible surface chemistry. These unique features have made dendrimers of increasing interest for design as imaging probes, drug delivery systems and theranostic vehicles for cancer treatment. This chapter mainly focuses on the recent developments in the field of nanoscale theranostic vehicles based on dendrimers for cancer treatment, and provides examples of these nanoscale theranostic carriers investigated in different cancer therapies. The novel developments of dendrimer based nanoscale theranostic vehicles described here provide great potential to achieve better therapeutic effects in cancer therapy guided in real-time.

12.1 Introduction

The quest for highly efficient and safe strategies has become the key factor for cancer treatments. Chemotherapy, one of the important types of cancer treatment, is effective for tumor suppression and cancer therapy. However, the

* Corresponding author: zwgu@scu.edu.cn

current clinical anticancer agents are mostly small molecules and have demonstrated some challenges which are needed to be addressed, such as the non-specific cytotoxicity, poor water-solubility, non-specific cellular uptake, unfavorable pharmacokinetics and drug resistance of cancers, which lead to low therapeutic index and high side effects and mortality. On the other hand, the cancer diagnosis is also important to the cancer treatment. Currently used molecular imaging modalities, such as magnetic resonance imaging (MRI), radionuclide-based imaging, computerized tomography (CT), ultrasound and optical imaging (bioluminescence and fluorescence),[1,2] provide the characterization at the cellular and subcellular levels at early stage of cancers. The imaging probes have been widely used in those imaging modalities for their predominant positive signal enhancement. The limitations in the clinical use of imaging probes are non-specificity, low relatively signal enhancing effect and fast elimination from the circulatory system due to the low molecular weight.[3] The development of highly efficient and safe imaging probes has also become one of the key factors for successful application of imaging modalities.

The emergence of nanomedicines provides the possibility to overcome the current shortcomings associated with cancer treatments.[4] Among the variety of nanoparticles, liposomes and polymeric nanoscale vehicles likely have the great potential clinical impact for the foreseeable future,[5] and they either have been already approved by FDA or are in clinical development for cancer treatments. Currently, nanoscale drug delivery vehicles have been widely investigated for cancer therapy,[6,7] suggesting enhanced therapeutic indexes, while simultaneously reducing side effects, owing to properties such as passively targeting tumors *via* the EPR effect and active cellular uptake.[6,8] It is worth noted that natural and synthetic polymers based nanoscale systems, such as polyethylene glycol (PEG)[9], polyglutamic acid (PGA)[10], polysaccharides[11], *N*-(2-hydroxypropyl)methacrylamide (HPMA) copolymer[12] and dendrimers[13], have been widely proved to be targeting nanoscale drug delivery vehicles for cancer diagnosis and therapy.

In the conventional cancer therapy, the therapeutic ability can be confirmed *via* imaging. However, the therapeutic progress of disease mediated by the administrated anticancer agents can hardly be monitored in real-time. The therapeutic agents and imaging probes are administrated separately. Therefore, this divided therapy and imaging strategy is not the optimal therapeutic time window for diseases, which is needed repeated administration and results in higher cost and suffering of patients. It is necessary to develop integrated platform combined with diagnostic and therapeutic agents for dynamically monitoring the administrated agent biodistribution *in vivo*, the progress of diseases and the therapeutic ability and efficacy.[14] With the development of biomaterials for cancer diagnosis and therapy, recently, the emerged concept termed as "theranostics", which integrates diagnostic and anticancer agents into a single system, is expected to play an important role in cancer therapy dynamically guided and monitored by imaging.[14,15]The nanoparticles, owing to the rich encapsulation and surface chemistry, are potential theranostic platforms simultaneously loading molecular imaging and therapeutic agents. For instance, iron oxide nanoparticles, being used as T_2 contrast agents, was functionalized with hydrophilic polymers and conjugated with chemotherapeutic agents, resulting in nanoscale systems for imaging guided cancer therapy.[14,16] Additionally, the iron oxide nanoparticles also can be encapsulated into micelle-based drug delivery systems, which have been investigated for dual imaging and drug delivery purposes.[17,18]

Dendrimers are hyperbranched and globular macromolecules with a multivalent surface (nanoscaffold), branched shells and a core attached with dendrons.[19,20] Since, Tomalia *et al.* successfully synthesized dendrimers poly amidoamine (PAMAM)[21], dendrimers with nanoscale sizes, such as PAMAM, poly(propyleneimine) (PPI) dendrimers and peptide dendrimers, have been investigated as biomaterials with attractive characteristics for nanomedicines, such as MRI contrast agents[22–24], drug/gene delivery systems[25–28]. Especially during the last few decades, dendrimers have been employed for drug delivery and imaging probes for cancer treatments, suggesting great potential alternative route to the development of nanoscale drug/imaging probes vehicles due to their features, such as precise nanostructure, low polydispersity, controllable molecular size, highly adaptable and flexible surface chemistry, and chemically stable molecular entities.[29–31] Importantly, the highly branched, multivalent surface and rich surface groups make them optimal candidates for a variety of functionalization for cancer treatment, such as drug conjugation, cancer targeting moieties attachment, imaging probes label. It is also reported that the dendrimers with unique structures can be employed to directly encapsulate anticancer agents that will be shuttled to the cytoplasm of cancer cells.[32] Compared to linear polymeric vehicles, the highly branched, globular architecture of these dendrimers gives rise to a number of interesting properties, such as increased solubility and very low intrinsic viscosities. The multivalent dendritic architectures with functional moieties have showed significantly higher efficient cellular uptake than linear polymeric drug delivery systems, such as PEG[33],and *N*-(2-hydroxypropyl)methacrylamide (HPMA) based vehicles.[34,35] Additionally, the dendritic drug delivery systems with optimal nanoscale sizes have demonstrated longer blood circulation time and increased drug accumulation in target site (tumor). Those attractive results can be attributed to the unique structural features, nanoscale size and compact spherical geometry in solution.[36] Owing to short clinical experience with dendrimers, however, a rational design of safe and nontoxic drug delivery system is still not possible.[37,38] Recently, peptide dendrimers synthesized from amino acids, such as lysine-based ones, have been used as drug/gene delivery systems and imaging probes both *in vitro* and *in vivo* studies, showing properties similar to proteins-water solubility, resistance to proteolytic digestion and good biocompatibility due to biodegradable property of the peptide branching units.[38,39]

In this chapter, we will concentrate on functional and multivalent dendrimers as targeting nanoscale theranostic drug delivery systems based on the EPR effect (passive targeting) and introduction of targeting moieties (active targeting) which have been used to delivery drug into tumors for cancer therapy guided by imaging, which are the most numerous and advanced.

12.2 Dendrimers for Cancer Diagnosis and Therapy

Owing to the properties of the endothelial lining of the blood vessel wall in many tumors,[40] such as the much more permeability than in the normal tissues,[41] the large gaps exist between endothelial cells in tumor blood vessels and tumor tissues provides selective extravasation and retention of nanoscale particles, which have showed great promise for enhancing the accumulation of nanoscale vehicles in tumors *via* EPR effects.[42] For dendrimer based drug delivery for cancer therapy, the anticancer agents can be encapsulated into the internal voids of dendrimers *via* physical incorporation based on host-guest chemistry, or by covalent conjugation onto the dendritic architecture.[43] Previous studies have showed that dendrimer/drug complexes and its conjugates can be applied *in vivo* by any administration such as oral, intravenous, intratumoral, intraperitoneal, ocular and transdermal ones.[13,44] On the other hand, the dendritic architectures were labeled with imaging moieties, or encapsulated with imaging nanoparticles, such as gold and SPIOs,[45–47] resulting in dendritic imaging probes for cancer diagnosis. Previous studies also have demonstrated that the cut-off size of the permeabilized vasculature varies from case to case. Therefore, the spontaneous "passive" targeting efficacy of nanoscale vehicles can be optimized *via* controlled size. It is well accepted that the diameters of nanoscale vehicles should be less than 200 nm to avoid uptake by the reticuloendothelial system (RES), and the sizes of nanoparticles should be controlled in the range of 30–150 nm, which are effective for passive targeted drug delivery or nanoscale imaging probes to solid tumors,[48] since the nanoscale vehicles with suitable sizes have been long-circulating in the blood to extend periods of time, resulting in a sufficient level of accumulation in the tumor tissue.

Therefore, for dendrimer based nanoscale vehicles for cancer treatments, the drug delivery and tumor targeting efficacy depend on the molecular weight of the dendrimers associated with the generations. The dendrimers with high generation showed higher drug encapsulation capability and can be assembled into nanoparticles suitable for passive targeting. However, for most of the dendrimers, such as PAMAM and PPI dendrimers, the size is below 10 nm. To address this issue, the generations can be increased to enhance drug loading and nanosizes. However, it's not simple to prepare dendrimer with high generation due to the high steric hindrance to chemical reaction accompanied with the increment of generation. The emergence of click chemistry, owing to its high efficiency, quantitative yields and technical simplicity, has provided great potential to address the challenges, which has been employed to prepare dendrimers with high generations.[49] Additionally, the nanoparticles with negative or neutral charge can prevent form protein adsorption in blood circulation, resulting in longer blood circulation. However, for the dendrimers PPI and PAMAM, the particles showed positive charge due to primary amine at the surface and tertiary amine groups in the branch. To address those shortcomings, the dendrimers were usually functionalized with certain negative charged moieties and water-soluble polymers, such as polyethylene glycol (PEG). The functionalization significantly enhanced the nanosizes of the dendritic vehicles as well as long blood circulation. For instance, we described the synthesis of Gd(Ⅲ)-based peptide dendrimers possessing highly controlled structures and single molecular weight and their potential of use as MRI contrast agents (Fig. 12.1).[3] These agents which have no obvious cytotoxicity and the PEGylated dendritic formulations showed the highest 9-fold increase in T_1 relaxivity to 39.2 Gd(Ⅲ) $mM^{-1}\cdot s^{-1}$ compared to clinical agent of Gd-DTPA. *In vivo* studies have shown that the mPEGylated Gd(Ⅲ)-based dendrimer provided much higher signal intensity (SI) enhancement in mouse kidney at 60 min post-injection, with 54.8% relatively enhanced SI. The higher accumulations of gadolinium in mouse liver and kidney were also detected for the mPEGylated dendrimer. Meanwhile, mPEGylated dendrimer showed much higher Gd(Ⅲ) concentration in blood with 38 μg Gd(Ⅲ)/g blood at 1 h post-injection comparing to other dendrimer formulations. Meanwhile, for the PEGylated dendrimers, the Gd(Ⅲ) concentration in blood also decreased, indicating that the Gd(Ⅲ) can be cleared from the body and potential good biosafety.[3] That is attributed to the biodegradablity of peptide dendrimers.[50] These findings provide an attractive alternative strategy to the design of multifunctional dendrimer based vehicles with controlled structures for cancer treatments.

The functional dendrimers, especially PEGylated ones, for drug loading by physical incorporation or covalent drug conjugations, have also showed great possibility as drug delivery vehicles for cancer therapy. Additionally, because PEG is highly water soluble and biocompatible, the PEGylated dendrimer showed no toxicity, lethality, or abnormality in blood chemistry.[51] PEG was typically conjugated to the surface of a hydrophobic dendrimer to provide a hydrophilic shell around a hydrophobic dendritic core to form a unimolecular micelle, or self-assembled into nanoparticles, which drastically increased drug loading and solubilization, as well as the blood circulation due to the nanoscale sizes. We described the preparation of mPEGylated peptide dendron-doxorubicin (dendron-DOX) conjugate and its potential as drug delivery system

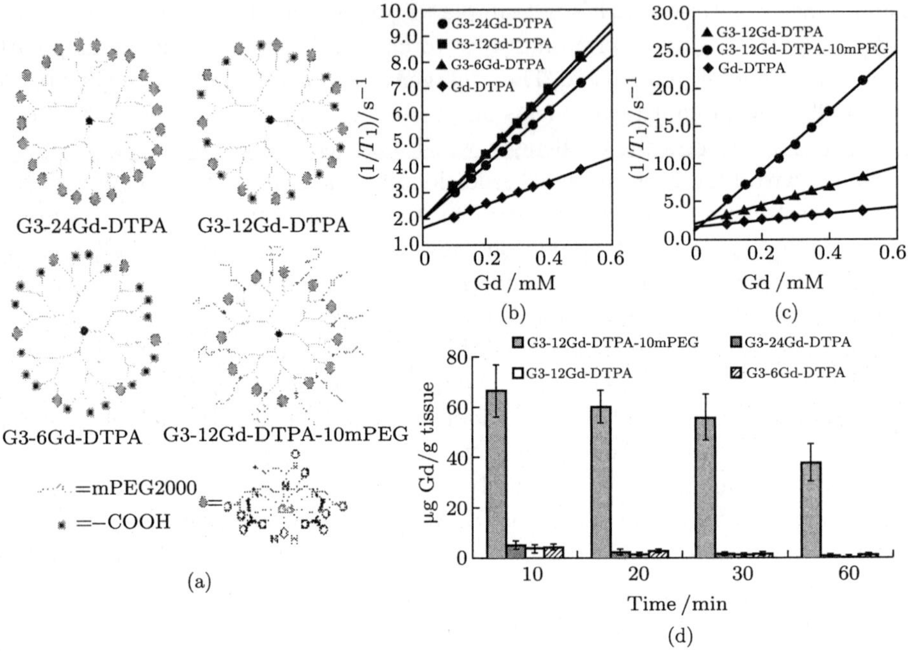

Fig. 12.1 The structures for gadolinium-based peptide dendrimers with controlled structures (a), longitudinal relaxation rate ($1/T_1$) vs the concentration of Gd(Ⅲ) in water of the synthesized peptide dendritic MRI contrast agents and Gd-DTPA at 1.5 T field (b,c), and the blood-contents time course of Gd(Ⅲ) in blood after administration of MRI contrast agents at a dose of 0.05 mmol Gd^{3+}/kg in BALB/C mice (n = 3–5 mice) (d). Reprinted from Ref. 3. Copyright (2011) with permission from Elsevier.

for breast tumor therapy.[52]The PEG was attached to the functional peptide dendron *via* click reaction; anticancer agent DOX was then conjugated to another tail of the PEGylated dendron through pH-sensitive hydrazone bond, resulting in mPEGylated peptide dendron-DOX conjugate with 14.0 wt% (weight percent) of DOX (Fig.12.2). The amphiphilic PEGylated dendron-DOX conjugate self-assembled into compact nanoparitcle with neutrally charged surface and diameter around 80 nm determinated by DLS, SEM and TEM. The nanoparticles showed pH-sensitive property. The drug release rates from the nanoparticle at pH 5.0 were much faster than those at pH 7.4 due to the pH-sensitive cleavage of the hydrazone bonds. As expected, the self-assembled nanoparticles demonstrated much higher antitumor activity and induced apoptosis on the 4T1 breast tumor model compared to the free drug DOX. Additionally, *in vivo* toxicity evaluation showed that drug-free dendron and drug-loading nanoparticle provided good biosafety in both healthy mice and tumor-bearing ones, as no significant systematic toxicity was observed *via* histological analysis. That was attributed to the biodegradability of peptide dendritic scaf-

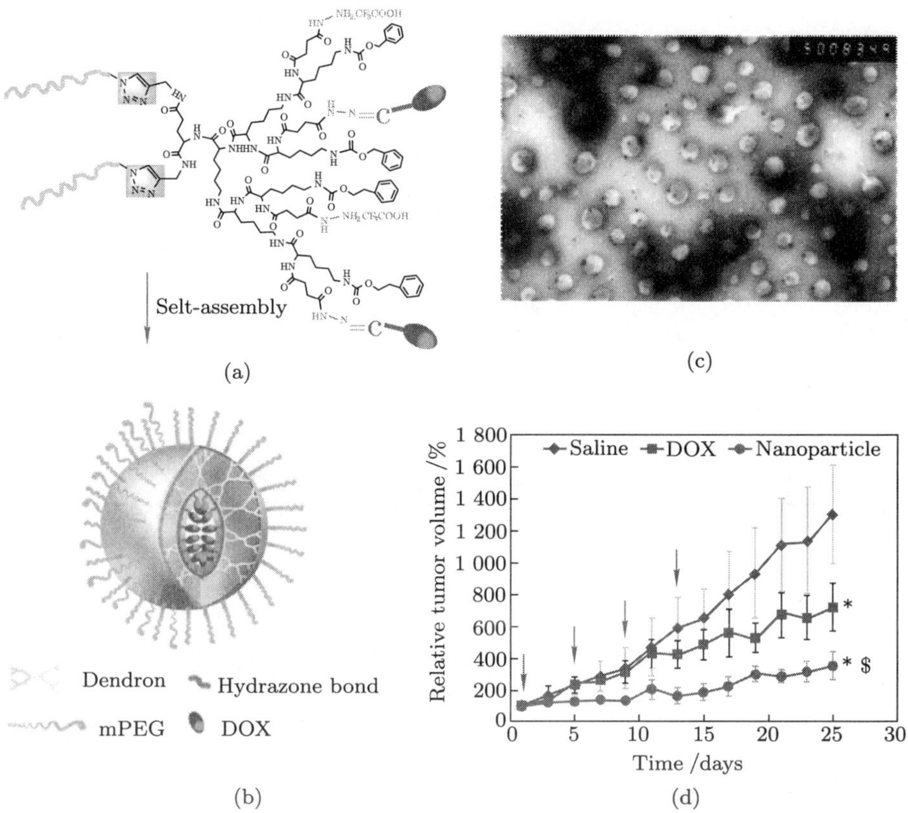

Fig. 12.2 Structures and illustration of mPEGylated peptide dendron-DOX conjugate (a), and its self-assembled nanoparticle (b). TEM image showed the size of the dendron-DOX conjugate based nanoparticle was around 80 nm (c). *In vivo* tumor growth inhibition studies showed the dendron-DOX conjugate based nanoparticle demonstrated significant tumor inhibition compared to free drug DOX (DOX) and saline in the breast tumor model ($n = 7$). ($^*p < 0.001$, compared to saline; $^\$p < 0.001$, compared to free drug DOX) (d). Reprinted from Ref. 52. Copyright (2013) with permission from Elsevier.

fold and lower side effects to normal tissues/cells mediated by nanoparticles. Therefore, the PEGylated and amphiphilic peptide dendron-DOX conjugate based nanoparticle showed promising high anticancer efficacy as well as good biosafety *in vitro* and *in vivo* by combination the features of peptide dendron, pH-sensitive linkages and nanomedicines. Another strategy to preparation of dendrimer based nanoparticles is dendronization of polymers,[53,54] and the resulted polymers with dendritic architecture were referred to as "dendronized polymers".[55,56]The amphiphilic dendronized polymers loading hydrophobic anticancer agents can self-assemble into nanoparticles taking advantage of dendrimer and linear polymer,[57,58] resulting in nanoparticles with suitable sizes for cancer therapy. This dendronized strategy can overcome the synthetic challenges associated with preparation of high generation dendrimers with satisfactory sizes for tumor passive targeting, owing to the significant increased molecular weights and nanoscale sizes (30–150 nm). For instance, the non-cytotoxic, biodegradable and water-soluble natural polysaccharide, heparin,[59,60] was dendronized with low generation peptide dendron (generation 2, G2) *via* click reaction; anticancer drug DOX was then attached to the surface of hydrophobic dendron *via* pH-sensitive hydrazone bond. This amphiphilic dendronized polymeric hybrid self-assembled into compact nanoparitcle with negatively charged surface, resulting in strong antitumor activity, high antiangiogenesis effects and induced apoptosis on the 4T1 breast tumor model, as well as no significant toxicity to normal organs of both tumor-bearing and healthy mice.[52]

Passive targeting approach is one of basis of cancer diagnosis and therapy, which is usually achieved by optimizing the size of the nanoscale vehicles. However, this strategy is still not very effective for cancer targeting, since only a small portion of the administrated drug or imaging delivery vehicles can accumulate in the solid tumor tissues. Some functional moieties, that can specifically recognize over-expression of the biomarkers in the tumor and cancer cells, have been introduced to the nanoscale carriers.[61] Therefore, this active targeting depends on the targeting moieties conjugated to the dendrimers to facilitate cell-receptor-mediated interactions or specific interactions between tissue and vehicles. Until now, the following substances, peptide, mono-, oligo- and polysaccharides, antibodies and their fragments, biotin, folate, and so on, have been employed as targeting moieties.[62,63] Owing to the nanoscale size and tumor targeting moieties, the functional nanoscale vehicles showed significant enhanced tumor targeting efficacy by combination of active targeting and passive targeting. Generally, the interactions of isolated targeting moiety-receptor are typically very weak. It is notable that, for dendrimers, multiple moieties conjugated to a single dendritic scaffold can coordinate the receptor binding in a multivalent way, and the selective interaction of moieties with receptors or cells can be significantly enhanced owing to the multivalency of dendrimer.[64–66] The significant increased targeting capability manifested by functional dendrimer has been termed as "cluster effect",[67–69] which is the feature of the functional dendrimer as targeting vehicles for cancer diagnosis and therapy. For instance, Gd-based multifunctional glycopeptide dendrimers bearing multiple galactose moieties on a single dendritic scaffold were designed and prepared as liver MRI probes.[69] These macromolecular Gd-ligand dendritic agents exhibited up to 3-fold increase in T_1 relaxivity compared to Gd-DTPA complexes.[70] The dendrimers functionalized with multiple galactosyl moieties demonstrated much higher cell uptake *in vitro* as in T_1-weighted scans. *In vivo* animal imaging studies showed the probes with galactosyl moieties provided better signal intensity (SI) enhancement in mouse liver mediated by asialoglycoprotein receptors (ASGP-R),[71] especially at 60 min post-injection, and among all the groups, the generation three (G3T) has shown the highest targeting efficacy.[70]

We have also reported functional dendrimer based drug delivery system for cancer therapy.[72,73] The dendrimers were synthesized from polyhedral oligomeric silsesquioxane (POSS) nanocubic core. The low generation dendrimers were then functionalized with targeting functions (biotin), and the anticancer drug DOX was covalently conjugated to the dendritic scaffold *via* pH-sensitive linkage. The multifunctional dendritic conjugate aggregated into compact nanoparticles with diameters around 100 nm, and showed pH-sensitive drug release property, as the drug release rates at pH 5.0 were much faster than those at pH 7.0. The *in vitro* confocal laser scanning microscopy (CLSM) and flow cytometry studies demonstrated that the cellular uptake by HeLa cells and internalization of the nanoparticles with biotin compared to the nanoparticles without biotin. Simultaneously, the biotin functionalized nanoparticles showed higher cytotoxicity to mice breast cancer 4T1 cells, as the IC50 (2.3 mg/mL) was lower compared to the nanoparticles without biotin (4.5 mg/mL). Importantly, the biotin functionalized nanoparticles resulted in high *in vivo* anticancer activity on mice xerograft breast cancer models with the tumor inhibition rate up to 60%, compared to the ones without biotin (43%) and free drug DOX (38%).[73] By combination of the EPR effect, the biotin functionalization may enhance the accumulation of the drug delivery systems and cell uptake, resulting in the highest anticancer activity.

12.3 Functional Dendrimers as Theranostic Nanoscale Vehicles for Cancer Treatment

12.3.1 *Dendrimer Based Photodynamic Therapy*

Photodynamic therapy (PDT) is now a well-recognized clinical therapeutic modality for the treatment of cancer, and great advances have been observed in the field in recent years. PDT involves three elements: a photosensitizer, light and oxygen. Porphyrins, their derivatives and porphyrin-inducing drugs are the most commonly used photosensitizers.[74] Compared to chemotherapy and radiotherapy, the PDT used in cancer therapy has showed some advantages due to the relative specificity and selectivity of PDT along with the absence of harmful side effects. Dendrimers, as proven drug delivery vehicles, were designed to carry PDT agents for cancer PDT.[75] For instance, a water-soluble, anionic surface functional and biocompatible dendrimer prepared *via* condensing phloroglucinol and succinic acid (PGSA) was employed to efficiently encapsulate a well known hydrophobic PDT agent of protoporphyrin IX (PpIX).[76] The cytotoxicity of the drug loaded dendrimer formulation towards dalton lymphoma ascites (DLA) cancer cell lines upon visible light treatment was studied and the cytotoxic reactive oxygen species (ROS) generation efficiency of the free PpIX and its dendrimer forms was evaluated. The results showed that, compared to free PpIX, the prepared PGSA dendrimer based systems reduced the toxicity of PpIX without light treatment. In contrast, treatment of DLA cells with drug loaded dendrimer form with light treatment resulted in significant enhancement in therapeutic efficacy of PpIX. Additionally, owing to the outstanding near-infrared (NIR) optical properties of PDT agents, PDT has been demonstrated to have great potential to be an efficient theranostic approach for cancer therapy and visualization of malignant tissue by fluorescence imaging.

Dendrimers, owing to its internal cavities of dendrimer structures features, can be utilized to encapsulate PDT agents. However, the drug loading efficiency depends on the generation of the dendrimer and its size. In order to enhance the drug loading and EPR effects of the dendrimers, herein, PEGylation has still been proved to be usable strategy. Kono *et al.* described two PEGylated dendrimers derived from PAMAM and PPI dendrimers (PEG-PAMAM and PEG-PPI), and their potential delivery of photosensitizers rose bengal (RB) and PpIX for PDT. The PEGyated dendrimers showed lower drug loading of PpIX molecules than RB, but the formed complexes were more stable under physiological conditions. Compared to PEG-PAMAM dendrimer, the PEG-PPI formed more stable complexes with photosensitizers due to their inner hydrophobicity. The complex of PpIX with PEG-PPI exhibited efficient cytotoxicity *in vitro* upon visible light treatment, compared with free PpIX, suggesting that PEGylated dendrimers are promising vehicles for PDT.[77] Recently, phthalocyanines (Pc), as photosensitizers, have promising potential as theranostic agents for fluorescence image-guided drug delivery and noninvasive treatment of deep tumors by PDT because of the outstanding NIR optical properties. To overcome some shortcomings of phthalocyanines, such as poor water solubility, aggregation and insufficient selectivity for cancer cells, PPI dendrimer (generation 4) were PEGylated and functionalized with luteinizing hormone-releasing hormone (LHRH) peptide for increasing biocompatibility and tumor-targeted delivery, while Pc molecule was modified with a hydrophobic linker for the significant enhancement of physical encapsulation.[78] The functionalized dendritic nanocarriers have an average diameter of 62.3 nm and narrow size distribution with a PDI of 0.100, which was suitable as drug delivery for cancer diagnosis and therapy *via* EPR effect (30–150 nm suitable for passive targeting). The high drug loading efficiency was observed (20% w/w), and the nanoscale drug delivery vehicles capsulated phthalocyanine derivative exhibited a distinct NIR absorption (700 nm) and fluorescence emission (710 and 815 nm), which are required for an efficient theranostic agent with both of PDT and fluorescence imaging (Fig.12.3). It was demonstrated that subcellular localization *in vitro* and organ distribution *in vivo* of the functional dendrimer-based nanocarrier can be determined due to the intrinsic fluorescence properties of phthalocyanine. The dendritic theranostic systems exhibited low cytotoxicity (IC50 = 28 μg/mL) without light treatment while significant PDT effects (IC50 = 0.9 μg/mL) was obtained through excessive generation of toxic reactive oxygen species with light treatment, suggesting significant potential of the functional dendrimer-based nanocarrier as efficient NIR theranostic vehicles.[78]

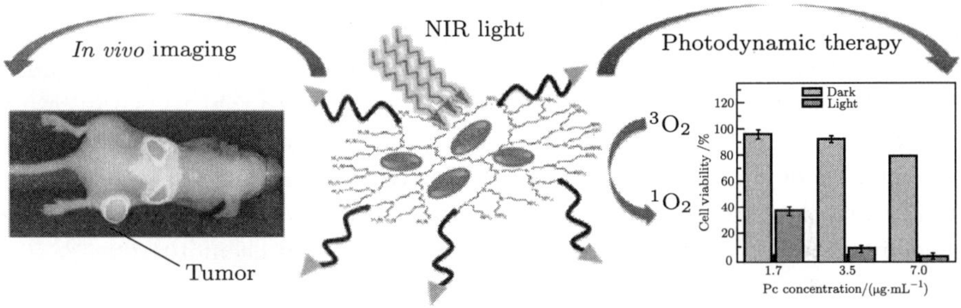

Fig. 12.3 The illustration of functinal dendrimer based theranostic vehicles and their *in vivo* images of tumor-bearing mouse, and the relative intracellular ROS levels in A2780/AD cells incubated with theranostic dendritic nanocarriers.[78] Reprinted from Ref. 78. Copyright (2013) American Chemical Society.

Although physical incorporation of PDT agents using dendrimers has some straightforward advantages, such as labile molecules degradation and eliminating solubility issue associated with hydrophobic compounds, there are still some problems associated with encapsulation approach, including low stability in terms of storage and burst release, leading to less pharmacokinetic or therapeutic indexes. To address the drawbacks caused by encapsulation, chemical conjugation serves to prepare dendrimer based delivery vehicles for PDT. PDT agents were covalent conjugated to dendritic scaffold to achieve increased drug loading. One example of such system is the 5-aminolaevulinic acid (ALA) based dendrimers. The endogenous protoporphyrin IX of ALA has been employed as PDT agent. However, the efficacy of ALA PDT agent is sub-optimal for thicker tumor due to its hydrophilic nature, which required efficient delivery vehicles to carry it to tumor. In order to improve the efficacy of the delivery of ALA to desired sites, the dendritic derivatives bearing several ALA molecules were prepared,[79–81] and their potential application for PDT *in vitro* and *in vivo* were investigated in a murine tumor model using systemic i.p. administration. The synthesized dendrimer was more efficient than free ALA for porphyrin synthesis at low concentrations, which was correlated well with higher cellular ALA dendrimer accumulation *in vitro*. *In vivo*, the porphyrin kinetics from free ALA showed an early peak in most tissues after 3 and 4 h administration. It's noted that dendrimer with several ALA molecules leaded to sustained porphyrin production for over 24 h.[82] Therefore, ALA based dendrimers may be synthesized and employed for metronomic PDT with low-dose ALA-PDT treatments. Another PDT agent used to prepare conjugated dendrimer for PDT is porphyrin, and it was often utilized as the core for dendrimers synthesis. The porphyrin based dendrimer can be used to encapsulate anticancer agents, and the theranostic vehicles for cancer combination therapy, resulting in significantly enhanced therapeutic indexes. Ma *et al.*[83] reported a novel porphyrin-poly(L-lysine) dendrons (PP-PLLD), where the porphyrin was as the core of dendrimers. The amphiphilic PP-PLLD dendrimer has a low critical micelle concentration in an aqueous solution, and can load anticancer drug DOX with a loading amount of 64 μg/mg. *In vitro* toxicity assay results showed that the PP-PLLD loading DOX demonstrated a higher cytotoxicity under the light condition than PP-PLLD or DOX alone, suggesting PP-PLLD and its drug loaded systems have a potential application as theranostic vehicles by combination of imaging, photodynamic therapy and chemotherapy.[83]

12.3.2 *Dendrimer/Magnetic Nanoparticle Hybrid Nanoparticles*

The studies on inorganic-organic hybrid for cancer chemotherapy and diagnostic imaging have attracted ever-increasing attention due to their unique physical characteristics and the capability of targeting cancer, whereas the inorganic parts may be metallic or inorganic nanoparticles and the organic shells which are hydrophilic polymers to improve the solubility and carry the chemotherapy and imaging agents, enhancing their potential performance in tumor for targeting diagnosis and therapy. Among the inorganic nanoparticles, imaging probe of gold and monodisperse magnetic nanoparticles, such as superparamagnetic iron oxide (SPIO) nanoparticles have been investigated for the cancer theranostics. There are two strategies to prepare dendrimer/inorganic nanoparticles hybrids for cancer diagnosis and therapy. Firstly, the unique structural feature of dendrimers allows them to be employed as templates for the preparation of various nanoscale vehicles *via* encapsulating inorganic nanoparticles or assembling with nanoparticles. Additionally, the rich surface chemistry of the dendrimers provides possibility to the surface functionalization with various biomoleculars, such as hydrophilic polymers, tumor targeting moieties, resulting in good colloidal stability, biocompatibility, and biofunctionality.[84]Secondly, tunable surface chemistry of nanoparticles allows their surface functionalization *via* conjugating dendrimers.[85,86] The formed nanoparticle core/dendrimer shell composite hybrid can be further employed as imaging probes, drug/gene delivery vehicles or theranostic nonocarriers.

SPIO have been investigated as theranostics vehicles owing to their current and potential usefulness as magnetic resonance imaging contrast agents, surface chemistry for drug loading, as well as colloidal mediators for cancer magnetic hyperthermia for targeting cancer therapy.[87–90] The SPIO based theranostics vehicles have demonstrated combined advantages such as T_2-type magnetic resonance imaging contrast enhancement, monitoring *in vivo* transport under external magnetic field, as well as hyperthermia cancer therapy *via* alternating magnetic field.[91,92] The surface functionalized SPIO based nanoparticles have been employed as MRI diagnostic imaging probes[93], biosensing systems, and targeted delivery therapeutic agents, such as chemotherapeutic drugs[14,94–97], DNA[98–100], and small interfering RNA (siRNA).[94,101–103]. For instance, Liu *et al.* reported the fabrication of SPIO/copolymer hybrid micelles.[17] The amphiphilic block copolymers poly(ε-caprolactone)-b-poly(glycerol monomethacrylate), PCL-b-PGMA, and PCL-b-P(OEGMA-co-FA) were modified with oligo(ethylene glycol) monomethyl ether methacrylate and folic acid-bearing moieties, respectively. The functional polymers were employed to physically encapsulate hydrophobic drugs paclitaxel (PTX) and SPIO nanoparticles within hydrophilic coronas, resulting in theranostics micelles. As MRI imaging probes, the clustering of SPIO nanoparticles within micellar coronas resulted significantly enhanced T_2 relaxivity ($r_2 = 121.1\ \text{s}^{-1}\cdot\ \text{mM}^{-1}$ Fe), which was much higher than that of surfactant-stabilized single SPIO nanoparticles ($r_2 = 28.3\ \text{s}^{-1}\cdot\text{mM}^{-1}$ Fe), suggesting that SPIO/polymers hybrid micelles can serve as a T_2-weighted MRI imaging contrast enhancer. Additionally, the controlled and sustained release of drug PTX from micelles was achieved due to the cumulative release of ~61% encapsulated drugs over ~130 h. For the SPIO micelles based nanoscale vehicles, the much higher anticancer efficacy can be achieved *via* modifying the surface with tumor-targeting moieties. Gao *et al.* report the preparation of multifunctional polymeric micelles loading with SPIO nanoparticles, anticancer drug DOX and cancer-targeting capability *via* $\alpha_v\beta_3$ integrins. The multifunctional polymeric micelles were developed to provide effective targeting to $\alpha_v\beta_3$-integrin receptors owing to the presence of peptide cRGD on the micelle surface, ultrasensitive detection by MRI, and growth inhibition in tumor SLK cells.[104] The amphiphilic block copolymer micelle embedded with SPIO nanoparticle clusters at the hydrophilic corona may be employed as multifunctional theranostics nanoscale vehicles for targeting drug delivery guided by MRI imaging.

Compared to micelles, the surface functionalized SPIO may be more suitable as biomaterials due to the better dispersibility and stability in aqueous solution,[105] since the functional ligands used to SPIO surface functionalization are often hydrophilic organic compounds/polymers. However, it is not simple to prepare surface functionalized organic/inorganic hybrid nanoparticles suitable for covalent transporting imaging and anticancer agents, which involved multistep synthesis. Two strategies, known as surface encapsulation and ligand-exchange, are commonly employed to prepare hydrophilic functionalized SPIO. The surface encapsulation often employs amphiphilic polymer to form micelle-like double-layer structures outside the hydrophobic nanoparticles. This strategy is suitable for large-scale preparation of functional SPIO based nanoparticles. However, several drawbacks for this approach have also been observed, such as instability under physiological conditions because of the weak interaction between hydrophobic double-layer structures, small size (often below 30 nm) resulting in shorter blood circulation time *in vivo*, and difficulty in manipulating the number of terminal functionalities for controllable bioconjugation due to the uncertainty in the number functional groups of polymer. As an alternative strategy, ligand-exchange is a novel method, since the functional ligands was conjugated to the SPIO *via* covalent bone and the original ligands can be displaced without changing the intrinsic properties of the iron core. The functionalized SPIO resulted from this approach are more stable due to the stronger coordinate bonds during the surfactant exchange reaction. Therefore, this method can address the inherent drawbacks of the surface-encapsulation one.

We have reported the preparation of peptide dendrimers functionalized SPIO using ligand-exchange method.[106] Firstly, amino acids L-lysine and L-glutamic acid based dendrimer ligands with a dopamine located at the focal point were synthesized. Secondly, the tail functionalized different generation dendrimers were conjugated to the surface of the SPIOs *via* ligand-exchange, resulting in functional SPIO nanoparticles with precisely controlled terminal functional groups and definite sizes. The two series of dendrimers functionalized nanoparticles were fully characterized by TEM, XRD, DLS, TGA, FTIR, VSM and zeta potential measurements. The nanoparticles modified with peptide dendrimers showed superior stability due to the water-soluble ligands. For the saturation magnetization of different functionalized SPIOs nanoparticles, no significant changes were observed. Recently, core-shell SPIO nanoparticles functionalized with fluorescein-modified dendrons were reported.[107] Compared to linear analogue, the dendronized molecules were confirmed to increase the surface functionalization, providing possibility of conjugating different biological molecules, such as imaging probes. Additionally, the alternative way of functionalization of core-shell SPIO nanoparticles by well-defined dendritic molecules allows the tunable functionalization of the MNPs' surface, where the rate of functionalization increases with the generation of the dendritic molecules. Therefore, the rich surface functional groups of SPIO nanoparticles functionalized with biocompatible dendrimers provide possibility of conjugating anticancer agents and imaging probes for cancer diagnosis and therapy.[108]

APTS-coated Fe_3O_4 nanoparticles were functionalized by arginine-based dendritic block *via* conventional growth approach of Michael addition/amidation reactions. The thus-obtained dendritic magnetite nanocarriers with an average size of 10 nm showed superparamagnetism with high magnetization values at room temperature. The aqueous colloidal suspension of the functionalized magnetite nanocarriers (10 mg/mL of Fe) exhibited a temperature rise up to

43 °C in 5 min and yielded a moderate specific absorption rate (SAR) value of 30 W/g of magnetite under the influence of AC magnetic field of 10 kA/m and 425 kHz frequency. Biosafety study on human cervical cancer cells indicated that the magnetite nanocarriers functionalized with dendritic polymer had no significant cytotoxicity. Additionally, the nanocarriers showed good capacity to encapsulate DOX due to the high drug loading (w/w, 65%). The magnetite nanocarriers demonstrated a pH-sensitive drug release of 54% at pH 5.0, while the release of DOX increased up to 80% on application of an AC magnetic field.[109]

As an alternative or complementary therapy in the treatment of cancer, gene therapy has been widely studied. The bottleneck to gene therapy is to delivery designed gene to desired site (tumor). Dendrimer have been employed as gene delivery, whereas the gene transfection efficiency with high generation dendrimer companied with high molecular weight turn to be higher. However, the high generation dendrimer showed high cytotoxicity. One strategy to address the drawbacks is to attach several low generation dendrimers to the same nanoparticle, such as SPIO,[110,111] which leads to an efficient DNA complexation and gene transfection.[112] Vallet-Regí *et al.* reported low generation PPI dendrimer functionalized SPIO nanoparticles as nonviral gene vehicles, where the dendrimers were attached to the SPIO through covalent bonds *via* a one-pot sol-gel synthetic route.[113] The core-shell hybrid nanocarriers were water soluble as colloidal ferrofluids which can be long-term stable in water at physiological pH. *In vitro* gene transfection magnified by the dendrimer functionalized SPIO nanoparticles were assayed with Saos-2 osteoblasts, using as reporter gene a plasmid DNA that codes for the green fluorescent protein. The gene transfection studies showed the transfection efficiency strongly depends on the presence of the magnetic field and the generation of dendrimer, whereas the generation 3 PPI dendrimer functionalized SPIO nanoparticles induced highest transfection efficiency. The nanosystems couple the DNA fragments and safe transport under magnetic stimulus, indicating a potential strategy to design gene vehicles delivering gene in a short time while keeping the structure integrity.[113]

12.3.3 *Gadolinium(Ⅲ)-labeled Dendrimers Based Drug Delivery Vehicles*

Gadolinium(Ⅲ)-chelates have been proved as the most commonly used T_1 MRI contrast agents, since they have the ability to generate a positive image contrast. The nanoparticles loaded with gadolinium(Ⅲ)-chelates, such as Gd-DTPA and Gd-DOTA, have showed significant potential as novel MRI contrast agents with upon increasing the longitudinal relaxation rates of surrounding water protons, leading to lower doses and higher positive single intensity. For instance, functional dendrimers with gadolinium(Ⅲ)-chelates at the surface have been widely investigated macromolecular MRI contrast agents, showing great possibility of application in cancer diagnosis. Combination of anticancer agents, gadolinium(Ⅲ)-chelates based nanoscale vehicles have been employed as theranostic systems for cancer treatment. Tomalia *et al.* has also proposed the nanoscale dendrimers can be as multi-purpose nanodevices for anticancer drug delivery and diagnostic MRI contrast agents.[114] Recently, Kono and coworkers[115] have reported a multi-functional liposomes based on temperature-sensitive copoly(EOEOVE-block-ODVE) and PAMAM dendrimers (Fig.12.4). The lipsomes have temperature-responsive drug release features and MR imaging functions (Gd-chelate-attached dendrimer). The anticancer drug DOX was then successfully loaded into the liposomes resulting in stable theranostic vehicles capable of retaining the therapeutic agent within their interior at physiological temperatures and releasing it at temperatures above 40 °C. The *in vivo* accumulation processes of the liposome was monitored *via* MR imaging, suggesting the size of liposome affected their accumulation efficiency in tumors (100 nm diameter were better than that of 50 nm diameter). Finally, the *in vivo* anticancer studies showed the successful DOX release from the DOX-loaded liposomes upon temperature increase, and tumor growth was strongly suppressed when the tumor was heated mildly at 44 °C for 10 min at 8 h after administration.[115]

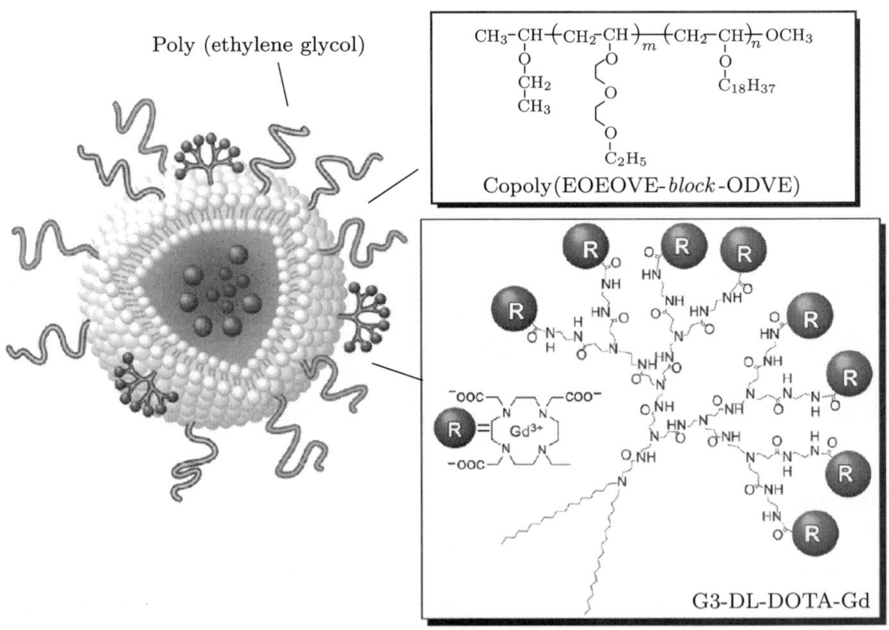

Fig. 12.4 Schematic illustration of design of copoly(EOEOVE-block-ODVE) and PAMAM dendrimer based liposome with temperature-controlled drug release feature and DOTA-Gd MRI imaging probes. Reprinted from Ref. 115. Copyright (2011) with permission from Elsevier.

12.4 Functional Dendrimers Labeled with Optical Probes as Drug Delivery Vehicles

For the anticancer agents without imaging property, the anticancer efficacy mediated by dendrimer *in vitro* and *in vivo*, and the biodistribution *in vivo* can be guided by fluorescence imaging.[116] After the dendrimer labeled with fluorescence imaging probes, such as fluoroisothiocynate (FITC), Cy5 and Cy5.5, the interaction with cells and their cellular entry can be visualized by flow cytometry, confocal microscopy and UV/visible spectroscopy.[29,117]For instance, NIR Cy5 dye has labeled drug delivery vehicles to monitor their accumulation in tumors, biodistribution, tumor targeting and therapy efficacy. Backer *et al.* reported the functional PAMAM dendrimers for boron neutron capture therapy (BNCT) of cancer.[118] The generation 5 dendrimer with 128 amino groups at the surface was firstly functionalized with 105 to 110 vascular endothelial growth factors (VEGF). To facilitate *in vivo* analysis, the boronated dendrimer was tagged with NIR dye Cy5, for NIR fluorescent imaging of their biodistribution and uptake in tumor-bearing mice. The boronated dendrimer conjugated with VEGF and Cy5 (VEGF-BD/Cy5) was proposed to target upregulate VEGF receptors (VEGFR), which are overexpressed on tumor neovasculature. The VEGF-BD/Cy5 bioconjugate showed no cytotoxicity to HEK293 cells. NIR fluorescent imaging was employed to monitor the *in vivo* activities in 4T1 mouse breast carcinoma, demonstrating selective accumulation of VEGF-BD/Cy5, but not the dendrimer without VEGF (BD/Cy5), particularly at the tumor periphery where angiogenesis was most active. The functional dendritic vehicle (VEGF-BD/Cy5) was designed to serve as a diagnostic tool as well as a potential therapeutic agent, suggesting potential as theranostic vehicle for BNCT of tumor neovasculature.

It is similar with nanoscale drug delivery with active targeting moieties, the multifunctional dendrimer based theranostic systems for cancer treatment were designed to enhance their accumulation in tumors and result in satisfactory anticancer efficacy. For instance, the primary amino groups on the surface of PAMAM dendrimer (generation 5) were firstly partly neutralized through partial acetylation, providing enhanced solubility of the dendrimer and preventing protein adsorption and nonspecific targeting interactions *in vivo*. The imaging probe FITC, targeting moiety folic acid (FA), and chemotherapeutic drug paclitaxel were conjugated to the acetylated dendrimer *via* the reaction with remaining nonacetylated primary amino groups. These multifunctional theranositc dendritic conjugates have been investigated *in vitro* for targeted delivery of paclitaxel to cancer cells, guided by fluorescence imaging.[119] It was also reported that the partially acetylated PAMAM dendrimer was conjugated with the targeting moiety biotin and the FITC, and the resulting theranostic dendritic conjugate exhibited much higher cellular uptake into HeLa cells than the conjugate without biotin as revealed by flow cytometry and confocal microscopy.[120] As gene delivery vectors for cancer treatment, the G5 PAMAM dendrimer functionalized with FITC and FA was self-assembled with complementary DNA oligonucleotides (5′-phosphate-modified 34-base-long oligonucleotides) to produce gene delivery system with "cluster effects" feature that target cancer cells that overexpress the high-affinity folate receptor. *In vitro* studies of the DNA/dendrimer nanoparticles showed specific binding to KB cells,[121] suggesting functional dendrimers with targeting moieties and imaging probes can be employed as theranostic vehicles for gene delivery.

Although dendrimers, such as PPI, PAMAM and peptide dendrimers, have been utilized as gene delivery vectors for gene therapy, the major limitation preventing its application is the low gene transfection efficacy. It should be noted that genetically engineered replication-selective adenoviruses represent efficient gene transfer vehicles. The clinical application of adenovirus-mediated gene therapy still faces hurdles that strongly limit efficient and safe application, such as immune and inflammatory responses, elimination by neutralizing antibodies, widespread expression of the coxsackieadenovirus receptor (CAR), and significant accumulation in the liver.[122] Recently, chemical modification of the adenovirus has been employed to overcome the limitations. Spitzweg *et al.* used PAMAM dendrimers with positively charged terminal amines to modify the negatively charged adenoviral capsid *via* virtue of electrostatic interaction, resulting in efficient internalization and transduction of tumor cells *in vitro*.[123] To further investigate this dendrimer/adenovirus-mediated gene delivery system, Spitzweg and his coworkers coated replication-deficient (Ad5-CMV/NIS) (CMV: cytomegalovirus) and replication-selective (Ad5-E1/AFP-E3/NIS) adenovirus serotype 5 with PAMAM dendrimers (generation 5), resulting in two nanoscale gene delivery carriers for carrying the hNIS gene. Additionally, these particles were coated virus particles by ^{123}I scintigraphy for systemic radiovirotherapy in a liver cancer xenograft mouse model. The sodium iodide symporter (NIS) as a theranostic gene was designed to guide their *in vivo* biodistribution, transduction efficacy and tumor treatment efficacy *via* combination of systemic oncolytic virotherapy and NIS-mediated radiotherapy. *In vitro*, the dendrimer/Ad5-CMV/NIS system showed partial protection from neutralizing antibodies and enhanced gene transfection efficacy in CAR-negative cells. For *in vivo* accumulation in the mice bearing subcutaneous HuH7 xenografts, radioiodine accumulation in tumor was approximately 13 % ID/g ^{123}I (biologic half-life, 3.5 h), while the gene delivery system was coated with dendrimer, the NIS-mediated radioiodine uptake in the liver was significantly reduced as compared with injection of the uncoated one after administration of dc300Ad5-CMV/NIS. After a single intravenous injection of the conditionally replicating Ad5-E1/AFP-E3/NIS, the dendrimer coated systems (dc300Ad5-E1/AFP-E3/NIS) showed a significant anticancer efficiency compared with the injection of uncoated adenovirus (Ad5-E1/AFP-E3/NIS). These results demonstrated efficient liver detargeting and tumor retargeting of adenoviral vectors can be achieved *via* dendrimer coating, resulting in a potential for systemic NIS gene therapy. Additionally, the dendrimers were not only employed as nonviral gene delivery systems, but also proposed to modify viral gene vectors.[124]

12.5 Conclusions

Dendrimers have showed several advantages over conventional polymers, including the fact that they are well-defined molecules with lower PDI, controlled nanoscale size and flexible chemistry, allowing these nanostructures as ideal vehicles for both passive and active drug delivery/diagnostic imaging applications for cancer treatments. Up to now, dendrimer based nanomedicines have achieved a number of encouraging successes, and some dendrimer-based nanomedicines are in the final stages of human clinical testing in the U.S. FDA approval process.[125,126] Although significant effort has been devoted to develop dendrimer-based drug delivery systems for cancer therapy, those functional nanoscale vehicles have been not yet in clinical trials or application, since some inherent problems, such as the vulnerability with hydrolysis or proteolysis, stability issues, biodegradability and biosafety, have been not addressed. Currently, most studies focused on the dendritic nanosacle drug delivery vehicles and imaging probes separately where both of the nanomedicines fully take advantages of the dendrimer features. In future, with further understanding of the unique physicochemical and biological features of the tumor and dendrimers, another area that will attract interests is the development of multifunctional nanoscale dendritic vehicle for cancer treatment, and their application for cancer therapy guided by imaging. To date, a few dendritic theranostic systems primarily have been designed to treat cancers in real-time, suggesting some obvious advantages. This kind of vehicles can directly reflect the therapeutic outcomes and the behaviors of the dendritic vehicles *in vivo*, such as biodegradability, biodistribution and accumulation in tumors. It should be pointed out that the correlation between imaging results and therapeutic efficacy needs to be investigated for reliability, which can be realized *via* optimizing the dendritic structures. Overall, the multifunctional dendrimer based nanoscale vehicles will achieve a general platform for cancer therapy guided by imaging.

References

1. Adams, S.; Baum, R. P.; Stuckensen, T.; Bitter, K.; Hör, G. Prospective comparison of 18F-FDG PET with conventional imaging modalities (CT, MRI, US) in lymph node staging of head and neck cancer. *Eur J Nucl Med* **1998**, *25*, 1255-1260.
2. Brindle, K. New approaches for imaging tumour responses to treatment. *Nat Rev Cancer* **2008**, *8*, 94-107.

3. Luo, K.; Liu, G.; She, W.; Wang, Q.; Wang, G.; He, B.; Ai, H.; Gong, Q.; Song, B.; Gu, Z. Gadolinium-labeled peptide dendrimers with controlled structures as potential magnetic resonance imaging contrast agents. *Biomaterials* **2011**, *32*, 7951-7960.
4. Heath, J. R.; Davis, M. E.; Hood, L. Nanomedicine targets cancer. *Sci Am* **2009**, *300*, 44-51.
5. Farokhzad, O. C.; Langer, R. Impact of nanotechnology on drug delivery. *ACS Nano* **2009**, *3*, 16-20.
6. Davis, M. E. Nanoparticle therapeutics: An emerging treatment modality for cancer. *Nat Rev Drug Discov* **2008**, *7*, 771-782.
7. Cho, K.; Wang, X.; Nie, S.; Shin, D. M. Therapeutic nanoparticles for drug delivery in cancer. *Clin Cancer Res* **2008**, *14*, 1310-1316.
8. Singh, R.; Lillard, Jr J. W. Nanoparticle-based targeted drug delivery. *Exp Mol Pathol* **2009**, *86*, 215-223.
9. Vicent, M. J.; Greco, F.; Nicholson, R. I.; Paul, A.; Griffiths, P. C.; Duncan, R. Polymer therapeutics designed for a combination therapy of hormone-dependent cancer. *Angew Chem Int Edit* **2005**, *117*, 4129-4134.
10. Li, C.; Yu, D. F.; Newman, R. A.; Cabral, F.; Stephens, L. C.; Hunter, N.; Milas, L.; Wallace, S. Complete regression of well-established tumors using a novel water-soluble poly(L-glutamic acid)-paclitaxel conjugate. *Cancer Res* **1998**, *58*, 2404-2409.
11. Hreczuk-Hirst, D.; Chicco, D.; German, L.; Duncan, R. Dextrins as potential carriers for drug targeting: Tailored rates of dextrin degradation by introduction of pendant groups. *Int J Pharm* **2001**, *230*, 57-66.
12. Duncan, R. Polymer conjugates as anticancer nanomedicines. *Nat Rev Cancer* **2006**, *6*, 688-701.
13. Gillies, E. R.; Frechet, J. M. Dendrimers and dendritic polymers in drug delivery. *Drug Discov Today* **2005**, *10*, 35-43.
14. Janib, S. M.; Moses, A. S.; MacKay, J. A. Imaging and drug delivery using theranostic nanoparticles. *Adv Drug Deliver Rev* **2010**, *62*, 1052-1063.
15. Sumer, B.; Gao, J. Theranostic nanomedicine for cancer. *Nanomedicine* **2008**, *3*, 137-140.
16. Xie, J.; Lee, S.; Chen, X. Nanoparticle-based theranostic agents. *Adv Drug Deliver Rev* **2010**, *62*, 1064-1079.
17. Hu, J.; Qian, Y.; Wang, X.; Liu, T.; Liu, S. Drug-loaded and superparamagnetic iron oxide nanoparticle surface-embedded amphiphilic block copolymer micelles for integrated chemotherapeutic drug delivery and MR imaging. *Langmuir* **2011**, *28*, 2073-2082.
18. Lai, J. R.; Chang, Y. W.; Yen, H. C.; Yuan, N. Y.; Liao, M. Y.; Hsu, C. Y.; Tsai, J. L.; Lai, P. S. Multifunctional doxorubicin/superparamagnetic iron oxide-encapsulated pluronic F127 micelles used for chemotherapy/magnetic resonance imaging. *J Appl Phys* **2010**, *107*, 09B318.
19. Fischer, M.; Vögtle, F. Dendrimers: From design to application — a progress report. *Angew Chem Int Edit* **1999**, *38*, 884-905.
20. Bosman, A.; Janssen, H.; Meijer, E About dendrimers: Structure, physical properties, and applications. *Chem Rev* **1999**, *99*, 1665-1688.

21. Tomalia, D.; Baker, H.; Dewald, J.; Hall, M.; Kallos, G.; Martin, S.; Roeck, J.; Ryder, J.; Smith, P. A new class of polymers: Starburst-dendritic macromolecules. *Polym J* **1985**, *17*, 117-132.
22. Kobayashi, H.; Brechbiel, M. W. Nano-sized MRI contrast agents with dendrimer cores. *Adv Drug Deliver Rev* **2005**, *57*, 2271-2286.
23. Konda, S. D.; Aref, M.; Wang, S.; Brechbiel, M.; Wiener, E. C. Specific targeting of folate-dendrimer MRI contrast agents to the high affinity folate receptor expressed in ovarian tumor xenografts. *Magn Reson Mater Phy Biol Med* **2001**, *12*, 104-113.
24. Kobayashi, H.; Brechbiel, M. W. Dendrimer-based nanosized MRI contrast agents. *Curr Pharm Biotechno* **2004**, *5*, 539-549.
25. Esfand, R.; Tomalia, D. A. Poly(amidoamine)(PAMAM) dendrimers: From biomimicry to drug delivery and biomedical applications. *Drug Discov Today* **2001**, *6*, 427-436.
26. Patri, A. K.; Kukowska-Latallo, J. F.; Baker Jr, J. R. Targeted drug delivery with dendrimers: Comparison of the release kinetics of covalently conjugated drug and non-covalent drug inclusion complex. *Adv Drug Deliver Rev* **2005**, *57*, 2203-2214.
27. Tang, M. X.; Redemann, C. T.; Szoka, F. C. *In vitro* gene delivery by degraded polyamidoamine dendrimers. *Bioconjugate Chem* **1996**, *7*, 703-714.
28. Dufès, C.; Uchegbu, I. F.; Schätzlein, A. G. Dendrimers in gene delivery. *Adv Drug Deliver Rev* **2005**, *57*, 2177-2202.
29. Kolhe, P.; Khandare, J.; Pillai, O.; Kannan, S.; Lieh-Lai, M.; Kannan, R. M. Preparation, cellular transport, and activity of polyamidoamine-based dendritic nanodevices with a high drug payload. *Biomaterials* **2006**, *27*, 660-669.
30. Han, L.; Huang, R.; Li, J.; Liu, S.; Huang, S.; Jiang, C. Plasmid pORF-hTRAIL and doxorubicin co-delivery targeting to tumor using peptide-conjugated polyamidoamine dendrimer. *Biomaterials* **2011**, *32*, 1242-1252.
31. Hu, J.; Su, Y.; Zhang, H.; Xu, T.; Cheng, Y. Design of interior-functionalized fully acetylated dendrimers for anticancer drug delivery. *Biomaterials* **2011**, *32*, 9950-9959.
32. Medina, S. H.; El-Sayed, M. E. Dendrimers as carriers for delivery of chemotherapeutic agents. *Chem Rev* **2009**, *109*, 3141-3157.
33. Khandare, J. J.; Jayant, S.; Singh, A.; Chandna, P.; Wang, Y.; Vorsa, N.; Minko, T. Dendrimer versus linear conjugate: Influence of polymeric architecture on the delivery and anticancer effect of paclitaxel. *Bioconjugate Chem* **2006**, *17*, 1464-1472.
34. Kopeček, J.; Kopečková, P.; Minko, T.; Lu, Z-R.; Peterson, C. Water soluble polymers in tumor targeted delivery. *J Control Release* **2001**, *74*, 147-158.
35. Kopeček, J.; Kopečová, P.; Minko, T.; Lu, Z-R. HPMA copolymer-anticancer drug conjugates: Design, activity, and mechanism of action. *Eur J Pharm Biopharm* **2000**, *50*, 61-81.
36. Tekade, R. K.; Kumar, P. V.; Jain, N. K. Dendrimers in oncology: An expanding horizon. *Chem Rev* **2008**, *109*, 49-87.
37. Samad, A.; Alam, M. I.; Saxena, K. Dendrimers: A class of polymers in the nanotechnology for the delivery of active pharmaceuticals. *Curr Pharm Design* **2009**, *15*, 2958-2969.

38. Sebestik, J.; Niederhafner, P.; Jezek, J. Peptide and glycopeptide gendrimers and analogous dendrimeric structures and their biomedical applications. *Amino Acids* **2011**, *40*, 301-370.
39. Gu, Z.; Luo, K.; She, W.; Wu, Y.; He, B. New-generation biomedical materials: Peptide dendrimers and their application in biomedicine. *Sci China Chem* **2010**, *53*, 458-478.
40. Hashizume, H.; Baluk, P.; Morikawa, S.; McLean, J. W.; Thurston, G.; Roberge, S.; Jain, R. K.; McDonald, D. M. Openings between defective endothelial cells explain tumor vessel leakiness. *Am J Pathol* **2000**, *156*, 1363-1380.
41. Bergers, G.; Benjamin, L. E. Tumorigenesis and the angiogenic switch. *Nat Rev Cancer* **2003**, *3*, 401-410.
42. Brigger, I.; Dubernet, C.; Couvreur, P. Nanoparticles in cancer therapy and diagnosis. *Adv Drug Deliver Rev* **2002**, *54*, 631-651.
43. Soliman, G. M.; Sharma, A.; Maysinger, D.; Kakkar, A. Dendrimers and miktoarm polymers based multivalent nanocarriers for efficient and targeted drug delivery. *Chem Comm* **2011**, *47*, 9572-9587.
44. Šebestík, J.; Reiniš, M.; Ježek, J. Dendrimers in drug delivery. In: Sebestik, J. *Biomedical Applications of Peptide-, Glyco-and Glycopeptide Dendrimers, and Analogous Dendrimeric Structures.* Springer, **2012**, 131-140.
45. Shi, X.; Wang, S.; Meshinchi, S.; Van Antwerp, M. E.; Bi, X.; Lee, I.; Baker, J. R. Dendrimer-entrapped gold nanoparticles as a platform for cancer-cell targeting and imaging. *Small* **2007**, *3*, 1245-1252.
46. Wang, H.; Zheng, L.; Peng, C.; Guo, R.; Shen, M.; Shi, X.; Zhang, G. Computed tomography imaging of cancer cells using acetylated dendrimer-entrapped gold nanoparticles. *Biomaterials* **2011**, *32*, 2979-2988.
47. Peng, C.; Zheng, L.; Chen, Q.; Shen, M.; Guo, R.; Wang, H.; Cao, X.; Zhang, G.; Shi, X. PEGylated dendrimer-entrapped gold nanoparticles for *in vivo* blood pool and tumor imaging by computed tomography. *Biomaterials* **2012**, *33*, 1107-1119.
48. Torchilin, V. P. Recent advances with liposomes as pharmaceutical carriers. *Nat Rev Drug Discov* **2005**, *4*, 145-160.
49. Luo, K.; Li, C.; Li, L.; She, W.; Wang, G.; Gu, Z. Arginine functionalized peptide dendrimers as potential gene delivery vehicles. *Biomaterials* **2012**, *33*, 4917-4927.
50. Boyd, B. J.; Kaminskas, L. M.; Karellas, P.; Krippner, G.; Lessene, R.; Porter, C. J. Cationic poly-L-lysine dendrimers: Pharmacokinetics, biodistribution, and evidence for metabolism and bioresorption after intravenous administration to rats. *Mol Pharm* **2006**, *3*, 614-627.
51. Chen, H. T.; Neerman, M. F.; Parrish, A. R.; Simanek, E. E. Cytotoxicity, hemolysis, and acute *in vivo* toxicity of dendrimers based on melamine, candidate vehicles for drug delivery. *J Am Chem Soc* **2004**, *126*, 10044-10048.
52. She, W.; Li, N.; Luo, K.; Guo, C.; Wang, G.; Geng, Y.; Gu, Z. Dendronized heparin-doxorubicin conjugate based nanoparticle as pH-responsive drug delivery system for cancer therapy. *Biomaterials* **2013**, *34*, 2252-2264.
53. Frauenrath, H. Dendronized polymers: Building a new bridge from molecules to nanoscopic objects. *Prog Polym Sci* **2005**, *30*, 325-384.

54. Schlüter, A. D.; Rabe, J. P. Dendronized polymers: Synthesis, characterization, assembly at interfaces, and manipulation. *Angew Chem Int Edit* **2000**, *39*, 864-883.
55. Gao, M.; Jia, X.; Kuang, G.; Li, Y.; Liang, D.; Wei, Y. Thermo- and pH-responsive dendronized copolymers of styrene and maleic anhydride pendant with poly(amidoamine) dendrons as side groups. *Macromolecules* **2009**, *42*, 4273-4281.
56. Laurent, B. A.; Grayson, S. M. Synthesis of cyclic dendronized polymers *via* divergent "graft-from" and convergent click "graft-to" routes: Preparation of modular toroidal macromolecules. *J Am Chem Soc* **2011**, *133*, 13421-13429.
57. Rudick, J. G.; Percec, V. Induced helical backbone conformations of self-organizable dendronized polymers. *Accounts Chem Res* **2008**, *41*, 1641-1652.
58. Percec, V.; Rudick, J. G.; Peterca, M.; Heiney, P. A. Nanomechanical function from self-organizable dendronized helical polyphenylacetylenes. *J Am Chem Soc* **2008**, *130*, 7503-7508.
59. Park, J. W.; Jeon, O. C.; Kim, S. K.; Al-Hilal, T. A.; Jin, S. J.; Moon, H. T.; Yang, V. C.; Kim, S. Y.; Byun, Y. High antiangiogenic and low anticoagulant efficacy of orally active low molecular weight heparin derivatives. *J Control Release* **2010**, *148*, 317-326.
60. Tang, D. W.; Yu, S. H.; Ho, Y. C.; Mi, F. L.; Kuo, P. L.; Sung, H. W. Heparinized chitosan/poly(γ-glutamic acid) nanoparticles for multi-functional delivery of fibroblast growth factor and heparin. *Biomaterials* **2010**, *31*, 9320-9332.
61. Arap, W.; Pasqualini, R.; Ruoslahti, E. Cancer treatment by targeted drug delivery to tumor vasculature in a mouse model. *Science* **1998**, *279*, 377-380.
62. Yu, B.; Zhao, X.; Lee, L. J.; Lee, R. J. Targeted delivery systems for oligonucleotide therapeutics. *The AAPS J* **2009**, *11*, 195-203.
63. Brannon-Peppas, L.; Blanchette, J. O. Nanoparticle and targeted systems for cancer therapy. *Adv Drug Deliver Rev* **2012**, *64*, 206-212.
64. Rolland, O.; Turrin, C. O.; Caminade, A. M.; Majoral, J. P. Dendrimers and nanomedicine: Multivalency in action. *New J Chem* **2009**, *33*, 1809-1824.
65. Chandrasekar, D.; Sistla, R.; Ahmad, F. J.; Khar, R. K.; Diwan, P. V. The development of folate-PAMAM dendrimer conjugates for targeted delivery of anti-arthritic drugs and their pharmacokinetics and biodistribution in arthritic rats. *Biomaterials* **2007**, *28*, 504-512.
66. Boswell, C. A.; Eck, P. K.; Regino, C. A.; Bernardo, M.; Wong, K. J.; Milenic, D. E.; Choyke, P. L.; Brechbiel, M. W. Synthesis, characterization, and biological evaluation of integrin $\alpha_v\beta_3$-targeted PAMAM dendrimers.*Mol Pharm* **2008**, *5*, 527-539.
67. Sampathkumar, S. G.; Yarema, K. J. Targeting cancer cells with dendrimers. *Chem Biol* **2005**, *12*, 5-6.
68. André, S.; Ortega, P. J. C.; Perez, M. A.; Roy, R.; Gabius, H. J. Lactose-containing starburst dendrimers: Influence of dendrimer generation and binding-site orientation of receptors (plant/animal lectins and immunoglobulins) on binding properties. *Glycobiology* **1999**, *9*, 1253-1261.
69. Luo, K.; Liu, G.; Zhang, X.; She, W.; He, B.; Nie, Y.; Li, L.; Wu, Y.; Zhang, Z.; Gong, Q. Functional L-lysine dendritic macromolecules as liver-imaging probes. *Macromol Biosci* **2009**, *9*, 1227-1236.

70. Luo, K.; Liu, G.; He, B.; Wu, Y.; Gong, Q.; Song, B.; Ai, H.; Gu, Z. Multifunctional gadolinium-based dendritic macromolecules as liver targeting imaging probes. *Biomaterials* **2011**, *32*, 2575-2585.
71. Roy, R. Syntheses and some applications of chemically defined multivalent glycoconjugates. *Curr Opin Struc Biol* **1996**, *6*, 692-702.
72. Yuan, H.; Luo, K.; Lai, Y.; Pu, Y.; He, B.; Wang, G.; Wu, Y.; Gu, Z. A novel poly(L-glutamic acid) dendrimer based drug delivery system with both pH-sensitive and targeting functions. *Mol Pharm* **2010**, *7*, 953-962.
73. Pu, Y.; Chang, S.; Yuan, H.; Wang, G.; He, B.; Gu, Z. The anti-tumor efficiency of poly(L-glutamic acid) dendrimers with polyhedral oligomeric silsesquioxane cores. *Biomaterials* **2013**, *34*, 3658-3666.
74. Zhang, Y.; Lovell, J. F. Porphyrins as theranostic agents from prehistoric to modern times. *Theranostics* **2012**, *2*, 905-915.
75. Ideta, R.; Tasaka, F.; Jang, W. D.; Nishiyama, N.; Zhang, G. D.; Harada, A.; Yanagi, Y.; Tamaki, Y.; Aida, T.; Kataoka, K. Nanotechnology-based photodynamic therapy for neovascular disease using a supramolecular nanocarrier loaded with a dendritic photosensitizer. *Nano Lett* **2005**, *5*, 2426-2431.
76. Kumar, M. S.; Babu, A.; Murugesan, R.; Jeyasubramanian, K. Novel water soluble dendrimer nanocarrier for enhanced photodynamic efficacy of protoporphyrin IX. *Nano Biomed Eng* **2012**, *4*, 132-138.
77. Kojima, C.; Toi, Y.; Harada, A.; Kono, K. Preparation of poly(ethylene glycol)-attached dendrimers encapsulating photosensitizers for application to photodynamic therapy. *Bioconjugate Chem* **2007**, *18*, 663-670.
78. Taratula, O.; Schumann, C.; Naleway, M. A.; Pang, A. J.; Chon, K. J.; Taratula, O. A multifunctional theranostic platform based on phthalocyanine-loaded dendrimer for image-guided drug delivery and photodynamic therapy. *Mol Pharm* **2013**, *10*, 3946-3958.
79. Di Venosa, G. M.; Casas, A. G.; Battah, S.; Dobbin, P.; Fukuda, H.; MacRobert, A. J.; Batlle, A. Investigation of a novel dendritic derivative of 5-aminolaevulinic acid for photodynamic therapy. *Int J Biochem Cell B* **2006**, *38*, 82-91.
80. Battah, S.; O'Neill, S.; Edwards, C.; Balaratnam, S.; Dobbin, P.; MacRobert, A. J. Enhanced porphyrin accumulation using dendritic derivatives of 5-aminolaevulinic acid for photodynamic therapy: An *in vitro* study. *Int J Biochem Cell B* **2006**, *38*, 1382-1392.
81. Battah, S. H.; Chee, C. E.; Nakanishi, H.; Gerscher, S.; MacRobert, A. J.; Edwards, C. Synthesis and biological studies of 5-aminolevulinic acid-containing dendrimers for photodynamic therapy. *Bioconjugate Chem* **2001**, *12*, 980-988.
82. Casas, A.; Battah. S.; Di Venosa, G.; Dobbin, P.; Rodriguez, L.; Fukuda, H.; Batlle, A.; MacRobert, A. J. Sustained and efficient porphyrin generation *in vivo* using dendrimer conjugates of 5-Ala for photodynamic therapy. *J Control Release* **2009**, *135*, 136-143.
83. Ma, D.; Liu, Z. H.; Zheng, Q. Q.; Zhou, X. Y.; Zhang, Y.; Shi, Y. F.; Lin, J. T.; Xue, W. Star-shaped polymer consisting of a porphyrin core and poly(l-lysine) dendron arms: Synthesis, drug delivery, and *in vitro* chemo/photodynamic therapy. *Macromol Rapid Comm* **2013**, *34*, 548-552.
84. Shen, M.; Shi, X. Dendrimer-based organic/inorganic hybrid nanoparticles in biomedical applications. *Nanoscale* **2010**, *2*, 1596-1610.

85. Abu Reziq, R.; Alper, H.; Wang, D.; Post, M. L. Metal supported on dendronized magnetic nanoparticles: Highly selective hydroformylation catalysts. *J Am Chem Soc* **2006**, *128*, 5279-5282.
86. Pan, B. F.; Gao, F.; Ao, L. M. Investigation of interactions between dendrimer-coated magnetite nanoparticles and bovine serum albumin. *J Magn Magn Mater* **2005**, *293*, 252-258.
87. Mornet, S.; Vasseur, S.; Grasset, F.; Duguet, E. Magnetic nanoparticle design for medical diagnosis and therapy. *J Mater Chem* **2004**, *14*, 2161-2175.
88. Zhu, L.; Wang, D.; Wei, X.; Zhu, X.; Li, J.; Tu, C.; Su, Y.; Wu, J.; Zhu, B.; Yan, D. Multifunctional pH-sensitive superparamagnetic iron-oxide nanocomposites for targeted drug delivery and MR imaging. *J Control Release* **2013**, *169*, 228-238.
89. Zou, P.; Yu, Y.; Wang, Y. A.; Zhong, Y.; Welton, A.; Galbán, C.; Wang, S.; Sun, D. Superparamagnetic iron oxide nanotheranostics for targeted cancer cell imaging and pH-dependent intracellular drug release. *Mol Pharm* **2010**, *7*, 1974-1984.
90. Chung, T. H.; Hsiao, J. K.; Hsu, S. C.; Yao, M.; Chen, Y. C.; Wang, S. W.; Kuo, M. Y. P.; Yang, C. S.; Huang, D. M. Iron oxide nanoparticle-induced epidermal growth factor receptor expression in human stem cells for tumor therapy. *ACS Nano* **2011**, *5*, 9807-9816.
91. Lewin, M.; Carlesso, N.; Tung, C. H.; Tang, X. W.; Cory, D.; Scadden, D. T.; Weissleder, R. TAT peptide-derivatized magnetic nanoparticles allow *in vivo* tracking and recovery of progenitor cells. *Nat Biotechnol* **2000**, *18*, 410-414.
92. Bamrungsap, S.; Chen, T.; Shukoor, M. I.; Chen, Z.; Sefah, K.; Chen, Y.; Tan, W. Pattern recognition of cancer cells using aptamer-conjugated magnetic nanoparticles. *ACS Nano* **2012**, *6*, 3974-3981.
93. Lee, J. H.; Huh, Y. M.; Jun, Y. W.; Seo, J. W.; Jang, J. T.; Song, H. T.; Kim, S.; Cho, E. J.; Yoon, H. G.; Suh, J. S. Artificially engineered magnetic nanoparticles for ultra-sensitive molecular imaging. *Nat Med* **2006**, *13*, 95-99.
94. Sun, C.; Lee, J. S.; Zhang, M. Magnetic nanoparticles in MR imaging and drug delivery. *Adv Drug Deliver Rev* **2008**, *60*, 1252-1265.
95. Dilnawaz, F.; Singh, A.; Mohanty, C.; Sahoo, S. K. Dual drug loaded superparamagnetic iron oxide nanoparticles for targeted cancer therapy. *Biomaterials* **2010**, *31*, 3694-3706.
96. Yu, M. K.; Jeong, Y. Y.; Park, J.; Park, S.; Kim, J. W.; Min, J. J.; Kim, K.; Jon, S. Drug-loaded superparamagnetic iron oxide nanoparticles for combined cancer imaging and therapy *in vivo*. *Angew Chem Int Edit* **2008**, *47*, 5362-5365.
97. Brulé, S.; Levy, M.; Wilhelm, C.; Letourneur, D.; Gazeau, F.; Ménager, C.; Le Visage, C. Doxorubicin release triggered by alginate embedded magnetic nanoheaters: A combined therapy. *Adv Mater* **2011**, *23*, 787-790.
98. Chen, G.; Chen, W.; Wu, Z.; Yuan, R.; Li, H.; Gao, J.; Shuai, X. MRI-visible polymeric vector bearing CD3 single chain antibody for gene delivery to T cells for immunosuppression. *Biomaterials* **2009**, *30*, 1962-1970.
99. Kamei, K.; Mukai, Y.; Kojima, H.; Yoshikawa, T.; Yoshikawa, M.; Kiyohara, G.; Yamamoto, T. A.; Yoshioka, Y.; Okada, N.; Seino, S. Direct cell entry of gold/iron-oxide magnetic nanoparticles in adenovirus mediated gene delivery. *Biomaterials* **2009**, *30*, 1809-1814.

100. Liu, G.; Swierczewska, M.; Lee, S.; Chen, X. Functional nanoparticles for molecular imaging guided gene delivery. *Nano Today* **2010**, *5*, 524-539.

101. Medarova, Z.; Pham, W.; Farrar, C.; Petkova, V.; Moore, A. *In vivo* imaging of siRNA delivery and silencing in tumors. *Nat Med* **2007**, *13*, 372-377.

102. Taratula, O.; Garbuzenko, O.; Savla, R.; Andrew, W. Y.; He, H.; Minko, T. Multifunctional nanomedicine platform for cancer specific delivery of siRNA by superparamagnetic iron oxide nanoparticles-dendrimer complexes. *Curr Drug Deliver* **2011**, *8*, 59-69.

103. Park, J. W.; Bae, K. H.; Kim, C.; Park, T. G. Clustered magnetite nanocrystals cross-linked with PEI for efficient siRNA delivery. *Biomacromolecules* **2010**, *12*, 457-465.

104. Nasongkla, N.; Bey, E.; Ren, J.; Ai, H.; Khemtong, C.; Guthi, J. S.; Chin, S. F.; Sherry, A. D.; Boothman, D. A.; Gao, J. Multifunctional polymeric micelles as cancer-targeted, MRI-ultrasensitive drug delivery systems. *Nano Lett* **2006**, *6*, 2427-2430.

105. Lee, H.; Lee, E.; Kim, D. K.; Jang, N. K.; Jeong, Y. Y.; Jon, S. Antibiofouling polymer-coated superparamagnetic iron oxide nanoparticles as potential magnetic resonance contrast agents for *in vivo* cancer imaging. *J Am Chem Soc* **2006**, *128*, 7383-7389.

106. Zhu, R.; Jiang, W.; Pu, Y.; Luo, K.; Wu, Y.; He, B.; Gu, Z. Functionalization of magnetic nanoparticles with peptide dendrimers. *J Mater Chem* **2011**, *21*, 5464-5474.

107. Heuzé, K.; Rosario, A. D.; Nlate, S.; Gaboyard, M.; Bouter, A.; Clérac, R. Efficient strategy to increase the surface functionalization of core-shell superparamagnetic nanoparticles using dendron grafting. *New J Chem* **2008**, *32*, 383-387.

108. Li, M.; Xu, L. Q.; Wang, L.; Wu, Y.; Li, J.; Neoh, K. G.; Kang, E. T. Clickable poly(ester amine) dendrimer-grafted Fe_3O_4 nanoparticles prepared *via* successive michael addition and alkyne-azide click chemistry. *Polym Chem* **2011**, *2*, 1312-1321.

109. Chandra, S.; Mehta, S.; Nigam, S.; Bahadur, D. Dendritic magnetite nanocarriers for drug delivery applications. *New J Chem* **2010**, *34*, 648-655.

110. Strable, E.; Bulte, J. W.; Moskowitz, B.; Vivekanandan, K.; Allen, M.; Douglas, T. Synthesis and characterization of soluble iron oxide-dendrimer composites. *Chem Mater* **2001**, *13*, 2201-2209.

111. Shi, X.; Thomas, T. P.; Myc, L. A.; Kotlyar, A.; Baker, Jr. J. R. Synthesis, characterization, and intracellular uptake of carboxyl-terminated poly(amidoamine) dendrimer-stabilized iron oxide nanoparticles. *Phys Chem Chem Phys* **2007**, *9*, 5712-5720.

112. Pan, B.; Cui, D.; Sheng, Y.; Ozkan, C.; Gao, F.; He, R.; Li, Q.; Xu, P.; Huang, T. Dendrimer-modified magnetic nanoparticles enhance efficiency of gene delivery system. *Cancer Res* **2007**, *67*, 8156-8163.

113. González, B.; Ruiz, H. E.; Feito, M. J.; de Laorden, C. L.; Arcos, D.; Ramírez, S. C.; Matesanz, C.; Portolés, M. T.; Vallet, R. M. Covalently bonded dendrimer-maghemite nanosystems: Nonviral vectors for *in vitro* gene magnetofection. *J Mater Chem* **2011**, *21*, 4598-4604.

114. Tomalia, D.; Reyna, L.; Svenson, S. Dendrimers as multi-purpose nanodevices

for oncology drug delivery and diagnostic imaging. *Biochem Soc Trans* **2007**, *35*, 61-67.

115. Kono, K.; Nakashima, S.; Kokuryo, D.; Aoki, I.; Shimomoto, H.; Aoshima, S.; Maruyama, K.; Yuba, E.; Kojima, C.; Harada, A. Multi-functional liposomes having temperature-triggered release and magnetic resonance imaging for tumor-specific chemotherapy. *Biomaterials* **2011**, *32*, 1387-1395.
116. Krasia-Christoforou, T.; Georgiou, T. K. Polymeric theranostics: Using polymer-based systems for simultaneous imaging and therapy. *J Mater Chem B* **2013**, *1*, 3002-3025.
117. Kolhe, P.; Misra, E.; Kannan, R. M.; Kannan, S.; Lieh, L. M. Drug complexation, *in vitro* release and cellular entry of dendrimers and hyperbranched polymers. *Int J Pharm* **2003**, *259*, 143-160.
118. Backer, M. V.; Gaynutdinov, T. I.; Patel, V.; Bandyopadhyaya, A. K.; Thirumamagal, B.; Tjarks, W.; Barth, R. F.; Claffey, K.; Backer, J. M. Vascular endothelial growth factor selectively targets boronated dendrimers to tumor vasculature. *Mol Cancer Ther* **2005**, *4*, 1423-1429.
119. Majoros, I. J.; Myc, A.; Thomas, T.; Mehta, C. B.; Baker, J. R. PAMAM dendrimer-based multifunctional conjugate for cancer therapy: Synthesis, characterization, and functionality. *Biomacromolecules* **2006**, *7*, 572-579.
120. Yang, W.; Cheng, Y.; Xu, T.; Wang, X.; Wen, L. P. Targeting cancer cells with biotin-dendrimer conjugates. *Eur J Med Chem* **2009**, *44*, 862-868.
121. Choi, Y.; Thomas, T.; Kotlyar, A.; Islam, M. T.; Baker, Jr. J. R. Synthesis and functional evaluation of DNA-assembled polyamidoamine dendrimer clusters for cancer cell-specific targeting.*Chem Biol* **2005**, *12*, 35-43.
122. Kreppel, F.; Kochanek, S. Modification of adenovirus gene transfer vectors with synthetic polymers: A acientific review and technical guide. *Mol Ther* **2007**, *16*, 16-29.
123. Vetter, A.; Virdi, K. S.; Espenlaub, S.; Rödl, W.; Wagner, E.; Holm, P. S.; Scheu, C.; Kreppel, F.; Spitzweg, C.; Ogris, M. Adenoviral vectors coated with PAMAM dendrimer conjugates allow CAR independent virus uptake and targeting to the EGF receptor. *Mol Pharm* **2013**, *10*, 606-618.
124. Grünwald, G. K.; Vetter, A.; Klutz, K.; Willhauck, M. J.; Schwenk, N.; Senekowitsch-Schmidtke, R.; Schwaiger, M.; Zach, C.; Wagner, E.; Göke, B.; Holm, P.S.; Ogris, M.; Spitzweg, C. Systemic image-guided liver cancer radiovirotherapy using dendrimer-coated adenovirus encoding the sodium iodide symporter as theranostic gene. *J Nucl Med* **2013**, *54*, 1450-1457.
125. Jiang, Y. H.; Emau, P.; Cairns, J. S.; Flanary, L.; Morton, W. R.; McCarthy, T. D.; Tsai, C. C. SPL7013 gel as a topical microbicide for prevention of vaginal transmission of SHIV89. 6P in macaques. *AIDS Res Hum Retrov* **2005**, *21*, 207-213.
126. McCarthy, T. D.; Karellas, P.; Henderson, S. A.; Giannis, M.; O'Keefe, D. F.; Heery, G.; Paull, J. R.; Matthews, B. R.; Holan, G. Dendrimers as drugs: Discovery and preclinical and clinical development of dendrimer-based microbicides for HIV and STI prevention. *Mol Pharm* **2005**, *2*, 312-318.

Chapter 13

Protein-based Nanoparticles for Tumor Targeting Theranostics

Yang Liu, Hao Wu, and Huihui Wang*
Division of Biotechnology, Dalian Institute of Chemical Physics, Chinese Academy of Sciences, Dalian 116023, China

Proteins are suitable carriers for drug delivery, imaging or therapy, which have been proved to be biocompatible and biodegradable. Protein-based nanoparticles, therefore, may be considered to be an ideal system for tumor targeting theranostics *in vivo*. Among different protein-based nanoparticle systems, albumin nanoparticles, especially human serum albumin (HSA) nanoparticles, have gained more attention, since they have great capacity of loading various anti-cancer drugs and imaging agents after targeted modifications without any serious side-effects *in vivo*. This chapter focuses on the discussion of albumin-based nanoparticles with respect to preparation techniques, formulation designs, targeted modifications, targeted drug delivery, imaging and photodynamic therapy for tumor theranostics.

13.1 Introduction

Targeted drug delivery and imaging play a key role in tumor therapy and diagnosis. Nanoparticles are widely used as a good drug delivery agent or imaging carrier in tumor targeting theranostics.[1–3] They can enhance tumor targeting, protect a drug from degradation, improve penetration and distribution in targeted tissues or cells, load or combine imaging contrast agents, *etc.*[2,4]

Many types of nanoparticles, including liposome, polymeric nanoparticles, solid lipid quantum nanoparticles, quantum dots, ceramic nanoparticles, nanomicelles, magnetic nanoparticles, nanocages and dendrimers, *etc*, have been used for various biomedical applications, such as drug delivery and imaging.[5,6]

Proteins are popular raw materials that can be used to fabricate polymeric nanoparticles.[7,8] Systems based on protein nanoparticles have been widely studied for drug delivery or imaging applications both *in vitro* and *in vivo*.[4,9–13] Proteins are absorbable, biodegradable with nonantigenicity, and their degradation end products are less toxic. They also have the advantages of being synthetic polymers at the same time.[9,14,15] In addition, proteins also have good functional characteristics, such as emulsification, gelation, *etc*, and they can go

∗ Corresponding author: liuyang@dicp.ac.cn

through the aqueous steric barrier.[4,16,17] Therefore, proteins are proved to be the ideal basic materials that are used as GRAS (generally regarded as safe) carriers in biomedical applications. Nanoparticles based on proteins also have similar advantages, including good stability, low toxicity, nonantigenicity and large-scale productivity, over other nanoparticle systems.[4,18,19].

Some animal proteins and plant proteins have been used as basic materials for protein-based nanoparticle fabrication. Animal proteins mainly include albumin, gelatin, collagen, elastin, casein, whey protein, *etc*, while gliadin and zein are the main types of plant proteins.[9,17] Among these different kinds of proteins, albumin, especially human serum albumin (HSA), is a typical protein used to prepare nanoparticles, because some studies indicate that there is no antigenicit when HSA-based nanoparticles are used by intravenous injections, which shows that HSA nanoparticles may be well tolerated *in vivo* without any potential toxicity.[4]

This chapter focuses on discussion of the albumin-based nanoparticles, with respect to formulation designs, targeted modifications, targeted drug delivery, imaging and photodynamic therapy for tumor theranostics.

13.2 Albumin Nanoparticles for Tumor Targeting Theranostics

13.2.1 *Albumin Nanoparticles*

Albumin with a molecular weight of 66.5 kDa is the main protein of human plasma (50%–60% of total plasma protein, 35–50 g/L human serum).[20,21] It is synthesized in the liver with a rate of about 10–15 g daily and HSA has a half-life of 19 days. The three-dimensional(3D) structure of HSA has been determined by X-ray crystallography.[22,23] The possible structure of HSA may be an ellipsoid that consists of three flexible spheres in a row, which is showed schematically in Fig.13.1.[21]

Albumin is relatively inexpensive and can be produced at large scale by recombinant protein expression,[24] and the recombinant albumin protein also shows good safety and efficacy in *in vivo* studies.[25] Because it is abundant, nontoxic, low immunogenic, and can cross through the vascular endothelia, albumin can be used as *in vivo* delivery carriers for drugs[21], photosensitizers[26], and imaging agents[27,28].

There are many sources from where we can get albumin, including egg white, bovine serum, human serum, soybeans, milk and grains,[4] which promotes the wide application of albumin.

Albumin nanoparticles are considered to be an attractive strategy for drug or imaging agent delivery *in vivo*.[4] For drug delivery, albumin nanoparticle matrix can incorporate a lot of drugs due to many binding sites in the albumin molecule.[29] Some positively or negatively charged molecules can also attached to the nanoparticles easily because of the high content of charged amino acids

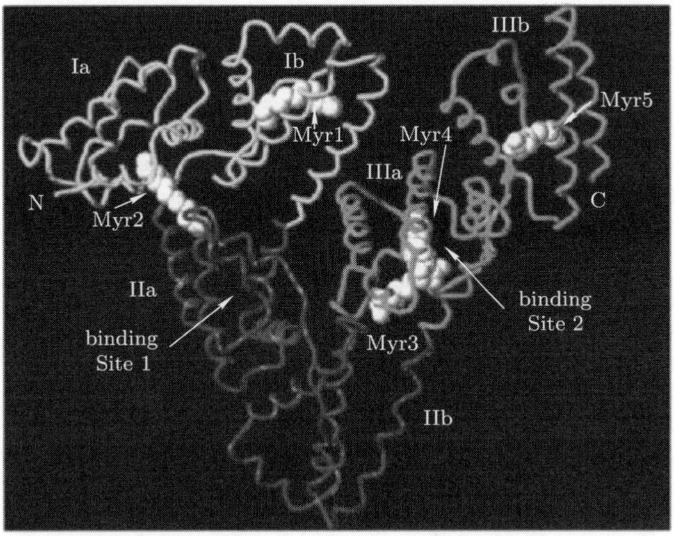

Fig. 13.1 X-ray structure of human serum albumin. Reproduced from Ref. 21. Copyright (2008) with permission from Elsevier. (color figure at the end of the book)

in albumin molecules omitting the requirement of other compounds.[30,31] Moreover, albumin nanoparticles with tumor targeting ligands can be easy to prepare by means of covalent reaction due to the presence of amino or carboxylic groups on the particle surface.[32,33] Clinical studies indicate that albumin nanoparticle may be well tolerated, which is proved by two kinds of HSA-based drug delivery carriers, including Albunex™ and Abraxane™.[34–37] For imaging purposes, albumin nanoparticle has been used in the fields of radionuclide-based imaging, magnetic resonance imaging, optical imaging or photodynamic therapy.[11,38]

In addition, albumin nanoparticles with better drug controlled release and targeted tumor imaging properties can be produced at large scale by means of some well known techniques, such as desolvation, emulsification, nano spray drying and self-assembly, which promote their commercial applications.[39]

13.2.2 *Preparation Techniques for Albumin Nanoparticles*

Elzoghby *et al.*[4] summarized the main techniques used to prepare albumin nanoparticles, including desolvation, emulsification, thermal gelation, nano spray drying, nab-technology and self-assembly, which have their own advantages and disadvantages, respectively.

13.2.2.1 *Desolvation*

In the process of desolvation, desolvating agent (ethanol) is continuously added into albumin aqueous solution drop by drop under stirring condition. During this process, the water-solubility of albumin is diminished gradually, and the solution appears turbid, which indicates albumin nanoparticles have been formed.[18]

Langer *et al.* prepared HSA nanoparticles with controllable particle size by means of the desolvation method.[18] The size of nanoparticles was varied from 100 to 300 nm with a narrow size distribution. In cell experiments, it was found that about 87% of cells took up these nanoparticles at a concentration of 0.025 mg/mL HSA nanoparticles, while nanoparticles were found in 98% of cells when 1 mg/mL nanoparticles was used (Fig.13.2).

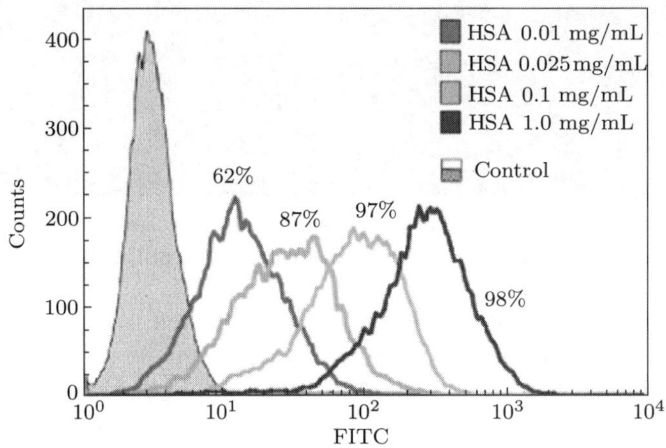

Fig. 13.2 Percentage of primary human macrophages internalized HSA nanoparticles when different concentrations of nanoparticles were used. Reprinted from Ref. 18. Copyright (2003) with permission from Elsevier. (color figure at the end of the book)

However, albumin nanoparticles might redissolve again when added into water, which showed their instability.[40] Therefore, glutaraldehyde could be used as a cross-linker to harden the coacervates by a simple crosslinking reaction.[31,41]

It was also found that the nanoparticle size could be affected by the amount of desolvating agent used or pH of HSA solution.[18,31] Nanoparticles with 150–280 nm diameter could be obtained by changing pH value. Higher pH values were helpful to generate smaller nanoparticles.[18] In addition, a narrower size distribution could be achieved after washing nanoparticles by differential centrifugation or by the addition of a desolvating agent (ethanol) to the albumin solution.[18,42]

Weber *et al.*[31] found that the lowest required glutaraldehyde concentration was about 40% with a reaction time of 24 h to generate stable nanoparticles. Moreover, methyl polyethylene glycol (PEG) modified oxidized dextran (Dextranox-MPEG) could be used to crosslink HSA nanospheres instead of glutaraldehyde.[43] To terminate denaturation, sodium metabisulfite or lysine solution could be used successfully.[44,45] Fine nanoparticle powders could be obtained after ethanol elimination, purification *via* centrifugation, and lyophilization.[15]

13.2.2.2 *Emulsification*

Thermal or chemical treatment are the main methods used to stabilize albumin nanoparticles generated by emulsification process.[46] It was found that heating was helpful to stabilize the oil phase containing albumin droplets.[47] When it was cooled, the mixture was diluted with ethyl ether to reduce the oil viscosity, which promoted the separation. In addition, albumin solution also could be emulsified in oil at room temperature, then denatured by resuspension in ether containing crosslinking agents, such as 2,3-butadiene and formaldehyde by means of chemical treatment.[46,47]

It was reported that cefamandolenafate-loaded BSA nanoparticles could be prepared by adding BSA aqueous solution dropwise to a continuous phase consisted of cyclohexane containing glutaraldehyde.[48] Yang *et al.*[49] added the w/o single emulsion dropwise to castor oil at 140±5 °C to stabilize BSA nanoparticles entrapping 10-hydroxycamptothecin (HCPT). It was found that NaOH concentration and stirring time had an important effect on the percentage of lactone formed after emulsification (Fig.13.3).

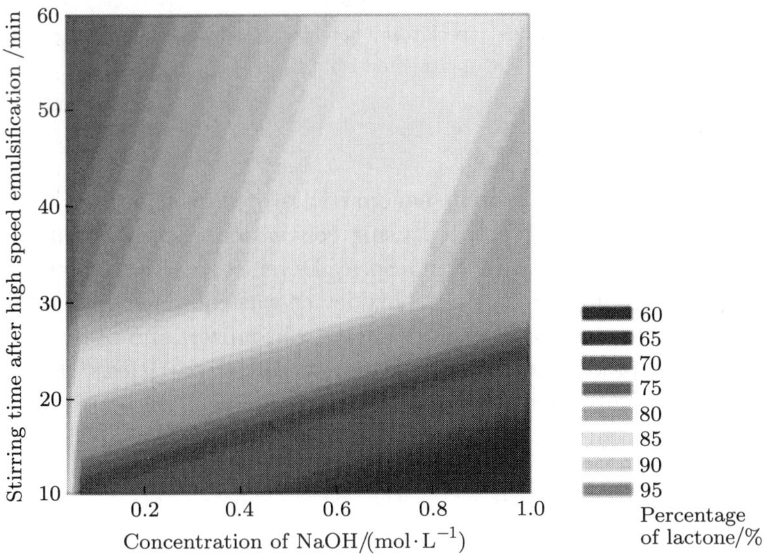

Fig. 13.3 Effect of NaOH concentration and stirring time on the percentage of lactone formed after emulsification. Reprinted from Ref. 49. Copyright (2007) with permission from Elsevier. (color figure at the end of the book)

13.2.2.3 *Thermal Gelation*

Thermal gelation is a sequential process consisting of heat-induced unfolding and many kinds of protein-protein interactions, such as hydrogen bonding and electrostatic.[50] It was found that BSA has the gelation property upon heating.[51] The BSA-dextran-chitosan nanoparticles were fabricated by heating the mixture of chitosan and BSA-dextran conjugates by means of the electrostatic attraction between BSA and chitosan as well as the thermal gelation of BSA (Fig.13.4).[50] These nanoparticles were spherical in shape with 130–230 nm diameter and doxorubicin could be effectively loaded.

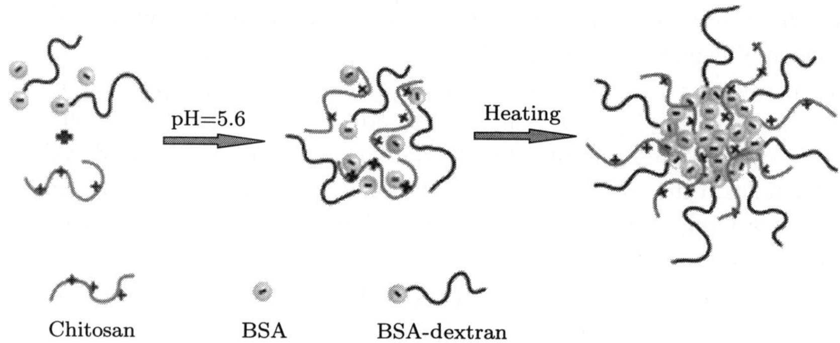

Fig. 13.4 Formation of BSA-dextran-chitosan nanoparticles by thermal gelation. Reprinted from Ref. 50. Copyright (2010) with permission from Elsevier.

13.2.2.4 *Nano Spray Drying*

It is difficult to generate protein nanoparticles of dimension less than 2 μm with high yield in a single step by using conventional spray drying method. Lee *et al.*[52] developed a novel Nano Spray Dryer B-90 (Fig.13.5) to produce BSA nanoparticles in a single step. This device was equipped with a vibrating mesh technology instead of conventional rotary atomizers and pressure nozzles. An electrostatic precipitator was also used to collect particles.[52] Preparation parameters of nanoparticles were also optimized, including spray mesh size, BSA solution concentration, surfactant concentration, drying air flow rate and inlet temperature.

13.2.2.5 *Nanoparticle Albumin-bound Technology*

To encapsulate lipophilic drugs into protein nanoparticles effectively, nab-technology has been developed by American Bioscience Inc. The mixed solution containing drugs and HSA passed through a jet under high pressure to generate HSA nanoparticles with 100-200 nm diameter.[53] Abraxane® is beneficial to the treatment of metastatic breast cancer, which has been approved

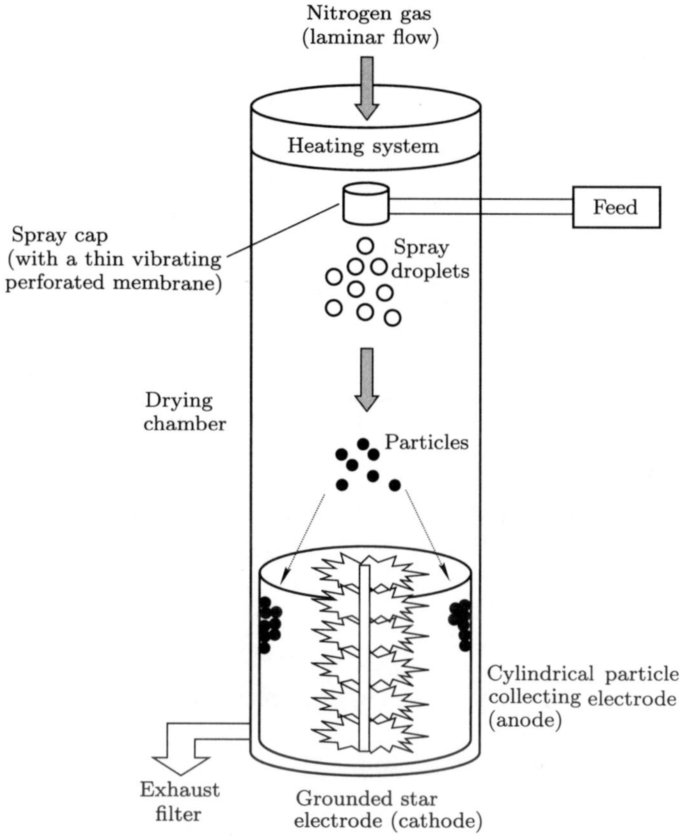

Fig. 13.5 Schematic diagram of Nano Spray Dryer B-90. Reprinted from Ref. 52. Copyright (2011) with permission from Elsevier.

by FDA.[54,55] Other nab drugs are also under development, including ABI-008 and ABI-009.[53]

Kim *et al.*[56] prepared HSA nanoparticles containing curcumin with 130–150 nm diameters by nab-technology. These nanoparticles showed excellent water solubility, which was 300-fold more than that of free curcumin (Fig.13.6). It was found that 50%–66% tumor growth was inhibited when these nanoparticles were injected into tumor xenograft HCT116 models, while only about 18% tumor growth inhibition was achieved when free curcumin was used.

13.2.2.6 *Self-assembly*

HSA micelles can also be generated by HSA self-assembly when the hydrophobicity of albumin is increased, which can be promoted by addition of a lipophilic drug or diminishment of primary amine groups on protein surface. Gong *et al.* synthesized octyl-modified BSA micelles containing paclitaxel.[57]A high drug-

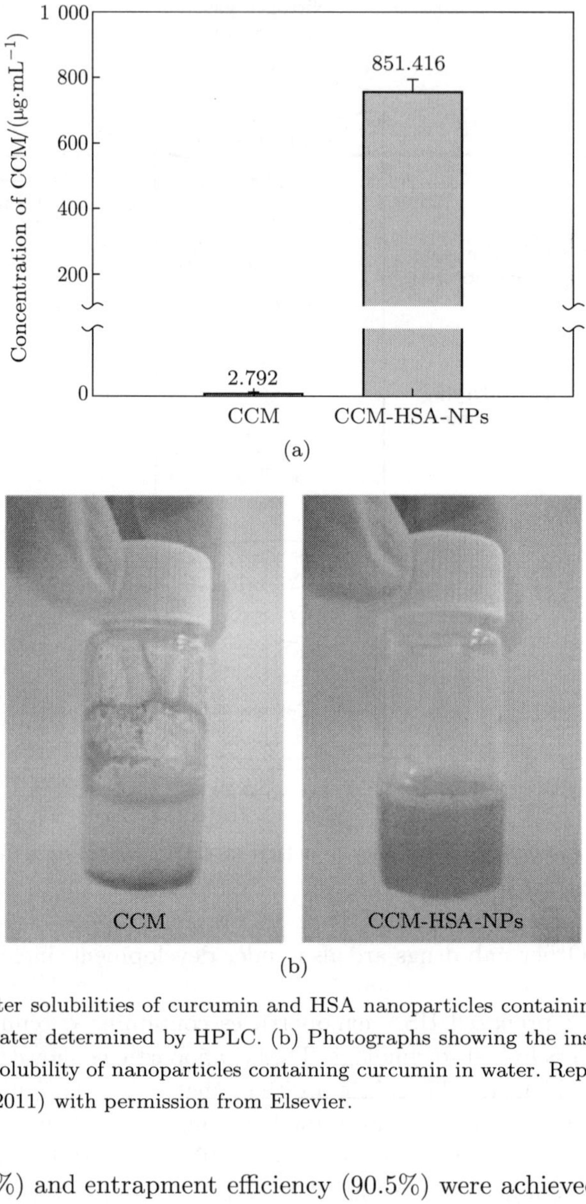

Fig. 13.6 Water solubilities of curcumin and HSA nanoparticles containing curcumin. (a) Solubilities in water determined by HPLC. (b) Photographs showing the insolubility of curcumin and the solubility of nanoparticles containing curcumin in water. Reprinted from Ref. 56. Copyright (2011) with permission from Elsevier.

loading (33.1%) and entrapment efficiency (90.5%) were achieved due to both micellar encapsulation and the binding interaction between paclitaxel and micelles. It was found that the size of micelles containing paclitaxel was smaller than blank micelles (Fig.13.7). The octyl-modified BSA was proved to be safer than Tween-80 and Cremophor EL as an injectable pharmaceutic adjuvant for paclitaxel.

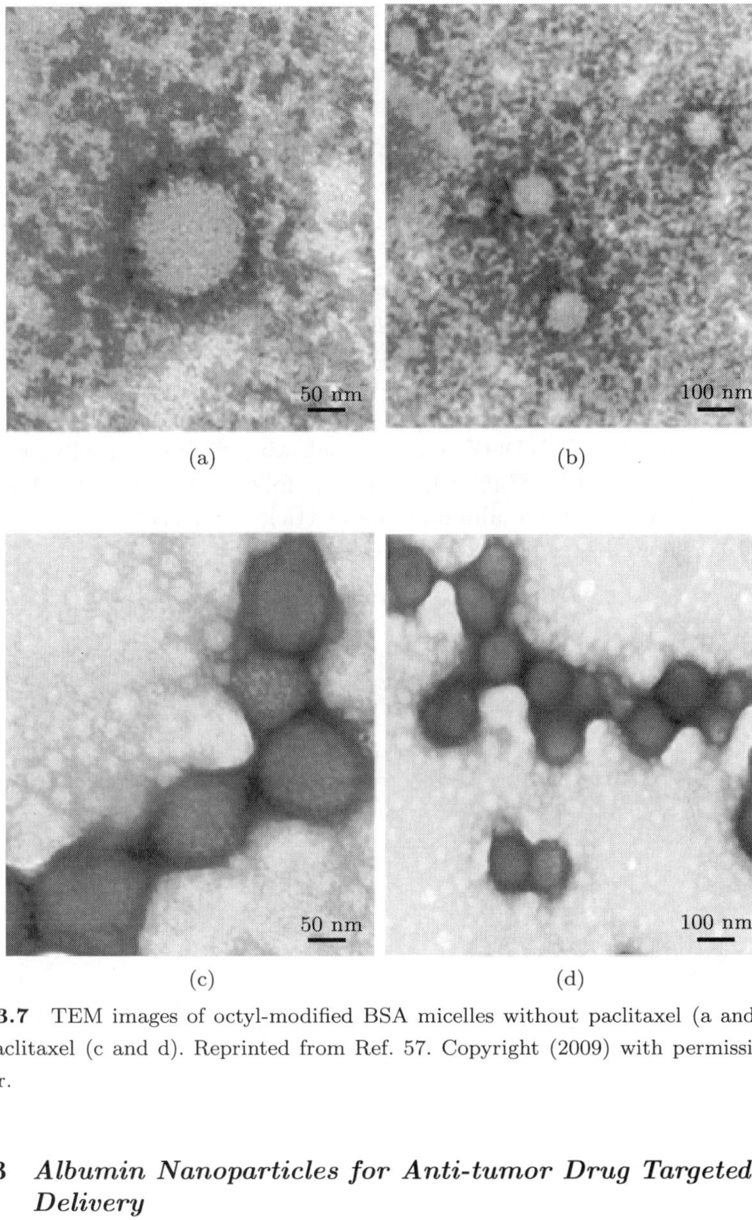

Fig. 13.7 TEM images of octyl-modified BSA micelles without paclitaxel (a and b) and with paclitaxel (c and d). Reprinted from Ref. 57. Copyright (2009) with permission from Elsevier.

13.2.3 *Albumin Nanoparticles for Anti-tumor Drug Targeted Delivery*

There are some functional groups, such as carboxylic and amino groups, on the surface of albumin-based nanoparticles, which is helpful for surface modification.[31] Several techniques, such as covalent reaction, surface coating and electrostatic adsorption, are usually used to conjugate ligands to the surface of albumin nanoparticles. Albumin can load drug as a biodegradable delivery carrier while ligands are used as targeting molecules for tumor cells. The typical targeting ligands include folate, peptide, apolipoprotein, transferrin, monoclonal antibodies, *etc.*[4]

13.2.3.1 *Folate*

The receptor of folic acid is usually over expressed in cancer cells, which has been thought to be an important tumor marker, while it is highly restricted in most of the normal tissues.[58]

Folic acid can be conjugated to albumin nanoparticles by covalent reaction due to the presence of the carboxylic group in folic acid and the amino groups on the surface of albumin nanoparticles. Shen *et al.*[45] prepared DOX-loaded folic acid-conjugated albumin nanoparticles (FA-DOX-AN) (Fig.13.8). It was found that folic acid modification accelerated the incorporation of nanoparticles into tumor cells. Moreover, FA-DOX-AN could selectively kill HeLa cells (with folic acid receptor) but not AoSMC cells (without folic acid receptor), which indicated that folate-conjugated albumin nanoparticles represented a drug delivery system that showed specificity for cancer cells.[59]

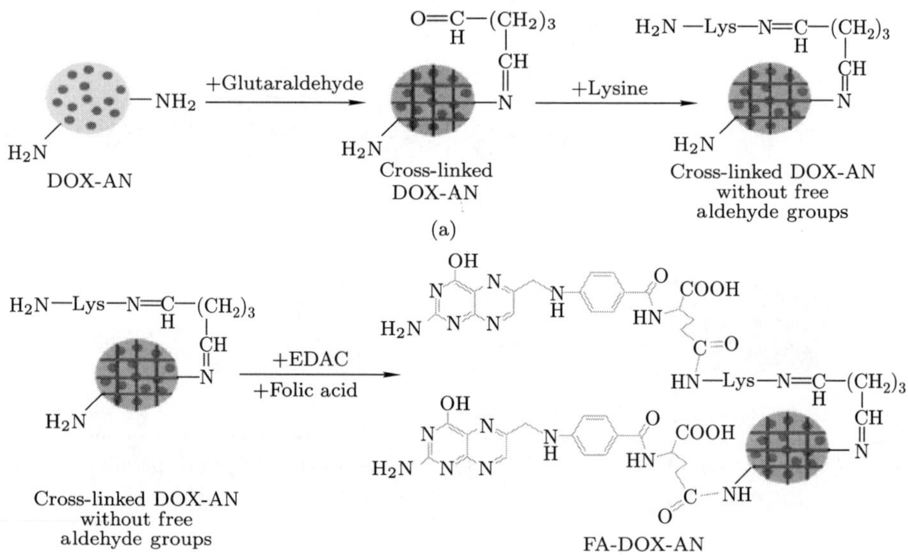

Fig. 13.8 Preparation process for the production of FA-DOX-AN. Reprinted from Ref. 45. Copyright (2011) with permission from Elsevier.

13.2.3.2 *Peptides*

$\alpha_v\beta_3$ integrin is a membrane receptor of some extracellular matrix (ECM) ligands, such as vitronectin and fibronectin, which is expressed on the surface of various tumor cells at high level.[60] So $\alpha_v\beta_3$ integrin can be efficiently used to target tumor cells.

Arginine-glycine-aspartic acid (RGD) peptide is a ligand of $\alpha_v\beta_3$ integrin with high binding affinity. Dubey *et al.*[61] prepared RGD peptide-anchored sterically stabilized BSA nanospheres (RGD-SN) loading fluorouracil with the size of 100 nm. It was found that RGD-SN could bind to endothelial cells more effectively than nanospheres without RGD, and also could prevent the metastasis and angiogenesis of tumor cells, which indicated that RGD-modified nanospheres could be used as an effective drug delivery carrier with the ability of targeting tumor cells.[61]

Similarly, Xu *et al.*[28] prepared HSA-based micelles with 30 nm diameter to deliver DOX. The inner core of micelles was formed by the interaction of HSA and doxorubicin. While PEGs were used to generate a hydrophilic outer layer, cyclic RGD peptide was conjugated on the surface of micelles, which was helpful for delivering drugs to the cells with $\alpha_v\beta_3$ integrin (Fig.13.9). It was found that these micelles could release drug continuously in the mimic intracellular environment and more drugs were uptaken by human melanoma cells by using RGD-modified micelles than unmodified micelles or free drug.

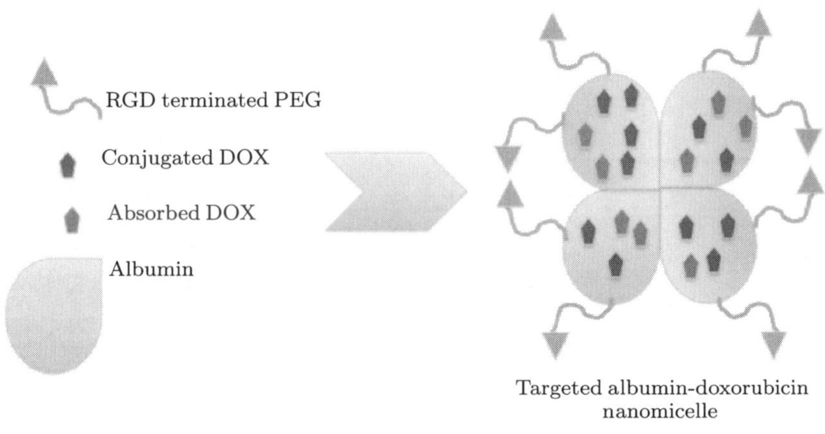

Fig. 13.9 Schematic of targeted HSA-doxorubicin nanomicelle. Reproduced with permission from Ref. 28. Copyright (2011) American Chemical Society.

13.2.3.3 *Apolipoprotein*

Apolipoprotein E (Apo E) is a class of apolipoprotein. In the central nervous system, Apo E is mainly produced by astrocytes and microglia, and it transports cholesterol to neurons *via* Apo E receptors. Protein nanoparticles conjugated with Apo E can promote the drug delivery across the blood-brain barrier.[4]

Zensi *et al.*[62] manufactured Apo E-modified HSA nanoparticles. To realize the modification, sulfhydryl groups were introduced into the apolipoproteins, then the thiolated lipoprotein solutions were added to activate HSA nanoparticles and stirred. Both Apo E-modified and unmodified HSA nanoparticles were injected intravenously into SV 129 mice. It was found that only Apo E-modified HSA nanoparticles were observed in brain capillary endothelial cells and neurones, whereas no uptake into the brain was observed with unmodified HSA nanoparticles (Fig.13.10). The uptake of Apo E-modified nanoparticles into endothelial cells *in vitro* was also determined. These results indicated that Apo E-modified HSA nanoparticles could be taken up into the cerebral endothelium by an endocytic mechanism followed by transcytosis into brain parenchyma.

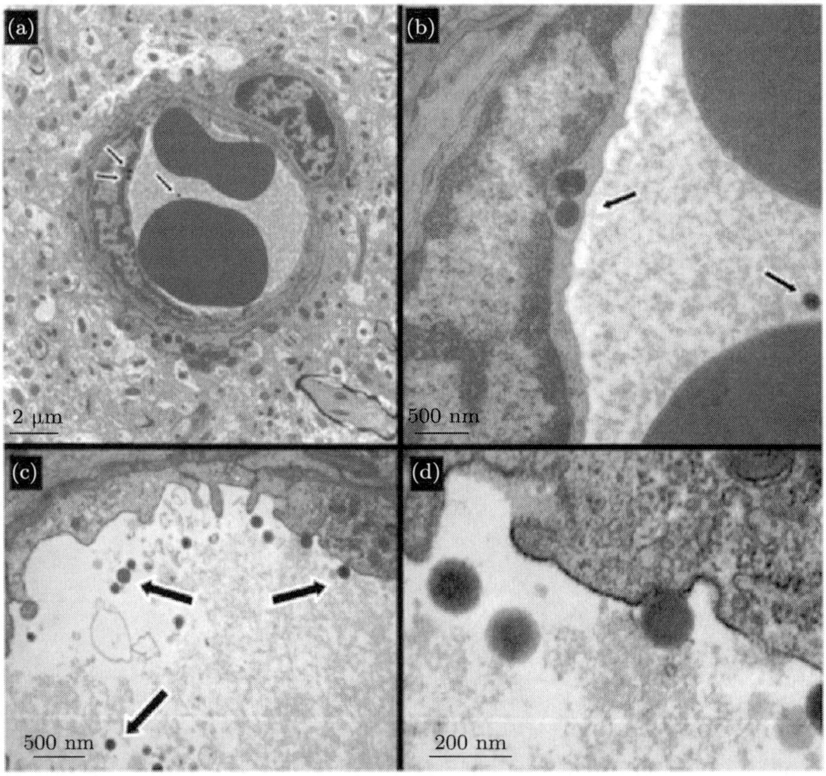

Fig. 13.10 Micrographs of the cortex and hippocampus region of SV 129 mice after the injection of Apo E-modified HSA nanoparticles. (a) A cross section of a blood vessel in the cortex region with two nanoparticles in the endothelium cell and one nanoparticle still in the lumen of the vessel. (b) A higher magnification of the same micrograph. (c) A cross section of a vessel in the hippocampus region of a mouse 15 min after nanoparticle injection. Some nanoparticles (arrows) can still be seen in the lumen of the vessel. (d) A higher magnification of the same vessel with the endothelium cell membrane forming a coated pit adjacent to a nanoparticle. Reprinted from Ref. 62. Copyright (2009) with permission from Elsevier.

13.2.3.4 *Transferrin*

Transferrins are iron-binding blood plasma glyco proteins, which can be produced in brain or liver. Mishra *et al.*[63] prepared HSA nanoparticles loading azidothymidine conjugated with transferrin on the surface of particles to realize brain targeting. To complete the modification, nanoparticles were activated with NHS-PEG-MAL-5000, then thiolated solution of transferrin was added to sulfhydryl-reactive HSA nanoparticles, and incubated under shaking for 18 h at room temperature. It was found that transferrin- modified HSA nanoparticles significantly enhanced the brain uptake, which indicated that transferrin ligand modified on nanoparticles was helpful for brain targeting.

Bae *et al.*[64] conjugated two kinds of ligands, including tumor necrosis factor (TNF)-related apoptosis-inducing ligand (TRAIL) and transferrin, on the surface of HSA nanoparticles loading DOX. Dimethylmaleic anhydride was first used to protect the amines on the surface of HSA, and then HSA nanoparticles were generated by using desolvation method. After that, TRAIL or transferrin was grafted on the nanoparticle surface. Figure 13.11 showed the whole preparation process of ligand-modified nanoparticles. It was found that TRAIL/transferrin-modified nanoparticles had remarkable cytotoxic and

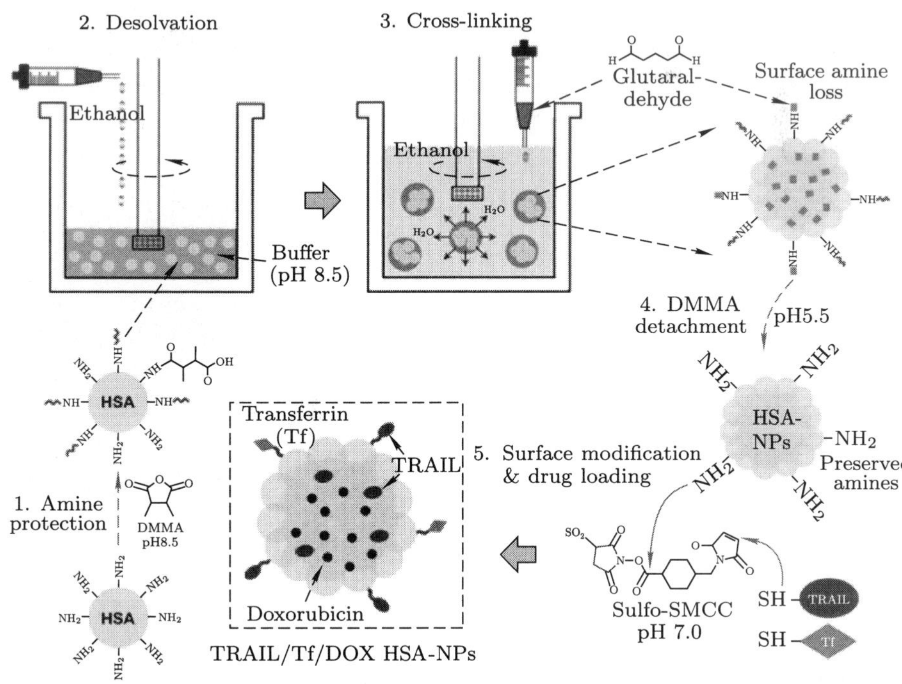

Fig. 13.11 Schematic of the preparation of TRAIL/transferrin-modified HSA nanoparticles loading doxorubicin. Reprinted from Ref. 64. Copyright (2012) with permission from Elsevier.

apoptotic activities in different types of tumor cells. These modified nanoparticles could also be detected in tumor tissues in an HCT 116-xenografted nu/nu mouse. These results show that TRAIL/transferrin-modified DOX HSA nanoparticles loading doxorubicin can be used as a useful targeting drug delivery system.

13.2.3.5 *Monoclonal Antibodies*

It has been reported that some malignancies, including head, breast, colorectal, neck, glioma and prostate cancers, show an over expression of epidermal growth factor receptor (EGFR). For example, HER2 has been investigated to be an important tumor targeting marker for metastatic breast cancer.[65] Kouchakzadeh *et al.*[66] conjugated a monoclonal antibody (1F2) against HER2 on the surface of PEGylated HSA nanoparticles. It was found that 1F2-modified nanoparticles could bind on nearly all of HER2 receptors of BT474 cells, which showed that 1F2-modified HSA nanoparticles was a potential system for the targeting delivery of drugs or imaging agents into tumor cells expressing HER2 receptors.

Moreover, a humanized IgG1 monoclonal antibody(cetuximab) that targeted EGFR was approved by FDA in 2004 to treat colorectal cancer.[67] Loew *et al.*[68] reported a cetuximab-modified HSA nanoparticles containing doxorubicin (Fig.13.12). The conjugation occurred due to the covalent reaction between thiolated antibody and sulfhydryl-reactive nanoparticles. It was found that cetuximab-modified nanoparticles could significantly bind on the surface of colon carcinoma cells with higher EGFR-expression and also could be uptaken by cells.

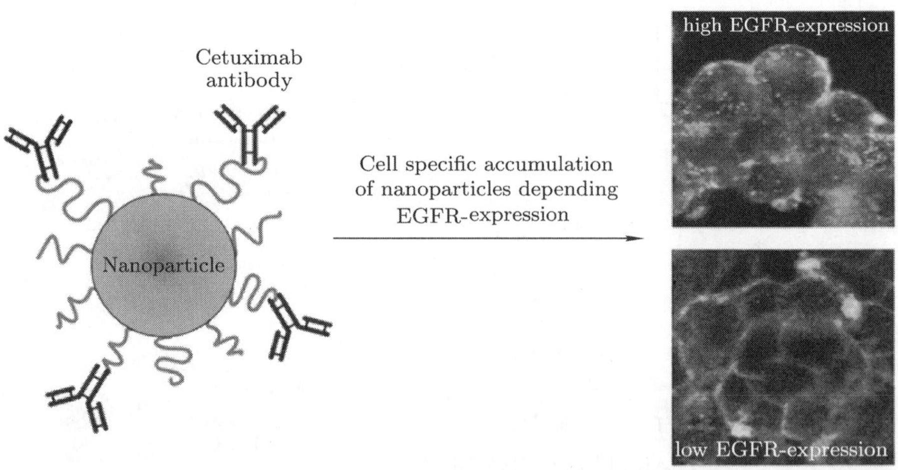

Fig. 13.12 Schematic of cetuximab-modified HAS nanoparticles containing doxorubicin. Reprinted from Ref. 68. Copyright (2011) with permission from Elsevier. (color figure at the end of the book)

13.2.4 *Albumin Nanoparticles for Targeted Tumor Imaging*

Molecular imaging has a wide potential in clinical diagnosis, because it can produce specific contrast in specified tissues and body invasion can also be minimized. Many types of modalities are involved in molecular imaging, such as nuclear imaging, magnetic resonance imaging (MRI) and optical imaging. Nuclear imaging has high sensitivity and high-contrast, but its spatial resolution is restricted.[38] Although MRI has higher spatial resolution, it is difficult to detect weak signals, leading to a low sensitivity. Therefore, it is necessary to develop new contrast reagents as signal amplifiers to promote the application of molecular imaging by using MRI.[69] On the contrary, optical imaging has poor tissue penetration. Near-infrared (NIR) window is thought to be suitable for *in vivo* imaging and nanoparticles are the basic and important platform for molecular imaging. Nanoparticles used for clinical diagnosis must be safe, efficient and standard. Albumin-based nanoparticles are found to be one of the proper carriers due to its useful characteristics.[38]

13.2.4.1 *Radionuclide-based Imaging*

Ozgur *et al.*[39] prepared radio labeled fluorescent dye conjugated BSA nanoparticles (BSANPs) with the size of 190–210 nm (Fig.13.13). BSA nanoparticles were generated by using desolvation technology. The modification was achieved by using gluteraldehyde. BSANPs were loaded pheophorbide-a (PH-A), followed by ^{99m}Tc labeling. It was found that the radio labeling efficiency of such systems were very high, about 90% and ^{99m}Tc-PH-A-BSANPs also showed a higher uptake in the breast and uterus than ^{99m}Tc-PH-A, which indicated that BSA NPs could increase the uptake of PH-A in breast cancer cells. This study

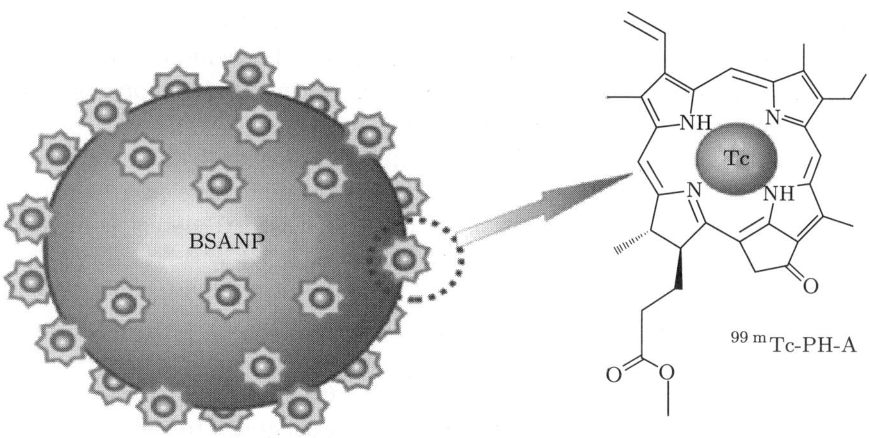

Fig. 13.13 Schematic of ^{99m}Tc-PH-A complex (right) attached BSA nanoparticles (left). Reprinted from Ref. 39. Copyright (2012) with permission from Elsevier.

showed that ^{99m}Tc labeled PH-A-BSANPs was proper to be used for nuclear imaging.

13.2.4.2 *Magnetic Resonance Imaging*

MRI signal has a close relationship with the precession of water hydrogen nuclei in a magnetic field. After application of radiofrequency pulses, the nuclei will return to the original aligned state, and this relaxation process can be used to produce an image.[70] Some contrast agents can be used to shorten the relaxation parameters of water in order to enhance the differentiation between tissues *in vivo*. Therefore, it is necessary to develop contrast agents based on nanoparticles.[37]

Stollenwerk *et al.*[38] prepared HSA-PLA nanoparticles with the size of 20–40 nm, and then verified their physicochemical properties as contrast agents for MRI. Gadolinium (Gd) was binded to the HSA in the form of DTPA-Gd chelates. It is known that Gd is a well-established signal enhancer in MRI, while DTPA is considered to be the chelator for Gd. In these nanoparticles, HSA retained its tertiary structure. It was also found that MRI imaging of liquid columns was feasible at low nanoparticle concentrations (below 10 mg/mL) and these particles were proved to be non-cytotoxic, non-thrombogenic and non-immunogenic.[38]

13.2.4.3 *Optical Imaging*

Optical imaging detects photons emitted from bioluminescent or fluorescent probes that are used for *in vivo* imaging. However, optical imaging suffers from poor tissue penetration (only 0–2 cm) and high background due to tissue autofluorescence, light absorption by proteins, heme groups, and even water. But NIR (700–900 nm) window is assumed to be suitable for *in vivo* imaging, because autofluorescence and tissue scattering can be reduced and the tissue penetration depth can be increased.[37] NIR fluorescent HSA nanoparticles hold wide potential as contrast agents for tumor diagnosis.

Bunschoten *et al.*[71]prepared self-assembled complexes by the non covalent interaction between HSA and peptides *via* NIR cyanine dyes (Fig.13.14). Fluorescence resonance energy transfer (FRET) was used to detect complex assembly and disassembly, because FRET was satisfied with the interactions between HSA-Cy5 and ICG, IR783-CO_2H and three IR783-labeled targeting peptides. Nanoparticles formed by the interactions between HSA and IR783-labeled peptides were coated with peptides that target some receptors, including $\alpha_v\beta_3$-integrins, CXCR4 and somatostatin. CXCR4-targeted nanoparticles showed great potential in sentinel lymph node procedures. In their work, novel targeted nanoparticles were formed by binding NIR-dye labeled peptides to the clinically approved HSA nanoparticles, which might be easier to use in clinical diagnosis.

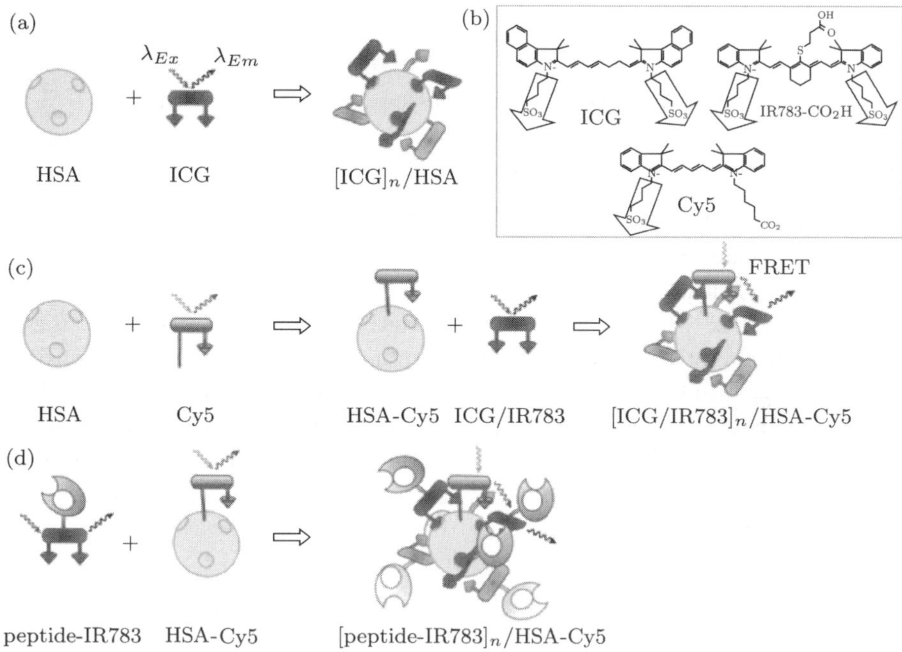

Fig. 13.14 (a) Formation of a fluorescent noncovalent complex of HSA/ICG. (b) Chemical structures of ICG, IR783-CO_2H and Cy5. (c) Covalent labeling of HSA with Cy5 and subsequent formation of a noncovalent quenched complex of HSA-Cy5 with the NIR-dyes ICG or IR783-CO_2H. (d) Formation of targeted nanoparticles based on the interaction of IR783-labeled peptides with HSA. Reprinted from Ref. 71. Copyright (2012) with permission from Elsevier.

Cohen *et al.*[72] manufactured new NIR fluorescent HSA nanoparticles by loading an NIR fluorescent dye in HSA nanoparticles. It was found that the photostability of the dye was significantly increased. In addition, targeting ligands, such as peanut agglutinin (PNA) and anti-carcinoembryonic antigen antibodies (anti-CEA) could be conjugated to the surface of NIR fluorescent nanoparticles. Fluorescent signal in tumor tissues was significantly increased due to the upregulated receptors. These nanoparticles could also detect specific colon tumor, which was confirmed by using a chicken embryo or rat models.[72]

In addition, rare-earth-doped nanoparticles are thought to be not ideal for clinical application due to their insolubility in aqueous solution, lack of functional surface groups for modification, and even cytotoxicity. Naczynski *et al.*[73] reported new nanoshell complexes, luminescent rare-earth-doped NaYF4 nanoparticles (REs), with good biocompatibility. Targeting ability of these particles was excitable with 920–980 nm near-infrared light, which was helpful for *in vivo* imaging applications. NaYF4 nanoparticles containing Yb and Er were encapsulated in HSA nanoshells to increase dispersibility in water and biological function. It was found that these particles were stable in aqueous solution for over 4 weeks. The HSA shell could confer cytoprotection and significantly enhanced the biocompatibility of REs. Composite particles conjugated with RGD could target tumor cells that expressed $\alpha_v\beta_3$ integrin receptors.

13.2.5 *Albumin Nanoparticles for Photodynamic Therapy*

Photodynamic therapy (PDT) is a promising method that can be used for cancer treatment by inducing tumor cell apoptosis or necrosis. Photosensitizers have PDT property. In the absence of light, photosensitizers have no cytotoxicity, but when a specific wave-length light is used, photosensitizers can generate cytotoxic oxygen that kills tumor cells. However, photosensitizers have poor solubility in water, and have no tumor-targeting ability, which made them restricted in clinical applications.[26,74] Therefore, it is important to develop nanoparticles to overcome the above limitations of photosensitizers for PDT. Jeong *et al.* prepared photosensitizer-conjugated HSA nanoparticles with 88 nm diameter.[74] The nanoparticles were formed by self-assembly between HSA and photosensitizer, chlorin e6 (Ce6). Ce6-conjugated HSA nanoparticles (Ce6-HSA-NPs) were nontoxic in the absence of light, but when the appropriate wavelength of light was used, nanoparticles could produce singlet oxygen and kill tumor cells *in vitro*. When these nanoparticles were injected through the tail vein into tumor-bearing mice, Ce6-HSA-NPs could enhance the distribution in tumor tissues compared with free Ce6, and good treatment results were achieved after following laser irradiation.

In addition, Yu *et al.*[75] first manufactured Au_2Se/Au core-shell nanostructures with strong NIR absorbance properties. These nanostructures were then embedded inside BSA nanospheres by using a spray dryer equipped with an ultrasonic atomizer followed by thermal denaturation. The size of these BSA nanospheres was between 450 and 600 nm. Au_2Se/Au-loaded BSA nanospheres could load zinc phthalocyanine (ZnPc) photosensitizer. It was found that after irradiation, the temperature of Au_2Se/Au-loaded BSA nanospheres increased over 20 °C and the release of ZnPc was enhanced compared to the ones without irradiation. Moreover, after incubation with human esophageal carcinoma, these nanospheres showed the ability of photothermal and photodynamic therapy when laser irradiation was used.

13.3 Outlook

In this chapter, albumin-based nanoparticles are described for their potential applications in tumor targeting theranostics, including drug delivery, imaging and therapy.

One problem of natural proteins is the big variation between different batches, which might limit the large-scale production of nanoparticle for clinical applications. To overcome this problem, it is possible to use recombinant protein, such as recombinant HSA and recombinant gelatin.[4,9] Nonetheless, new tumor-targeting ligands should be developed to enhance the targeting capacity of nanoparticles when used *in vivo*. Additionally, it is also necessary to find new safe crosslinkers of protein nanoparticles instead of the toxic ones. Moreover, basic research on the mechanisms of protein particle action and of the protein-drug interactions at the molecular level is very important for the design of novel drug nanocarriers and can promote the clinical applications of them in tumor targeting theranostics.[9]

Acknowledgments

This work was supported by National Natural Science Foundation of China (No. 31271055,31470944).

References

1. van Vlerken, L. E.; Amiji, M. M. Multi-functional polymeric nanoparticles for tumour-targeted drug delivery. *Expert Opin Drug Del* **2006**, *3* (2), 205-216.
2. Peng, X. H.; Qian, X. M.; Mao, H.; Wang, A. Y.; Chen, Z.; Nie, S. M.; Shin, D. M. Targeted magnetic iron oxide nanoparticles for tumor imaging and therapy. *Int J Nanomed* **2008**, *3* (3), 311-321.
3. Rhyner, M. N.; Smith, A. M.; Gao, X. H.; Mao, H.; Yang, L. L.; Nie, S. M. Quantum dots and multifunctional nanoparticles: New contrast agents for tumor imaging. *Nanomedicine* **2006**, *1* (2), 209-217.
4. Elzoghby, A. O.; Samy, W. M.; Elgindy, N. A. Albumin-based nanoparticles as potential controlled release drug delivery systems. *J Control Release* **2012**, *157* (2), 168-182.
5. Liu, Y. Y.; Miyoshi, H.; Nakamura, M. Nanomedicine for drug delivery and imaging: A promising avenue for cancer therapy and diagnosis using targeted functional nanoparticles. *Int J Cancer* **2007**, *120* (12), 2527-2537.
6. Sahoo, S. K.; Jain, T. K.; Reddy, M. K.; Labhasetwar, V. Nano-sized carriers for drug delivery. In: Shoseyov, O.; Levy, L. *NanoBioTechnology: BioInspired Devices and Materials of the Future*. Humana Press Inc., **2008**, 329-348.
7. Farrugia, C. A.; Groves, M. J. The activity of unloaded gelatin nanoparticles on murine melanoma B16-F0 growth *in vivo*. *Anticancer Res* **1999**, *19* (2A), 1027-1031.
8. Farrugia, C. A.; Groves, M. J. Gelatin behaviour in dilute aqueous solution: Designing a nanoparticulate formulation. *J Pharm Pharmacol* **1999**, *51* (6), 643-649.
9. Elzoghby, A. O.; Samy, W. M.; Elgindy, N. A. Protein-based nanocarriers as promising drug and gene delivery systems. *J Control Release* **2012**, *161* (1), 38-49.

10. Kuijpers, A. J.; van Wachem, P. B.; van Luyn, M. J. A.; Brouwer, L. A.; Engbers, G. H. M.; Krijgsveld, J.; Zaat, S. A. J.; Dankert, J.; Feijin, J. *In vitro* and *in vivo* evaluation of gelatin-chondroitin sulphate hydrogels for controlled release of antibacterial proteins. *Biomaterials* **2000**, *21* (17), 1763-1772.
11. Abdelmoez, A. A.; Thurner, G. C.; Wallnoefer, E. A.; Klammsteiner, N.; Kremser, C.; Talasz, H.; Mrakovcic, M.; Froehlich, E.; Jaschke, W.; Debbage, P. Albumin-based nanoparticles as magnetic resonance contrast agents: II. Physicochemical characterisation of purified and standardised nanoparticles. *Histochem Cell Biol* **2010**, *134* (2), 171-196.
12. Yan, F.; Zhang, Y.; Yuan, H.-K.; Gregas, M. K.; Vo-Dinh, T. Apoferritin protein cages: A novel drug nanocarrier for photodynamic therapy. *Chem Commun* **2008**, (38), 4579-4581.
13. Li, K.; Zhang, Z.-P.; Luo, M.; Yu, X.; Han, Y.; Wei, H.-P.; Cui, Z.-Q.; Zhang, X.-E. Multifunctional ferritin cage nanostructures for fluorescence and MR imaging of tumor cells. *Nanoscale* **2012**, *4* (1), 188-193.
14. Vandelli, M. A.; Rivasi, F.; Guerra, P.; Forni, F.; Arletti, R. Gelatin microspheres crosslinked with D,L-glyceraldehyde as a potential drug delivery system: Preparation, characterisation, *in vitro* and *in vivo* studies. *Int J Pharmaceut* **2001**, *215* (1-2), 175-184.
15. Sahin, S.; Selek, H.; Ponchel, G.; Ercan, M. T.; Sargon, M.; Hincal, A. A.; Kas, H. S. Preparation, characterization and *in vivo* distribution of terbutaline sulfate loaded albumin microspheres. *J Control Release* **2002**, *82* (2-3), 345-358.
16. Elzoghby, A. O.; El-Fotoh, W. S. A.; Elgindy, N. A. Casein-based formulations as promising controlled release drug delivery systems. *J Control Release* **2011**, *153* (3), 206-216.
17. MaHam, A.; Tang, Z. W.; Wu, H.; Wang, J.; Lin, Y. H. Protein-based nanomedicine platforms for drug delivery. *Small* **2009**, *5* (15), 1706-1721.
18. Langer, K.; Balthasar, S.; Vogel, V.; Dinauer, N.; von Briesen, H.; Schubert, D. Optimization of the preparation process for human serum albumin (HSA) nanoparticles. *Int J Pharmaceut* **2003**, *257* (1-2), 169-180.
19. Kommareddy, S.; Amiji, M. Preparation and evaluation of thiol-modified gelatin nanoparticles for intracellular DNA delivery in response to glutathione. *Bioconjugate Chem* **2005**, *16* (6), 1423-1432.
20. He, X. M.; Carter, D. C. Atomic-structure and chemistry of human serum-albumin. *Nature* **1993**, *364* (6435), 362.
21. Kratz, F. Albumin as a drug carrier: Design of prodrugs, drug conjugates and nanoparticles. *J Control Release* **2008**, *132* (3), 171-183.
22. Carter, D. C.; He, X. M.; Munson, S. H.; Twigg, P. D.; Gernert, K. M.; Broom, M. B.; Miller, T. Y. 3-dimensional structure of human-serum albumin. *Science* **1989**, *244* (4909), 1195-1198.
23. Carter, D. C.; Ho, J. X. Structure of serum-albumin. *Adv Protein Chem* **1994**, *45*, 153-203.
24. Cregg, J. M.; Cereghino, J. L.; Shi, J. Y.; Higgins, D. R. Recombinant protein expression in Pichia pastoris. *Mol Biotechnol* **2000**, *16* (1), 23-52.
25. Chuang, V. T. G.; Otagiri, M. Recombinant human serum albumin. *Drug Today* **2007**, *43* (8), 547-561.

26. Chen, K.; Preuss, A.; Hackbarth, S.; Wacker, M.; Langer, K.; Roeder, B. Novel photosensitizer-protein nanoparticles for photodynamic therapy: Photophysical characterization and *in vitro* investigations. *J Photoch Photobio B* **2009**, *96* (1), 66-74.
27. Chaumet-Riffaud, P.; Martinez-Duncker, I.; Marty, A. L.; Richard, C.; Prigent, A.; Moati, F.; Sarda-Mantel, L.; Scherman, D.; Bessodes, M.; Mignet, N. Synthesis and application of lactosylated, Tc-99m chelating albumin for measurement of liver function. *Bioconjugate Chem* **2010**, *21* (4), 589-596.
28. Xu, R.; Fisher, M.; Juliano, R. L. Targeted albumin-based nanoparticles for delivery of amphipathic drugs. *Bioconjugate Chem* **2011**, *22* (5), 870-878.
29. Patil, G. V. Biopolymer albumin for diagnosis and in drug delivery. *Drug Develop Res* **2003**, *58* (3), 219-247.
30. Irache, J. M.; Merodio, M.; Arnedo, A.; Camapanero, M. A.; Mirshahi, M.; Espuelas, S. Albumin nanoparticles for the intravitreal delivery of anticytomegaloviral drugs. *Mini-Rev Med Chem* **2005**, *5* (3), 293-305.
31. Weber, C.; Kreuter, J.; Langer, K. Desolvation process and surface characteristics of HSA-nanoparticles. *Int J Pharmaceut* **2000**, *196* (2), 197-200.
32. Steinhauser, I. M.; Langer, K.; Strebhardt, K. M.; Spankuch, B. Effect of trastuzumab-modified antisense oligonucleotide-loaded human serum albumin nanoparticles prepared by heat denaturation. *Biomaterials* **2008**, *29* (29), 4022-4028.
33. Ulbrich, K.; Hekmatara, T.; Herbert, E.; Kreuter, J. Transferrin- and transferrin-receptor-antibody-modified nanoparticles enable drug delivery across the blood-brain barrier (BBB). *Eur J Pharm Biopharm* **2009**, *71* (2), 251-256.
34. Geny, B.; Mettauer, B.; Muan, B.; Bischoff, P.; Epailly, E.; Piquard, F.; Eisenmann, B.; Haberey, P. Safety and efficacy of a new transpulmonary echo contrast agent in echocardiographic studies in patients. *J Am Coll Cardiol* **1993**, *22* (4), 1193-1198.
35. Ibrahim, N. K.; Desai, N.; Legha, S.; Soon-Shiong, P.; Theriault, R. L.; Rivera, E.; Esmaeli, B.; Ring, S. E.; Bedikian, A.; Hortobagyi, G. N.; Ellerhorst, J. A. Phase I and pharmacokinetic study of ABI-007, a Cremophor-free, protein-stabilized, nanoparticle formulation of paclitaxel. *Clin Cancer Res* **2002**, *8* (5), 1038-1044.
36. Hawkins, M. J.; Soon-Shiong, P.; Desai, N. Protein nanoparticles as drug carriers in clinical medicine. *Adv Drug Deliver Rev* **2008**, *60* (8), 876-885.
37. Janib, S. M.; Moses, A. S.; MacKay, J. A. Imaging and drug delivery using theranostic nanoparticles. *Adv Drug Deliver Rev* **2010**, *62* (11), 1052-1063.
38. Stollenwerk, M. M.; Pashkunova-Martic, I.; Kremser, C.; Talasz, H.; Thurner, G. C.; Abdelmoez, A. A.; Wallnoefer, E. A.; Helbok, A.; Neuhauser, E.; Klammsteiner, N.; Klimaschewski, L.; von Guggenberg, E.; Froehlich, E.; Keppler, B.; Jaschke, W.; Debbage, P. Albumin-based nanoparticles as magnetic resonance contrast agents: I. Concept, first syntheses and characterisation. *Histochem Cell Biol* **2010**, *133* (4), 375-404.
39. Ozgur, A.; Lambrecht, F. Y.; Ocakoglu, K.; Gunduz, C.; Yucebas, M. Synthesis and biological evaluation of radiolabeled photosensitizer linked bovine serum albumin nanoparticles as a tumor imaging agent. *Int J Pharmaceut* **2012**, *422* (1-2), 472-478.

40. Li, F. Q.; Hu, J. H.; Lu, B.; Yao, H.; Zhang, W. G. Ciprofloxacin-loaded bovine serum albumin microspheres: Preparation and drug-release *in vitro*. *J Microencapsul* **2001**, *18* (6), 825-829.
41. Meziani, M. J.; Sun, Y. P. Protein-conjugated nanoparticles from rapid expansion of supercritical fluid solution into aqueous solution. *J Am Chem Soc* **2003**, *125* (26), 8015-8018.
42. Paik, S.-Y.-R.; Hoang Hai, N.; Ryu, J.; Che, J.-H.; Kang, T. S.; Lee, J. K.; Song, C. W.; Ko, S. Robust size control of bovine serum albumin (BSA) nanoparticles by intermittent addition of a desolvating agent and the particle formation mechanism. *Food Chem* **2013**, *141* (2), 695-701.
43. Lin, W.; Coombes, A. G. A.; Garnett, M. C.; Davies, M. C.; Schacht, E.; Davis, S. S.; Illum, L. Preparation of sterically stabilized human serum-albumin nanospheres using a novel dextranox-mpeg cross-linking agent. *Pharmaceut Res* **1994**, *11* (11), 1588-1592.
44. Zimmer, A. K.; Chetoni, P.; Saettone, M. F.; Zerbe, H.; Kreuter, J. Evaluation of pilocarpine-loaded albumin particles as controlled drug-delivery systems for the eye. II. Coadministration with bioadhesive and viscous polymers. *J Control Release* **1995**, *33* (1), 31-46.
45. Shen, Z.; Li, Y.; Kohama, K.; Oneill, B.; Bi, J. Improved drug targeting of cancer cells by utilizing actively targetable folic acid-conjugated albumin nanospheres. *Pharmacol Res* **2011**, *63* (1), 51-58.
46. Sundar, S.; Kundu, J.; Kundu, S. C. Biopolymeric nanoparticles. *Sci Technol Adv Mat* **2010**, *11*, 014104-014116.
47. Jahanshahi, M.; Babaei, Z. Protein nanoparticle: A unique system as drug delivery vehicles. *Afr J Biotechnol* **2008**, *7* (25), 4926-4934.
48. Crisante, F.; Francolini, I.; Bellusci, M.; Martinelli, A.; D'Ilario, L.; Piozzi, A. Antibiotic delivery polyurethanes containing albumin and polyallylamine nanoparticles. *Eur J Pharm Sci* **2009**, *36* (4-5), 555-564.
49. Yang, L.; Cui, F.; Cun, D.; Tao, A.; Shi, K.; Lin, W. Preparation, characterization and biodistribution of the lactone form of 10-hydroxycamptothecin (HCPT)-loaded bovine serum albumin (BSA) nanoparticles. *Int J Pharmaceut* **2007**, *340* (1-2), 163-172.
50. Qi, J.; Yao, P.; He, F.; Yu, C.; Huang, C. Nanoparticles with dextran/chitosan shell and BSA/chitosan core: Doxorubicin loading and delivery. *Int J Pharmaceut* **2010**, *393* (1-2), 176-184.
51. Dissanayake, M.; Kelly, A. L.; Vasiljevic, T. Gelling properties of microparticulated whey proteins. *J Agr Food Chem* **2010**, *58* (11), 6825-6832.
52. Lee, S. H.; Heng, D.; Ng, W. K.; Chan, H.-K.; Tan, R. B. H. Nano spray drying: A novel method for preparing protein nanoparticles for protein therapy. *Int J Pharmaceut* **2011**, *403* (1-2), 192-200.
53. Desai, N. P.; Trieu, V.; Hwang, L. Y.; Wu, R.; Soon-Shiong, P.; Gradishar, W. J. Improved effectiveness of nanoparticle albumin-bound (nab) paclitaxel versus polysorbate-based docetaxel in multiple xenografts as a function of HER2 and SPARC status. *Anti-Cancer Drugs* **2008**, *19* (9), 899-909.
54. Damascelli, B.; Cantu, G.; Mattavelli, F.; Tamplenizza, P.; Bidoli, P.; Leo, E.; Dosio, F.; Cerrotta, A. M.; Di Tolla, G.; Frigerio, L. F.; Garbagnati, F.;

Lanocita, R.; Marchiano, A.; Patelli, G.; Spreafico, C.; Ticha, V.; Vespro, V.; Zunino, F. Intraarterial chemotherapy with polyoxyethylated castor oil free paclitaxel, incorporated in albumin nanoparticles (ABI-007)-Phase I study of patients with squamous cell carcinoma of the head and neck and anal canal: Preliminary evidence of clinical activity. *Cancer* **2001**, *92* (10), 2592-2602.

55. Desai, N.; Trieu, V.; Yao, Z. Increased antitumor activity, intratumor paclitaxel concentrations, and endothelial cell transport of cremophor-free, albumin-bound paclitaxel, ABI-007, compared with cremophor-based paclitaxel. *Clin Cancer Res* **2006**, *12* (12), 3869.
56. Kim, T. H.; Jiang, H. H.; Youn, Y. S.; Park, C. W.; Tak, K. K.; Lee, S.; Kim, H.; Jon, S.; Chen, X.; Lee, K. C. Preparation and characterization of water-soluble albumin-bound curcumin nanoparticles with improved antitumor activity. *Int J Pharmaceut* **2011**, *403* (1-2), 285-291.
57. Gong, J.; Huo, M.; Zhou, J.; Zhang, Y.; Peng, X.; Yu, D.; Zhang, H.; Li, J. Synthesis, characterization, drug-loading capacity and safety of novel octyl modified serum albumin micelles. *Int J Pharmaceut* **2009**, *376* (1-2), 161-168.
58. Atkinson, S. F.; Bettinger, T.; Seymour, L. W.; Behr, J. P.; Ward, C. M. Conjugation of folate *via* gelonin carbohydrate residues retains ribosomal-inactivating properties of the toxin and permits targeting to folate receptor positive cells. *J Biol Chem* **2001**, *276* (30), 27930-27935.
59. Ulbrich, K.; Michaelis, M.; Rothweiler, F.; Knobloch, T.; Sithisarn, P.; Cinatl, J.; Kreuter, J. Interaction of folate-conjugated human serum albumin (HSA) nanoparticles with tumour cells. *Int J Pharmaceut* **2011**, *406* (1-2), 128-134.
60. Bello, L.; Zhang, J. P.; Nikas, D. C.; Strasser, J. F.; Villani, R. M.; Cheresh, D. A.; Carroll, R. S.; Black, P. M. $\alpha_v\beta_3$ and $\alpha_v\beta_5$ integrin expression in meningiomas. *Neurosurgery* **2000**, *47* (5), 1185-1195.
61. Dubey, P. K.; Singodia, D.; Verma, R. K.; Vyas, S. P. RGD modified albumin nanospheres for tumour vasculature targeting. *J Pharm Pharmacol* **2011**, *63* (1), 33-40.
62. Zensi, A.; Begley, D.; Pontikis, C.; Legros, C.; Mihoreanu, L.; Wagner, S.; Buechel, C.; von Briesen, H.; Kreuter, J. Albumin nanoparticles targeted with Apo E enter the CNS by transcytosis and are delivered to neurones. *J Control Release* **2009**, *137* (1), 78-86.
63. Mishra, V.; Mahor, S.; Rawat, A.; Gupta, P. N.; Dubey, P.; Khatri, K.; Vyas, S. P. Targeted brain delivery of AZT *via* transferrin anchored pegylated albumin nanoparticles. *Drug Target* **2006**, *14* (1), 45-53.
64. Bae, S.; Ma, K.; Kim, T. H.; Lee, E. S.; Oh, K. T.; Park, E.-S.; Lee, K. C.; Youn, Y. S. Doxorubicin-loaded human serum albumin nanoparticles surface-modified with TNF-related apoptosis-inducing ligand and transferrin for targeting multiple tumor types. *Biomaterials* **2012**, *33* (5), 1536-1546.
65. Wartlick, H.; Michaelis, K.; Balthasar, S.; Strebhardt, K.; Kreuter, J.; Langer, K. Highly specific HER2-mediated cellular uptake of antibody-modified nanoparticles in tumour cells. *Drug Target* **2004**, *12* (7), 461-471.
66. Kouchakzadeh, H.; Shojaosadati, S. A.; Tahmasebi, F.; Shokri, F. Optimization of an anti-HER2 monoclonal antibody targeted delivery system using PEGylated human serum albumin nanoparticles. *Int J Pharmaceut* **2013**, *447* (1-2), 62-69.

67. Wagner, S.; Rothweiler, F.; Anhorn, M. G.; Sauer, D.; Riemann, I.; Weiss, E. C.; Katsen-Globa, A.; Michaelis, M.; Cinatl, J. Jr.; Schwartz, D.; Kreuter, J.; von Briesen, H.; Langer, K. Enhanced drug targeting by attachment of an anti α_v integrin antibody to doxorubicin loaded human serum albumin nanoparticles. *Biomaterials* **2010**, *31* (8), 2388-2398.
68. Loew, K.; Wacker, M.; Wagner, S.; Langer, K.; von Briesen, H. Targeted human serum albumin nanoparticles for specific uptake in EGFR-expressing colon carcinoma cells. *Nanomed-Nanotechnol* **2011**, *7* (4), 454-463.
69. Laurent, S.; Vander Elst, L.; Fu, Y. J.; Muller, R. N. Synthesis and physicochemical characterization of Gd-DTPA-B(sLe(x))A: A new MRI contrast agent targeted to inflammation. *Bioconjugate Chem* **2004**, *15* (1), 99-103.
70. Richard, C.; de Chermont, Qle, M.; Scherman, D. Nanoparticles for imaging and tumor gene delivery. *Tumori* **2008**, *94* (2), 264-270.
71. Bunschoten, A.; Buckle, T.; Kuil, J.; Luker, G. D.; Luker, K. E.; Nieweg, O. E.; van Leeuwen, F. W. B. Targeted non-covalent self-assembled nanoparticles based on human serum albumin. *Biomaterials* **2012**, *33* (3), 867-875.
72. Cohen, S.; Margel, S. Engineering of near IR fluorescent albumin nanoparticles for *in vivo* detection of colon cancer. *J Nanobiotechnol* **2012**, *10*, 36.
73. Naczynski, D. J.; Andelman, T.; Pal, D.; Chen, S.; Riman, R. E.; Roth, C. M.; Moghe, P. V. Albumin nanoshell encapsulation of near-infrared-excitable rare-earth nanoparticles enhances biocompatibility and enables targeted cell imaging. *Small* **2010**, *6* (15), 1631-1640.
74. Jeong, H.; Huh, M.; Lee, S. J.; Koo, H.; Kwon, I. C.; Jeong, S. Y.; Kim, K. Photosensitizer-conjugated human serum albumin nanoparticles for effective photodynamic therapy. *Theranostics* **2011**, *1*, 230-239.
75. Yu, C.; Wo, F.; Shao, Y.; Dai, X.; Chu, M. Bovine serum albumin nanospheres synchronously encapsulating "gold selenium/gold" nanoparticles and photosensitizer for high-efficiency cancer phototherapy. *Appl Biochem Biotech* **2013**, *169* (5), 1566-1578.

Chapter 14

Conclusions and Future Perspectives

Shanmin Gao[1,2] *and Narayan S. Hosmane*[1,3*]

1 Department of Chemistry and Biochemistry, Northern Illinois University, DeKalb, IL 60115, USA

2 School of Chemistry and Materials Science, Ludong University, Yantai 264025, China

3 Visiting Professor of Chinese Academy of Sciences for International Senior Scientists at Ningbo Institute of Materials Technology and Engineering (NIMTE), Ningbo 315201, China

This book, edited by Mingqian Tan and Aiguo Wu, summarizes some of the cancer theranostic applications of nanomaterials with focused topics in 11 chapters highlighting magnetic nanomaterials, semiconductor quantum dots, mesoporous silica nanoparticles, surface enhanced Raman scattering nanomaterials, polymeric nanosystems, carbon nanomaterials, lipopolyplex nanoparticles, microbubbles, upconversion nanomaterials, functional dendrimers, protein-based nanoparticles, among other topics of great interest in nanomedicine.

14.1 Topical Background and Conclusions

Cancer is one of the leading causes of death and it kills thousands of people every day. Various types of treatment methods have been developed to treat and cure cancer. In cancer therapy, nanotechnology may provide a better way to cure cancer.

The rapid development in nanoscience and nanotechnology has brought great opportunities to cancer treatment in various directions. Developments in the field of nanotechnology have brought together distinct fields such as chemistry, engineering, biology, and medicine, and have paved the way to design novel molecular contrast probes for imaging and theranostics. Nanomaterials are considered as a medical boon for the diagnosis, treatment and prevention of cancer.

Nanomaterials are the moieties that have undergone extensive explorations in recent years for biomedical applications, especially in cancer treatment. With

* Corresponding author: hosmane@niu.edu

respect to cancer, nanomaterials are studied for diagnosis, drug delivery, gene delivery, bioseparation, hyperthermia, phototherapy, chemotherapy, imaging mechanisms, among other uses.

Over the past decade, concepts and tools, derived from nanotechnology, have been applied to overcome the problems of conventional techniques for advanced diagnosis and therapy in cancer treatment area. Advances in nanotechnology have created new paradigms for theranostics, which is defined as the combination of therapeutic and diagnostic agents within a single platform. A variety of emerging nanomaterials and technology, including magnetic nanomaterials, quantum dots, surface enhanced Raman scattering (SERS) technology, carbon nanomaterials, lipopolyplex nanoparticles, nano/microbubbles, protein-based nanomaterials, mesoporous silica nanoparticles and neutron capture therapy have been developed for tumor imaging and therapy. Such nanomaterials possess unique optical, electrical, magnetic and/or electrochemical properties.

This book summarizes some of the cancer theranostic applications of nanomaterials. The following targeted topics are the highlights of the book and the individual chapters will provide the detailed information on these topics that will take you to the boundary of your imagination where the nanomedicine will undoubtedly be the future for diagnosis and treatments.

14.1.1 *Magnetic Nanomaterials*

Magnetic resonance imaging (MRI) contrast agents based on magnetic nanomaterials, including paramagnetic and superparamagnetic nanomaterials, have been widely used in clinical medicine. The applications of the magnetic nanomateritals in cancer research mainly concentrate on magnetic resonance imaging, hyperthermia, photodynamic therapy, photothermal therapy and drug/gene delivery. In particular, superparamagnetic nanoparticles offer unique properties including quantum effects, large surface area and easy surface chemistry. The superparamagnetic nanoparticles may yield better MRI contrast agents compared to conventional transition metal ions. However, surface functionalizations and toxicity have been the issues that have limited their use in clinical settings. Understanding the interplay of saturation magnetization, coercivity, and magnetocrystalline anisotropy is critical for optimizing magnetic characteristics needed for effective imaging and therapeutics.

14.1.2 *Semiconductor Quantum Dots (QDs)*

Semiconductor quantum dots exhibit unique optical and electronic properties, such as broad absorption spectra, narrow emission spectra and excellent photostability. These properties rendering them to be promising candidates for a variety of biomedical and medicinal research applications. The flexibility and versatility of quantum dots may provide the keys to answer important biological questions and ultimately improve diagnostics and therapy of cancer. The surface of quantum dots can be engineered with functional biomolecules to selectively target specific sites *in vitro* and *in vivo*. A synergistic combination of quantum dots with biophotonics provides unprecedented opportunities for addressing many of the current challenges in cancer diagnosis and therapy.

14.1.3 *Mesoporous Silica Nanoparticles (MSNs)*

Mesoporous silica nanoparticles have attracted widespread interest for biomedical use due to their easy accessibility and uniformly sized pores, stable structure, easy modification of the internal and external surface, extremely high surface area and pore volume, and good biocompatibility. MSNs possess three topological domains, framework, nanochannels, and outermost surface, each of which can be functionalized sequentially with distinct moieties, making MSN a unique theranostic nanoplatform. The modified functional groups on the surface allow MSNs to be conjugated with diverse targeting molecules, peptides, and aptamers for efficient uptake and accumulation by specific target sites. Furthermore, to release the incorporated drug payload, nanochannels in MSNs can be controlled by varying external and/or internal stimuli. Multifunctionalized MSNs have proven to be a key platform for theranostics that orchestrates traceable imaging, targeted delivery, and selective therapeutics for cancer treatments.

14.1.4 *Surface Enhanced Raman Scattering (SERS)*

Raman spectroscopy and surface enhanced Raman scattering (SERS) have been employed for imaging cancer cells. SERS is an analytical method that is well known for its high sensitivity, specificity, flexibility, and is capable of providing multiplicity. SERS not only serves as one of the most powerful tools for exploring material properties at a single molecular level, but also plays very important role in applied sciences. In biomedical sciences, SERS based methods were reported to be able to monitor the chemical reactions inside a single cell, or probe different cancer markers and cells.

14.1.5 *Polymeric Nanostructures*

Polymeric nanostructures can be engineered into different forms of micelles, nanospheres, nanogels, vesicles, and dendrimers, *etc.* Their sizes can be tailored and optimized for specific biomedical diagnosis and treatment. Polymeric nanostructures are rich in functional groups, such as amino acid, glucose unit, and nucleotides and, therefore, they are particularly versatile in structural architecture. Polymeric nanostructures have found promising applications in cancer imaging and therapies. The nanomedicinal systems, based on such species, hold significant potential for improving the treatment of cancer. However, adverse effects resulting from the interactions among all components must be taken into consideration in designing and developing these polymeric nanosystems.

14.1.6 *Carbon Nanomaterials*

Nano-carbons, including fullerenes, carbon dots, carbon nanotubes, graphene, and nanodiamonds, are important nanostructures that are attracting tremendous interests during the past two decades. Due to their apparent low toxicity, carbon-based nanoparticles found important applications, particularly in biomedicine. Some of the carbon nanomaterials show fluorescence in the visible wavelength range, particularly graphene and carbon dots, but, being nanosized, they all have common corresponding properties. The main theranostics applications of nano-carbons are for improving tumor imaging, allowing earlier detection, and specific drug delivery allowing high treatment efficient and reduced side effects.

14.1.7 *Lipoplexes and Polyplexes Nanoparticles*

Cationic lipids (lipoplexes) and cationic polymers (polyplexes) emerge as a great potential alternative to viral vectors in the last decade. The power of this technique is that it can obtain information about gene delivery and preclinical and clinical image diagnosis simultaneously, including optional imaging, MRI and positron emission tomography (PET). Additionally, therapeutic effect is useful on an extensive screening of gene carriers in preclinical processes. This targeted gene theranostic method has found broad application in gene therapy science. Nonetheless, a substantial progress in this field is still warranted. Enhanced understanding of specific targeted gene delivery and combination of diagnosis and treatment of lipopolyplex-mediated gene delivery would be beneficial.

14.1.8 *Nano/Microbubbles Technology*

Contrast-enhanced ultrasound agents in the form of nano/microbubbles can preferentially extravasate into tumor tissue, and be activated through tumor-directed ultrasound for imaging and controlled delivery purposes. Ultrasound contrast agents are not only effective in ultrasonic imaging, but also as important tools for drug or gene delivery by disrupting nano/microbubbles locally using ultrasound beams. Nano/microbubbles can enhance cell permeability through cavitation and increase the cytotoxicity of delivered agents. But some challenges still remain for the application of this approach to cancer therapy and imaging, particularly due to nonhomogeneity of nanodroplet distribution within the tumor.

14.1.9 *Upconversion Nanomaterials*

Upconversion nanomaterials are a relatively new class of compounds that are being developed as agents for *in-vivo* fluorescence imaging. For bio-applications, upconversion nanomaterials are superior to traditional fluorescent agents since they provide remarkably high optical contrast because of absence of autofluorescence and decreased light scattering. The near-infrared (NIR) to visible or NIR to NIR upconversion fluorescence is also a good solution to the photo damage and possible phototoxicity in living organisms, caused by UV radiation. However, upconversion may not be an advantage if the fluorescence is in the visible region, which suffers from the penetration issue, but may be satisfactory for imaging thin tissue sections.

14.1.10 *Functional Polymeric Dendrimers*

Functional polymeric dendrimers are hyperbranched nanostructures that can be controlled in size, molecular weight and chemical composition of the final product by controlling the number of polymerization generations. Compared to other polymeric materials, dendrimers are nanosized and symmetrical, with precise structures, low polydispersity, as well as highly adaptable and flexible surface chemistry. These unique features have made dendrimers as materials of increasing interest for design as imaging probes, boron neutron capture therapy agents, drug delivery systems and theranostic vehicles in cancer treatment.

14.1.11 *Protein-based Nanoparticles*

Proteins are suitable carriers for drug delivery, imaging or therapy and have been shown to be nontoxic, non-immunogenic, biocompatible and biodegradable. Therefore, protein-based nanoparticle may be an ideal system for *in vivo* tumor targeting theranostics. Among different protein-based nanoparticle systems, albumin nanoparticles, especially human serum albumin (HSA) nanoparticles, have gained more attention, because they have great capacity of loading various anti-cancer drugs and imaging agents after targeted modifications without any serious *in vivo* side-effects.

14.2 Future Perspectives of Theranostics

The unique characteristics of nanomaterials, such as high surface-to-volume ratio, broad optical properties, and facile surface chemistry and functionalization hold pledge in the clinical field for cancer therapeutics. Despite these advances, it is still too early to predict their success as theranostics agents that must overcome many challenges including biocompatibility, pharmacokinetics, *in vivo* targeting efficacy, and cost-effectiveness.The development is still in progress, and many barriers must be crossed before they can be translated into the clinic. Here are those focal points that need to be addressed before theranostics become the future of nanomedicine.

(1) Before these nanomaterials are tested in cancer patients, the detailed preclinical studies should be conducted, especially to investigate the pharmacokinetics, *in vivo* tumor targeting, and therapeutic effects of these nanomaterials. Furthermore, the clearance of the nanoparticles from the body and their acute and long-term toxicity need to be examined. Nonetheless, the nanomaterials must meet the safety, effectiveness, and quality control standards of any new drug development.

(2) The major approaches of nanotechnology in tumor treatment must include the development of nanomaterials with less or no tissue-resistance, their biocompatibility, ability as nanocarriers for drug delivery, and enhanced energy deposition in tissue with or without the external influence of microwave, light, magnet, *etc.*

(3) The most crucial aspect in future development of nanomaterials medicament should focus on the ability for multifunctionalization and successful engineering and fabrication of multimodal nanotheranostic designs. The ultimate goal should be to maximize the amount of diagnostic information and therapeutic efficacy, while minimizing the time frame for early diagnosis, and reducing the degree and frequency of invasive interventions.

(4) In order to increase the range of light, thermal, ultrasound and magnetically sensitive compounds that fulfill the requirements of generally recognized safe products, new designs and synthetic protocols for unique biocompatible nanomaterials are warranted. Different techniques could be used to prepare multifunctional nanoparticles and to modify nanoparticle surfaces that are required for different applications. Multimodality and multifunctionality should be rendered to these nanoprobes that offer the potential to facilitate early detection, accurate diagnosis and efficient therapeutic intervention.

To sum up, the development of novel theranostic approaches in cancer treatments, based on nanomaterials, has been and would remain an area of great demand in cancer therapeutics. In order to realize their clinical applications for diagnosis and therapy, it is important to control and tune the fine-size, shape, morphology and surface coating of nanomaterials to satisfy the requirements of biocompatibility, long-term stability and *in vivo* targeting ability. More generally, it is also important to evaluate the toxicity of nanomaterials and potential risks for environment and human health.

A major goal in nanomedicine is the coherent implementation of multifunctional platforms within a single targeted system that would simultaneously perform diagnosis, targeted delivery and efficient therapy. The ideal theranostic nanomaterials should include several important features: ① selectivity and rapid accumulation in tumor tissue; ② identification of biochemical and morphological characteristics of the surface area; ③ an effective delivery of therapeutics; and ④ must be safe and biodegradable to form nontoxic byproducts. Thus, it is a challenge to develop complex theranostic systems that are responsive to biochemical signals or biomarkers typically present in a range of less than nanomolar concentrations. Such systems need a complex, hierarchical organization of the responsive nanomaterials to accommodate various possible amplification mechanisms. Another challenge in the design of theranostic agents is to develop systems that can respond to several external stimuli in an intelligent way. Much research remains to be done before practical clinical applications become viable even in a remote part of the globe. It is the public who need to alert and then convince the law-makers in the government to invest and investigate the future prospects of nanotechnology-based theranostics and/or nanomedicine.

Abbreviation Index

A

ABC	ATP-binding cassette
Abraxane®	albumin-bound paclitaxel
AC	alternating current
ADTPA	Gd(III) aminobenzyl-diethylene triamine pentacetic acid
AF	Alexa Fluor
AF-PBCA-MP	air-filled polybutylcyanoacrylate microparticle
Ag-FON	silver film over nanosphere
ALA	5-aminolaevulinic acid
aMØ	alveolar macrophage
AN	albumin nanoparticle
anti-CEA	anti-carcinoembryonic antigen antibodies
Apo E	apolipoprotein E
APS	aminopropyl-silanes
Apt	aptamer
APTS	aminopropyltriethoxysilane
ASGP-R	asialoglycoprotein receptors
ASON	anti-survivin antisense oligonucleotides
A549	human lung adenocarcinoma epithelial cell

B

BBB	blood-brain barrier
BCNU	1,3-bis(2-chloroethyl)-1-nitrosourea
bFGF	basic fibroblast growth factor
BL	bioluminescence
BNCT	boron neutron capture therapy
BOLD	blood oxygenation level-dependent
BPEI	branched polyethylenimine
BSA	bovine serum albumin
BSANP	BSA nanoparticle

C

CA	contrast agent
CAR	coxsackieadenovirus receptor
CD	carbon dot
CEHDA	coaxial electrohydrodynamic atomization
Ce6	chlorin e6
Ce6-HSA-NP	Ce6-conjugated HSA nanoparticle
CL	cationic liposome
CLSM	confocal laser scanning microscopy
CMC	carboxymethyl chitosan
CMV	cytomegalovirus
CNT	carbon nanotube
COIN	composite-organic-inorganic-nanoparticle
CPT	camptothecin
c[RGDfC]	Cyclo(Arg-Gly-Asp-D-Phe-Cys)
CS	chitosan
CSC	cancer stem cell
CT	computerized tomography; charge transfer
CTAB	cetyltrimethyl ammonium bromide
CTC	circulating tumor cell
Curc-OEG	curcumin with oligo(ethylene glycol)
CY	cyclosporine

D

DAPI	$4', 6$-diamidino-2-phenylindole
DC	dendritic cell
DCA	dendritic contrast agent
DCE	dynamic contrast enhanced
DDS	drug delivery system
Dextranox-MPEG	methyl polyethylene glycol modified oxidized dextran
DHLA	dihydrolipoic acid
DiI	$1, 1'$-dioctadecyl-$3, 3, 3', 3'$-tetramethyl-indocarbocyanine perchlorate
DiO	$3, 3'$-dioctadecyloxacarbocyanine perchlorate
DLS	dynamic light scattering
DMNPE	4,5-dimethoxy-2-nitroacetophenone
DNA	deoxyribonucleic acid
DNR	daunorubicin
DOPC	dioleoyl phosphatidyl choline
DOPE	dioleoyl phosphatidyl ethanolamine

DOTA	1,4,7,10-tetraazacyclododecane-1,4,7,10-tetraacetic acid
DOX	doxorubicin
DO3A	1,4,7,10-tetraazacyclododecane-1-(2-hydroxypropyl)-4,7,10-triacetic acid
DPA	D-penicillamine
ds	double-stranded
DTPA	diethylene triamine pentaacetic acid
DTTC	3, 3′-diethylthiatricarbocyanine

E

EC	endothelial cell
EDC	1-ethyl-3-(3-dimethylaminopropyl) carbodiimide
EF	enhancement factor
EGF	epidermal growth factor
EGFR	epidermal growth factor receptor
EM	electromagnetic
EMFL	endohedral metallofullerenes
EPC	endothelial progenitor cell
EPR	enhanced permeability and retention
ESA	excited-state absorption
ETO	etopoxide
ETU	energy transfer upconversion

F

FA	folic acid
FA-DOX-AN	doxorubicin-loaded folic acid-conjugated albumin nanoparticle
FDA	Food and Drug Administration
FGF	fibroblast growth factor
FGFR	fibroblast growth factor receptors
FI	fluorescent imaging
FISH	fluorescence *in situ* hybridization
FITC	fluoroisothiocynate
FL	fullerene
fMRI	functional MRI
FRET	fluorescence resonance energy transfer
FRα	folate receptor alpha
FWHM	full width at half maximum

G

Gd	gadolinium
Gd-albumin	gadolinium-labeled-albumin
Gd-BOPTA	Gd-benzyloxypropionictetraacetate
Gd-DTPA-BMA	Gd-DTPA-bismethylamide
Gd-EOB-DTPA	Gd-ethoxybenzyl-DTPA
GEM	gemcitabine
GFP	green fluorescent protein
GI	gastrointestinal
GO	graphene oxide
GON	gadolinium oxide nanocrystal
GON-AN	GON-loaded AN
GP	graphene
GQD	graphene quantum dot
GRAS	generally regarded as safe

H

HA	hyaluronic acid
HAD	hexanedioic acid
HCPT	10-hydroxycamptothecin
HEK	human epidermal keratinocytes
HEK293	human embryonic kidney
HeLa cell	human cervical cancer cell
HepG2 cell	human hepatocellular carcinoma cell
HER2	human epidermal growth factor receptor 2
HFF	human foreskin fibroblasts
hMSC	human mesenchymal stem cell
HPC	hydroxypropylcellulose
HPPH	3-(1′-hexyloxyethyl)-3-devinyl pyropheophorbide
HSA	human serum albumin
HSV-TK	herpes simplex virus thymidine kinase
HT	hyperthermia treatment
HUVEC	human umbilical vein endothelial cell

I

ICG	indocyanine green
IHC	immunohistochemistry
IL-12	interleukin-12
IONP	iron oxide nanoparticle
IR	infrared

IV intravenous

L

L-Arg L-arginine
LBL layer-by-layer
LCST lower critical solution temperature
LHRH luteinizing hormone-releasing hormone
LLC1 Lewis lung carcinoma
LMNP lipid-micellar nanoparticle
LPD lipopolyplex
LSP localized surface plasmon
LSS liquid, solid and solution phases

M

MAA mercaptoacetic acid
Mal-PEG-NHS α-Maleinimido-ω-carboxysuccinimidyl ester poly (ethylene glycol)
Mb myoglobin
MB methylene blue; microbubble
MCD mesoporous carbon dot
MCF-7 Michigan Cancer Foundation -7
MC540 merocyanine 540
MDR multidrug resistance
MIT Massachusetts Institute of Technology
MMP matrix metalloproteinase
MnO-NP manganese oxide nanoparticle
MNP magnetic nanoparticle
MPA mercaptopropionic acid
MPA-DOX PEGylated albumin-DOX conjugate attached 2-mercaptoethanol
MPS mercaptopropy-ltris(methyloxy)silane
MPTS 3-mercaptopropyl trimethoxysilane
MRA MR angiography
MRI magnetic resonance imaging
mRNA micro RNA
MSC mesenchymal stem cells
MSN mesoporous silica nanoparticle
MTX methotrexate
MUA mercaptoundecanoic acid
MW molecular weight
MWCNT multi-walled carbon nanotube

N

ND	nanodiamond
NI	nuclear imaging
NIR	near-infrared
NIS	sodium iodide symporter

O

OAS	octa(3-aminopropyl)silsesquioxane
OI	optical imaging
OPC	opal photonic crystal

P

PA	photon avalanche
PAA	poly (acrylic acid)
PAMAM	polyamidoamine
PAMAM-PEG-T7	peptide HAIYPRH (T7)-conjugated PEG-modified polyamidoamine dendrimer
PANC-1	pancreatic cancer cell-1
PArg	polyarginine
PAsp	polyasparagine
PBS	phosphate buffer solution
Pc	phthalocyanines
Pc4	Pc photosensitizer
PDGF	platelet derived growth factor
PDMAEMA	poly[2-(N,N-dimethylamino)ethyl methacrylate
pDNA	plasmid DNA
PDT	photodynamic therapy
PE	phosphatidyl ethanolamine
PEG	polyethylene glycol
PEG-b-PLA	poly(ethylene glycol-block-(dl)lactic acid) block copolymers
PEGDA	polyethylene glycol diacrylate
PEG-PCL	PEG-block-poly (caprolactone)
PEG-PLLA	PEG-block-poly(L-lactide)
PEG-thioPLL	poly(ethylene glycol)-block-poly(L-lysine)
PEI	polyethylenimine
PESDA	perfluorobutane-exposed sonicated dextrose albumin
PET	positron emission tomography
PFP	perfluoropentane
PGA	polyglutamic acid

PGSA	phloroglucinol and succinic acid
PH-A	pheophorbide-a
PHFL	polyhydroxy fullerenes
PLGA	poly(lactic-co-glycolic acid)
PLL	polylysine
PLL-PEG-PLL	poly(L-leucine)-block-poly(ethylene glycol)-block-poly(L-leucine)
PMMA	poly(methyl methacrylate)
PMMAVP	hydrophobic copolymer
PNA	peptide nucleic acids
PNB	plasmonic nanobubbles
PNIPAM	poly(N-isopropylacrylamide)
P (NIPAM-co-MAA)	poly (N-isopropylacrylamide-co-methacrylic acid)
POSS	polyhedral oligomeric silsesquioxane
PPI	poly(propyleneimine)
PP-PLLD	porphyrin-poly(L-lysine) dendrons
PpIX	protoporphyrin IX
PS	photosensitizer
PSA	prostate specific antigen
PSMA	prostate specific membrane antigen
PTT	photothermal therapy
PTX	paclitaxel
PVA	polyvinyl alcohol
PVP	poly(vinyl pyrrolidone)

Q

QD	quantum dot
QD-Apt-DOX	QD-aptamer-doxorubicin conjugate

R

RB	rose bengal
RE	rare earth
RES	reticuloendothelial system
RGD	Arg-Gly-Asp
RGD-SN	RGD peptide-anchored sterically stabilized BSA nanospheres
RGO	reduced graphene oxide
Rh6G	Rhodamine 6G
ROI	reactive oxygen intermediates; region of interest
ROS	reactive oxygen species

RPA-DOX	RGD-PEG-albumin-DOX conjugates
RRS	resonance Raman scattering

S

SAR	specific absorption rate
SCCHN	squamous cell carcinoma of the head and neck
SELEX	systematic evolution of ligands by exponential enrichment
SEM	scanning electron microscope
SERS	surface enhanced Raman scattering
SESORS	surface enhanced spatially offset Raman spectroscopy
SI	signal intensity
siRNA	small interfering RNA
SLN	sentinel lymph nodes
SLP	specific loss power
SMCC	4-(N-maleimidomethyl)-cyclohexanecarboxylicacid N-hydroxysuccinimideester
SNR	signal-to-noise ratio
SP	surface plasmon
SPDP	N-succinimidyl-3-(2-pyridylthio) propionate
SPECT	single photon emission computed tomography
SPIO	superparamagnetic iron oxide
SPM	superparamagnetic
SPR	surface plasmon resonance
SR	specific receptors
ss	single-stranded
ssDNA	single stranded DNA
SWCNT	single-walled carbon nanotube

T

TEM	transmission electron microscopy
TEOS	tetraethyl orthosilicate
TET	triplet energy transfer
TF/Tf	transferrin
TGA	thioglycolic acid
TGF-β1	transforming growth factor-β1
TNF	tumor necrosis factor
TOPO	trioctylphosphine oxide
TOPSe	trioctylphosphine selenide
TPA-PDT	2-photon activated-photodynamic therapy

TRAIL	TNF-related apoptosis-inducing ligand

U

UC	upconversion
UCL	upconversion luminescence
UCNC	upconversion nanocrystal
UCNP	upconversion nanoparticle
UI	ultrasound imaging
US	ultrasound
USI	ultrasound imaging
UTMD	ultrasound targeted microbubble destruction
UV	ultraviolet

V

VEGF	vascular endothelial growth factor
VEGFR	VEGF receptors
VIS	visible
VMAT2	vesicular monoamine transporter
VPTT	volume phase transition temperature

W

WGA	wheat germ agglutinin

X

X-CT	X-ray computed tomography

Z

ZnPc	zinc phthalocyanine

3D	three-dimensional
4T1	mouse breast cancer cells
5-FU	5-Fluorouracil

β-CD	β-cyclodextrin

Color Figures

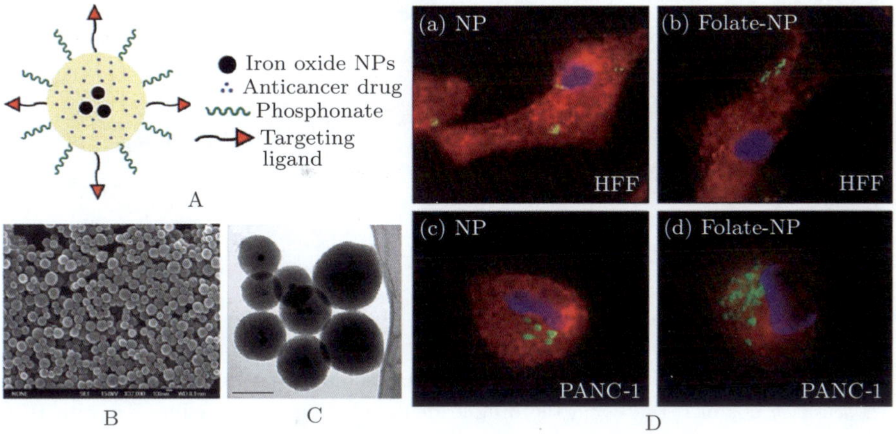

Fig. 1.5 A. Schematic illustration of multifunctional MSNs doped with iron oxide NPs, hydrophobic anticancer drugs, and surface modifications with phosphonate and FA. SEM (B) and TEM (C) images of the iron oxide NPs incorporated MSNs. D. Fluorescence microscopy images of the cancer cells PANC-1 and human foreskin fibroblasts (HFF) treated with camptothecin (CPT)-loaded FA-NP and NPs (green fluorescence). HFF cells treated with (a) NPs and (b) folate-NP, PANC-1 cells treated with (c) NPs and (d) folate-NP. The blue fluorescence shows the cell nuclei stained with DAPI, and red fluorescence indicates the membranes stained with WGA. Image is adapted with permission from Ref. 124. Copyright (2008) American Chemical Society.

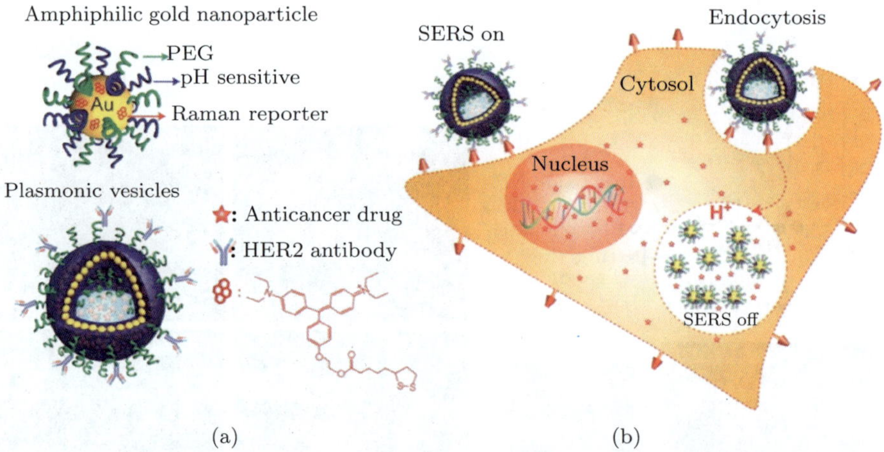

Fig. 1.6 A. (a) Schematic illustration of gold NPs modified with Raman reporter, PEG and pH-sensitive polymer grafts and the drug-loaded plasmonic vesicle conjugated with HER2 antibody for cancer cell targeting. (b) Schematic illustration of cellular binding, uptake, and intra-organelle disruption of SERS-encoded pH-sensitive plasmonic vesicles. B. (a) SERS spectra of SKBR-3 cells treated with HER2-targeted vesicles (1) and untargeted vesicle (2), MCF-7 cells treated with targeted vesicles (3), and SKBR-3 control cells (4). (b) SERS spectra of SKBR-3 cells treated incubated with targeted vesicles at 30 min, 60 min and 90 min, respectively. C. Dark-field (a, e, i), fluorescence (c, g, k), and the overlaid images (b, d, f, h, j, l) of SKBR-3 cells treated with DOX-loaded pH-sensitive plasmonic vesicles after 30 min incubation (a–d) and the post incubation images of the cells at 60 min (e–h) and 90 min (i–l). The red fluorescence shows DOX, blue fluorescence indicates cell nuclei. Image is adapted with permission from Ref. 33. Copyright (2012) American Chemical Society.

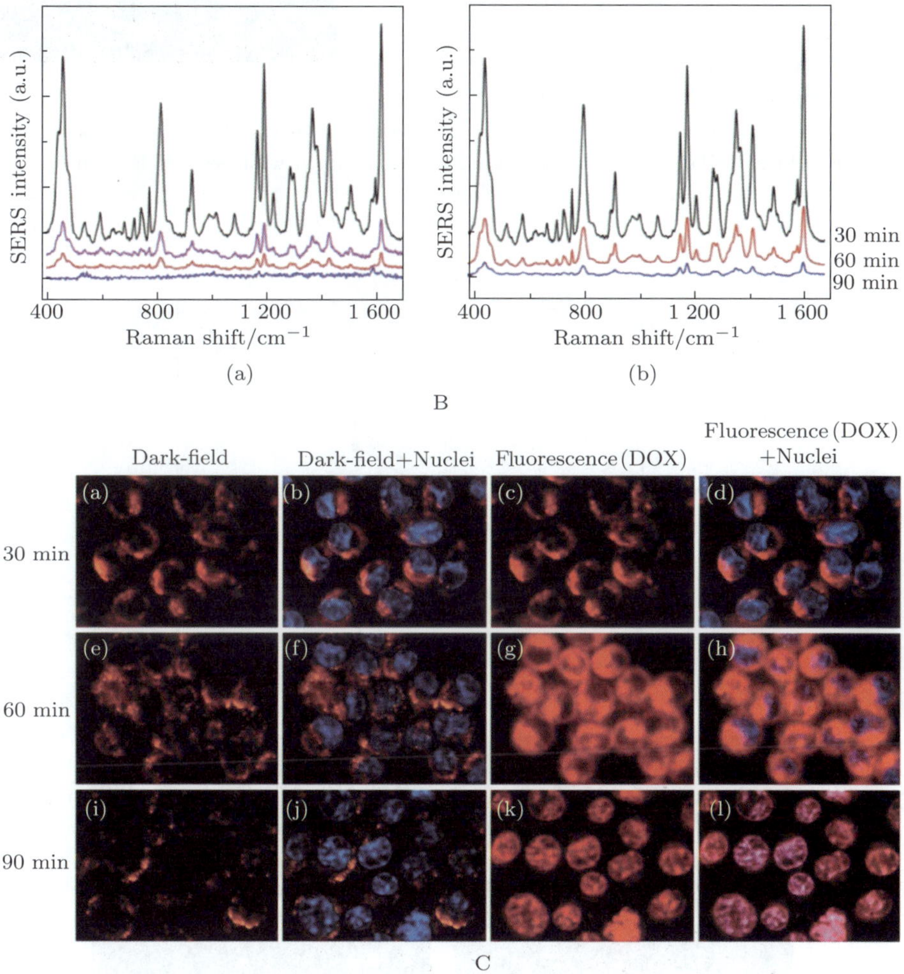

Fig. 1.6 (Continued).

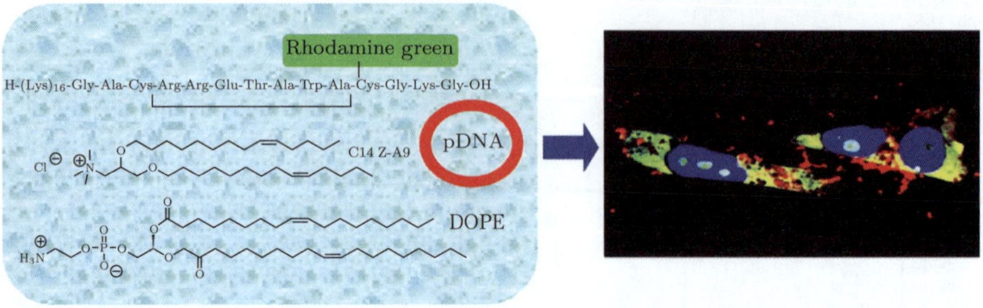

Fig. 1.9 Lipopolyplex vectors consisting of lipid-peptide-pDNA for *in vitro* DNA transfection in breast MDA-MB-231cancer cells. Image is adapted with permission from Ref. 36. Copyright (2010) American Chemical Society.

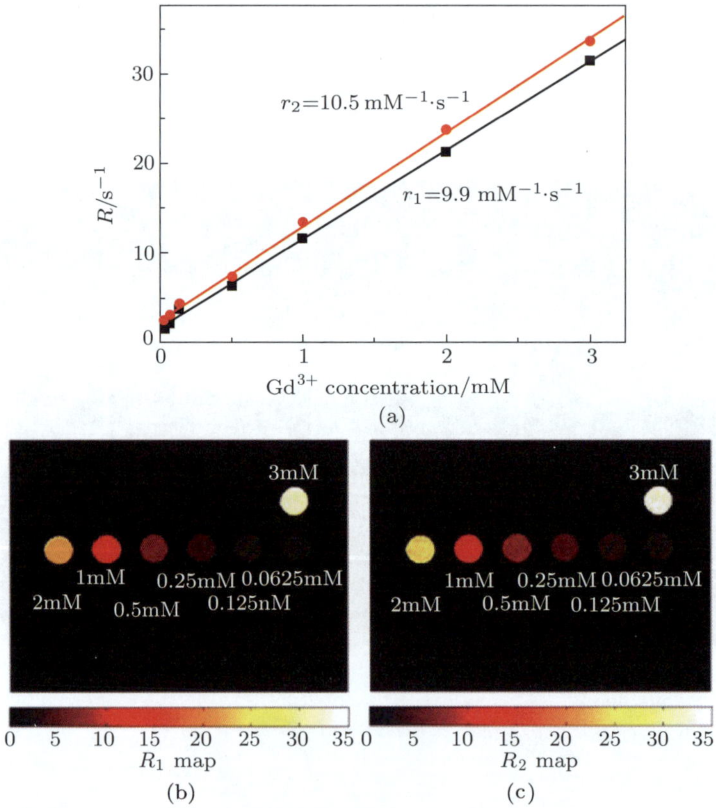

Fig. 3.1 (a) Plots of the R_1 and R_2 versus Gd^{3+} concentration. r_1 and r_2 are calculated from the slopes. (b) R_1 and (c) R_2 map images as a function of Gd^{3+} concentration. Image is adapted with permission from Ref. 45. Copyright (2009) American Chemical Society.

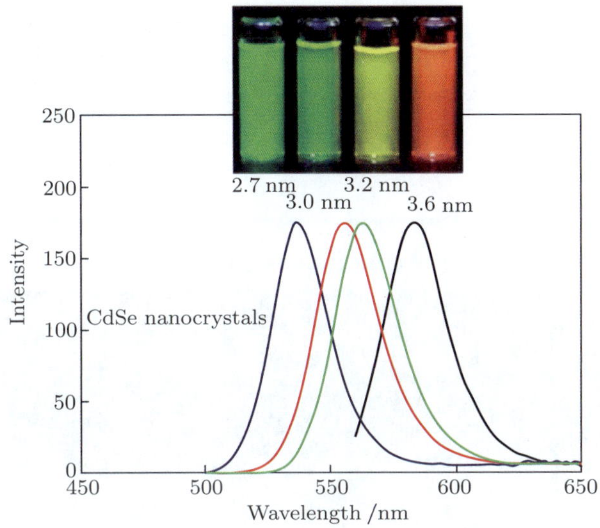

Fig. 4.1 Emission maxima shift to longer wavelength as sizes of QDs increase. All spectra are typically around 30 to 50 nm (full width at half maximum). The inset shows the photo of QDs with different size upon UV irradiation.

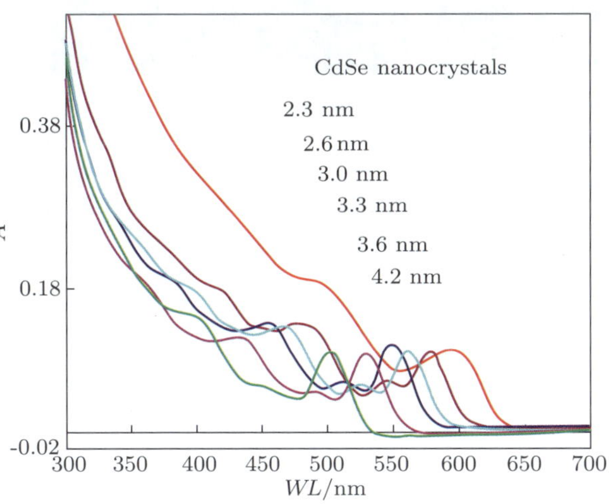

Fig. 4.3 Absorption spectra of CdSe QDs with different sizes.

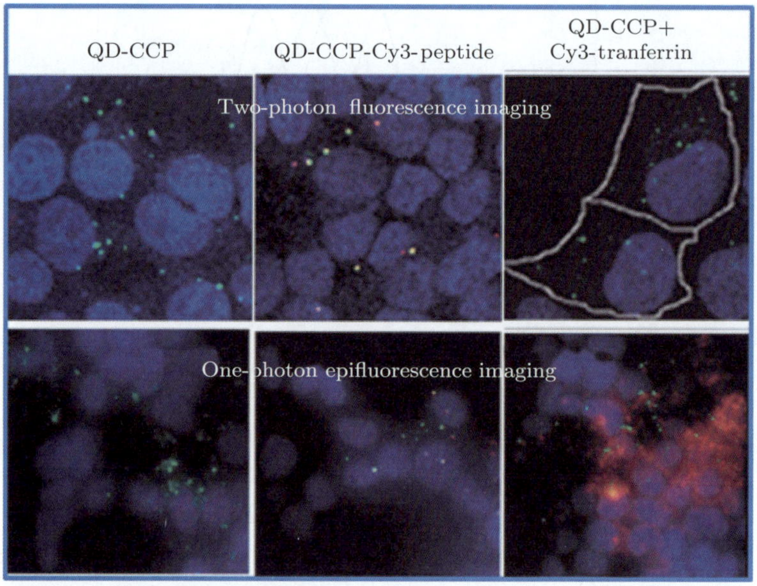

Fig. 4.7 Two-and one-photon excitation of QD-Cy3 FRET system. QDs are efficiently excited by both methods, enabling fluorescence of conjugated Cy3 molecules *via* FRET. QD is located outside the cell nuclei. However, only two-photon excitation precludes non-FRET excitation of Cy3 dye, reducing significant background produced by conventional one-photon fluorescence. Reproduced from Ref. 139 with permission of Wiley-VCH Verlag GmbH & Co. KGaA, Copyright © 2007.

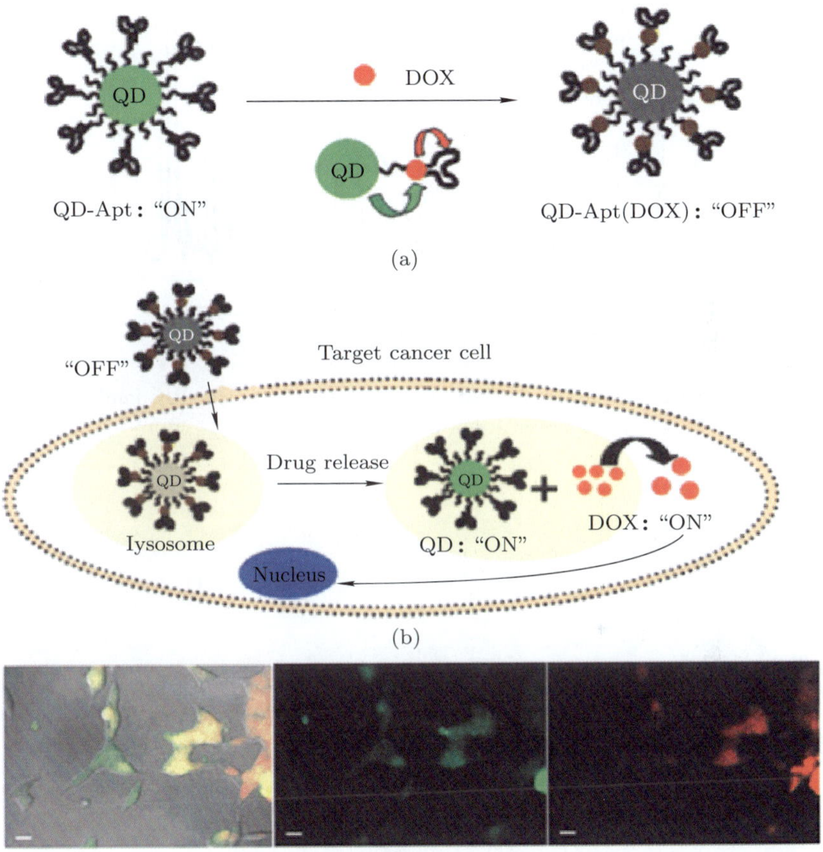

Fig. 4.11 Schematic images of (a) QD-Apt-DOX FRET system. QD was functionalized with Apt and then loaded DOX, leading to the fluorescence quench, (b) tumor-targeted delivery of QD-Apt-DOX. The release of DOX from the QD-Apt-DOX conjugates induces fluorescence recovery, thereby detecting the intracellular delivery of drug and QD localization, as well as killing of cancer cells. (c) Images of target cells after incubation with QD-Apt-DOX. DOX and QD are shown in red and green, respectively. (scale bar: 20 μm). Image is adapted with permission from Ref. 39. Copyright (2007) American Chemical Society.

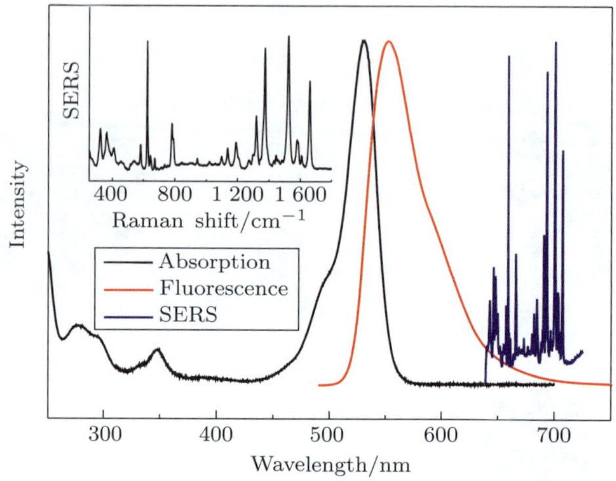

Fig. 6.5 Absorption, fluorescence, and SERS of Raman reporter Rhodamine 6G(Rh6G). Inset shows the "normal" plot of SERS of Rh6G.

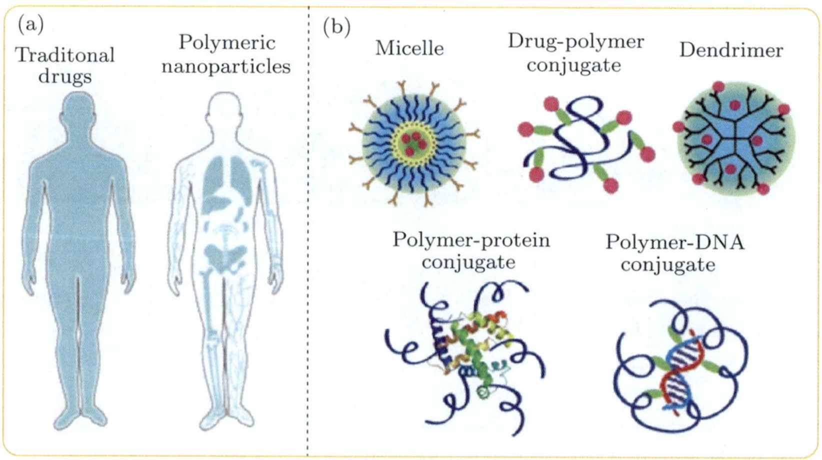

Fig. 7.1 Schematic diagrams showing (a) the difference between conventional medicine and nanotechnology and (b) various types of nanocarrier systems.

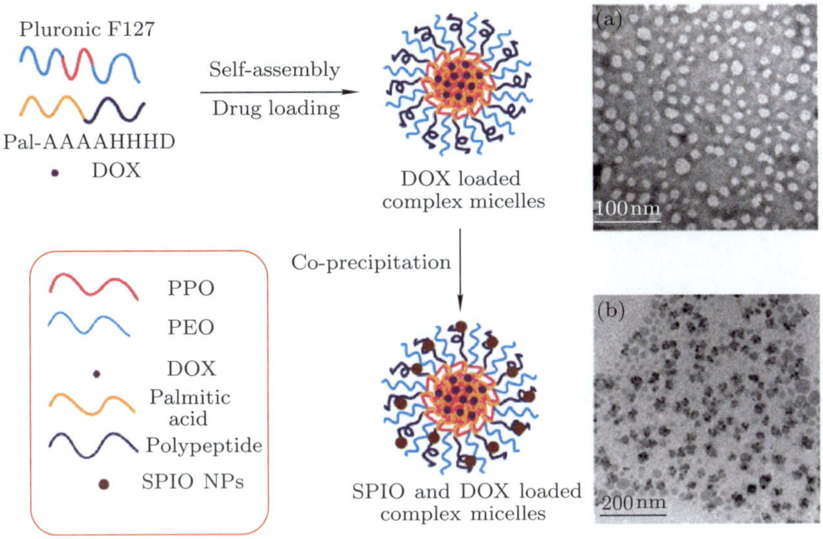

Fig. 7.2 Schematic illustration of the synthesis of the hybrid complex micelles with shell embedded magnetic nanoparticles and TEM images of (a) complex micelles and (b) hybrid micelles.[75]

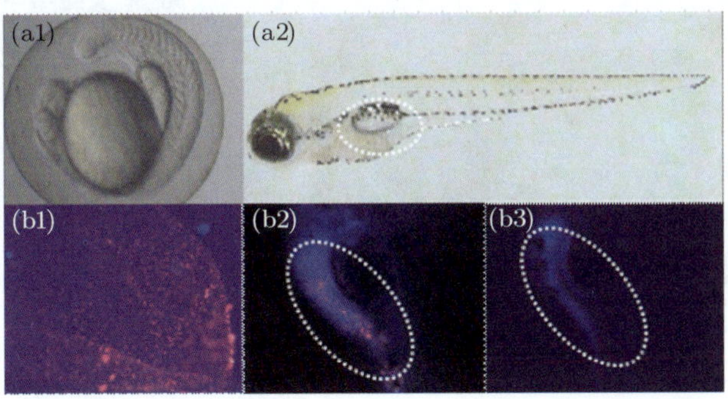

Fig. 7.3 Images of zebrafish embryo (a1), the full body shot of a fish (a2), zebrafish embryo exposure to micelles (b1), larva after treated with micelles for 6 days (b2) and then with fresh water for 3 days (b3). The images of (b1), (b2) and (b3) were obtained by excitation under UV light with a blue background.

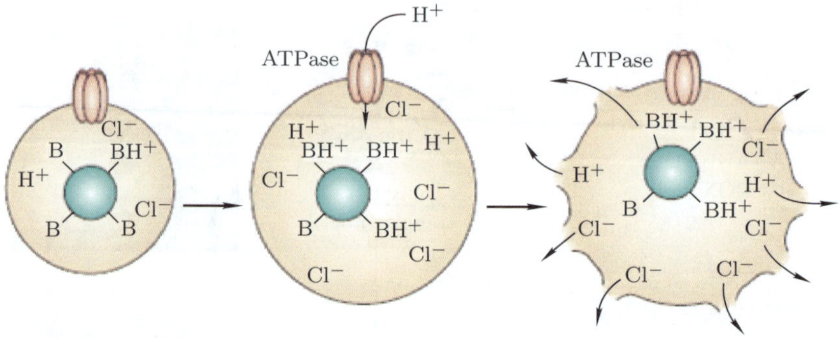

Fig. 9.5 Schematic of the proton-sponge mechanism. Protonation of the proton-sponge polymer (green) causes increased influx of protons (and counter-ions) into endocytic vesicles. Increasing osmotic pressure causes the vesicle to swell and rupture. Image is adapted with permission from Ref. 12. Copyright (2005), rights managed by Nature Publishing Group.

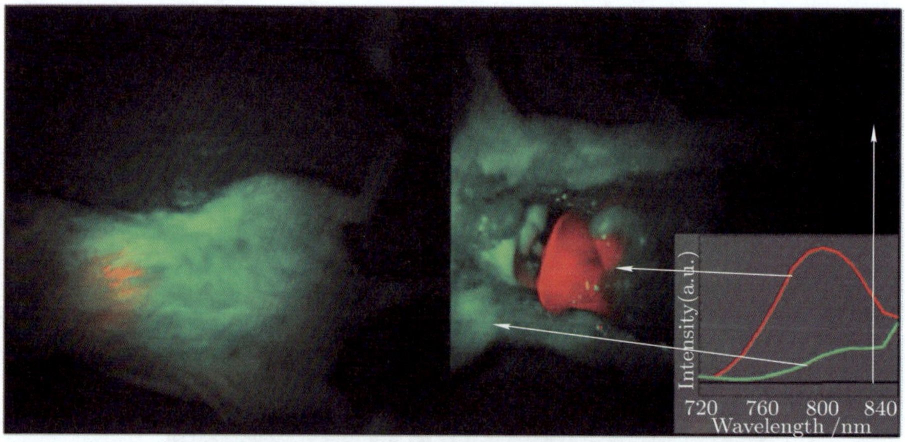

Fig. 11.2 Whole body images of mouse injected i.v. with UCNPs; intact mouse (left), same mouse after dissection (right). The red color indicates emission from UCNPs, green and black show background as indicated by the arrows. The inset presents the PL spectra corresponding to the spectrally unmixed components of the multispectral image obtained with the Maestro system. Reprinted with permission from Ref. 8. Copyright (2008) American Chemical Society.

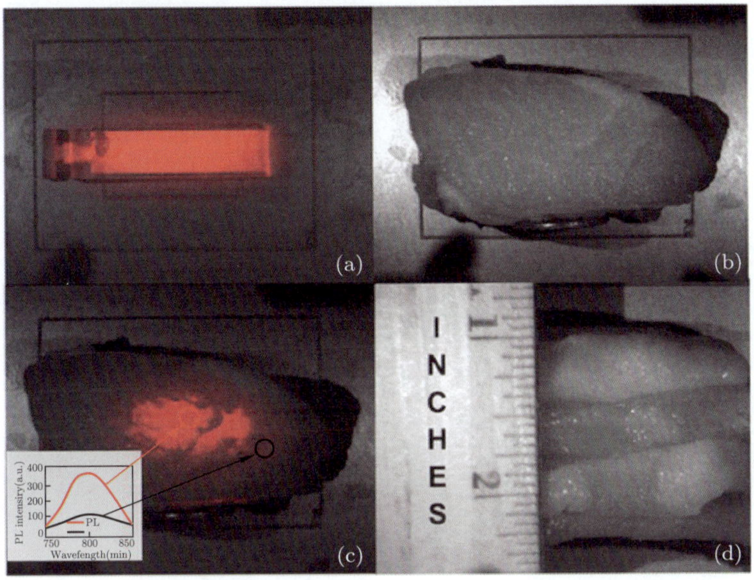

Fig. 11.3 (a) UC PL bright-field image of a cuvette filled with a suspension of the core/shell nanoparticles. (b) Bright-field image of a cuvette covered with pork tissue with a quarter coin stood aside showing its thickness. (c) Merged UC PL/bright-field image of the cuvette covered with pork tissue. (d) Bright-field image of the pork tissue (side view). The inset in (c) shows the spectra obtained from the circled areas. Reprinted with permission from Ref. 29. Copyright (2012) American Chemical Society.

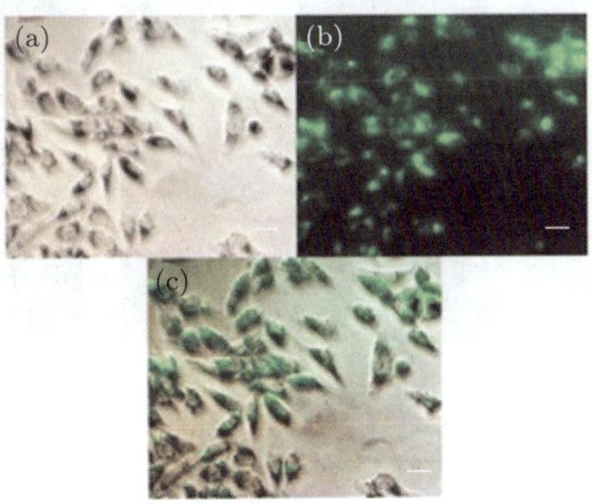

Fig. 11.4 Optical microscopy images of HepG2 cells incubated with ZnS:Tb,Gd nanoparticles: (a) phase contrast image, (b) fluorescence image and (c) overlay of fluorescence and phase contrast images. Scale bar = 20 μm. Reprinted with permission from Ref. 32. Copyright (2013) The Royal Society of Chemistry.

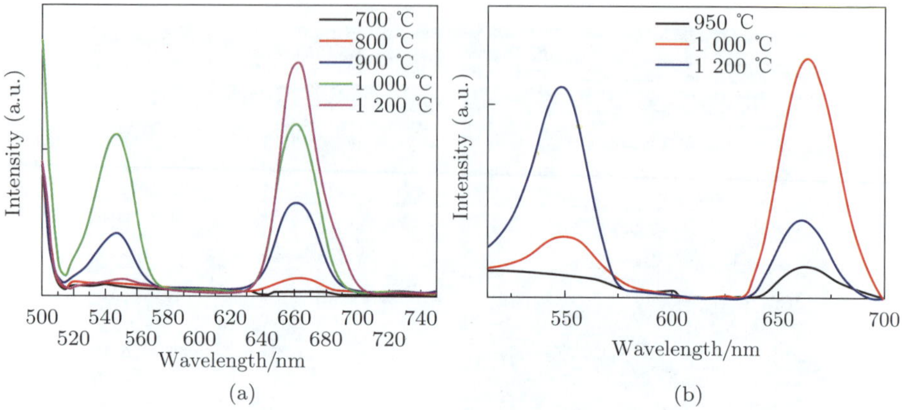

Fig. 11.5 PL spectra of (a) ZnO:Er,Yb,Gd and (b) ZnS:Er,Yb,Gd nanoparticles with varied sintering temperatures (λ_{EX} = 980 nm). Adapted with permission from Ref. 32. Copyright (2013) The Royal Society of Chemistry.

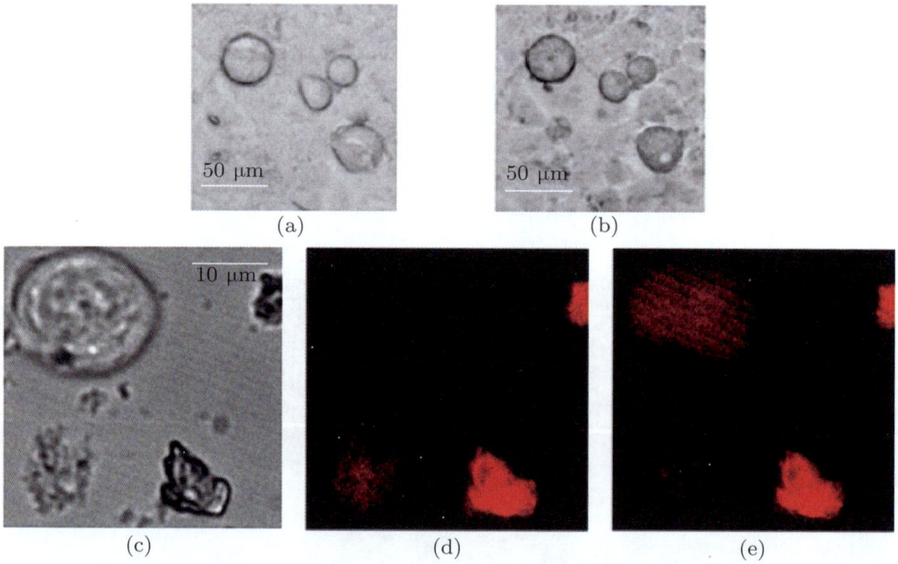

Fig. 11.7 (a) MCF-7/AZ cells incubated with Trypan Blue and antibody-functionalized MC540-coated PUNPs modified with antibody on surface. (b) MCF-7/AZ cells in (a) after 974 nm laser irradiation for 36 min. (c) Optical image of MCF-7/AZ cells incubated with PI and MC 540-coated PUNPs modified with antibody on surface. (d) Fluorescence image of MCF-7/AZ cells in (c) before 974 nm laser irradiation. (e) Fluorescence image of MCF-7/AZ cells in (c) after 974 nm laser irradiation for 43 min. Reprinted with permission from Ref. 78. Copyright (2007) American Chemical Society.

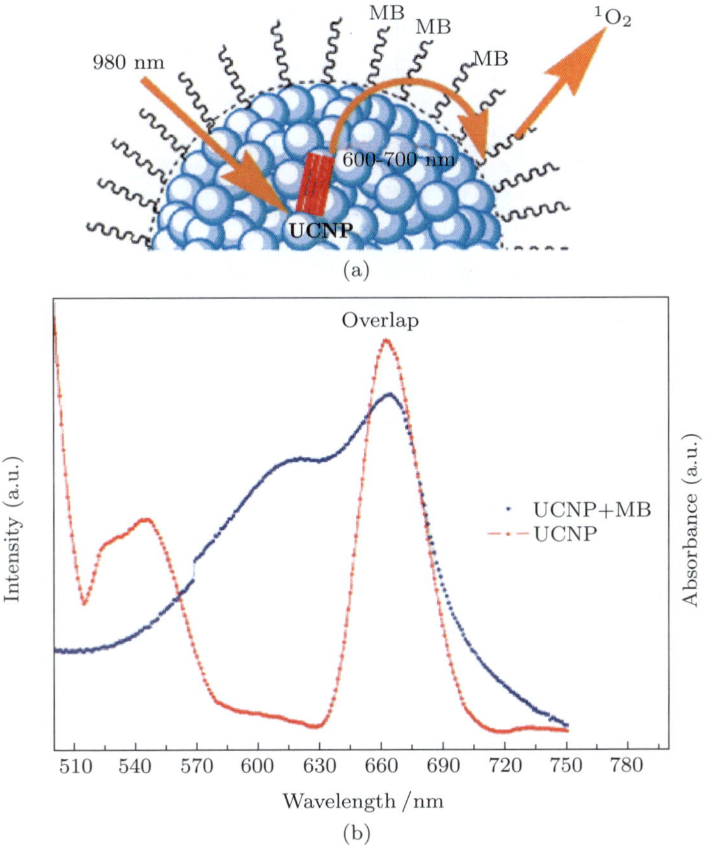

Fig. 11.8 (a) Schematic illustration showing how PDT works using UCNPs-MB. On exposure to NIR light, the nanoparticles convert NIR light to visible light, which will activate the MB to produce reactive oxygen species to kill cancer cells. (b) Overlap between ZnO:Er,Yb,Gd-MB absorption and ZnO:Er,Yb,Gd UCNPs emission: red-band emission of upconversion fluorescence spectrum of ZnO:Er,Yb,Gd under NIR-laser excitation (red line); UV/Vis absorption spectrum of ZnO:Er,Yb,Gd-MB in water located at 665 nm (blue line). Reprinted with permission from Ref. 87. Copyright (2013) American Chemical Society.

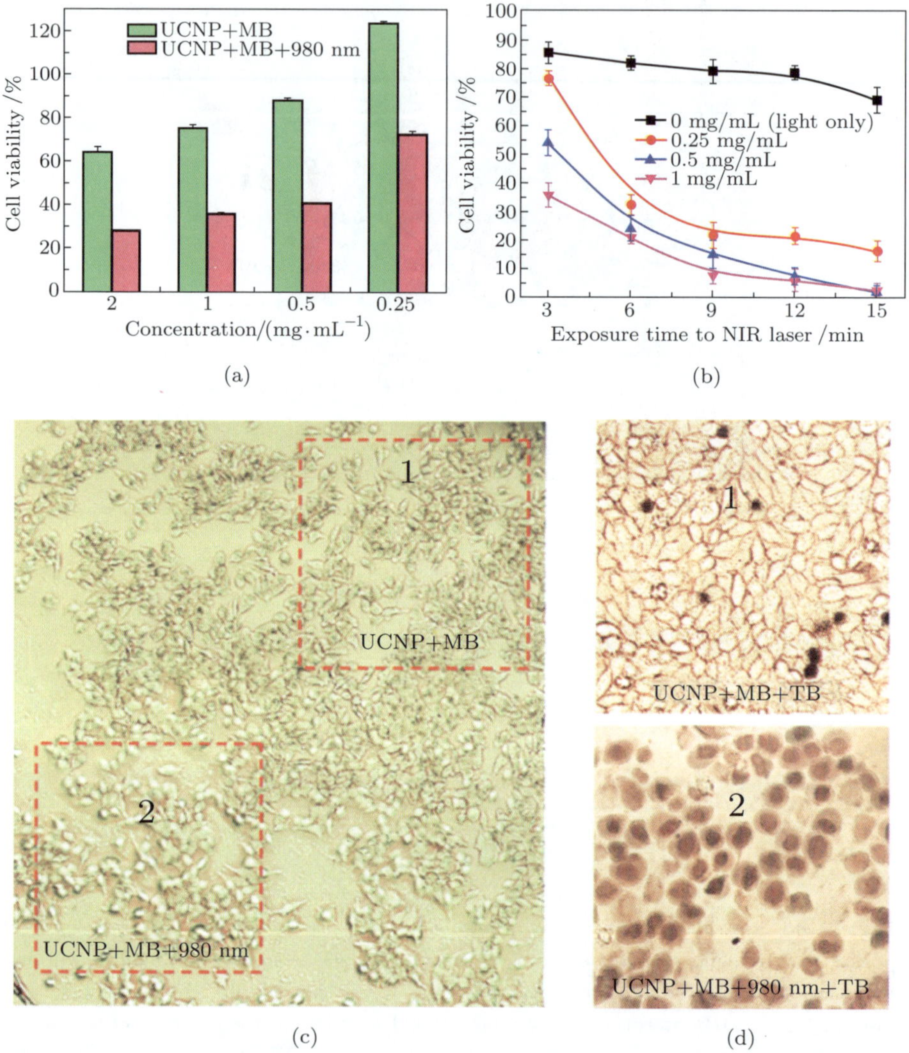

Fig. 11.9 (a, b) Cell viability data of HepG2 cells after various treatments indicated as evaluated by the standard MTT assay. (c) Optical microscopy image of HepG2 cells incubated with ZnO:Er,Yb,Gd-MB particles for 12 h (zone 1) and further exposed to NIR laser for a period of time (zone 2). (d) The corresponding TB stained cells image of (c). Reprinted with permission from Ref. 87. Copyright (2013) American Chemical Society.

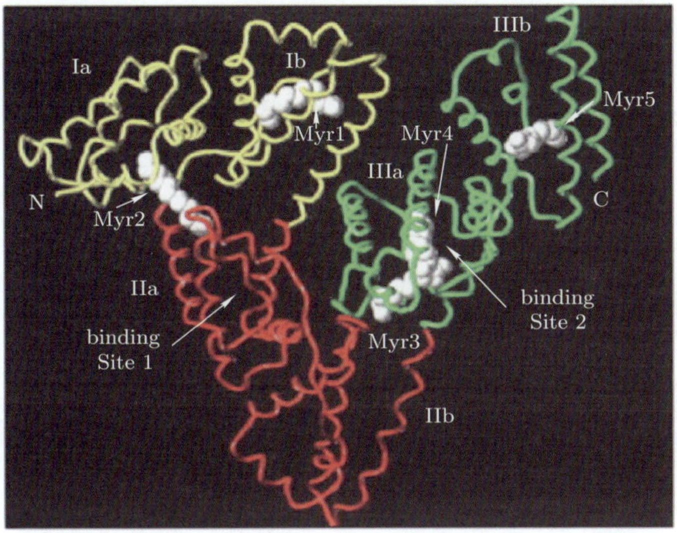

Fig. 13.1 X-ray structure of human serum albumin. Reproduced from Ref. 21. Copyright (2008) with permission from Elsevier.

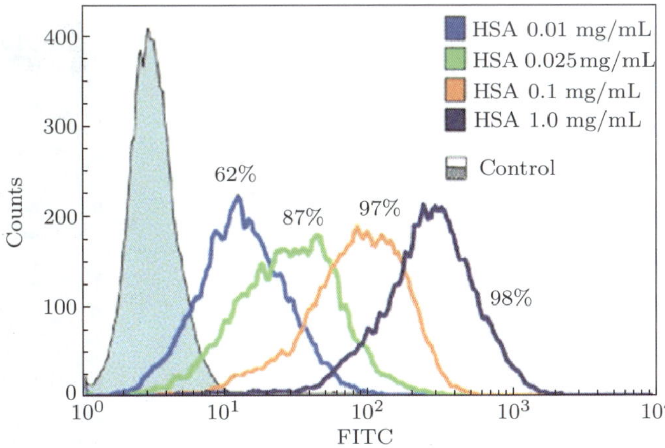

Fig. 13.2 Percentage of primary human macrophages internalized HSA nanoparticles when different concentrations of nanoparticles were used. Reprinted from Ref. 18. Copyright (2003) with permission from Elsevier.

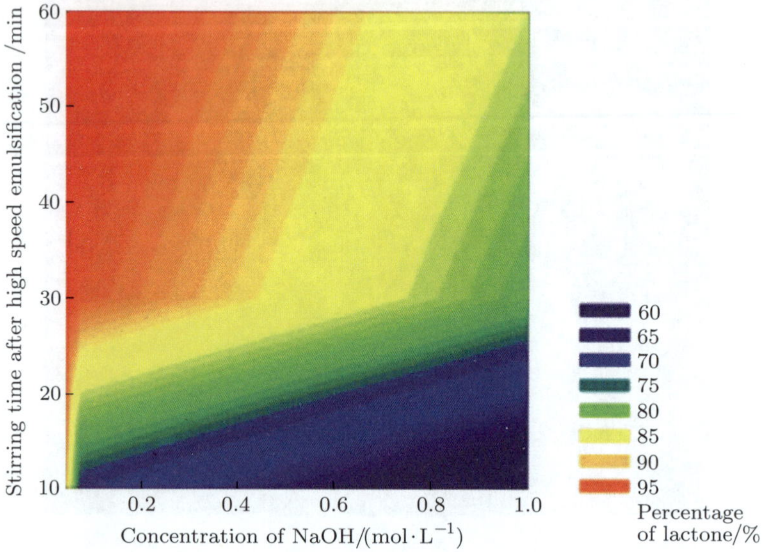

Fig. 13.3 Effect of NaOH concentration and stirring time on the percentage of lactone formed after emulsification. Reprinted from Ref. 49. Copyright (2007) with permission from Elsevier.

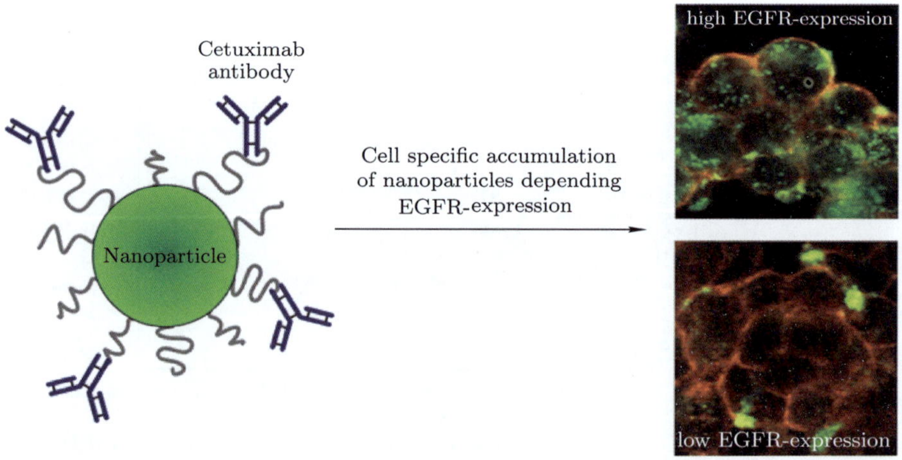

Fig. 13.12 Schematic of cetuximab-modified HAS nanoparticles containing doxorubicin. Reprinted from Ref. 68. Copyright (2011) with permission from Elsevier.

HEP NANO

纳米科学与技术著作系列
HEP Series in Nanoscience & Nanotechnology

> 已出书目

□ 印刷电子学：材料、技术及其应用

崔铮 等 编著

ISBN 978-7-04-034129-4

□ 时间分辨光谱基础

郭础

ISBN 978-7-04-036009-7

□ 有序介孔分子筛材料

赵东元、万颖、周午纵 著

ISBN 978-7-04-036543-6

□ 微纳米加工技术及其应用（第三版）

崔铮 著

ISBN 978-7-04-036974-8

□ 纳米科技基础（第二版）

陈乾旺 编著

ISBN 978-7-04-038650-9

■ Nanomaterials for Tumor Targeting Theranostics: A Proactive Clinical Perspective

肿瘤靶向诊治纳米材料：前瞻性临床展望

（英文版，与 World Scientific 合作出版）

谭明乾、吴爱国 主编

ISBN 978-7-04-042924-4

> 即将出版

□ Printed Electronics: Materials, Technologies and Applications

印刷电子学：材料、技术及其应用（英文版，与 Wiley 合作出版）

Zheng Cui, et al.

□ 微纳磁珠分离分析技术及其应用

刘清珺 主编